Conservation Agriculture and Climate Change
Impacts and Adaptations

Ritesh Saha
Dhananjay Barman
Madhusudan Behera
Gouranga Kar

ICAR-Central Research Institute for Jute and Allied Fibres
Barrackpore, West Bengal – 700 121, India

CRC Press
Taylor & Francis Group
Boca Raton London New York

CRC Press is an imprint of the
Taylor & Francis Group, an **informa** business

NIPA GENX ELECTRONIC RESOURCES & SOLUTIONS P. LTD.
New Delhi-110 034

First published 2023
by CRC Press
4 Park Square, Milton Park, Abingdon, Oxon, OX14 4RN

and by CRC Press
6000 Broken Sound Parkway NW, Suite 300, Boca Raton, FL 33487-2742

© 2023 New India Publishing Agency

CRC Press is an imprint of Informa UK Limited

The right of the contributors to be identified as authors of this work has been asserted in accordance with sections 77 and 78 of the Copyright, Designs and Patents Act 1988.

All rights reserved. No part of this book may be reprinted or reproduced or utilised in any form or by any electronic, mechanical, or other means, now known or hereafter invented, including photocopying and recording, or in any information storage or retrieval system, without permission in writing from the publishers.

For permission to photocopy or use material electronically from this work, access www.copyright.com or contact the Copyright Clearance Center, Inc. (CCC), 222 Rosewood Drive, Danvers, MA 01923, 978-750-8400. For works that are not available on CCC please contact mpkbookspermissions@tandf.co.uk

Trademark notice: Product or corporate names may be trademarks or registered trademarks, and are used only for identification and explanation without intent to infringe.

Print and electronic editions not for sale in South Asia (India, Sri Lanka, Nepal, Bangladesh, Pakistan, Afghanistan and Bhutan).

British Library Cataloguing-in-Publication Data
A catalogue record for this book is available from the British Library

ISBN: 9781032428680 (hbk)
ISBN: 9781032428697 (pbk)
ISBN: 9781003364665 (ebk)

DOI: 10.4324/9781003364665

Typeset in Times New Roman
by NIPA, Delhi

About the Editors

Dr. Ritesh Saha, born on 15th January 1974 in Kolkata West Bengal, is presently working as Principal Scientist in the Division of Crop Production, ICAR-CRIJAF, Barrackpore, Kolkata. He started his professional career as Scientist (ARS) in 1999 and served at various research institutes like ICAR-CIFRI, Vadodara centre; ICAR Research complex for NEH Region, Meghalaya; ICAR-IIHR, Bhubaneswar Centre and ICAR-IISS, Bhopal. He is a Soil Scientist with Specialization in Soil Physics and Soil Water Conservation.

He has 22 years of experience in research and training in the field of soil physics, soil quality and physical health, rainwater harvesting, water conservation, land configuration and conservation agriculture. Notable contributions of Dr. Saha include integrated assessment of zero tillage and organic residue management on soil physical behaviour, water productivity and sustainability under rice and jute based cropping system, analysis of carbon stock density under long-term experiment/field trials in various agro-ecosystem; Development of farming system approach with special intervention of agroforestry for maintaining soil quality against the traditional land use practice of shifting cultivation; Identification of perennial grasses based on hydro-physical behaviour and organic carbon dynamics for improving soil quality in degraded land; Development and standardization a new method of low-cost micro rainwater harvesting technology (Jalkund) for enabling diversified use of stored water for various farm activities like crop, livestock and fish production during water-stress period under hilly terrain; and Evaluation of raised and sunken bed system (RSB) for increasing water economy and total system productivity under rice-based cropping system.

Dr. Saha has to his credit more than 70 research papers published in journals of international and national repute. Besides this, he also authored 3 books, 25 book chapters, 11 research bulletins, 5 extension leaflets, 10 technical articles, 11 review articles and 7 popular articles. He has also presented over more than 60 research papers in both national and international conference. His outstanding work done in the field of scientific research with special emphasis on soil quality/resilience, rainwater harvesting, conservation agriculture, C sequestration has brought many laurels/prestigious awards like Golden Jubilee Commemoration Young Scientist Award of Indian Society of Soil Science, New Delhi (2012), Associate Fellowship from National Academy of Agricultural Sciences (2014) and Dr B C Deb memorial award for popularization of science by Indian Science Congress Association, Kolkata (2015), SURE Scientist of the Year award by Society for Upliftment of Rural Economy, Varanasi (2014), BIOVED Young Scientist Associate award (2016), and Dr. J S P Yadav Memorial Award for Excellence in Soil Science (Team award) by, ISSS, New Delhi (2016).

Dr. Dhananjay Barman, born on 7th May 1982 in Dinhata, Cooch Behar district, West Bengal, is presently working as Senior Scientist in the Division of Crop Production, ICAR-CRIJAF, Barrackpore, Kolkata. He started his professional career as Scientist (ARS) in 2008 at ICAR-IISWC, Research Centre, Koraput, Odisha. He is a Soil Scientist with Specialization in Remote Sensing, GIS and Soil Physics.

He has 13 years of experience in research and training in the field of Agro-climatology, Remote Sensing and GIS, Soil Physics, Soil and Water Conservation, Watershed Management and Agroadvisory Services. Notable contributions of Dr. Barman include his development of JuteMet®, a web-based Agrometeorological Database Management System (ADBMS)-cum-Agroadvisory System, and obtained Trademark for that. He also developed models for soil temperature prediction from air temperatures in Indo-Gangetic plain. He has experience in eddy covariance technique and using this technique he measured CO_2 sequestration potential of jute-based agroecosystem. He successfully used GIS and remote sensing-based approach for mapping of potential areas for mango-sisal intercropping system in red and laterite zone of West Bengal, and land use/land cover dynamics of shifting cultivated areas in Eastern Ghats Highlands of India. Dr. Barman has to his credit 35 research papers published in national and international journals. He also authored 8 book chapters, 3 research bulletins, and many technical articles.

Dr. Barman has been bestowed with Young Scientist Award (2016) in NRM, and Outstanding Scientist Award (2019) in Agricultural Physics conferred by Venus International Foundation; Young Scientist Award (2019) conferred by Indian Association of Soil and Water Conservationists, Dehradun; and Best Young Scientist Award (2021) conferred by ICAR-CRIJAF, Barrackpore.

Dr. Madhusudan Behera was born in Keonjhar district of Odisha in 1961. He obtained his Ph.D degree from Odisha University of Agriculture and Technology in 2009. Dr. Behera started his professional experience as a Research Associate followed by Junior Agronomist at OUAT, Bhubaneswar and served various Institutes like Rubber Research Institute of India, Kottayam as Junior Scientist and ICAR-Indian Institute of Water Management, Bhubaneswar as Chief Technical Officer. He is presently working as Principal Scientist (Agronomy) in Division of Crop Production at Central Research Institute for Jute and Allied Fibre Barrackpore with specialization in Agronomy.

He has 33 years of experience and implemented 32 nos of Multi-disciplinary projects as PI and Co-PI in the field of Integrated Farming System, Cropping Systems, Crop Diversification, Micro-irrigation, Integrated Nutrient Management, Conservation Agriculture Practices, Organic farming, and Watershed management.

To disseminate these technologies in farmers field, he has organized and delivered lectures at Farmers Trainings, Trainer's Trainee Programmes, Farmers Fair, Field Day, Front Line Demonstrations under SCSP, NEH, TSP, Jute ICARE, Govt. Of India, ICAR, SWPA, FPARP, SAU, NATP, NABARD including radio talks, articles in newspaper and publication of more than 100 book chapters, research bulletins, leaflets and folders and Training Manuals. Based on the outcomes of the projects, he has 39 nos of publications in various national and international journals along with 65 nos of abstracts in different national and international conferences and seminar. He is also involved in reviewing various research papers for National and International Journals. He is associated with affiliated Professional Societies and bodies as active member. His research papers were acknowledged at various platforms and forums and awarding papers with Dr. JSP Yadav Best paper Award, Krishi Gaurav Samman, Best Paper Award, Best Poster Awards at various Conferences and Institutional Proficiency Awards for his services towards the Institution. Moreover, Dr Behera is the receiver of the prestigious Bharat Jyoti Award for his meritorious service, outstanding performance and remarkable contribution in his field in the year 2014.

Dr. Gouranga Kar, born on 1st September, 1968 in Bankura, West Bengal, is presently the Director of ICAR-Central Research Institute for Jute & Allied Fibres (CRIJAF), Barrackpore, Kolkata. He did his M.Sc and Ph.D in the discipline of Agricultural Physics in 1993 and 1996, respectively, from Indian Agricultural Research Institute (IARI), New Delhi. He started his professional career as Scientist (ARS) in Aug., 1996 at ICAR-Indian Institute of Water Management (IIWM), Bhubaneswar, Odisha and served there at various capacities till 2020.

He has 25 years of research experience in the field of Land use and cropping system characterization using remote sensing and GIS, crop growth modelling, water and watershed management, climate change research, mitigation and adaptation strategies. He has handled many externally funded projects as Principal investigator during his stay at Bhubaneswar. Some of the important projects are National Innovation for Climate Resilient Agriculture (NICRA) funded by Ministry of Agriculture and Farmers' Welfare, GoI; Integrated watershed management project (IWMP) sponsored by Odisha Watershed Development Mission, Bhubaneswar; Farmers' Participatory Action Research Programme (2nd Phase) funded by Ministry of Water Resources, GoI; Natural resource mapping and hydrologic study of watershed using remote sensing and GIS sponsored by DST, GoI and Developing low cost technology for acid soil management funded by TIFAC, GoI. For pursuing the frontier research, he was awarded with Norman Borlouge Fellowship in 2008 and Fulbright-Nehru Senior Research Fellowship, USA during 2011-12. His research work has been published in different leading international and national journals with high repute. He contributed more than 300 publications which includes referred journals, reviews, book/book chapters, research bulletin, training manual/leaflets, proceedings of seminar/conferences etc.

Dr. Kar is recipient of several national and international awards/honours, namely, ICAR-Hari Om Ashram Trust Award (2008), ICAR-Vasant Rao Naik Award (2005), Rajbhasa Gaurav Puraskar, Ministry of Home Affairs, GoI (2019, NAAS Recognition Award (2012), ICAR-Swami Sahajananda Saraswati Award (2019), Fulbright-Nehru Senior Research Fellowship, USA (2011-12), IARI-Hooker Award (2018), IARI-Sukumar Bose Memorial Award (2007), Dr K.G. Tejwani Award of IASWC (2008), Associates of NAAS (2008), Young Scientist Award of Indian Science Congress Association, ISCA (2001), Young Scientist Golden Jubillee Commemoration Award of ISSS (2006), Ekamra Shree Award of Govt of Odisha (2016), Rajdhani Gaurav Samman of Govt of Odisha (2015), Ekamra Shree Award of Govt of Odisha (2017). He is fellow of National Academy of Agricultural Sciences; FNAAS (2011), Indian Association of Soil & Water Conservationists; FIASWC (2012), Association of Agrometeorologists; FAAM (2010), Indian Society of Soil Science; FISSS (2018), West Bengal Academy of Science and Technology; FWAST (2017) and Indian Society of Coastal Agricultural Research; FISCAR (2020).

Dr. H. S. Sen

Former Director,
ICAR- Central Research Institute
for Jute & Allied Fibres
Barackpore, West Bengal
Blog: hssen-hssen-coastalmanagement.blogspot.com

2/74, Naktala, P.O. : Naktala
Kolkata - 700 047
Phone : +91-33-2411 2381
Mobile : +91 9874189762, +91 8902784681
E-mail : hssen.india@gmail.com
hssen2000@hotmail.com

Foreword

There is global consensus that current climate is changing and it is for real. Agriculture is one sector of the economy that is exposed directly and considerably affected by climate change and climate variability. Rural communities are struggling with the realities, together with declining farm productivity and rising cost of cultivation associated with depleting and degrading natural resources. Climate resilient agriculture therefore requires a major shift in the way land and water are managed to ensure that these resources are used more efficiently.

Conservation agriculture (CA) is a farming system approach that promotes minimum soil disturbance (some people loosely call it as 'no' tillage) together with maintenance of a permanent soil cover, and diversification of plant species reinforcing ecosystem services through a number of interrelated pathways. The production of food in CA reinforces the ecosystem services through nutrient recycling, soil formation habitat derived from crop rotation, minimum soil disturbance together with plant residue retention, have also the minimal impact on climate change. Despite some challenges associated with CA, if implemented judiciously, CA could ensure current food security and nutrition for all without compromising the economic, social, and environmental bases for future generations. There is a need to sensitize the resource poor farmers/stakeholders about CA through innovative participatory approaches.

This book entitled "Conservation Agriculture and Climate Change: Impacts and Adaptations" will provide a comprehensive understanding of the subject with topics related to climate change mitigation strategies, approaches and impact of conservation agriculture on natural resource management. I hope this book will be useful to researchers, teachers and students to understand CA as climate smart agricultural practice for sustainable agriculture in present day scenario.

I congratulate Dr. Ritesh Saha, Principal Scientist (ICAR-CRIJAF) and other Editors for publishing this book in academic spirit.

Dated the 22nd April, 2021
Kolkata

(H.S. Sen)

Preface

Conventional tillage and burning crop residues has degraded the soil resource base and intensified soil degradation with concomitant decrease in crop production capacity. The emerging issue of global warming coupled with green house gases emissions has further aggravated the scenario. Conservation agriculture (CA) helps in reducing many negative effects of conventional agriculture such as soil erosion, soil organic matter (SOM) decline, water loss, soil physical degradation, and fuel use. CA helps to improve biodiversity in the natural and agro-ecosystems. Complemented by other good agricultural practices (GAPs) including the use of quality seeds, integrated pest, nutrient and water management etc., CA provides a base for sustainable intensification of the agricultural production system. Moreover, the yield levels in CA systems are comparable and even higher than traditional intensive tillage systems with substantially less production costs.

The conservation agriculture (CA) practiced over an estimated 100 M ha area worldwide and across a variety of climatic, soil and geographic zones, has proved to be energy and input efficient, besides addressing the emerging environment and soil health problems. The CA technologies involving no- or minimum tillage with direct seeding and bed planting, residue management (mainly residue retention) and crop diversification have potential for improving productivity and soil quality, mainly by soil organic matter (SOM) build-up. This bring many possible benefits including reduced water and energy use (fossil fuels and electricity), reduced greenhouse gas (GHG) emissions, soil erosion and degradation of the natural resource base, increased yields and farm incomes, and reduced labor shortages.

We place on record our sincere thanks to all the authors for their support and contribution. We wish to express our deep sense of gratitude to Dr. T. Mahapatra, DG (ICAR) and Secretary (DARE) for his constant encouragement to us.

We are also thankful to M/s NIPA, New Delhi for their support in compilation, designing and publishing the book.

We do hope, this book will be of immense use to researchers, scientists, students and policy makers for efficient and sustainable management of natural resources under climate change scenario.

Dated the 15th April, 2021 **Editors**

Contents

Foreword... *ix*

Preface.. *xi*

List of Contributors .... *xvii*

Introduction to Conservation Agriculture Under Climate Change Scenario

1. Climate Change and Its Impact in Agriculture and
   Water Resources 1
   Gouranga Kar

2. Climate Smart Agriculture: Special Reference to Conservation
   Agriculture...21
   P. Bhattacharyya

3. Relevance of Conservation Agriculture in Climate Change
   Perspective: An Issue for Climate Resilient Agriculture....................33
   P.K. Ghosh, C.P. Nath, Debarati Datta and K.K. Hazra

4. Conservation Agriculture: Issues, Challenges and Prospects
   in India..53
   A.K. Biswas, J. Somasundaram, K.M. Hati, Pramod Jha
   A.K. Viswakarma, R.S. Chaudhary and A.K. Patra

5. Conservation Agriculture *vis-a vis* Resource Conservation
   Technologies (RCTs): Why and How? 75
   R. Saha, J. Mitra and Alka Paswan

6. Conservation Agriculture and Resource Conservation
   Technologies in Indo-Gangetic Plains: Status and
   Challenges Ahead 93
   D. K. Kundu

7. Impact of Climate Change on Insect Pest Dynamics and
   its Mitigation...107
   S. Satpathy, B.S. Gotyal and V. Ramesh Babu

Approaches of Conservation Agriculture as Climate Change Adaptation Strategies

8. **Crop Diversification in CA /RCTs under Climate Change Perspective: Special Reference to Jute & Allied Fibres**.......119

   A.K. Ghorai and Debarati Datta

9. **Resource Conservation through Crop Residue Management for Sustainable Agriculture**.................135

   M.S. Behera, Laxmi Sharma and Pradipta Samanta

10. **Agroecological Dynamics and Strategic Cultural Management of Weeds in Conservation Agriculture**.................143

    S. Sarkar and B. Majumdar

11. **Prospects of Organic Farming as Resource Conservation Technology**.................157

    Brij Lal Lakaria, Satish Bhagwatrao Aher, Pramod Jha A.B. Singh, B.P., Meena, S. Ramana and J. K. Thakur

12. **Cover Crops: Potential and Prospects in Conservation Agriculture**..167

    Debarati Datta, Sourav Ghosh, R. Saha and C. P. Nath

13. **Biodynamic Farming and Organic Farming: Traditional Approach for Resource Conservation**.................189

    Mahua Banerjee and R. Saha

14. **Environmental Friendly Insect Pest Management under Resource Conservation Technologies**.................211

    B.S. Gotyal, S. Satpathy and V. Ramesh Babu

15. **Scope and Potential of Precision Farming in Conservation Agriculture for Improving Input Use Efficiency** 223

    K.M. Hati, J. Somasundaram, R.S. Chaudhary, R.K. Singh A.K. Biswas, N.K. Sinha and M. Mohanty

16. **Role of GIS, Remote Sensing and Agro Advisory in Conservation Agriculture**.................233

    D. Barman, R. Saha, Tania Bhowmick, Abhishek Bagui Girindrani Dutta and Shikhasri Das

17. **Overview of Crop Growth Models as Support System to Conservation Agriculture**.................249

    Saon Banerjee and Soumen Mondal

Impact of Conservation Agriculture for Natural Resource Management

18. Natural Resource Management Through Conservation Agriculture Under Climate Change Scenario 263

   Debashis Mandal

19. Conservation Agriculture: An Approach Towards Sustainability of Soil Physical Health 283

   A. Kundu, S. Mukherjee, R. Nandi and P. K. Bandyopadhyay

20. Nutrient Dynamics Through Conservation Agriculture Under Climate Change Scenario 309

   A.R. Saha, S.P. Mazumdar and Alka Paswan

21. Conservation Agriculture and Its Impact on Soil Quality in Climate Change Scenario 319

   S.P. Mazumdar, S. Sasmal and Ria Bhattacharya

22. Impact of Conservation Agriculture Practices on Soil Microbial Diversity 335

   B. Majumdar, S. Sarkar, Lipi Chattopadhyay and Shrestha Barai

23. Impact of Conservation Agriculture Practices on Soil Water Dynamics 351

   S. Mitra, R. Saha and N.M. Alam

24. Conservation Agriculture for Enhancing Soil Health and Crop Production 361

   J. Somasundaram, A.O. Shirale, N.K.Sinha, B.P. Meena, K.M. Hati M. Mohanty, A.K. Naorem, A.K. Biswas and A.K. Patra

25. Energy Budgeting and Farm Mechanization in Conservation Agriculture 373

   R.K. Naik and Alka Paswan

Climate Change Mitigation Strategies & Socio Economic Impact of Conservation Agriculture

26. Mining Genetic Resources for Plant Traits Suited to Changing Climatic Conditions 385

   Pratik Satya, Suman Roy, Laxmi Sharma, Soham Ray, Amit Bera Srinjoy Ghosh

27. Scope of Agroforestry Systems for Climate Change Adaptation 405

R. Saha, D. Barman, Suman Roy and Pradipta Samanta

28. Conservation Agriculture Approaches for Reducing Carbon Footprint..417

A.K. Singh

29. Greenhouse Gas Estimation: Techniques and Mitigation Technologies for Reducing Carbon Footprint in Agriculture 433

P. Bhattacharyya, S. R. Padhy and P. K. Dash

30. Prospects and Constraints in Adoption of Conservation Agriculture Practices 449

Shamna. A, S.K. Jha, S. Kumar and M.L. Roy

31. Retrospect and Prospects of Resource Conservation Technologies in Indo-Gangetic Plains 457

S.K. Jha, S. Kumar, Shamna A., M.L. Roy and T. Samajdar

List of Contributors

Aher, S.B., *Scientist, ICMR- National Institute for Research in Environmental Health. Bhauri Bhopal -462 030, Madhya Pradesh*

Alam, N.M., *Senior Scientist, All India Network Project- Jute & Allied Fibres. ICAR- Central Research Institute for Jute & Allied Fibre, Barrackpore . Kolkata-700 121, West Bengal*

Bagui, A., *Senior Research Fellow, Crop Production Division, ICAR- Central Research Institute for Jute & Allied Fibre, Barrackpore , Kolkata-700 121, West Bengal*

Bandyopadhyay, P.K., *Professor, Division of Soil Science & Agricultural Chemistry, Bidhan Chandra Krishi Viswavidyalaya, Mohanpur -741 252, West Bengal*

Banerjee, M. *Assistant Professor, Department of Agronomy, Institute of Agriculture Visva-Bharati, Sriniketan, Birbhum- 731 236, West Bengal*

Banerjee, S., *Professor, Department of Agril. Meteorology & Physics, Bidhan Chandra Krishi Viswavidyalaya, Mohanpur -741 252, West Bengal*

Barai, S., *Senior Research Fellow, Crop Production Division, ICAR- Central Research Institute for Jute & Allied Fibre, Barrackpore, Kolkata-700 121, West Bengal*

Barman, D., *Senior Scientist, Crop Production Division, ICAR- Central Research Institute for Jute & Allied Fibre, Barrackpore, Kolkata-700 121, West Bengal*

Behera. M.S., *Principal Scientist, Crop Production Division, ICAR- Central Research Institute for Jute & Allied Fibre, Barrackpore, Kolkata-700 121, West Bengal*

Bera, A., *Senior Scientist, Crop Improvement Division, ICAR- Central Research Institute for Jute & Allied Fibre, Barrackpore, Kolkata-700 121, West Bengal*

Bhattacharya, R., *Senior Research Fellow, Crop Production Division, ICAR- Central Research Institute for Jute & Allied Fibre, Barrackpore, Kolkata-700 121, West Bengal*

Bhattacharyya, P., *National Fellow & Principal Scientist, Crop Production Division ICAR-National Rice Research Institute, Cuttack -753 006, Odisha*

Bhowmick, T., *Senior Research Fellow, Crop Production Division, ICAR- Central Research Institute for Jute & Allied Fibre, Barrackpore, Kolkata-700 121, West Bengal*

Biswas, A.K., *Head & Principal Scientist, Soil Chemistry & Fertility Division, ICAR-Indian Institute of Soil Science, Nabi Bagh, Bhopal - 462 038, Madhya Pradesh*

Chattopadhyay, L., *Senior Research Fellow, Crop Production Division, ICAR- Central Research Institute for Jute & Allied Fibre, Barrackpore, Kolkata-700 12, West Bengal*

Chaudhary, R.S., *Head & Principal Scientist, Soil Physics Division, ICAR-Indian Institute of Soil Science, Nabi Bagh, Bhopal - 462 038, Madhya Pradesh*

Das, S., *Project Assistant, Crop Production Division, ICAR- Central Research Institute for Jute & Allied Fibre, Barrackpore, Kolkata-700 121, West Bengal*

Dash, P.K., *Research Scholar, Crop Production Division, ICAR-National Rice Research Institute, Cuttack -753 006, Odisha*

Datta, D., *Scientist, Crop Production Division, ICAR- Central Research Institute for Jute & Allied Fibre, Barrackpore, Kolkata-700 121, West Bengal*

Dutta, G., *Young Professional, Crop Production Division, ICAR- Central Research Institute for Jute & Allied Fibre, Barrackpore, Kolkata-700 121, West Bengal*

Ghorai, A.K., *Principal Scientist, Crop Production Division, ICAR- Central Research Institute for Jute & Allied Fibre, Barrackpore, Kolkata-700 121, West Bengal*

Ghosh, P.K., *Director, ICAR–National Institute of Biotic Stress Management Raipur - 493 225, Chhattisgarh*

Ghosh, Sourav, *Scientist, Crop Production Division, ICAR- Directorate of Onion and Garlic Research Rajgugunagar - 410 505, Maharashtra*

Ghosh, Srinjoy, *Young Professional, Crop Production Division,ICAR- Central Research Institute for Jute & Allied Fibre, Barrackpore, Kolkata-700 121, West Bengal*

Gotyal, B.S., *Senior Scientist, Crop Protection Division, ICAR- Central Research Institute for Jute & Allied Fibre, Barrackpore, Kolkata-700 121, West Bengal*

Hati, K.M., *Principal Scientist, Soil Physics Division, ICAR-Indian Institute of Soil Science, Nabi Bagh, Bhopal - 462 038, Madhya Pradesh*

Hazra, K.K., *Scientist, Crop Production Division, ICAR–Indian Institute of Pulses Research, Kanpur – 208 024, Uttar Pradesh*

Jha, P., *Principal Scientist, Soil Chemistry & Fertility Division, ICAR-Indian Institute of Soil Science, Nabi Bagh, Bhopal - 462 038, Madhya Pradesh*

Jha, S.K., *Principal Scientist & In-charge, Agricultural Extension Section, ICAR- Central Research Institute for Jute & Allied Fibre, Barrackpore, Kolkata-700 121, West Bengal*

Kar, G., *Director, ICAR- Central Research Institute for Jute & Allied Fibre, Barrackpore, Kolkata-700 121 West Bengal*

Kumar, S., *Principal Scientist, Agricultural Extension Section, ICAR- Central Research Institute for Jute & Allied Fibre, Barrackpore, Kolkata-700 121, West Bengal*

Kundu, A., *Post Graduate Scholar, Division of Soil Science & Agricultural Chemistry, Bidhan Chandra Krishi Viswavidyalaya, Mohanpur -741 252, West Bengal*

Kundu, D.K., *Former Head, Crop Production Division, ICAR- Central Research Institute for Jute & Allied Fibre, Barrackpore, Kolkata-700 121, West Bengal*

Lakaria, B.L., *Principal Scientist, Soil Chemistry & Fertility Division, ICAR-Indian Institute of Soil Science, Nabi Bagh, Bhopal - 462 038, Madhya Pradesh*

Majumdar, B., *Principal Scientist, Crop Production Division, ICAR- Central Research Institute for Jute & Allied Fibre, Barrackpore, Kolkata-700 121, West Bengal*

Mandal, D., *National Fellow & Principal Scientist, Soil Science & Agronomy Division ICAR- Indian Institute of Soil and Water Conservation, Dehradun -248 195, Uttarakhand*

Mazumdar, S.P., *Senior Scientist, Crop Production Division, ICAR- Central Research Institute for Jute & Allied Fibre, Barrackpore, Kolkata-700 121, West Bengal*

Meena, B.P., *Scientist, Soil Chemistry & Fertility Division, ICAR-Indian Institute of Soil Science, Nabi Bagh, Bhopal - 462 038, Madhya Pradesh*

Mitra, J., *Head & Principal Scientist, Crop Improvement Division, ICAR- Central Research Institute for Jute & Allied Fibre, Barrackpore, Kolkata-700 121, West Bengal*

Mitra, S., *Principal Scientist & In-Charge, All India Network Project- Jute & Allied Fibres ICAR- Central Research Institute for Jute & Allied Fibres, Barrackpore, Kolkata-700 121 West Bengal*

Mohanty, M., *Principal Scientist, Soil Physics Division, ICAR-Indian Institute of Soil Science Nabi Bagh, Bhopal - 462 038, Madhya Pradesh*

Mondal, S., *Research Scholar, Department of Agril. Meteorology & Physics, Bidhan Chandra Krishi Viswavidyalaya, Mohanpur -741 252, West Bengal*

Mukherjee, S., *Post Graduate Scholar, Division of Soil Science & Agricultural Chemistry Bidhan Chandra Krishi Viswavidyalaya, Mohanpur -741 252, West Bengal*

Naik, R.K., *Senior Scientist, Crop Production Division, ICAR- Central Research Institute for Jute & Allied Fibre, Barrackpore, Kolkata-700 121, West Bengal*

Nandi, R., *Post Graduate Scholar, Division of Soil Science & Agricultural Chemistry Bidhan Chandra Krishi Viswavidyalaya, Mohanpur -741 252, West Bengal*

Naorem, A.K., *Scientist, Regional Research Station – Kukma, Bhuj, ICAR-Central Arid Zone Research Institute, Bhuj- 370 105, Gujarat*

Nath, C.P., *Scientist, Crop Production Division, ICAR–Indian Institute of Pulses Research Kanpur – 208 024, Uttar Pradesh*

Padhy, S.R., *Research Scholar, Crop Production Division, ICAR-National Rice Research Institute, Cuttack -753 006, Odisha*

Paswan, A., *Young Professional, Crop Production Division, ICAR- Central Research Institute for Jute & Allied Fibre, Barrackpore, Kolkata-700 121, West Bengal*

Patra, A.K., *Director, ICAR-Indian Institute of Soil Science, Nabi Bagh, Bhopal - 462 038 Madhya Pradesh*

Ramana, S., *Principal Scientist, Environmental Soil Science Division, ICAR-Indian Institute of Soil Science, Nabi Bagh, Bhopal - 462 038, Madhya Pradesh*

Ramesh Babu, V., *Scientist, Crop Protection Division, ICAR- Central Research Institute for Jute & Allied Fibre, Barrackpore, Kolkata-700 121, West Bengal*

Ray, S., *Scientist, ICAR-Indian Agricultural Research Institute, Pusa, New Delhi – 110 012*

Roy, M.L., *Senior Scientist, Agricultural Extension Section, ICAR- Central Research Institute for Jute & Allied Fibre, Barrackpore, Kolkata-700 121, West Bengal*

Roy, S., *Scientist, Crop Production Division, ICAR- Central Research Institute for Jute & Allied Fibre, Barrackpore, Kolkata-700 121, West Bengal*

Saha, A.R., *Principal Scientist & In-Charge, Crop Production Division, ICAR-Central Research Institute for Jute & Allied Fibre, Barrackpore, Kolkata-700 121, West Bengal*

Saha, R., *Principal Scientist, Crop Production Division, ICAR- Central Research Institute for Jute & Allied Fibre, Barrackpore, Kolkata-700 121, West Bengal*

Samanta, P., *Research Assistant, ICAR-Central Research Institute for Jute & Allied Fibre, Barrackpore, Kolkata-700 121, West Bengal*

Sarkar, S., *Principal Scientist, Crop Production Division, ICAR- Central Research Institute for Jute & Allied Fibre, Barrackpore, Kolkata-700 121, West Bengal*

Sasmal, S., *Senior Research Fellow, Crop Production Division, ICAR-Central Research Institute for Jute & Allied Fibre, Barrackpore, Kolkata-700 121, West Bengal*

Satpathy, S., *Head & Principal Scientist, Crop Protection Division, ICAR- Central Research Institute for Jute & Allied Fibre, Barrackpore, Kolkata-700 121, West Bengal*

Satya, P., *Principal Scientist, Crop Improvement Division, ICAR- Central Research Institute for Jute & Allied Fibre, Barrackpore, Kolkata-700 121, West Bengal*

Shamna, A., *Senior Scientist, Agricultural Extension Section, ICAR- Central Research Institute for Jute & Allied Fibre, Barrackpore, Kolkata-700 121, West Bengal*

Sharma, L., *Scientist, Crop Production Division, ICAR- Central Research Institute for Jute & Allied Fibres Barrackpore Kolkata-700 121 West Bengal*

Shirale, A.O., *Scientist, Soil Chemistry & Fertility Division, ICAR-Indian Institute of Soil Science, Nabi Bagh, Bhopal - 462 038, Madhya Pradesh*

Singh, A.B., *Principal Scientist, Soil Biology Division, ICAR-Indian Institute of Soil Science Nabi Bagh, Bhopal - 462 038, Madhya Pradesh*

Singh, A.K., *Principal Scientist, Crop Production Division, ICAR- Central Research Institute for Jute & Allied Fibre, Barrackpore, Kolkata-700 121, West Bengal*

Singh, R.K., *Principal Scientist, Soil Physics Division, ICAR-Indian Institute of Soil Science Nabi Bagh, Bhopal - 462 038, Madhya Pradesh*

Sinha, N.K., *Scientist, Soil Physics Division, ICAR-Indian Institute of Soil Science, Nabi Bagh Bhopal - 462 038, Madhya Pradesh*

Somasundaram, J., *Principal Scientist, Soil Physics Division, ICAR-Indian Institute of Soil Science, Nabi Bagh, Bhopal - 462 038, Madhya Pradesh*

Thakur, J.K., *Scientist, Soil Biology Division, ICAR-Indian Institute of Soil Science, Nabi Bagh, Bhopal - 462 038, Madhya Pradesh*

Viswakarma, A.K., *Principal Scientist, Soil Chemistry & Fertility Division, ICAR-Indian Institute of Soil Science, Nabi Bagh, Bhopal - 462 038, Madhya Pradesh*

Introduction to Conservation Agriculture Under Climate Change Scenario

1

Climate Change and Its Impact in Agriculture and Water Resources

Gouranga Kar

ICAR-Central Research Institute for Jute and Allied Fibres, Barrackpore
West Bengal – 700 121, India

Introduction

Water and food security are the key challenges under climate change as both are highly vulnerable to continuously changing climatic patterns. Climate change has resulted in increases in globally averaged mean annual air temperature and variations in regional precipitation and these changes are expected to continue and intensify in the future. The projected changes in climate patterns over India include increase in surface temperature, variations in rainfall, increasing occurrence of extreme weather events like floods and droughts, rise in sea levels and impact on the Himalayan glaciers. Sectors of the Indian economy are most likely to be impacted by such changes in climate are those dependent on natural resources, namely agriculture, water and forestry. The likely impacts such as reduction in food production, water scarcity, loss of forest biomass Climate change is expected to bring more intense and more frequent extreme weather events including droughts and floods in India. The global atmospheric concentration of carbon di-oxide CO_2 for increased from preindustrial value of about 280 ppm to about 410 ppm at the present (2021). The global increase in CO_2 concentration is primarily due to fossil fuel use and land use change. Atmospheric methane was 1803 ppb in 2011, this is 150% greater than before 1750. Atmospheric nitrous oxide (N_2O) was 324 ppb in 2011, this is 150% greater than before 1750 (Table 1). These increase in GHGs have resulted in warming of the climate system by 0.74 °C between 1906 and 2005.

The challenges faced by agriculture in the coming decades are daunting. The two-way relationship of climate change and agriculture is of utmost significance and wider ranging in developing countries due to their large dependence on agricultural practice for livelihoods. The scale of modification needed for meeting the sustainable development goals, including those of zero poverty, hunger and the imperative action required for addressing climate change will necessitate the transformation of global production systems. Agriculture, the world's most symbolic driver of environmental change (Godfray and Garnett, 2014) is vulnerable to the ever-changing climatic conditions, which are related to diverse emissions and pollutions. Agriculture contributes to alteration in climate through emissions of greenhouse gases (GHGs) such as carbon dioxide, methane and nitrous oxide which are aftermaths of emissions coming directly from fossil fuel use, crop residue burning, tillage practices, fertilized agricultural soils, deforestation and livestock manure in large proportion.

Climate change trend

As per the fifth assessment report of the IPCC (IPCC-AR-5), increase of global mean surface temperatures for 2081–2100 relative to 1986–2005 is projected to likely be in the ranges derived from the concentration-driven CMIP5 model simulations, that is, 0.3°C to 1.7°C (RCP2.6), 1.1°C to 2.6°C (RCP4.5), 1.4°C to 3.1°C (RCP6.0), 2.6°C to 4.8°C (RCP8.5) (Table 1). The IPCC (IPCC-AR-5) concluded the warming of earth's climate system based on direct observation of changes in temperature, sea level and snow cover in the northern hemisphere during 1850 to the present (Table 2) (IPCC, 2014). The accelerated increase in the greenhouse gases (GHG) concentration in the atmosphere is a major cause for climate change. As per the IPCC (2007) report, the maximum growth in the emission of greenhouse gases (GHG) has occurred between 1970 and 2004, i.e. 145% increase from energy supply sector, 120% from transport, 65% from industry, 40% from change in land use patterns and during this period global population increases by 69%. As per the WMO (2013), the world experienced unprecedented high-impact climate extremes during the 2001–2010 decade that was the warmest since the start of modern measurements in 1850.

Table 1: Abundance and lifetime of green house gases in the atmosphere

Parameters	CO_2	CH_4	N_2O	CFCs
Average concentration 100 years ago (ppbv)	290	900	270	0
Current concentration (ppbv) (2007)	390	1774	324	3-5
Projected concentration in the year 2030	400-500	2800-3000	400-500	3-6
Atmospheric life time (year)	5-200	9-15	114	75
Global wanning potential (100 years relative to CO_2)	1	25	298	4750-10900

Source: IPCC (2014)

Table 2: Projected change in global mean surface air temperature and global mean sea level rise for the mid- and late 21st century relative to the reference period of 1986-2005.

		2046-2065		2081-2100	
Global Mean Surface Temperature Change (°C)	Scenario	Mean	Likely range	Mean	Likely range
	RCP2.6	1.0	0.4 to 1.6	1.0	0.3 to 1.7
	RCP4.5	1.4	0.9 to 2.0	1.8	1.1 to 2.6
	RCP6.0	1.3	0.8 to 1.8	2.2	1.4 to 3.1
	RCP8.5	2.0	1.4 to 2.6	3.7	2.6 to 4.8
Global Mean Sea Level Rise (m)	Scenario	Mean	Likely range	Mean	Likely range
	RCP2.6	0.24	0.17 to 0.32	0.40	0.26 to 0.55
	RCP4.5	0.26	0.19 to 0.33	0.47	0.32 to 0.63
	RCP6.0	0.25	0.18 to 0.32	0.48	0.33 to 0.63
	RCP8.5	0.30	0.22 to 0.38	0.63	0.45 to 0.82

So, far much attention has been given to climate change adaptation as an anticipatory and planned process, managed through new policies, technological innovations and development interventions. But these policies and strategies need to be implementation because most of the fresh water resources are depleting at a very fast rate due to unprecedented escalation in demand from domestic, irrigational and industrial sectors. Impact of climate change such as depletion of water resources (Shallow & deep aquifer depletion) and decline in agricultural production has increased and has escalated food inflation globally. The condition is extremely severe in continents like Africa, where most of the northern portion is extremely dry. Western India, Middle East and Arab Countries, where most of the domestic, irrigational and industrial demands are met by Surface and groundwater are facing severe crisis due to depletion of water resources.

Sources of major Greenhouse Gases from agricultural activities

The three major GHGs are carbon dioxide, methane and nitrous oxide which emit from different agricultural activities. A brief description about their sources and sinks is given below.

Carbon Dioxide

The main sources of carbon dioxide emission are burning of fossil fuels, deforestation and land-use changes, decay of organic matter, forest fires, eruption of volcanoes. Within agriculture, soil is the main contributor with factors such as soil texture, temperature, moisture, pH, and available C and

N, influencing CO_2 emission from soil. Emission of CO_2 is more from a tilled soil than from an undisturbed soil (no till). Temperature has a marked effect on CO_2 evolution from soil by influencing root and soil respiration. It may be mentioned that plants, oceans and atmospheric reactions are the major sinks of carbon dioxide.

Methane

Methane is about 25-times more effective as a heat trapping gas than that of CO_2. The main sources of methane are: wetlands, organic decay, termites, natural gas and oil extraction, biomass burning, rice cultivation, cattle and refuse landfills. The primary sources of methane from agriculture include animal digestive processes, rice cultivation and manure storage and handling. The removal in the Stratosphere and soil are the main sinks of methane. In ruminant animals, methane is produced as a by-product of the digestion of feed in the rumen under anaerobic condition. Methane emission is related to the composition of animal diet (grass, legume, grain and concentrates) and the proportion of different feeds (e.g., soluble residue, hemicellulose and cellulose content). Mitigation of methane emitted from livestock is approached most effectively by strategies that reduce feed input per unit of product output.

Nutritional, genetic and management strategies to improve feed efficiency increase the rate of product (milk, meat) output per animal. Because most CH_4 is produced in the rumen by fermentation, practices that speed the passage of feed from the rumen can also reduce methane formation.

Methane is also formed in soil through the metabolic activities of a small but highly specific bacterial group called 'methanogens. Their activity increases in the submerged, anaerobic conditions developed in the wetland rice fields, which limit the transport of oxygen into the soil, and the microbial activities render the water-saturated soil practically devoid of oxygen. The upland, aerobic soil does not produce methane. Water management, therefore, plays a major role in methane emission from soil.

Altering water management practices, particularly mid-season aeration by short-term drainage as well as alternate wetting and drying can greatly reduce methane emission from rice cultivation. Improving organic matter management by promoting aerobic degradation through composting or incorporating soil during off-season drain-period is another promising technique.

Globally, 70–80% anthropogenic land uses are taken up by livestock, especially ruminants (FAO 2009; Bellarby *et al.*, 2013) which consume about 35% of agricultural produce (Foley *et al.*, 2011). This sector is in direct competition with potential complementary land uses, such as bioenergy crop production and natural land conservation (Smith *et al.*, 2010; Phelps and Kaplan, 2017). Methane is produced from enteric fermentation of ruminant animals while denitrification and nitrification processes in soils are major sources of N_2O. Methane (30–39% of total emissions for dairy), nitrous oxide (17–22% of total emissions for dairy), carbon dioxide from energy consumption (about 15%) and CO_2 fluxes from land-use changes (14–38%) are the major shareholder in GHG emission in livestock farming (Weiss and Leip, 2012).

The CO_2 from the atmosphere is fixed through photosynthesis during crop growth and emitted as CO_2 *via* respiration of plant and soil microbes and manure decomposition. Animal respiration, manure storages and barn floors are major sources of CO_2 on dairy farms (Sejian *et al.*, 2018). Methane is emitted from enteric fermentation and released by eructation and respiration, manure storages and manure deposited by animals inside barns or on pasture (Sejian *et al.*, 2018). Denitrification and nitrification processes in soils cultivated for fodder, surface of slurry manure storage, stacked manure, bedded pack manure on barn floors, and on manure-laden dry lot surfaces act as N_2O sources in dairy farm.

Nitrous Oxide

As a greenhouse gas, nitrous oxide is 298-times more effective than CO_2. Forests, grasslands, oceans, soils, nitrogenous fertilizers, and burning of biomass and fossil fuels are the major sources of nitrous oxide, while it is removed by oxidation in the Stratosphere. Soil contributes to the largest amount of nitrous oxide emission. The major sources are soil cultivation, fertilizer and manure application, and burning of organic material and nitrous oxide emission from soil represents a loss of soil nitrogen, reducing the nitrogen-use efficiency.

Appropriate crop management practices, which lead to increased N-use efficiency, hold the key to reduce nitrous oxide emission. Site-specific nutrient management, fertilizer placement and proper type of fertilizer supply nutrients in a better accordance with plant demands, thereby reduce nitrous oxide emission.

Agriculture contributes around 80% of the N_2O to the atmosphere annually by human activities out of which 60% is a resultant of soil management (Pires *et al.*, 2015). Total N_2O emission from agriculture sector was about 2.11 Pg in 2008 (World Bank 2012). Nitrous oxide has a high global warming potential (310 times that of CO_2) (UNFCCC, 2004), which makes it a large contributor to GHG budgets. Environmental factors like precipitation and potential evapotranspiration during N fertilization affect intensity of these emissions (Carlson *et al.*, 2017; Millar *et al.*, 2018). For instance, direct leaching and emissions of N_2O due to N fertilization are proportional to the ratio of precipitation to potential evapotranspiration (Tongwane *et al.*, 2016). Land application of manures and fertilizers contributes in almost half of agricultural GHG emission (Ren *et al.*, 2017). Type of manure, methods and rates of application determine the intensity and magnitude of GHG emission footprint. Solid manure having recalcitrant carbon and nitrogen, has lower potential GHG emission footprint than liquid manures (Aguirre-Villegas and Larson, 2017). Fertilization rate has a direct positive correlation with nitrous oxide fluxes (Zhang *et al.*, 2016) and incorporation of manure in soil results in less vulnerable emission than surface application.

According to a report, considering crop cultivation and dairying, rice cultivation (10%), use of synthetic fertilizer (12.6%), burning crop residue and pasture (4.7%), left over crop residues (4.7%), enteric emissions (39.7%), manure left on pastures (16.1%), poor manure management (6.7%) and manure application (3.7%) contribute significantly to total GHG emission (Patra and Babu, 2017).

Impacts of Climate Change on Agriculture and Allied Sectors

Impacts of climate change on crop

Global climatic changes can affect agriculture through their direct and indirect effects on the crops, soils, livestock and pests. An increase in atmospheric carbon dioxide level will have a fertilization effect on crops with C3 photosynthetic pathway and thus will promote their growth and productivity. The increase in temperature, depending upon the current ambient temperature, can reduce crop duration, increase crop respiration rates, alter photosynthate partitioning to economic products, affect the survival and distribution of pest populations, hasten nutrient mineralization in soils, decrease fertilizer-use efficiencies, and increase evapo-transpiration rate. Indirectly, there may be considerable effects on land use due to snow melt, availability of irrigation water, frequency and intensity of inter- and intra-seasonal droughts and floods, soil organic matter transformations, soil erosion, changes in pest profiles, decline in arable areas due to submergence of coastal lands, and availability of energy. Equally important determinants of food supply are socio-economic environment, including government policies, capital availability, prices and returns, infrastructure, land reforms, and inter and international trade that might be affected by the climatic change. In brief the impacts of climate change on agriculture and allied sectors are given below:

- Yield of major cereals crops, especially wheat is likely to be reduced due to decrease in grain filling duration, increased respiration, and / or reduction in rainfall/irrigation supplies.

- Increase in extreme weather events such as floods, droughts, cyclones and heat waves will adversely affect agricultural productivity.

- Reduction in yields in the rainfed areas due to changes in rainfall pattern during monsoon season and increased crop water demand.

- Incidence of cold waves and frost events may decrease in future due to global warming and it would lead to a decreased probability of yield loss associated with frost damage in northern India in crops such as mustard and vegetables.

- Quality of fruits, vegetables, tea, coffee, aromatic, and medicinal plants may be affected.

- Incidence of pest and diseases of crops to be altered because of more enhanced pathogen and vector development, rapid pathogen transmission and increased host susceptibility.

- Agricultural biodiversity is also threatened due to the decrease in rainfall and increase in temperature, sea level rise, and increased frequency and severity of droughts, cyclones and floods.

Simulating the effects of elevated temperature and CO_2 on growth and productivity of winter maize- a case study

The climate change impacts on crop growth and yield were assessed with weather series representing both the current (average of 24 years, 1985-2008) and changed climates (Current maximum and minimum temperatures + 1°C and + 2°C). The weather series for simulations in the changed climate was modified accordingly and run under both present (370 ppm) and enhanced CO_2 level (550 ppm). The impact of change in CO_2 concentration i.e. from 370 ppm to 550 ppm, under current climatic condition (average of 24 years, 1985-2008) on phenology, grain yield and total dry matter of maize was also studied (Table 3 and 4). In this investigation, DSSAT 4.5 model was calibrated and evaluated to study the effect of elevated temperature and carbon dioxide and their interaction on growth and productivity of maize crop. Study revealed that 1.0°C increase in temperature from the current (average of 1985-2008) reduced crop duration of November sown crop by 8 days under both 370 ppm (current) and 550 ppm CO_2 concentrations. Higher days of reduction in crop duration was observed (13 days) when 2.0°C temperature was increased from the current. According to model simulations with 2.0°C increase in current temperature, yield of maize will be reduced up to 13.8% under 370 ppm CO_2 concentration. The increase in temperature had negative effects on leaf area index (LAI), total dry matter and yield; however, these effects were found lesser under 550 ppm CO_2 concentrations as compared to 370 ppm. 4.2 Impact of climate change on water resources Changing global climatic patterns

8 Conservation Agriculture and Climate Change

coupled with declining per capita availability of surface and ground water resources have made sustainable agricultural a great challenge in India.

Table 3: Summary of observed and simulated results during model calibration with crop data of 2007-08 under current climatic condition

Crop parameters	Unit	Observed	Simulated	RMSE	Error (%)	d-index
Anthesis	days	55	54	0.42	- 1.81	0.96
Maturity	days	113	112	0.58	-0.88	0.87
Maximum LAI	-	6.4	6.0	0.36	-6.25	0.80
Grain yield	kg ha^{-1}	5847	5805	198.7	-0.72	0.94
Total Biomass	kg ha^{-1}	13440	13250	245.6	- 1.41	0.92

Table 4: Impact of temperature and CO_2 scenarios on phenology, growth, yield and yield component of maize (cv. Novjyot)

Temperature	Current (av. of 1985-2008)	1.0 °C	Difference (absolute)	2.0 °C	Difference (absolute)
CO_2 Concentration : 370 ppm					
Crop duration (days)	114	106	8	101	13
Maximum leaf area index	6.1	5.6	0.5	5.45	0.65
Total above ground dry biomass (kg ha^{-1})	13600	12306	1294	11515	2085
No. of grains m^{-2}	2357	2340	17	2314	43
Grain yield (kg ha^{-1})	5800	5318	482	4996	804
CO_2 Concentration : 550 ppm					
Crop duration (days)	114	106	8	101	13
Maximum leaf area index	6.3	5.8	0.5	5.65	0.65
Total above ground dry biomass (kg ha^{-1})	14304	12805	1499	12374	1930
No. of grains m^{-2}	2445	2345	100	2325	120
Grain yield (kg ha^{-1})	6094	5504	590	5298	796

Warming of the climate system in recent decades is unequivocal, as is now evident from observations of increases in global average air and ocean temperatures, widespread melting of snow and ice and rising global sea level (IPCC, 2014). The Inter-Governmental Panel on Climate Change (IPCC) in its 5th Assessment Report (2014) has observed the following impacts of climate change on water resources.

- The global water cycle will be affected in response to the warming.

- The contrast in precipitation between wet and dry regions and seasons will increase.

- The ocean will continue to warm causing heat to penetrate deeper to affect ocean circulation.
- Global mean sea level will continue to rise.
- Arctic and Antarctic ice cover will continue to shrink.
- Global glacier volume will further decrease.
- Some of the impacts of climate change on water resources are given in table-5.

Table 5: Impact of climate change on India's water resources during the next century.

Region/location	Impact
Indian subcontinent	Increase in monsoonal and annual run-off in the central plains No substantial change in winter run-off Increase in evaporation and soil wetness during monsoon and on an annual basis
Odisha and West Bengal	Sea-level rise by 1 m inundating 1700 km^2 of prime agricultural land
Indian coastline	One metre sea-level rise on the Indian coastline is likely to affect a total area of 5763 km^2 risking 7.1 million people
All-India	Increases in potential evaporation across India
Central India	Basin located in a comparatively drier region is more sensitive to climatic changes
Kosi Basin	Decrease in discharge of the Kosi River and decrease in run-off by 2-8%
Southern and Central India	Soil moisture increases marginally by 15-20% during monsoon months
Chenab River	Increase in discharge in the Chenab River
River basins of India	General reduction in the quantity of the available run-off, increase in Brahmini basin
Mahanadi river basin	Increasing run-off and intensities of flood
Damodar Basin	Decreased river flow
Rajasthan	Increase in evapo-transpiration
Kasnabati river basin	Increase in potential evapo-transpiration and lateral flow
Lower Brahmaputra	Increased peak flow with less frequent low flows
Sutluj Basin	Little change in total stream flow but substantial change in the distribution of stream flow

Impact of elevated temperature on crop evapotranspiration and water footprints of some winter season crops

Climate change due to increase in temperature rise will demand higher amount of water for irrigation. At the same time the higher temperature will change the crop physiology and shorten the crop growth period which in turn will reduce the irrigation days. These contradictory phenomena will change the total irrigation water demand which is required to quantify for long-term water resources planning and management. Assuming the same crop duration, as a case study the crop evapotranspiration of some winter season crops of Dhenkanal, Orissa under current and projected climate change scenario (RCP 8.5) were determined and are given in Table 5. Irregularities in onset of monsoon, drought, flood and cyclone Indian agriculture is highly dependent on the onset, retreat and magnitude of monsoon precipitation, particularly in the rainfed areas of east, north-east and south India. Climate modelers and IPCC documents have projected possibilities of increasing variability in Asian Monsoon circulation in a warmer world. Despite expansion of area under irrigation, droughts, caused by inadequate and uneven distribution of rainfall, continue to be the most important climatic aberrations, which influence the agricultural production in India. The severity of a drought will be intensified in a warmer climate change would be a major problem in the Indian subcontinent. Rise in sea level In South, South East and East Asia about 10% of the regional rice production, which is enough to feed 200 million people, is from the areas that are susceptible to 1 m rise in the sea level. Direct loss of land combined with less favourable hydraulic conditions may reduce rice yields by 4% if no adaptation measures are taken, endangering the food security of at least of 75 million people. Saltwater intrusion and soil salinization are other concerns for agricultural productivity.

Table 6: Impact of elevated temperature on crop evapo-transpiration (ETc, mm) of some winter season crops under scenario of RCP 8.5 (Study area: Dhenkanal, Odisha: Sowing - last week of November)

Crops	Current ETc (mm) 2010	2050 RCP (8.5)	% Increase from Current	2070 RCP (8.5)	% Increase from Current	2095 RCP (8.5)	% Increase from Current
Potato	409.4	429.8	4.7	443.4	8.3	459.3	12.2
Blackgram	332.5	342.7	3.1	362.4	9.0	378.1	13.7
Sunflower	426.7	444.7	4.7	461.7	8.2	476.2	11.6
Wheat	396.0	418.3	5.3	434.0	9.6	447.1	12.9
Chickpea	377.6	395.8	4.8	410.1	8.6	419.5	11.1
Safflower	414.1	434.8	4.5	450.1	8.7	470.0	13.5
Mustard	375.1	398.1	5.9	409.2	9.1	417.1	11.2
Linseed	281.2	295.4	4.7	309.9	10.2	315.5	12.2
Rapeseed	331.4	348.7	4.5	360.2	8.7	366.5	10.6
Tomato	514.6	545.4	5.1	565.0	9.8	574.3	11.6
Cabbage	500.5	526.2	4.3	551.6	10.2	557.6	11.4
Cauliflower	496.1	523.7	4.4	539.8	8.8	564.1	13.7
Okra	420.7	441.4	3.9	461.9	9.8	474.1	12.7
Carrot	398.0	420.8	4.8	442.6	11.2	447.4	12.4

Temporary deficits of water resources are still possible because of increased fluctuations of stream flow caused by higher volatility of precipitation and increased evaporation during all seasons and of seasonal cutbacks because of lower accumulation of snow and ice. Clean water supply may also decrease due to a warmer environment inducing lower water quality. Algae-producing toxins could damage the quality of sources such as lakes. Such overall decline in renewable water supply will intensify competition for water among agriculture, ecosystems, industry, and energy production, affecting regional water, energy, and food security. The increased temperature leads to increase in potential evapotranspiration which will result in more water demand for irrigation and ultimately lowering of groundwater table at some places. It has been estimated that the water footprint in wheat under Indo-Gangetic plains is 800-1700 m^3/ton during 2016, which will be around 1100-2000 m^3/ton 2090 (Fig.1). Decrease in rice yield by 20% due to water shortage under climate change conditions is predicted in India. There will be short term increased availability of water in the rivers and their tributaries due to melting of glaciers in the Himalayas; but in the long run, water availability will decrease considerably. Increased erratic nature of precipitation may increase runoff losses and floods in the wet season. The water balance indifferent parts of India will be disturbed and the quality of groundwater along the coastal track will be affected more due to intrusion of sea waters (Pathak *et al.*, 2013).

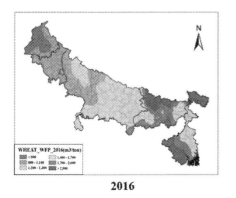
2016
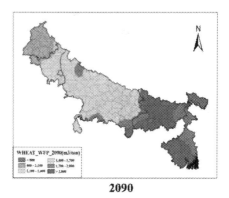
2090

Fig. 1: Impact of climate change (RCP 4.5) on water footprints of wheat in Indo-Gangetic region

Effect on Livestock

Climate change will affect fodder production and nutritional security of livestock. Increased temperature would enhance lignification of plant tissues, reducing the digestibility. Increased water scarcity would also decrease production of feed and fodder.

- Major impacts on vector-borne diseases will be through expansion of vector populations in the cooler areas. Changes in rainfall pattern may also influence expansion of vectors during wetter years, leading to large outbreaks of diseases.
- Global warming would increase water, shelter, and energy requirement of livestock for meeting the projected milk demands.
- Climate change is likely to aggravate the heat stress in dairy animals, adversely affecting their reproductive performance.

Effect on Fisheries

- Increasing temperature of sea and river water is likely to affect breeding, migration and harvests of fishes.
- Impacts of increased temperature and tropical cyclonic activity would affect the capture, production and marketing costs of the marine fish.
- Coral bleaching is likely to increase due to higher sea surface temperature.

Impact of climate change on soil fertility and microbes

- Soil temperature affects the rates of organic matter decomposition and release of nutrients. At high temperatures, though nutrient availability will increase in the short-term, in the long-run organic matter content will diminish, resulting in a decline in soil fertility.
- Organic matter content, which is already quite low in Indian soils, would become still lower. Quality of soil organic matter may be affected.
- The residues of crops under the elevated CO_2 concentrations will have higher C:N ratio, and this may reduce their rate of decomposition and nutrient supply.
- Rise in soil temperature will increase N mineralization, but its availability may decrease due to increased gaseous losses through processes such as volatilization and denitrification.
- There may be a change in rainfall volume and frequency, and wind may alter the severity, frequency and extent of soil erosion.
- Rise in sea level may lead to salt-water ingression in the coastal lands, turning them less suitable for conventional agriculture.

Impact of climate change on pests and diseases

As temperature increases, the insect-pests will become more abundant through a number of interrelated processes, including range extensions and phonological changes, as well as increased rates of population development, growth, migration and over-wintering. The climate change is likely to alter the balance between insect pests, their natural enemies and their hosts. The rise in temperature will favour insect development and winter survival.

Rising atmospheric carbon dioxide concentrations may lead to a decline in food quality for plant feeding insects, as a result of reduced foliar nitrogen levels. The epidemiology of plant diseases will be altered. The prediction of disease outbreaks will be more difficult in periods of rapidly changing climate and unstable weather. Environmental instability and increased incidence of extreme weather may reduce the effectiveness of pesticides on targeted pests or result in more injury to nontarget organisms.

The role of weather in outbreak of plant diseases has been realized from early 20th century. In the last years of the Second World War (1943), Bengal had to face a serious famine due to outbreaks of *Helminthosporium* in rice (Brown spot). The high yielding, short duration and more fertilizer responsive varieties have been found more susceptible to diseases and have caused great havoc in time to time in many parts of the country. The incidence of wheat rust in Madhya Pradesh in 1947, ergot disease of bajra (pearl millet) in Haryana in 1976. Downy mildew in Rajasthan in 1994, Tungro virus of rice in U.P and Bihar in 1966, Virus diseases of rabi in Karnataka and Andhra Pradesh are worth mentioning. The studies of weather in relationship with disease infestation are necessary to take preventive measures. Early detection of inoculums of infection is very often necessary in determining whether a given or anticipated weather situation will cause an outbreak of a disease in epidemic form. Short and medium range forecasts on disease infestation based on eminent weather will be useful for farmers to take protective measures. In case of air borne diseases where the infective spores get emigrated from distant regions, aero mycological observations are required to issue warning about the arrival of inoculum.

Irregularities in Onset of Monsoon, Drought, Flood and Cyclone

Indian agriculture is highly dependent on the monsoon precipitation; irregularities in its onset, erratic nature, drought and frequent dry spell. Increase in extreme weather events such as floods, droughts, cyclones and heat waves will adversely affect agricultural productivity and biodiversity. Floods are expected to occur more frequently on more than half of the earth's surface. During winter, snowfalls are expected to decrease in mid-latitudes, resulting in less significant snowmelt during spring. In the past decade, climate extremes such as droughts have intensified in warmer zones resulting in episodes of price hikes for food. Severe droughts will put additional pressure on existing water supply systems of dry areas but could be manageable in wetter areas, assuming adaption measures are timely implemented. Kar *et al.*, (2019) projected that 8.5 to 14.3% water footprints of rice and wheat are to be increased in different districts of Indo-Gangetic region at the end of this century.

Impact of climate change on Groundwater in India

Impact of climate change on the groundwater regime is expected to be severe. The most optimistic assumption suggests that an average drop in groundwater level by one meter would increase India's total carbon emissions by over 1%, because the time of withdrawal of the same amount of water will increase fuel consumption. A more realistic assumption reflecting the area projected to be irrigated by groundwater, suggests that the increase in carbon emission could be 4.8% for each meter drop in groundwater levels (Mall *et al.*, 2006). Climate change is likely to affect groundwater due to changes in precipitation and evapotranspiration.

Rising sea levels may lead to increased saline intrusion into coastal and island aquifers, while increased frequency and severity of floods may affect groundwater quality in alluvial aquifers. Sea level rise leads to intrusion of saline water into the fresh groundwater in coastal aquifers and thus adversely affects groundwater resources. For two small and flat coral islands at the coast of India, the thickness of freshwater lens was computed to decrease from 25 m to 10 m and from 36 m to 28 m, respectively, for a sea level rise of only 0.1 m (Mall *et al.*, 2006). It is projected that most irrigated areas in India would require more water around 2025 and global net irrigation requirements would increase relative to the situation without climate change by 3.5–5% by 2025 and 6–8% by 2075 (Kumar, 2012). In India, roughly 52% of irrigation consumption across the country is extracted from groundwater; therefore, it can be an alarming situation with decline in groundwater and increase in irrigation requirements due to climate change. Change in climate will affect the soil moisture, groundwater recharge and frequency of flood or drought episodes and finally groundwater level in different areas. In a number of studies, it is projected that increasing temperature and decline in rainfall may reduce net recharge and affect groundwater levels. By taking the impact of climate change on groundwater, to meet the demand of water, it requires to recharge the water for enhancing the level of groundwater in the aquifer.

Rise in Sea Level

In Asia about 10% of the regional rice production, which is enough to feed 200 million people, is from the areas that are susceptible to 1 m rise in the sea level. Direct land loss together with unfavourable hydraulic conditions may reduce rice yields by 4% if no adaptation measures are taken, endangering the food security of at least of 75 million people. Saltwater envision in fresh water bodies will be the other prime issues promoting water scarcity conditions (Mimura, 2013). Forecasted ocean level rise will threaten crucial food-producing areas along the coasts of India and Bangladesh, the major rice producers.

Need to Develop Climate Resilient Agriculture

The Indian agricultural production system faces the daunting task of having to feed 17.5 percent of the global population with only 2.4 per cent of land and 4 per cent of the water resources at its disposal. With the continuously degrading natural resource base compounded further by global warming and associated climate changes resulting in increased frequency and intensity of extreme weather events, "business as usual" approach will not be able to ensure food and nutrition security to the vast population as well as environmental security (the need of the hour). The challenge is formidable because more has to be produced with reduced carbon and water footprints. To achieve this task of paving the way for climate smart agriculture we need to take several measures that will have enabling policies, institutions and infrastructure in place and the farming community be better informed and empowered with necessary resources.

Climate resilient agriculture (CRA), encompassing adaptation and mitigation strategies and the effective use of biodiversity at all levels - genes, species and ecosystems-is thus an essential prerequisite for sustainable development in the face of changing climate CRA means the incorporation of adaptation, mitigation and other practices in agriculture which increases the capacity of the system to respond to various climate related disturbances by resisting damage and recovering quickly. Such perturbations and disturbances can include events such as drought, flooding, heat/cold wave, erratic rainfall pattern, long dry spells, insect or pest population explosions and other perceived threats caused by changing climate. In short it is the ability of the system to bounce back.

Climate resilient agriculture includes an in-built property in the system for the recognition of a threat that needs to be responded to, and also the degree of effectiveness of the response. CRA will essentially involve judicious and improved management of natural resources viz., land, water, soil and genetic resources through adoption of best bet practices.

Mitigation Strategies to Climate Change

The strategies for mitigating methane emission from rice cultivation could be alteration in water management, particularly promoting mid-season aeration by short-term drainage; improving organic matter management by promoting aerobic degradation through composting or incorporating it into soil during off-season drained period; use of rice cultivars with few unproductive tillers, high root oxidative activity and high harvest index; and application of fermented manures like biogas slurry in place of unfermented farmyard manure (Pathak and Wassmann, 2007). Methane emission from ruminants can be reduced by altering the feed composition, either to reduce the percentage which is converted into methane or to improve the milk and meat yield.

The most efficient management practice to reduce nitrous oxide emission is site-specific, efficient nutrient management (Pathak, 2010). The emission could also be reduced by nitrification inhibitors such as nitrapyrin and dicyandiamide (DCD). There are some plant-derived organics such as neem oil, neem cake and karanja seed extract which can also act as nitrification inhibitors.

Mitigation of CO_2 emission from agriculture can be achieved by increasing carbon sequestration in soil through manipulation of soil moisture and temperature, setting aside surplus agricultural land, and restoration of soil carbon on degraded lands. Soil management practices such as reduced tillage, manuring, residue incorporation, improving soil biodiversity, micro aggregation, and mulching can play important roles in sequestering carbon in soil. Some technologies such as intermittent drying, site-specific N management, etc. can be easily adopted by the farmers without additional investment, whereas other technologies need economic incentives and policy support (Wassmann and Pathak, 2007).

Conclusion

Climate change is a key political issue and its consequences, such as food insecurity are already generating conflict in vulnerable regions around the globe. Global warming projections suggest that climate change impacts will vary greatly among regions, and happen on different time scales. The expected impacts of climate change such as extreme temperatures, flooding, droughts, rising ocean levels will not only exacerbate existing tensions but will also be a major challenge for livelihood security. It is important to keep in mind that a myriad of interrelations exist among communities worldwide and thus, effect of climate change occurring in a particular region may trigger ripple effects around the globe *via* internationally connected systems like the economy. Moreover, climate change may modify migration patterns of human beings, thus triggering collateral consequences elsewhere, even in remote areas. Therefore, sustainable solutions to address reduction in water resources, food security issues, or energy challenges should consider this relationship to avoid having unintended collateral consequences in other areas. Integration of agricultural production has a huge potential to enhance nutrient recycling or recovery. Sustainable agricultural systems aim to produce high-quality and affordable food in sufficient quantity and at the same time leave low impact on environment. The agronomical tactics, such as crop diversification, higher N use efficiency, intensified crop rotation by reducing summer fallow, soil carbon sequestration, crop residue retention, minimum soil disturbance and integrated farming approach, are the torch bearers for reducing GHG emission. Integration of these improved farming practices together enables to reduce the use of mineral fertilizers, increase the system productivity and lower the carbon footprint. Strategies like anti-methanogenic chemical, increasing diet digestibility, improving health and welfare, increasing reproductive efficiency and breeding for higher productivity have been identified as GHG emission reduction strategies from the livestock sector. Further, algae have enormous potential to reduce CH_4 emission form ruminant and recover nutrient from wastewater. Nations capable of producing green food products and feed ingredients would benefit financially in the future. Widespread adoption of eco-friendly agriculture would be essential to mitigate GHGs and increase soil carbon storage; thus, accomplishing carbon neutrality in farm sector as a realistic goal. Eventually, by implementation of appropriate policies in all sectors of farming, along with the adoption of improved technologies, food security with minimum C footprint can be achieved effectively, and economically.

References

Aguirre-Villegas, H.A. and Larson, R.A. (2007). Evaluating greenhouse gas emissions from dairy manure management practices using survey data and lifecycle tools. *Journal of Cleaner Production*, 143:169–179.

Bellarby, J., Tirado, R., Leip, A., Weiss, F., Lesschen, J.P., Smith, P. (2013). Livestock greenhouse gas emissions and mitigation potential in Europe. *Global Change Biology*, 19: 3–18.

Carlson, K. M., Gerber, J. S., Mueller, N.D., Herrero, M., Mac Donald, G.K., Brauman, K.A., Havlik, P., O'Connell C.S., Johnson, J.A., Saatchi, S. and West, P.C. (2017). Greenhouse gas emissions intensity of global croplands. *Nature Climate Change*, 7: 63.

FAO (2009). The state of food and agriculture. Livestock in balance. FAO, Rome.

Foley, J.A., Ramankutty, N., Brauman, K.A., Cassidy, E.S., Gerber, J.S., Johnston, M., Mueller, N. D., O'Connell C, Ray, D.K., West, P.C., Balzer, C. (2011). Solutions for a cultivated planet. *Nature*, 478: 337.

Godfray, H.C.J. and T. Garnett, (2014): Food security and sustainable intensification. *Philosophical Transactions of the Royal Society B: Biological Sciences.*, 369: 20120273–20120273, doi:10.1098/rstb.2012.0273.

Mall, R.K., Singh, R., Gupta, A., Srinivasan, G. and Rathore, L.S. (2006). Impact of climate change on Indian agriculture: a review. *Climate Change*, 78(2): 445-478. .

IPCC (2014). Inter-Governmental Panel on Climate Change Fifth Assessment Report Synthesis.

IPCC (2012). Managing the Risks of Extreme Events and Disasters to Advance Climate Change Adaptation. [Field, C.B., V. Barros, T.F. Stocker, D. Qin, D.J. Dokken, K.L. Ebi, M.D. Mastrandrea, K.J. Mach, G.-K. Plattner, S.K. Allen, M. Tignor, and P.M. Midgley (eds.)]. A Special Report of Working Groups I and II of IPCC Intergovernmental Panel on Climate

18 Conservation Agriculture and Climate Change

Change. Cambridge University Press, Cambridge, United Kingdom and New York, USA, 594 pp.

IPCC (2018). An IPCC Special Report on the Impacts of Global Warming of 1.5_C Above Pre-Industrial Levels and Related Global Greenhouse Gas Emission Pathways, in the Context of Strengthening the Global Response to the Threat of Climate Change, Sustainable Development, and Efforts to Eradicate Poverty. Summary for Policymakers. Geneva: Intergovernmental Panel on Climate Change, Available at: http://report.ipcc.ch/ sr15/pdf/ sr15_spm_final.pdf.

Kar, G., Mohany, S., Anand, P.S.B., Panda. D.K., Raviraj, A., Rank, R.D., Singh, P.K. (2019). *National Innovations for Climate Resilient Agriculture, Annual Report*, 2018-19: p14-18. Published by Indian Institute of Water Management, Bhubaneswar.

Kumar, C.P. (2012). Climate Change and Its Impact on Groundwater Resources. *International Journal of Engineering and Science*, 1(5): 43-60.

Mall, R.K., Singh, R., Gupta, A., Srinivasan, G. and Rathore, L. S. (2006). Impact of climate change on Indian agriculture: a review. *Climatic Change*, 78: 445–478.

Millar, N., Urrea, A., Kahmark, K., Shcherbak, I., Robertson, G. P. and Ortiz Monasterio, I. (2018). Nitrous oxide flux responds exponentially to nitrogen fertilizer in irrigated wheat in the Yaqui Valley, Mexico. *Agriculture, Ecosystems & Environment*, 261: 125–132.

Mimura Nobuo (2013). Sea-level rise caused by climate change and its implications for society. *Proceedings of the Japan Academy, Series B.*, 89(7): 281- 301.

MoEFCC (2018). India: Second Biennial Update Report to the United Nations Framework Convention on Climate Change. Ministry of Environment, Forest and Climate Change, Government of India.

Nicola, L., (2019). Impact of climate change on animal health and welfare. *Animal Frontiers*, 9(1): 26-3.

OECD (2015). Agriculture and Climate Change. Trade and Agriculture Directorate. (http://www.oecd.org/tad/sustainable-agriculture/agriculture-and-climate change.html).

Pathak, H. (2010). Mitigating greenhouse gas and nitrogen loss with improved fertilizer management in rice: Quantification and economic assessment. *Nutrient Cycling in Agroecosystems*, 87: 443-45.

Pathak, H. and Wassmann, R. (2007). Introducing greenhouse gas mitigation as a development objective in rice-based agriculture: I. Generation of technical coefficients. *Agricultural Systems*, 94: 807-825.

Pathak, H., Sankhyan, S., Dubey, D.S., Bhatia, A., and Jain, N. (2013). Dry direct-seeding of rice for mitigating greenhouse gas emission: Field experimentation and simulation. *Paddy* and *Water Environment*, 11(1): 593–601. https://doi.org/10.1007/s10333-012-0352-0.

Patra, N.K. and Babu, S.C. (2017). Mapping Indian agricultural emissions. Available from http://www.indiaenvironmentportal.org.in/files/file/Mapping%20Indian%20Agricultural%20 Emissions.pdf.

Phelps, L.N., Kaplan, J.O. (2017). Land use for animal production in global change studies: Defining and characterizing a framework. *Global Change Biology*, 23: 4457–4471.

Pires, M.V., da Cunha, D.A., de Matos Carlos S and Costa, M.H. 2015. Nitrogen use efficiency, nitrous oxide emissions, and cereal production in Brazil: current trends and forecasts. *PLoS One*, 10: e0135234.

Ren, F., Zhang, X., Liu, J., Sun, N., Wu, L., Li, Z. and Xu, M. (2017). A synthetic analysis of greenhouse gas emissions from manure amended agricultural soils in China. *Scientific Reports*, 7: 8123. DOI:10.1038/ s41598-017-07793-6.

Sejian, V., Prasadh, R.S., Lees, A.M., Lees, J.C., Al-Hosni, Y.A., Sullivan, M.L. and Gaughan, J.B. (2018). Assessment of the carbon footprint of four commercial dairy production systems in Australia using an integrated farm system model. *Carbon Management*, 9: 57–70.

Smith, P., Gregory, P.J., Vuuren, D. van, Obersteiner, M., Havlı́k, P., Rounsevell, M., Woods, J., Stehfest, E. and Bellarby, J. (2010). Competition for land. *Philosophical Transactions of the Royal Society B: Biological Sciences*, 365: 2941–2957.

Syakila, A. and Kroeze, C. (2011). The global nitrous oxide budget revisited. *Greenhouse gas measurement and management*, 1(1): 17-26.

Tongwane, M, Mdlambuzi, T., Moeletsi, M., Tsubo, M., Mliswa, V. and Grootboom, L. (2016). Greenhouse gas emissions from different crop production and management practices in South Africa. *Environmental Development*, 19: 23–35.

UNFCCC. (2004). United Nations Framework Convention on Climate Change. Available at: http://unfccc.int/2860.php.

Wassmann, R. and Pathak, H. (2007). Introducing greenhouse gas mitigation as a development objective in rice-based agriculture: II. Cost-benefit assessment for different technologies, regions and scales. *Agricultural Systems*, 94: 826-84.

Weiss, F. and Leip, A. (2012). Greenhouse gas emissions from the EU livestock sector: a life cycle assessment carried out with the CAPRI model. *Agriculture, Ecosystems & Environment*, 149: 124–13.

World Bank, (2012). Carbon sequestration in agricultural soils. Washington, DC: The World Bank.

Zhang, J., Jiang, J. and Tian, G. (2016). The potential of fertilizer management for reducing nitrous oxide emissions in the cleaner production of bamboo in China. *Journal of Cleaner Production*, 112: 2536–2544.

2

Climate Smart Agriculture: Special Reference to Conservation Agriculture

P. Bhattacharyya

*ICAR-National Rice Research Institute (NRRI), Cuttack
Odisha-753 006, India*

Introduction

Climate smart agriculture (CSA) refers to the incorporation of adaptation and mitigation practices in agriculture that enhances the system-resistance and recovery-mechanism against climatic hazards. Climatic hazards / disturbances include drought, flood, heat / cold wave, erratic precipitation, sudden dry spell *etc.* In nutshell, CSA is the ability of the system to bounce back. It can also be explained as an approach to build-up resilience of agricultural systems, to increase adaptive capacities of farming communities to climate change and variability, and to ensure food security (Prasad *et al.*, 2015).

Climate change can be defined as a statistically significant variation in either the mean state of the climate or in its variability, persisting of an extended period, typically three decades or longer (Pathak *et al.*, 2014; NAAS, 2013, Bhattacharyya *et al.*, 2016). The surface air temperature, sea level rise, occurrence of extreme events (heat wave, cyclone, flash flood) is the direct evidences of climate change. The annual average surface temperature has increased by 0.87°C in 100 years (IPCC, 2018). Recent report revealed that more than 90% of Northern Hemisphere land areas outside the tropics showed at least 1°C above average, whereas temperatures were less extreme in Southern Hemisphere. Importantly, temperatures were above normal over most of the ocean areas. Global sea level rose to about 15 mm, in between November (2014) and February (2016), as a result of ElNino, which was well above the post-1993 trend of 3.0-3.5 mm year[1]. However, Indian summer monsoon rainfall since 1901 to 2012 showed no long-term trends, with some regional changes. Importantly, night temperatures have increased sharply during recent years, which couple with erratic precipitation could significantly affect agricultural production. The fifth IPCC (IPCC, 2014) reports clearly brought out global and regional impacts of climate change on Agriculture, Forestry and Other Land Use (AFOLU). South Asia has been characterized as one of the most vulnerable regions. Impacts of climate change witnessed all over the world; however, countries like India is more vulnerable because of its huge population primary dependent on agriculture for livelihood, predominance of small and marginal farmers, and fragmented land holding.

Climate change impacted agriculture both directly and indirectly. Direct effects include changes in productivity (both quantity and quality), and changes in resource management in (namely, water, soil, farm inputs, machineries *etc.*). Specifically, shortening of crop duration and increasing temperature hasten crop maturity, thus, adversely affecting productivity of annual crops. Direct effects on reproduction include spikelet sterility, poor pollination and fertilization processes in crops like rice, wheat, sunflower *etc.* (NAAS, 2013). Indirect impacts of climate change on agriculture include changes in water availability due to erratic precipitation, pest-disease outbreaks due to temperature enhancement, *etc.* (Gadgil, 2012; NAAS, 2013). However, some favorable effects have also been observed as a consequence of climate change. In India, Attri and Rathore (2003) and Roy *et al.* (2012) reported an increase in wheat and rice yields by 16-37 and 20%, respectively, under elevated CO_2 condition. However, an increase in temperature by 3-4°C or more could cancel out the positive effect of CO_2. Various field studies in India have inferred that direct effect of climate change would be small on *kharif* crop rather than *rabi*. But *kharif* agriculture may become vulnerable due to increased incidence of extreme events like rainy days, flash flood, terminal drought, humidity aberration, pest virulence *etc. Rabi* production may become comparatively more vulnerable due to higher increase in temperature during reproductive stage of crops, asymmetry of diurnal temperature, and higher uncertainty of rainfall (Peng *et al.*, 2004; Roy *et al.*, 2012; Bhattacharya *et al.*, 2014).

Technological Options for CSA

Technological options for CSA include 'innovative package of practices, processes, products with optimum blending of indigenous technical and traditional knowledge of communities having decision support tools, and livelihood options suitable for site-specific situation which encourages tackling climatic risk in agriculture'. Different technological options for CSA can be broadly grouped into four categories; (i) natural resources management, (ii) managing crop improvement and production, (iii) livestock and fisheries production systems, and (iv) enabling support systems.

Technological options / interventions under natural resource management include, rain water harvesting and recycling; *in-situ* moisture conservation; drainage networking in flood-prone areas; site-specific zero/ minimum tillage; ground water recharging; mulching; crop residue management; soil quality and health management; and carbon sequestration. On the other hand the technological options for crop improvement include, stress tolerant cultivars to drought, flood, submergence, salinity, high temperature, cold, pest and disease; resilient cropping and farming systems; contingent crop sequence; crop diversification; direct seeded rice; staggered community nursery techniques for paddy; advancement of planting dates of *rabi* crops to protect terminal heat stress or frost; site-specific nutrient management; and integrated pest-disease management.

In livestock and fisheries front, stress tolerant breeds, improved feed and mineral supplementation, unseasonal fodder production, preventive vaccination, modernized heat shock resistant shelter, innovative fish spawn protection mechanism could be viable options. However, the discussion on livestock and aquaculture aspects is out of the scope of this chapter.

Very importantly, village level support systems need to be strengthened which must include the mechanism and management of custom hiring of farm implements, seed bank (participatory or co-operative), and fodder bank. Small commodity group (horizontal) for maintenance of village level infrastructures, collective marketing chain, weather index / vulnerability based insurance scheme, real time advisory services are feasible and effective options.

Conservation Agriculture and CSA

Majority of the RCTs in crop production systems aimed at higher utilization efficiency of resources by adopting residue management, reduced tillage practices, crop diversification, cropping system approach, nutrient-water and energy conservation and thereby cut down the GHGs emissions and increasing resilience of the systems. The application of suitable RCTs could also retain carbon in soil which is useful for carbon sequestration.

Some of the popular RCTs in crop production systems includes, (i) residue management, (ii) crop diversification, (iii) reduced or zero tillage practices, (iv) nutrient conservation techniques (green manuring, brown manuring, INM etc), (v) water conservation for rice (direct seeding, bed planting, aerobic rice etc), (vi) energy, labour saving technologies by farm mechanization.

Therefore, it could be noticed that many of the technological options followed in RCTs are also practiced in CSA. The mitigation of GHGs emission must be addressed along with energy use efficiencies in RCTs to make them an integral part of CSA. On the other hand, CSA approaches should also take care of conservation of the natural resources which is the intrinsic part of RCTs. In the following sections we would be discussing the CSA with special reference to RCTs and GHGs emissions.

Natural Resources Management in CSA

Harvesting and recycling of rainwater by renovating the farm-pond at drought affected Nacharam Village of Khamman, Telangana in 2014 could provide two critical irrigations to cotton, chilies and fodder. As a consequence, farmers realized an additional yield and income from cotton (250 kg ha⁻¹, Rs. 10,000/-), chilies (100 kg ha⁻¹, Rs. 9,000/-) and fodder grass (4 t ha⁻¹, Rs. 10,900/-). Similarly, harvested rainwater was used through drip in order to increase the water use efficiency and yield at Harigao village, Aurangabad during 2014-15 in chickpea, resulting an increase of 69%. Not only that, rainwater harvesting promotes crop diversification and overall land productivity as evident in Hirehalli village (drought affected) of Tumkur, where farmers took additional crops like carrot, tomatoand aster in *rabi* with harvested water and earned a total income of Rs. 67,400/- including *kharif* crops like ground nut and finger millet. Construction of new water harvesting structures or village pond / renovation of existing structure demands high initial investment, which is very difficult to maintain by single or couple of farmers. Therefore, participatory community tanks / water harvesting structures (WHSs) are recommended under CSA. It not only reduces the burden for individual farmer, but at the same time increases the sustainability of system. Another very useful tool of water harvesting is check dam creation which in one hand saves productive soil from erosion and on the other hand recharges the ground water. A suitable example of that is community tank in Vadavathur village, Tamil Nadu, which helped farmers to take high value crops and fetched additional productivity (20-30% more) of onion during 2014. After harvesting the rainwater through check dam tube well could be recharged very efficiently and outrage it helps in decreasing salinity of groundwater for early irrigation of wheat / *rabi* crops. This technology is also useful in the salinity belts of Rajasthan, Haryana and Punjab. Trapping of sub-surface water source through small check dam (one side perforated) or small water harvesting tanks is very useful technology for hilly areas ('*Jalkund*').

Conservation agriculture, zero tillage wheat, using of Happy Seeder (second generation farm machineries) for wheat seeding and direct seeding of rice are the tools of CRA which got good success in north western region of India, where soils are light textured, farm holding is large and mechanization enters into the agricultural systems effectively.

Crop Improvement / Production Technologies in CSA

Introduction of stress tolerant (drought, flood, water stagnation, salinity, heat, cold, frost or biotic) cultivars is one of the major technique of CSA under crop improvement technologies. Another possible intervention of CSA is contingent cropping. As for example, contingent crops like sesame (c.v. Madhuri) and sunflower (c.v. PKV 559) can be adopted for delayed planting (August) due to delayed monsoon arrival, and these have been found to perform better and produce higher yields, compared to late planted soybean in Takali village, Amarabati district of Maharastra in 2014. Similarly, growing of short duration pigeon pea (c.v. BRG-2) for delayed onset of monsoon, aerobic paddy (c.v. CR-204), direct sown short duration and drought tolerant rice (c.v. BVD-110, Sahabhagi) are the other options which can increase resilience in system under climatic risk. Resilient intercropping systems such as cotton + green gram and cotton + black gram (1:1); soybean + pigeon pea (4:2) are very useful combinations in drylands of central India (states like Maharastra). Crop diversification is the inherent resilient mechanism which is true for avoiding / resisting climate risk also. Short duration foxtail millet (c.v. SIA-3085 and Suryanandi) in sole cotton system in Andhra Pradesh is an ideal example of that in dry land. The same holds true in case of Jharsuguda (drought prone district) of Odisha where maize was adopted instead of upland paddy and farmers could gain higher B:C ratio than conventional upland paddy. Participatory seed bank at village level by producing and storing seeds for short duration drought and flood tolerant varieties would be a viable option in contingency situation. Similarly, fodder bank also could be executed in community level.

RCTs in Relation to GHGs Emissions and CSA

The suitable RCTs could help to sequester soil organic carbon and reduce emission of GHGs. Particularly, zero tillage systems combined with adequate water management can reduce the release of CH_4 and N_2O (Gao, 2006).

(a) *Residue and nutrient management and GHGs emission*

The effect of urea alone and in combination with rice straw and green manure on the emission of CH_4, CO_2 and N_2O were quantified in tropical submerged rice system in India. On a seasonal basis, cumulative emission of CH_4 was highest (122.7 kg ha⁻¹) in rice straw + green manure. Cumulative seasonal emissions of CO_2-C ranged from 1100.3 kg ha⁻¹ in the control (unfertilized treatment) to 1858.5 kg ha⁻¹ in the rice straw + green manure treatment. Seasonal N_2O-N emissions were in the order of urea (1.0 kg ha⁻¹) > rice straw + urea (0.84 kg ha⁻¹) > rice straw + green manure (0.72 kg ha⁻¹) > control (0.23 kg ha⁻¹). The global warming potential (GWP) on CO_2 equivalent basis was in the order of rice straw + green manure (10,188 kg CO_2 equivalent ha⁻¹) > rice straw + urea (9418 kg CO_2 equivalent ha⁻¹) > urea (8084 kg CO_2 equivalent ha⁻¹) > control (5862 kg CO_2 equivalent ha⁻¹) (Table 1). The combination of an inorganic fertilizer, such as urea, with rice straw on a 1:1 nitrogen basis resulted in a significant build-up of soil carbon, enhancement of crop yield and lower GHG emission when compared to rice straw and green manure and it could be a viable option to mitigate global warming and maintain soil health (Bhattacharyya et al., 2012a).

Table 1: Greenhouse gas emissions and GWP on seasonal basis from flooded rice soil

Treatment	CH_4 emission (Kg ha⁻¹)	CO_2-C emission (Kg ha⁻¹)	N_2O-N emission (Kg ha⁻¹)	GWP (Kg CO_2 ha⁻¹)
Control	69.7	1100.3	0.23	5862
Urea	92.6	1447.7	1.00	8084
Rice straw + Urea	115.4	1680.6	0.84	9418
Rice straw + green manure	122.7	1858.5	0.72	10188

*Dose of N: 60 kg ha⁻¹; (cv. Gayatri)

The impact of long term organic amendments (RCTs) on the soil carbon (C) storage in relation to greenhouse gas (GHG) emission from rice field was quantified in a tropical Aeric Endoaquept in Cuttack, India. The treatments were unamended control, farmyard manure (FYM), green manure (GM) (*Sesbania aculeata*), FYM + GM and rice straw (RS) + GM combination. The trend of cumulative seasonal CH_4 emissions was in the order of FYM + GM (162 kg ha⁻¹) > FYM (143 kg ha⁻¹) > GM (109 kg ha⁻¹) > RS + GM (83 kg ha⁻¹) > Control (65 kg ha⁻¹) (Table 2). Cumulative N_2O–N emissions were in the order of GM (0.72 kg ha⁻¹) > RS + GM (0.66 kg ha⁻¹) > FYM + GM (0.60 kg ha⁻¹) > FYM (0.57 kg ha⁻¹) > Control (0.22 kg ha⁻¹) (Table 2). Whereas Cumulative CO_2-C emission followed the order FYM + GM (1910 kg ha⁻¹) > FYM (1480 kg ha⁻¹) > GM (1430 kg ha⁻¹) > RS + GM (1290 kg ha⁻¹) > Control (1000 kg ha⁻¹) (Table 2). Among the five treatments, the FYM + GM treatment has increased the global warming potential (GWP) by 110 % as well decreased the C efficiency ratio by 24 % in comparison to control. On the other hand, under RS + GM treatment the soil organic C and total C contents were significantly higher in the order of 34 and 53 %, respectively. Soil C storage was impacted to a maximum in the resource conservation techniques, like application of RS + GM at 1:1 [Nitrogen (N) basis] in rice-fallow cropping system in the tropical flooded soil planted to rice. From the environmental sustainability point of view, the implementation of RS + GM amendment techniques is the most adoptable option which gives relatively higher yield, reduced GHG emissions and high capacity to store C in the soil (Bhattacharyya et al., 2012b).

Table 2: Greenhouse gas emission and global warming potential in an under different organic manure treated soil (after 10 years of treatment effect)

Treatment	CH$_4$ emission (Kg ha^{-1})	CO$_2$-C emission (Kg ha^{-1})	N$_2$O-N emission (Kg ha^{-1})	GWP (Kg CO$_2$ ha^{-1})
Control	65	1000	0.22	5370
FYM	143	1480	0.57	9210
GM	109	1430	0.72	8260
FYM + GM	162	1910	0.60	11300
RS + GM	83	1290	0.66	7080

*rice *cv.* Geetanjali

The effect of N fertilizer and rapeseed (*Brassica napus* L.) straw on CO$_2$ emission was assessed in a paddy field of Subtropical China (Iqbal *et al.*, 2009). To understand the effect of residue incorporation (rapeseed straw) and N on CO$_2$ flux in rice, the CO$_2$ emission were measured during the growth stages of rice (*Oryza sativa* L.) from row, inter-row and bare soil. Soil CO$_2$ fluxes from row (797–1214 g C m^{-2} season^{-1}) were significantly higher than from inter-row (289–403 g C m^{-2} season^{-1}) and bare soil (148–241 g C m^{-2} season^{-1}), due to the contribution of rhizosphere respiration (Fig. 1). Among different treatments, N fertilization significantly increased the CO$_2$ flux from row with the highest being observed from NPK + straw treatment and lowest from control (no fertilizer). No significant differences among different treatments were observed from inter-row and bare soil.

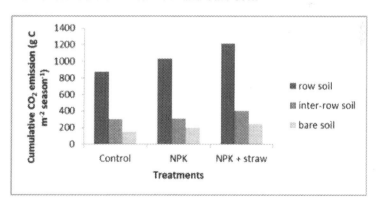

Fig 1. Cumulative CO$_2$ emission (g C m^{-2} season^{-1}) from the paddy field.

(b) *Effect of tillage practices on GHGs emission*

The impacts of four tillage practices in rice-wheat cultivation system on fluxes of GHGs (CH$_4$, N$_2$O and CO$_2$) and yield of rice were assessed in eastern Indo-Gangetic Plains of India (Pandey *et al.*, 2012). The four tillage treatments comprise of (i) conventional tilling and puddling before transplanting of rice and conventional tilling before sowing of wheat (RCT-WCT), (ii) conventional tilling and puddling before rice transplanting and no tilling before sowing of wheat (RCT-WNT), (iii) no tilling before rice sowing and conventional tilling before wheat sowing (RNT-WCT), and (iv) no tilling before sowing of both rice and wheat (RNT-WNT). No tillage significantly reduced the CH_4 and N_2O emission, but increased CO_2 fluxes than conventional tillage. No tillage practices in rice were more effective to reduce CH_4 and N_2O emission than no tillage in wheat (Fig. 2).

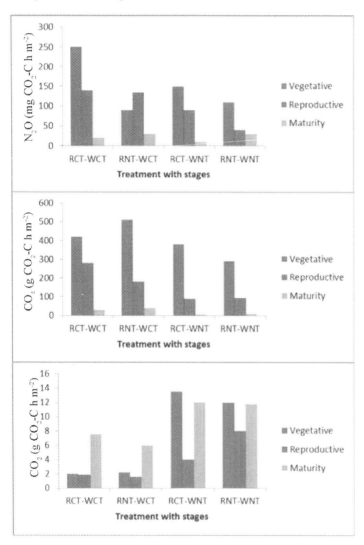

Fig. 2: Impact of tillage practices on GHGs emissions

(c) Effect of water management on NEP, GPP and RE

Net ecosystem exchange (NEE) was found to be negative during the daytime and positive during the night time for both flooded and aerobic rice fields in Philippines. The higher gross primary production (GPP), ecosystem respiration (RE) and net ecosystem production (NEP) were observed under flooded condition than aerobic system in rice indicated that flooded rice ecology behaved as better CO2-C sink than aerobic rice (Alberto et al., 2012, Bhattacharyya et al., 2014) (Table 3).

Table 3: Integrated values of GPP (g C m^{-2}), RE (g C m^{-2}), NEP (g C m^{-2}) and average grain yields (t ha^{-1}) in flooded and aerobic rice fields during 5 cropping seasons from 2008-2010

Cropping period	GPP		RE		NEP		Grain yield	
	flooded	aerobic	flooded	aerobic	flooded	aerobic	flooded	aerobic
2008 DS	882	620	718	606	164	14	5.6	2.1
2008 WS	914	856	619	774	295	82	6.1	1.9
2009 DS	873	951	522	733	351	218	7.4	2.5
2009 WS	762	500	537	461	225	39	5.1	1.9
2010 DS	765	688	578	514	187	174	7.5	2.3

(Note: DS: Dry season; WS: Wet season)

(d) Comparative assessment of different RCTs on GWP

The efficiency and performance of different RCTs like mid-season drainage, aerobic rice production system, System of Rice Intensification (SRI), direct seeded rice (DSR), sprinkler irrigation, zero tillage, integrated nutrient management, organic rice, crop diversification were tested against transplanted rice for assessing the GWP in upper and lower Indo-Gangetic Plains (IGPs). It was estimated with the help of InfoRCT (Information on Use of Resource-Conserving Technologies) model. The RCTs like mid-season drainage, aerobic rice cultivation, SRI, DSR, sprinkler irrigation, zero tillage and crop diversification approach proved to have low emission potential than transplanted rice grown in puddled soil with continuous flooding following conventional agricultural practices in Upper and Lower IGPs (Table 4) (Pathak et al., 2011).

Table 4: Total GWP in rice with different technological options in Upper and Lower IGP

Technology	GWP in Upper IGP	GWP in Lower IGP
Transplanted rice	3957	2934
Mid-season drainage	3625	2357
Aerobic rice	3141	1741
System of rice intensification	1542	1034
Direct seeded rice	2623	979
Sprinkler irrigation	2494	735
Zero till	637	346
Integrated nutrient management	5707	6089
Organic rice	6129	8569
Green manuring	5699	6086
Crop diversification	2118	529

Conclusion

Agriculture has the potential to mitigate emissions of GHGs by adopting technologically viable and economically feasible RCTs. However, the opportunities, constraints, interventions required for promoting low carbon technologies in different production systems must be identified. Easily available and locally adopted RCTs are successfully applied on field under the conservation agricultural practices in different rice-based cropping systems around the world keeping in mind stable agricultural production, maintaining better soil health and quality as well as reduced GHGs emissions. However, more research is needed to combine geographic information system, yield-emission models, relevant site-specific crop-soil information and agricultural management practices in relation to CSA for devising adoption strategies for curbing gaseous-C and N fluxes in agriculture.

To conclude, the way forward of CSA is to climate smart agriculture. This shift requires standardized methodologies for vulnerability assessment, enhancement of density of weather observations, efficient management of weather related information, real time diffusion of drought, heat and submergence tolerant crop varieties, adoption of location-specific conservation agriculture, management of climate risk through weather-based agro-advisory and affordable weather insurance product. Above all, Government along with NGOs and private sectors should join hand in hand for timely relief payment and promoting insurance for dealing with climate vagaries.

References

Alberto, M.C.R., Hirano, T., Miyata, A., Wassmann, R., Kumar, A., Padre, A. and Amante, M. (2012). Influence of climate variability on seasonal and inter-annual variations of ecosystem CO_2 exchange in flooded and non-flooded rice fields in the Philippines. *Field Crops Research*, 134: 80-94.

Attri, S.D. and Rathore, L.S. (2003). Simulation of impact of projected climate change on wheat in India. *International Journal of Climatology*, 23: 693-705.

Bhattacharyya, P. and Mohapatra, T. (2013). Soil organic carbon dynamics vis a vis anticipatory climatic changes and crop adaptation strategies. In Final Report of NAIP (C-4), "Soil organic carbon dynamics vis-à-vis anticipatory climatic changes and crop adaptation strategies", CRRI, Cuttack, pp 71.

Bhattacharyya, P., Roy, K.S., Dash, P.K., Neogi, S., Shahid, Md., Nayak, A.K., Raja, R., Karthikeyan, S., Balachandar, D., Rao, K.S. (2014). Effect of elevated carbon dioxide and temperature on phosphorus uptake in tropical flooded rice (*Oryza sativa* L.). *European Journal of Agronomy*, 53: 28-37.

Bhattacharyya, P., Roy, K.S., Nayak, A.K. (2016). Greenhouse Gas Emission from Agriculture: Monitoring, Quantification & Mitigation. Narendra Publishing House. New Delhi. ISBN 13-9789384337964.

Bhattacharyya. P., Roy, K.S., Neogi, S., Adhya, T.K., Rao, K.S. and Manna, M.C. (2012a) Effects of rice straw and nitrogen fertilization on greenhouse gas emissions and carbon storage in tropical flooded soil planted with rice. *Soil and Tillage research*, 124: 119-130.

Bhattacharyya. P., Roy, K.S., Neogi, S., Chakravorti, S. P., Behera, K. S., Das, K. M., Bardhan, S. and Rao, K. S. (2012b). Effect of long term application of organic amendment on C storage in relation to global warming potential and biological activities in tropical flooded soil planted to rice. *Nutrient Cycling in Agroecosystems,* 94: 273-285.

Gadgil Sulochana (2012). Monsoon variability, climate change and agriculture, *Yojana*, 56 (July issue): 19-23.

Gao, H. (2006). The impact of conservation agriculture on soil emissions of nitrous oxide. Draft report. Beijing, China: Asian and Pacific Centre for Agricultural Engineering and Machinery.

Intergovernmental Panel on Climate Change, 2018.Intergovernmental Panel on Climate Change. Global Warming of 1.5° C: An IPCC Special Report on the Impacts of Global Warming of 1.5° C Above Pre-Industrial Levels and Related Global Greenhouse Gas Emission Pathways, in the Context of Strengthening the Global Response to the Threat of Climate Change, Sustainable Development, and Efforts to Eradicate Poverty. Rosenfeld, A., Dorman, M., Schwartz, J., Novack, V., Just, A.C. and Kloog, I., (2017). Estimating daily minimum, maximum, and mean near surface air temperature using hybrid satellite models across Israel. *Environmental Research*, 159: 297-312.

IPCC (2014): *Climate Change 2014: Impacts, Adaptation, and Vulnerability. Part A: Global and Sectoral Aspects. Contribution of Working Group II to the Fifth Assessment Report of the Intergovernmental Panel on Climate Change* [Field, C.B., V.R. Barros, D.J. Dokken, K.J. Mach, M.D. Mastrandrea, T.E. Bilir, M. Chatterjee, K.L. Ebi, Y.O. Estrada, R.C. Genova, B. Girma, E.S. Kissel, A.N. Levy, S. MacCracken, P.R. Mastrandrea, and L.L. White (eds.)]. Cambridge University Press, Cambridge, United Kingdom and New York, NY, USA, 1132 pp.

Iqbal, J., Lin, S., Hu, R. and Feng, M. (2009). Temporal variability of soil-atmospheric CO_2 and CH_4 fluxes from different land uses in mid-subtropical China. *Atmospheric Environment*, 43(37): 5865-5875.

NAAS (2013). Climate Resilient Agriculture in India. Policy Paper No. 65, National Academy of Agricultural Sciences, New Delhi: 20 p.

Pandey, D., Agrawal, M. and Bohra, J.S. (2012). Greenhouse gas emissions from rice crop with different tillage permutations in rice–wheat system. *Agriculture, Ecosystems & Environment,* 159: 133-144.

Pathak, H. (2014). Carbon Economy in Agriculture: Negotiations at the UNFCC, Presentation made at the Brainstorming Session on Carbon Economy in Indian Agriculture at NAAS on 01 February, 2014.

Pathak, H., Saharawat, Y.S., Gathala, M. and Ladha, J.K. (2011). Impact of resource-conserving technologies on productivity and greenhouse gas emission in rice-wheat system. *Greenhouse Gas Science and Technology*, 1: 261-277.

Peng S, Huang J, Sheehy J.E., Laza, R.C., Visperas, R.M., Zhong X, Centeno G. S., Khush, G.S, and Cassman, K.G. (2004). Rice yields 12 Climate Change and its Impact on Indian Agriculture decline with higher night temperature from global warming, *Proceedings of National Academy of Science*, 101: 9971-9975.

Prasad, YG., Srinivasa Rao, Ch., Prasad, JVNS., Rao, KV., Ramana, DBV., Gopinath, KA., Srinivas, I., Reddy, BS., Adake, R., Rao. VUM., Maheswari, M., Singh, AK and Sikka, AK. (2015). Technology Demonstrations: Enhancing resilience and adaptive capacity of farmers to climate variability. National Innovations in Climate Resilient Agriculture (NICRA) Project, ICAR-Central Research Institute for Dryland Agriculture, Hyderabad.109 p.

Roy, K.S., Bhattacharyya. P., Neogi, S., Rao, K. S., Adhya, T.K. (2012). Combined effect of elevated CO_2 and temperature on dry matter production, net assimilation rate, C and N allocations in tropical rice (Oryza sativa L.). *Field Crops Research*, 139: 71-79.

3

Relevance of Conservation Agriculture in Climate Change Perspective: An Issue for Climate Resilient Agriculture

P.K. Ghosh, C.P. Nath[1], Debarati Datta[2] and K.K. Hazra[1]

ICAR–National Institute of Biotic Stress Management, Raipur
Chhattisgarh - 493 225, India
[1]ICAR–Indian Institute of Pulses Research, Kanpur, U.P - 208 024, India
[2]ICAR–Central Research Institute for Jute and Allied Fibers
Barrackpore, W.B - 700 121, India

Introduction

Climate change is a statistically significant variation in either the mean state of the climate or in its variability, persisting for an extended period (typically decades or longer). Warming of the climate system is unequivocal, as is now evident from observations of increases in global average air and ocean temperatures. Eleven years from 1995-2006 rank among the twelve warmest years in the instrumental record of global surface temperature (since 1850). The 100-year linear trend (1906-2005) of 0.74 [0.56 to 0.92] °C is larger than the corresponding trend of 0.6 [0.4 to 0.8] °C (1901-2000) and over the 21st century average temperature of earth surface is likely to go up by an additional of 1.8-4°C (IPCC, 2007). This temperature increase can be attributed to the altered energy balance of the climate system resulting from changes in atmospheric concentrations of the green house gases (GHGs). Among the principal components of radiative forcing of climate change, CO_2 has the highest positive forcing leading to warming of climate. CO_2 has the least global warming potential among the major green house gases but due to its much higher concentration in the atmosphere; it is the major contributor towards global warming and climate change. Agriculture sector in India contributes 28% of the total GHG emissions (NATCOM, 2004). The global average from agriculture is only 13.5% (IPCC, 2007). In future, the percentage emissions from agriculture in India are likely to be smaller due to relatively much higher growth in emissions in energy-use transport and industrial sectors. The emissions from agriculture are primarily due to methane emissions from rice fields, enteric fermentation in ruminant animals and nitrous oxides from application of manures and fertilizers to agricultural soils (NATCOM, 2004).

Causes of Climate Change

Influence of humans

Humans have been influencing the climate since the start of the Industrial Revolution. Since then, the average world temperature has risen by approximately 0.8^0C. In North-West Europe (including the Netherlands) the average temperature has risen by 1.5 degrees. The sea level has risen by around 20 cm and most of the glaciers have shrunk dramatically. Up to 1950 the influence of nature was more important than human influence. After that, the pattern in the average world temperature can only be explained by factoring in the human influence. Even so, a slight decline in temperature did appear from the mid-1940s to the mid-1970s. It was linked to a dramatic increase in cooling aerosols from the post-war industrialization in the western world. It was also caused by a mild decline in solar activity and some major volcanic eruptions in the second half of this period. According to the latest IPCC report, it is more than likely (more than 90% probability) that most of the global warming in recent decades is attributable to the observed increase in greenhouse gases.

CO_2 and climate change

The most well-known and the most important greenhouse gas is CO_2. The concentration of CO_2 in the atmosphere is subject to variation even without human intervention. The carbon cycle causes an exchange of CO_2 between the biosphere and the oceans on the one hand and the atmosphere on the other. Vast amounts of CO_2 are also released by the burning of fossil fuels. There is incontrovertible evidence that the CO_2 concentration in the air has never been so high in 800,000 years (probably even 60 million years) as it is now. The trend suggests that CO_2 emissions will continue to rise globally, although the economic crisis did prevent a rise in 2009. The Netherlands (per head of population) is high on the list of CO_2 emitters in the world. Besides CO_2, methane (CH_4), nitrous oxide (N_2O), fluorinated gases, ozone (O_3) and water vapour are important greenhouse gases. Residue burning is also a major problem which emits the GHGs in atmosphere. Open field burning of crop residue leads to emission of trace gases like CH_4, CO, N_2O, NO_X and other hydrocarbons. It also emits large amount of particulates that are composed of wide variety of organic and inorganic species.

Aerosols

Aerosols are less well-known than greenhouse gases. Aerosols are dust particles which, in addition to CO_2, are released into the atmosphere in large quantities when wood and fossil fuels are burned. Some aerosols have a cooling effect on the climate, others have a warming effect. On balance they have a cooling rather than a warming effect, but no-one can give a clear idea of the magnitude, because we still do not understand how aerosols influence the occurrence and characteristics of clouds. Natural phenomena, greenhouse gases and aerosols create an imbalance in the incoming and outgoing radiation in the atmosphere. This process is known as radiative forcing. When the Earth heats up, the short-wave radiation from the sun that enters the atmosphere is greater than the long-wave radiation that exits the atmosphere. The temperature changes on Earth will not stop until the radiation balance is restored. Given the immense capacity of oceans to absorb heat, it will take a long time to strike a new balance.

Uncertainty

The extent of global warming in the future is swathed in uncertainty; first, because we have no idea of how much of an increase to expect in greenhouse gases (depending on economic growth), and secondly, because we do not know exactly how our climate system will respond (climate sensitivity).

The Greenhouse Gases (GHGs) and their Source

The most important GHGs directly emitted by humans include CO_2, CH_4, N_2O and several others. The sources and recent trends of these gases are detailed below.

Carbon dioxide

Carbon dioxide is the primary greenhouse gas that is contributing to recent climate change. CO_2 is absorbed and emitted naturally as part of the carbon cycle, through plant and animal respiration, volcanic eruptions, and ocean-atmosphere exchange. Human activities, such as the burning of fossil fuels and changes in land use, release large amounts of CO_2, causing concentrations in the atmosphere to rise. Atmospheric CO_2 concentrations have increased by more than 40% since pre-industrial times, from approximately 280 parts per million by volume (ppmv) in the 18th century to over 400 ppmv in 2015. The heat from burning cereal straw can penetrate into soil up to 1 cm elevating the temperature to as high as 33.8-42.2^{0}C. About 32-76 % of the straw weight and 27-73% N are lost in burning. Bacterial and fungal populations are decreased immediately and substantially only on top 2.5 cm upon burning. Repeated burning in the field permanently diminishes the bacterial population by more than 50%. Burning immediately increased the exchangeable NH_4-N and bicarbonate extractable-Phosphorus content but there is no build up of nutrients in the profile. Long term burning reduces total N and C and potentially mineralized N in the 0-15 cm soil layer.

Fig. 1: Residue burning is the potential source GHGs emission in atmosphere

Methane

Methane is produced through both natural and human activities. For example, natural wetlands, agricultural activities, and fossil fuel extraction and transport all emit CH4. Methane is more abundant in Earth's atmosphere now than at any time in at least the past 800,000 years. Due to human activities, CH_4 concentrations increased sharply during most of the 20th century and are now more than two-and-a-half time's pre-industrial levels. In recent decades, the rate of increase has slowed considerably. The agricultural activities for emitting methane are:

- Ruminants
- Rice field
- Manure management
- Residue burning

Nitrous oxide

Nitrous oxide is produced through natural and human activities, mainly through agricultural activities and natural biological processes. Fuel burning and some other processes also create N_2O. Concentrations of N_2O have risen approximately 20% since the start of the Industrial Revolution, with a relatively rapid increase toward the end of the 20th century. Overall, N_2O concentrations have increased more rapidly during the past century than at any time in the past 22,000 years.

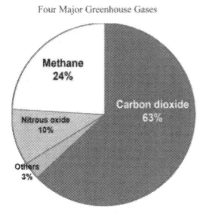

Fig. 2: Contribution of different GHGs

Table 1: The global warming potential of six major greenhouse gases

Gas	Global warming potential	Atmospheric life (Years)
CO_2	1	5 to 200
CH_4	21	12
N_2O	310	114
HFC	140 to 11,700	1.4 to 260
PFC	6,500 to 9,200	10,000 to 50,000+
SF_6	23,900	3200

Source: IPCC (2007).

Impact of Climate Change on Agriculture

Agriculture is an important sector of the Indian economy. Agriculture and fisheries are highly dependent on the climate. Increases in temperature and CO_2 can increase some crop yields in some places. But to realize these benefits, nutrient levels, soil moisture, water availability, and other conditions must also be met. Changes in the frequency and severity of droughts and floods could pose challenges for farmers and ranchers and threaten food safety. Meanwhile, warmer water temperatures are likely to cause the habitat ranges of many fish and shellfish species to shift, which could disrupt ecosystems. Overall, climate change could make it more difficult to grow crops, raise animals, and catch fish in the same ways and same places as we have done in the past. The effects of climate change also need to be considered along with other evolving factors that affect agricultural production, such as changes in farming practices and technology.

Impacts on crops

- Higher CO_2 levels can affect crop yields. Some laboratory experiments suggest that elevated CO_2 levels can increase plant growth. However, other factors, such as changing temperatures, ozone, and water and nutrient constraints, may counteract these potential increases in yield. For example, if temperature exceeds a crop's optimal level, if sufficient water and nutrients are not available, yield increases may be reduced or reversed. Elevated CO_2 has been associated with reduced protein and nitrogen content in alfalfa and soybean plants, resulting in a loss of quality. Reduced grain and forage quality can reduce the ability of pasture and rangeland to support grazing livestock.

- More extreme temperature and precipitation can prevent crops from growing. Extreme events, especially floods and droughts, can harm crops and reduce yields.

- Dealing with drought could become a challenge in areas where rising summer temperatures cause soils to become drier. Although increased irrigation might be possible in some places, in other places water supplies may also be reduced, leaving less water available for irrigation when more is needed.

- Many weeds, pests, and fungi thrive under warmer temperatures, wetter climates, and increased CO_2 levels. The ranges and distribution of weeds and pests are likely to increase with climate change. This could cause new problems for farmers' crops previously unexposed to these species.

- Though rising CO_2 can stimulate plant growth, it also reduces the nutritional value of most food crops. Rising levels of atmospheric carbon dioxide reduce the concentrations of protein and essential minerals in most plant species, including wheat, soybeans, and rice. This direct effect of rising CO_2 on the nutritional value of crops represents a potential threat to human health. Human health is also threatened by increased pesticide use due to increased pest pressures and reductions in the efficacy of pesticides.

Impacts on livestock

- Heat waves, which are projected to increase under climate change, could directly threaten livestock. Exposure to high temperature events can cause heat-related losses to agricultural producers. Heat stress affects animals both directly and indirectly. Over time, heat stress can increase vulnerability to disease, reduce fertility, and reduce milk production.

- Drought may threaten pasture and feed supplies. Drought reduces the amount of quality forage available to grazing livestock. Some areas could experience longer, more intense droughts, resulting from higher summer temperatures and reduced precipitation. For animals that rely on grain, changes in crop production due to drought could also become a problem.

- Climate change may increase the prevalence of parasites and diseases that affect livestock. The earlier onset of spring and warmer winters could allow some parasites and pathogens to survive more easily. In areas with increased rainfall, moisture-reliant pathogens could thrive.

- Potential changes in veterinary practices, including an increase in the use of parasiticides and other animal health treatments, are likely to be adopted to maintain livestock health in response to climate-induced changes in pests, parasites, and microbes. This could increase the risk of pesticides entering the food chain or lead to evolution of pesticide resistance, with subsequent implications for the safety, distribution, and consumption of livestock and aquaculture products.

- Increases in CO_2 may increase the productivity of pastures, but may also decrease their quality. Increases in atmospheric CO_2 can increase the productivity of plants on which livestock feed. However, the quality of some of the forage found in pasturelands decreases with higher CO_2. As a result, cattle would need to eat more to get the same nutritional benefits.

Impacts on fisheries

- Many aquatic species can find colder areas of streams and lakes or move north along the coast or in the ocean. Nevertheless, moving into new areas may put these species into competition with other species over food and other resources.

- Some marine disease outbreaks have been linked with changing climate. Changes in temperature and seasons can affect the timing of reproduction and migration. Many steps within an aquatic animal's lifecycle are controlled by temperature and the changing of the seasons. For example, in the Northwest warmer water temperatures may affect the lifecycle of salmon and increase the likelihood of disease. Combined with other climate impacts, these effects are projected to lead to large declines in salmon populations.

- In addition to warming, the world's oceans are gradually becoming more acidic due to increases in atmospheric CO_2. Increasing acidity could harm shellfish by weakening their shells, which are created by removing calcium from seawater. Acidification also threatens the structures of sensitive ecosystems upon which some fish and shellfish rely.

In nutshell climate change can impact agriculture in following ways:

- Soil: Drier, reduced productivity
- Irrigation: Increased demand, reduced supply
- Pests: Increased ranges and populations
- Production: Reduced crop yield, particularly in south Asia
- Livestock: Increased diseases and heat stress
- Fishery: Affected abundance and spawning
- Economic impact: Reduced agricultural output
- Agricultural productivity in India was estimated to decrease by 2.5 to 10% by 2020 to 5 to 30% by 2050 (IPCC assessment).

Introduction to Conservation Agriculture

Conservation agriculture is practiced globally on an estimated 155 million hectares in all continents and agricultural ecologies. North and South America have the largest area under CA (about 100 M ha), while Africa and Europe have the least (about 3 M ha). In India, there are divergent views on the area of land under CA (Derpsch *et al.*, 2010) estimated that CA is practiced on about 1.5 M ha in the Indo-Gangetic Plains (IGP), and is otherwise known through resource conservation technologies (RCTs). The spread of CA is concentrated in the rice–wheat system in the Indian IGP. Indian IGP comprises of Trans (GP), upper (GP), middle (GP) and the lower (GP). The TGP is the food hub of South Asia's 'Green Revolution'. No doubt, it heralded the introduction, opportunities and challenges of CA. The IGP of South Asia includes India, Nepal, Bangladesh and Pakistan. The aggregate of no-till and reduced till wheat area in the IGP is about 2 million hectares in 2004–2005. Recent assessments of CA in the IGP across India, Pakistan, Bangladesh and Nepal and in the rice–wheat cropping system with large adoption of no-till wheat is about 5 million ha, but only marginal adoption of permanent no-till systems and full CA (Friedrich *et al.*, 2012). The CA based cropping system practiced in the IGP is rarely a full conservation agriculture but rather a stepwise adoption or periodic CA which involve reduce or minimum tillage including crop residue and rotation in one season i.e., in wheat crop and not in rice crop, grown in the succeeding season.

Table 2: Global spread (M ha) of conservation agriculture in different countries

Country	Area (2008-09)	Area (2015-16)	% increase
USA	26.5	43.2	63.0
Brazil	25.5	32.0	25.5
Argentina	19.7	31.0	57.4
Canada	13.4	19.9	48.5
Australia	12.0	22.3	85.8
Paraguay	2.4	3.0	25.0
Kazakhstan	1.3	2.5	92.3
China	1.3	9.0	592.3
India	-	1.5	-
Others	4.4	16.1	265.9
Global total	106.5	180.5	69.5

Source: Kassam *et al.* (2019)

Genesis

The CA is one of the major drivers of sustainable agricultural intensification in the IGP vis-à-vis the increasing soil carbon depletion, declining groundwater table, increasing air pollution and the stagnating or low yields of the rice–wheat system. The ZT wheat after rice is the most widely adopted resource conserving technology in South Asia and in the Indian IGP. Thus it has become the predominant CA based cropping system. Zero-till wheat is aided by significant costs savings and potential yield increases (Erenstein and Laxmi, 2008). In these systems, ZT is only applied to the wheat crop–and does not essentially involve the retention of crop residue as mulch or the use of crop rotations. Also, the subsequent rice crop continuously puddled and transplanted. This anomaly in CA practices (in one season and not in the other) present a serious inadequacy from the ecologically-sustainable intensification outlook, as the benefits accumulated in the wheat season is lost in the subsequent puddled and transplanted rice (Nath *et al.*, 2017). Even in zero tillage wheat, farmers usually do not intentionally retain mulch and often burn the preceding rice straw–although the anchoring straw remained in the soil after burning may be enough to satisfy the requirements of residue mulch in CA. This suggests that farmers decides what components of CA practices that satisfies their aspiration and are available or easy to use vis-à-vis the prospects, limitations and trade-offs they face. Moving towards a full conservation agriculture calls for an improved management of crop residue and its retention and the shift towards direct-seed aerobic rice and crop diversification. Regional inequality in terms of agricultural productivity seems to favour the less intensified eastern IGP areas than the highly intensified north-western (NW-IGP), in terms of yield gains and cost savings from ZT practice. Socio-economic and system benefits of ZT in India is not a function of farm size. Smallholders have taken the advantage of ZT-drill in contract services. However, the reduced labour savings may boomerang against employment generation for farm labours who depend on land preparation for their livelihoods. The use of ZT in wheat unwraps the opportunity for the adoption of a full CA cropping system, and to other crops (pulses and vegetables) in the IGP. It also opens the scope for triple cropping in rice–wheat systems thereby increasing cropping intensity and diversity. For the expansion and scaling up of CA, it will have to deal with the impeding trade-offs or short falls associated with this anomaly.

Need for CA in the present agricultural scenario

The conservation agriculture is the need of present agricultural scenario to curb the challenges in present agriculture viz.

- Deteriorating soil health
- Declining factor productivity (water, nutrient, energy)
- Declining/stagnating yields and farm income
- Receding groundwater table
- Climate change and global warming

Options for adoption of conservation agriculture

- Availability of new machinery
- Herbicides for weed control
- Non-availability of labour
- Increasing costs, energy crisis, erosion losses, pollution hazards
- Residue burning – recycling
- Increase in cropping intensity

Principles of Conservation Agriculture

Soil disturbance regulation, surface residue management and crop rotation are fundamentals (core pillars) of CA (FAO, 2012). Controlled traffic or minimum physical soil disturbance on cropland–zero or reduced tillage enhances soil natural processes and recycling. This ensures that soil life, aggregates and structural quality is preserved, which promotes ecosystems sustainability. Permanent soil cover regulates erosion and temperature effect on surface soil, provides substrate for microorganism existence. Soils under diverse cropping systems by and large have a higher SOC pool than monocultures. Exclusion of summer fallow and growing a winter cover crop augments soil quality through SOC sequestration. Crop diversification through rotations, cover- and intercrops contributes to recycling nutrients, disrupt weed, pest and disease cycles, enhance biological nitrogen fixation (BNF) when legumes are included and ensure diversify food diets. Agroecosystems sustainability can be enhanced by changing from monoculture to rotation cropping. The CA integrated with best management practices can give higher yield, lower irrigation use, increase in irrigation water productivity; and reduction in energy use than conventional tillage. Also, net returns can be increased with reduction in production cost in CA with best management based systems.

Table 3: Key elements of conservation agriculture for ecosystem sustainability

Conventional agriculture	Conservation agriculture
Cultivating land, using science and technology to dominate nature	Least interference with natural processes
Excessive mechanical tillage and soil erosion	No till / drastically reduced tillage (Biological tillage)
Residue burning or incorporation	Surface retention of residues
Use of ex-situ FYM/composts	Use of in-situ organics/ composts
Green manuring (incorporated)	Brown manuring (surface drying)
Free-wheeling of farm machinery	Controlled traffic
Crop-based management	Cropping system-based management
Single or sole crops	Intercropping / relay cropping
Uneven field levels	Precision laser land leveling

Potential Benefit of Conservation Agriculture

To be widely adopted, all new technology needs to have benefits and advantages that attract a broad group of farmers who understand the differences between what they are doing and what they need. In the case of CA these benefits can be grouped as:

- **Economic benefits** that improve production efficiency.

- **Agronomic benefits** that improve soil productivity.

- **Environmental and social benefits** that protect the soil and make agriculture more sustainable.

Economic benefits

Three major economic benefits can result from CA adoption are (a) time saving and thus reduction in labour requirement, (b) reduction of costs, e.g. fuel, machinery operating costs and maintenance, as well as a reduced labour cost and (c) higher efficiency in the sense of more output for a lower input. The positive impact of CA on the distribution of labour during the production cycle and, even more important, the reduction in labour requirement are the main reasons for farmers to adopt CA, especially for farmers who rely fully on family labour. Manual labour for soil preparation is back-breaking and unnecessary. Should the supply of labour be reduced, through sickness or migration, then the system can quickly become unsustainable.

Agronomic benefits

Adopting CA leads to improvement of soil productivity through organic matter increase, in-situ soil water conservation, improvement of soil structure, and thus rooting zone. The constant addition of crop residues leads to an increase in the organic matter content of the soil. In the beginning this is limited to the top layer of the soil, but with time this will extend to deeper soil layers. Organic matter plays an important role in the soil: fertilizer use efficiency, water holding capacity, soil aggregation, rooting environment and nutrient retention, all depend on organic matter.

Environmental benefits

The CA reduces the soil erosion, and thus reduces road, dam and hydroelectric power plant maintenance costs. It also improve water and air quality, increase biodiversity and carbon sequestration. Residues on the soil surface reduce the splash-effect of the raindrops, and once the energy of the raindrops has dissipated the drops proceed to the soil without any harmful effect. This results in higher infiltration and reduced runoff, leading to less erosion. The residues also form a physical barrier that reduces the speed of water and wind over the surface. Reduction of wind speed reduces evaporation of soil moisture. Soil erosion is reduced close to the regeneration rate of the soil or even adding to the system due to the accumulation of organic matter. Soil erosion fills surface water reservoirs with sediment, reducing water storage capacity. Sediment in surface water increases wear and tear in hydroelectric installations and pumping devices, which result in higher maintenance costs and necessitates earlier replacement. More water infiltrates into the soil with conservation agriculture rather than running off the soil surface. Streams are then fed more by subsurface flow than by surface runoff. Thus, surface water is cleaner and more closely resembles groundwater in conservation agriculture than in areas where intensive tillage and accompanying erosion and runoff predominate. Some few profit from the change and turn into pests. However, most organisms are negatively affected and either they disappear completely or their numbers are drastically reduced.

Conservation Agriculture and Crop Residues Utilization

In cereal-based cropping systems, huge volume of crop residues are produced and are used as animal feed, thatching homes and domestic fuels. Large portion of unused crop residues are spread on wheat field by combine-harvester which are not suitable to feed the cattle because soil particles are attached with them. The conventional land preparation requires several passes of disc and harrow to incorporate crop residues with aim to create a suitable seed bed. Additional tillage leads to a long turnaround period, resulting 2–3 weeks delay in sowing of succeeding crop to avoid N- immobilization (Singh *et al.*, 2004). India being an agriculture-dominant country produces more than 500 million tons of crop residues annually. Besides using as animal feed, for thatching of homes, and as a source of domestic and industrial fuel, a large portion of unused crop residues are burnt in the fields primarily to clear the left-over straw and stubbles after the harvest. Non-availability of labour, high cost of residue removal from the field and increasing use of combines in harvesting the crops are main reasons behind burning of crop residues in the fields.

Burning of crop residues causes environmental pollution, is hazardous to human health, produces greenhouse gases causing global warming and results in loss of plant nutrients like N, P, K and S. Management of crop residues in combination with zero tillage is a proven strategy for soil and water conservation and enhancement of soil biological processes. Residue application through soil incorporation has been found beneficial to soil health, crop productivity, nutrient-use efficiency. Therefore, appropriate management of crop residues assumes a great significance. Surface residue management is one of the fundamental principles of CA. Permanent soil cover regulates erosion and temperature effect on surface soil, provides substrate for microorganism existence. farm management options to address the issue of burning as well as maintaining soil health and long-term sustainability of RWCS. Permanent crop cover with recycling of crop residues is a pre-requisite and integral part of conservation agriculture.

Fig. 3: Zero-till wheat with rice residues

Carbon Sequestration and Reduced GHGs Emission

India has the potential to cut its annual GHG emissions from agriculture and livestock which is equal to 85.5 mega tonnes of CO_2 equivalent per year. Climate change mitigation generally involves reduction in anthropogenic emissions of greenhouse gases. Examples of climate change mitigation include switching to renewable and nuclear energy, which is low carbon energy sources, reforestation to increase the capacity of carbon sinks to remove greater amounts of carbon dioxide from the atmosphere. Under CA, it could be achieved with minimal cost by efficient use of fertiliser, adoption of zero-tillage,direct seeding of rice, improved cultivations such as direct seeding of the corn, the wheat and maize FIRB system; no-tillage system; precise land levelling; precision water management approaches including drip and sprinkler systems; precision agriculture sensor-based reliable nutrient management (Srinivasarao *et al.*, 2016). By shifting water management from the current practice to intermittent flooding in all the irrigated rice-growing areas of the country, the CH4 flux from irrigated rice fields could be reduced by 40% (Bhatia *et al.*, 2012). It was also stated that by modifying water management from continuous flooding to alternate flooding or applying urea alone rather than urea coupled with farmyard manure would reduce global warming potential by 15 and 29%, respectively. Similarly, DSR decrease CH4 emissions drastically having a considerable potential to reduce the global warming potential by about 35-75% over the conventional puddled transplanted rice. Such options capture the atmospheric carbon and store in the soil; are important not only for global warming mitigation but also can contribute towards improving the fertility status of the soil (Nath *et al.*, 2019). Moreover, the GHGs emissions can be abridged by the substitution of fossil fuels by the agricultural feedstocks (e.g. crop residues, dung, and dedicated energy crops) for energy produce. Carbon sequestration can contribute to mitigating climate change through biomass production and conservation practices. In effect, improving agricultural management practices through CA enriches soil carbon content via sequestration (Lu *et al.*, 2011). While over the earlier period (2015-2045) CA mono-cropping and double-cropping systems increase sequestration rates by 0.4 and 0.8 ton per hectare per year respectively, at the later period these rates halve. They project a total carbon sequestration of 420 T g C over both simulated periods across the 16 million hectares in northeast China. Acknowledging the carbon sequestration potential of CA in Australia, Rochecouste *et al.* (2015), used a series of system models to forecast an atmospheric CO_2 reduction of about 3 tons yearly had Australian dryland grain farmers adopted conservation practices. Besides predictive studies on the potential of CA to sequester carbon, other direct investigations reveal the immediate impact of conservation practices to increase soil organic carbon via sequestration. Levels of SOC rise when there is more organic matter inputs coupled with lower rates of organic matter decomposition. This occurs whenever there is a diversified crop rotation, an increased crop productivity, enhanced return of crop residues, use of organic and green manure, use of cover crops, agro-forestry, improved irrigation, mulching, reduced soil tillage, reduced bare fallow periods and erosion control. No-till/ reduced till (minimum soil disturbance) leads to lower exposure to oxidation of organic soils and lower atmospheric CO_2 emissions particularly in comparison to tilled soils. Various conservation agriculture management practices are followed in irrigated and rainfed based cropping system for reduction of GHG.

Conservation Agriculture Based Technologies for Climate Change Mitigation and Adaptation

(a) Laser land leveling

Laser land leveling alters fields having a constant slope of 0 to 0.2% using laser equipped drag buckets and gives a smooth land surface (± 2 cm). Large horsepower tractors and soil movers equipped with global positioning systems (GPS) and/or laser-guided instrumentation help to move soil either by cutting or filling to create the desired slope. Laser leveling provides a very accurate, smooth and graded field, which helps in saving of irrigation water up to 20% and improves the use-efficiency of applied N.

(b) Zero/minimal tillage

Conservation tillage is the collective umbrella term, commonly given to no-tillage, direct-drilling, minimum-tillage and/or ridge-tillage, to denote that the specific practice has a conservation goal of some nature. Usually, the retention of 30% surface cover by residues characterizes the lower limit of classification for conservation tillage, but other conservation objectives for the practice include conservation of time, fuel, earthworms, soil water, soil structure and nutrients (Das et al., 2020).

(c) Bed planting (narrow/broad beds)

In bed planting, crops are grown on the raised beds alternated by furrows. Beds are usually made at 0.6-1.0 m wide, and 2-3 rows of crops are sown on the beds. The furrow irrigated raised-bed system (FIRBS) of wheat cultivation has been shown to result in saving of seed by 25-40%, water by 25-40% and nutrients by 25%, without affecting the grain yield.

Fig. 4: Bed planting

(d) Direct-seeded rice

Direct dry seeding of rice with subsequent aerobic soil conditions reduces overall water demand, saves labour, fuel and time, and gives similar yield to transplanted rice, if weeds are effectively controlled. The technology does not affect rice quality and can be practised in different ecologies such as upland, medium and lowland, deep water and irrigated areas. Soil health is maintained or improved, and fertilizer and water use efficiencies increase. Therefore, it can be a feasible alternative to conventional puddled transplanted rice.

(e) Sesbania brown manuring

In brown manuring, rice and Sesbania are sown together and allowed to grow for 25-30 days before knocking down Sesbania crop with 2,4-D ester salt at a rate of 0.40-0.50 kg/ha. The technology smothers weed, reduces herbicide use, lowers irrigation application, supplies 15-20 kg N/ha with a fresh biomass of 10-12 t/ha, facilitates better emergence of rice where soil crusting occurs, conserves moisture with brown mulch, improves soil C content and increases farmers' income.

(f) Residues retention for mulch

Cropland offers a huge potential for sequestering C, especially when crop residues are managed properly. Permanent or semi-permanent crop/plant residues cover on soil, which can be a growing crop or dead mulch, has a role to protect soil physically from sun, rain and wind and to feed soil biota/ micro-organisms that take over the tillage function and soil nutrient balancing. Crop residues significantly influence soil physical, chemical and biological properties. It helps in water conservation through enhanced water infiltration, and reducing evaporation, and wind and water erosion.

(g) Crop diversification

Crop rotations especially with legumes and use of organic fertilizers have an affirmative influence on soil microbial populations. Carbohydrates and proteins are profuse in crop-residues and act as excellent substrates for the growth of soil microbes. Easy and fast decomposable carbon is more effective for the growth of soil microbes (Ghosh *et al.*, 2020). Alternative crops can increase the yield of subsequent wheat crops by depriving soil-borne wheat pathogens of a host and are often referred to as break crops. The rice-wheat cropping system of Indo-Gangetic plains remains fallow for 60-70 days between harvest of wheatand sowing of subsequent rice crop. The short duration mungbean varieties like SML 832 and SML 668 can be grown during this fallow period. Sisti *et al.* (2004) observed no significant difference in soil organic matter under zero-tillage and conventional tillage in wheat-soybean system, whereas the C and N stocks were significantly higher under zero tillage compared to conventional tillage in soybean-wheat-hairy vetch system. Moreover, the N can be added to the soil through the inclusion of short duration leguminous crops in the cropping system that helps in organic matter build-up (Adu-Gyamfi *et al*., 2007).

Aspects in Conservation Agriculture and Future Thrusts

- Traditional mindset of farmers hinder the transition from conventional farming to no-till farming
- Necessary equipment is costly and needs government support
- Heavier reliance on herbicides
- Prevalence of weeds, disease and other pests may shift in unexpected ways
- May initially require more N fertilizer
- Can slow germination

Policy and development needs

- Promoting CA in different states with supplying machineries for conservation agriculture on subsidized rates, promoting custom hiring systems
- Developing a crop residues management policy for each state defining clearly various competing uses.
- Developing and implementing appropriate legislation on prevention and monitoring of on-farm crop residues burnings through incentives and punishment.
- Capacity building of under and post-graduate students and training of farmers. Every agricultural university should have courses on crop residue management and conservation agriculture both at under- and postgraduate levels.
- Including the component of conservation agriculture in soil health card for proper monitoring of crop residues retention/burning and its impact on soil health.
- Familiarizing conservation agriculture technologies at KVKs and state agricultural departments for awareness generation among the farmers.

Conclusion

In order to bridge/fill yield gaps, it is of paramount importance to adopt a full-fledged CA in rice-wheat cropping system in the NW-IGPs. Management of crop residues with conservation agriculture is vital for long-term sustainability of Indian agriculture. Hence, burning of residues must be discouraged and utilized gainfully for conservation agriculture in improving soil health and reducing environmental pollution. Regions where crop residues are used for animal feed and other useful purposes, some amount of residues should be recycled into the soil. Development of appropriate farm machinery are needed to facilitate collection, volume reduction, transportation and application of crop residues, and sowing of the succeeding crop under a layer of residues on soil surface under CA.

References

Adu-Gyamfi, J.J., Myaka, F.A, Sakala, W.D., Odgaard, R., Vesterager, J.M. and Hogh-Jensen, H. (2007). Biological nitrogen fixation and nitrogen and phosphorus budgets in farmer-managed intercrops of maizeepigeonpea in semi-arid southern and eastern Africa. *Plant and Soil,* 295:127-136.

Bhatia, A., Agarwal P.K., Jain, N. and Pathak, H. (2012). Greenhouse gas emission from rice and wheatgrowing areas in India: Spatial analysis and upscaling. *Greenhouse Gas Science and Technology,* 2: 115– 125.

Das, T.K., Nath, C.P., Das, S., Biswas, S., Bhattacharyya, R., Sudhishri, S., Raj, R., Singh, B., Kakralia, S.K., Rathi, N. and Sharma, A.R. (2020). Conservation Agriculture in rice-mustard cropping system for five years: Impacts on crop productivity, profitability, water-use efficiency, and soil properties. *Field Crops Research,* 250: 107781.

Derpsch, R., Friedrich, T., Kassam, A., and Hongwen, L. (2010). Current status of adoption of no-till farming in the world and some of its main benefits. *International Journal of Agriculture and Biological Engineering,* 3: 1–26.

Erenstein, O. and Laxmi, V. (2008). Zero tillage impacts in India's rice–wheat systems: A review. *Soil and Tillage Research,* 100: 1–14.

FAO (2012). Food and Agriculture Organisation of the United Nation, 2012. Available online at http://www.fao.org/ag/ca/6c.html.

Friedrich, T., Derpsch, R. and Kassam, A.H. (2012). Global overview of the spread of Conservation Agriculture. *Field Actions Science Report,* 6: 1–7.

Ghosh, P.K., Hazra, K.K., Venkatesh, M.S., Praharaj, C.S., Kumar, N., Nath, C.P., Singh, U. and Singh, S.S. (2020). Grain legume inclusion in cereal–cereal rotation increased base crop productivity in the long run. *Experimental Agriculture,* 56(1): 142-158.

IPCC, (2007). In: Solomon, S., Qin, D., Manning, M., Chen, Z., Marquis, M., Averyt, K.B., Tignor, M., Miller, H.L. (Eds.), Climate Change 2007: The Physical Science Basis. Contribution of I to the Fourth Assessment Report of the Intergovernmental Panel on Climate Change. Cambridge University Press, Cambridge, United Kingdom and New York, NY, USA, pp. 996.

Kassam, A., Friedrich, T. and Derpsch, R. (2019). Global spread of conservation agriculture. *International Journal of Environmental Studies,* 76(1): 29-51.

Lu, M, Zhou, X, Luo, Y, Yang, Y, Fang, C, Chen, J and Li, B. (2011). 'Minor stimulation of soil carbon storage by nitrogen addition: A meta-analysis'. *Agriculture, Ecosystems and Environment,* 140(1): 234-44.

NATCOM (2004). India's intial national commission to the united nation's framework convention on climate change. National communication project ministry of environment and forests, Govt. of India.

Nath C.P., Das, T.K., Rana, K.S., Bhattacharyya, R., Pathak, H., Paul, S., Meena, M.C. and Singh, S.B. (2017). Weed and nitrogen management effects on weed infestation and crop productivity of wheat–mungbean sequence in conventional and conservation tillage practices. *Agricultural Research*, 6:33–46.

Nath, C.P., Hazra, K.K., Kumar, N., Praharaj, C.S., Singh, S.S., Singh, U. and Singh, N.P. (2019). Including grain legume in rice–wheat cropping system improves soil organic carbon pools over time. *Ecological Engineering,* 129: 144-153.

Rochecouste, J-F, Dargusch, P., Cameron, D and Smith, C. (2015). 'An analysis of the socio-economic factors influencing the adoption of conservation agriculture as a climate change mitigation activity in Australian dryland grain production'. *Agricultural Systems*, 135: 20-30.

Singh, Y., Singh, B., Ladha, J.K., Khind, C.S., Khera, T.S. and Bueno, C.S. (2004). Effect of residue decomposition on productivity and soil fertility in rice – wheat rotation. *Soil Science Society of America Journal,* 68: 854–64.

Sisti, C.P.J., Santos, H., dos P, Kohhann, R., Alves, B.J.R., Urquiaga, S., and Boddey, R. M. (2004). Change in carbon and nitrogen stocks in soil under 13 years of conventional or zero tillage in southern Brazil. *Soil and Tillage Research,* 76: 39-58.

Srinivasarao, Ch., Rani, Y.S., Veni, V.G., Sharma, K.L., Sankar, G.M., Prasad, J.V.N.S., Prasad, Y.G. and Sahrawat, K.L. (2016). Assessing village-level carbon balance due to greenhouse gas mitigation interventions using EX-ACT model. *International Journal of Environmental Science and Technology*, 13(1): 97-112.

4

Conservation Agriculture: Issues, Challenges and Prospects in India

A.K. Biswas, J. Somasundaram, K.M. Hati, Pramod Jha
A.K. Viswakarma, R.S. Chaudhary and A.K. Patra

ICAR- Indian Institute of Soil Science, Nabibagh, Berasia Road
Bhopal - 462 038, India

Introduction

Persistent use of conventional farming practices based on extensive tillage, especially when combined with removal or *in-situ* burning of crop residues, has magnified soil erosion losses and swiftly degraded soil resource base. One of the glaring examples for the aforementioned statement is the 'Dust Bowl' in the U.S during 1930s in Great plains, where 91 M ha of land was degraded by severe soil erosion (Hobbs, 2007).

One of the primary challenges of our time is to feed growing and more demanding world population with reduced external inputs and minimal environmental impacts (nature paper). Conservation Agriculture (CA) is a set of management practices for sustainable agricultural production without excessively disturbing the soils, while protecting it from the processes of soil degradation like erosion, compaction, aggregate breakdown, loss of organic matter, leaching of nutrients, and processes that are accentuating due to anthropogenic interactions in the presence of extremes of weather and management practices. The organic materials conserved through this practice are decomposed slowly, and much of materials are incorporated into the surface soil layer, thus reducing the liberation rate of carbon as CO_2 into the atmosphere. In the total balance, carbon is sequestered in the soil, and turns the soil into a net sink of carbon. So, CA enhance soil health by improving soil aggregation, reducing compaction through promotion of biological tillage, increasing surface soil organic matter and carbon content, and moderating soil temperature and weed suppression. CA reduces cost of cultivation, saves time, increases yield through timelier seeding/planting, reduces pest and diseases through stimulation of biological diversity, and reduces green house gas emissions.

Machinery development, refinement and adoption for a range of soil and cropping situations will be fundamental in any success to promote conservation agriculture practices. Agricultural machinery or tools, which support conservation agriculture generally refer to the cultivation systems with minimum or zero tillage and in-situ management of crop residues. Minimum tillage is aimed at reducing tillage to the minimum necessary that would facilitate favorable seedbed condition for satisfactory establishment of crop. Zero tillage is however an extreme form of minimum tillage. With the development of direct drilling machines, almost all research work was attempted to define the responses of direct-drilled seeds in relation to soil micro-environments. Different designs of direct drilling machines *viz.* zero till drill, no till plant drill, strip till drill, and rotary slit no till drill have been developed with controlled traffic measures for energy efficient and cost-effective seeding of crops without tillage. In this paper various issues, challenges and prospects of conservation agriculture in India has been discussed.

Conservation Tillage

CA is a slower-evolving agricultural revolution that began at the same time as the Green Revolution and emerged as a new paradigm to achieve goals of sustainable agricultural production. It is a major step towards transition to sustainable agriculture. The concept of CA is emerged from reduced tillage. Concepts for reducing tillage and keeping soil covered came up and the term conservation tillage was introduced to reflect such practices aimed at soil protection (Friedrich *et al.*, 2012; Wall, 2007). Seeding machinery developments allowed then, in the 1940s, to seed directly without any soil tillage. At the same time theoretical concepts resembling today's CA principles were elaborated by Edward Faulkner in his book "Ploughman's Folly" (Faulkner, 1945) and Masanobu Fukuoka with the "One Straw Revolution" (Fukuoka, 1975). It wasn't until herbicides became readily available in the late 1950s and early 1960s that the era of conservation tillage began. Baker *et al.* (2002) defined conservation tillage as - *it is a collective umbrella term commonly given to no-tillage, direct-drilling, minimum-tillage and/or ridge-tillage, to denote that the specific practice has a conservation goal of some nature* (Table 1). Usually, the retention of 30% surface cover by residues characterizes the lower limit of classification for conservation-tillage, but other conservation objectives for the practice include conservation of time, fuel, earthworms, soil water, soil structure and nutrients. Thus, residue levels alone do not adequately describe all conservation tillage practices.

"Conservation tillage is a set of practices that leave crop residues on the surface which increases water infiltration and reduces erosion. It is a practice used in conventional agriculture to reduce the effects of tillage on soil erosion. However, it still depends on tillage as the structure forming element in the soil. Nevertheless, conservation tillage practices such as zero tillage practices can be transition steps towards Conservation Agriculture." In other words, conservation tillage uses some of the principles of conservation agriculture, but has more soil disturbance. FAO has characterized conservation agriculture as follows:

"Conservation agriculture maintains a permanent or semi-permanent organic soil cover. This can be a growing crop or dead mulch. Its function is to protect the soil physically from sun, rain and wind and to feed soil biota. The soil micro-organisms and soil fauna take over the tillage function and soil nutrient balancing. Mechanical tillage disturbs this process. Therefore, zero or minimum tillage and direct seeding are important elements of CA. A varied crop rotation is also important to avoid disease and pest problems." (FAO, 2011).

Now it is clear that CA does not just mean not tilling the soil and then doing everything else the same. It is a holistic system with interactions among households, crops, and livestock since rotations and residues have many uses within households; the result is a sustainable agriculture system that meets the needs of farmers (Sayre and Hobbs, 2004).

Table 1: Descriptions of different tillage systems

Tillage systems	Descriptions (Conservation Tillage Information Centre, 2002)
Zero-Tillage/ No-Tillage	It is a form of conservation tillage, has been used for thousand years of indigenous cultures simply because humans did not have manual force to till a significant area. Soil is left undisturbed from planting to harvest except for nutrient injection. Planting or drilling is accomplished in a narrow seed-bed by zero-tillers. Weed control is accomplished primarily by herbicides. Cultivation may be used for emergency weed control.
Mulch Tillage	The soil is disturbed prior to planting. Tillage tools such as chisels, field cultivators, disks, sweeps, or blades are used. Weed control is accomplished with herbicides, cultivation or both.
Reduced Tillage (Minimum tillage)	Any seedbed preparation system that leave 15 to 30 % residue cover after planting or 500 to1000 kg/ha of small grain residue equivalent throughout the critical wind erosion period is considered a reduce tillage system. Reducing tillage operations/ trips when compared to conventional systems.
Ridge Tillage	The soil is disturbed prior to planting by tillage tools such as chisels, field cultivators, disks, sweeps or blades. Weed control is accomplished with herbicides and/or cultivation.
Conventional Tillage	Tillage types that leaves less than 15% residue cover after planting during critical wind/water erosion period. Generally, includes ploughing or other intensive tillage operations; residue removed/ burned during cultivation. Weed control is accomplished with herbicides, cultivation or both.

Key Principles of CA

The main threat to soil resource are soil erosion, loss of organic matter (OM), soil compaction, soil sealing etc. It has been realized worldwide that crop residue retention through CA is able to revert the soil degradation process.

The term, *CA, refers to the system of raising of crops in rotation without tilling the soil while retaining crop residues in the soil surface* (Abrol and Sunita, 2006) that has three key principles (Fig 1)

Fig. 1: Principles of Conservation Agriculture

- Maintenance of permanent vegetative cover (>30%) or mulch to protect soil surface
- Minimal soil disturbances enable through no-till/reduced tillage
- Diversified crop sequences/rotations (spatial and temporal crop sequencing).

These CA principles are applicable to a wide range of crop production systems from low-yielding, dry, rainfed conditions to high-yielding, irrigated conditions. The whole range of agricultural practices, including handling crop residues, sowing and harvesting, water and nutrient management, disease and pest control, etc. need to be evolved and evaluated through adaptive research with active farmers' involvement.

Of late, conservation agricultural practices are gaining increased attention worldwide. According to recent estimates, globally conservation agriculture practices are now being adopted on about 180 m ha area (Kassam *et al.*, 2018) (Table 2). South American countries (e.g. Brazil, Argentina, Colombia etc) practicing conservation agriculture reported to have a remarkable positive effects on water footprints of crops by improving soil water infiltration, increasing soil water retention and reducing runoff and contamination of surface and ground water. Unlike, in the rest of the world, CA technologies in

India are spreading mostly in the irrigated areas of the Indo-Gangetic plains (about 5 mha) where rice-wheat (RW) cropping system dominates (WCCA Report 2009). Evidently, yields in the rice-wheat (RW) system of the Indo-Gangetic Plains (IGP) of India are higher about 200-500 kg/ha with no-till because of timelier planting, better crop stands and also able to escape terminal heat at maturity stage.

Table 2: CA spread under different continents during 2015-16

Continent	Area (M ha)	Percent of total area under CA	% of cropland of the region
South America	69.90	38.7	63.2
North America	63.18	35.0	28.1
Australia & New Zealand	22.67	12.6	45.5
Asia	13.93	7.7	4.1
Russia & Ukraine	5.70	3.2	3.6
Europe	3.56	2.0	5.0
Africa	1.51	0.8	1.1
World total	180.44	100	12.5

Source: Kassam, A. *et al.* (2018)

Conventional Agriculture *vs* Conservation Agriculture

The conventional agriculture is mainly based on soil tillage as the main operation. The most widely known tool for this operation is the plough, which has become a symbol of agriculture (Fig. 2 & 3). Moving from conventional agriculture to conservational agriculture represents one of the great, global challenges in terms of changing habits and mind sets.

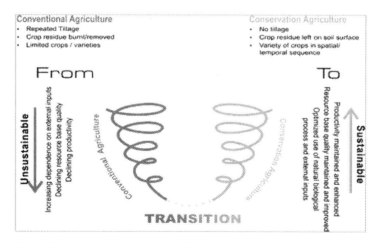

Fig. 2: Conventional Agriculture vs Conservation Agriculture

Soil tillage has in the past been associated with increased fertility, which originated from the mineralization of soil nutrients as a consequence of soil tillage. This process leads in the long term to a reduction of soil organic matter. Soil organic matter not only provides nutrients for the crop, but it is also, above all else, a crucial element for the stabilization of soil structure. Therefore, most soils degrade under prolonged intensive arable agriculture. This structural degradation of the soils results in the formation of crusts and compaction and leads to soil erosion (Fig. 3). The process is dramatic under tropical climatic situations but can be noticed all over the world. Mechanization of soil tillage, allowing higher working depths and speeds and the use of certain implements like ploughs, disk harrows and rotary cultivators have particularly detrimental effects on soil structure. Excessive tillage of agricultural soils may result in short term increases in fertility, but will degrade soils in the medium term. Structural degradation, loss of organic matter, erosion and falling biodiversity are all to be expected.

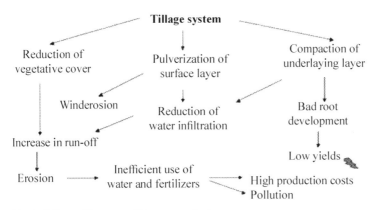

Fig. 3: Effect of intensive tillage on agriculture
Source: Friedrich *et al.* (2009)

However, CA systems have not been extensively tried or promoted in other major agro-ecoregions like rainfed semi-arid tropics, the arid regions and the mountain agro-ecosystems. In contrast to the homogenous growing environment of the IGP, the production systems in semi-arid and arid regions are quite heterogeneous in terms of land and water management and cropping systems (Kumar *et al.*, 2011). These include the core rainfed areas which cover up to 60-70% of the net sown area and the remaining irrigated production systems. The rainfed cropping systems are mostly single cropped in the Alfisols while in Vertisols, a second crop is generally taken on the residual moisture. Also in *rabi* black soils, farmers keep lands fallow during *kharif* and grow *rabi* crop on conserved moisture. Sealing, crusting, sub-surface hard pans and cracking are the key constraints which cause high erosion and obstruct infiltration of rainfall. Leaving crop residues on the surface in CA is a major concern in these rainfed areas due to its competing uses as fodder, leaving very little or no residues available for surface application. Agroforestry and alley cropping systems are other options for CA practices. This indicates that the concept of CA has to be adopted in a broader perspective in the arid and semi-arid areas.

Experiences at Indian Institute of Soil Science (IISS) showed that reduced tillage is a suitable option for growing soybean–wheat crops in Vertisols with saving of energy and labour (Subba Rao *et al.*, 2009). Soil organic carbon (SOC) is vital for sustainable soil quality and food production systems, which can be improved under conservation agriculture practices compared to conventional system. The soil properties like, organic carbon and pools, physical and biological properties were also improved under conservation agricultural practices.

Similarly, Central Research Institute for Dryland Agriculture (CRIDA), Hyderabad, has shown that in dryland ecosystems, it is possible to raise a second crop with residual soil moisture by covering the soil with crop residues. In a network project on tillage conducted since 1999 at various centers of the All India Coordinated Research Project for Dryland Agriculture (AICRPDA), it was found that rainfall and soil type had a strong influence on the performance of reduced tillage. In arid regions (<500 mm rainfall), low tillage was found on par with conventional tillage and weed problem was controllable in arid Inceptisols and Aridisols. In semi arid (500-1000 mm) region, conventional tillage was superior. However, low tillage + interculture were superior in semi-arid Vertisols and low tillage + herbicide was superior in Aridisols.

Issues in adoption of Conservation Agriculture

The CA systems are quite different from the conventional practices. The key challenges are

i) Development, standardization and adoption of farm machinery for seeding amidst of crop residues with minimum soil disturbance;

ii) Developing crop harvesting and management systems with residues maintained on soil surface; and developing and continuously improving site specific crop, soil and pest management strategies that will optimize the benefits of the new systems.

iii) Stratification of nutrients especially build-up of phosphorus and potassium in the surface layer and difficulty in application of N fertilizer under residue retained plots.

iv) Developing machinery like high clearance for standing crop as well as management of higher quantity of residues on the soil surface.

v) Training of man power for adoption of the new techniques which needs new skills and management strategies.

Challenges in Conservation Agriculture

There are challenges in the conservation agriculture like crop establishment, weed and residue management. These issues are discussed below in detail

i) Crop Residue Generation

Globally, the total crop residue production is estimated at 3.8 billion tones per year, of which 74% are from cereals, 8% from legumes, 3% from oilseeds, 10% from sugar crops and 5% from tuber crops (Lal, 2005). The total dry biomass generated annually was 682.61 million tons (MT) from the selected eleven crops in the three seasons. Out of this total annual crop biomass, 400.27 MT (58.57%) was generated during kharif season, 265.30 MT was generated during rabi season (38.92 %), and 17.05 MT was generated during summer season (2.50%) (Fig. 4). Among the States, maximum biomass generation was from Uttar Pradesh (17.99%) followed by Maharashtra (10.52%), Punjab (8.15%) and Gujarat (6.4%). During kharif season Uttar Pradesh generated maximum crop biomass (19.14%) followed by Maharashtra (16.24 %), Gujarat (8.75%) Punjab (7.51%), and Karnataka (7.49%). In rabi season 17.43% biomass was generated in Uttar Pradesh followed by Madhya Pradesh (12.70%), Rajasthan (10.65%) and Bihar (8.53%). The cereal crops (rice, wheat, maize) contribute 58% while rice crop alone contributes 33% of crop residues (Fig. 5 &6). Wheat ranks second with 21% of residues whereas fibre crops contribute 10% of residues generated from all crops. Among fibres, cotton generates maximum (53 Mt) of crop residues. Sugarcane occupying an area of 4% of the total cultivated area generates residues comprising tops and leaves to the extent of 17% The residue generated is utilized mainly as industrial/domestic fuel, fodder for animals, packaging, bedding, wall construction, *in- situ* incorporation and green manuring, thatching andleft in field for open burning. However, in case of combine harvesting almost all the residue generated is left in the field that finally ends up in burning (Gupta *et al.,* 2004). It is estimated that surplus crop residue of 91-141 Mt is being burnt on the field.

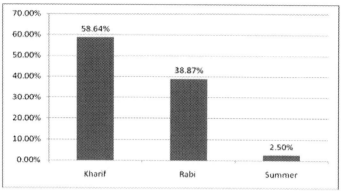

Fig. 4: Proportion of crop biomass generated during different crop season
Source: TIFAC (2018)

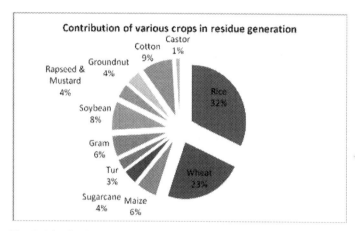

Fig. 5: Distribution of area under different crops
Source: TIFAC (2018)

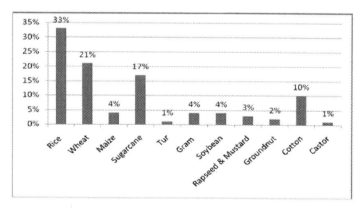

Fig. 6: Share of different crops in residue generation
Source: TIFAC (2018)

ii) Crop Residue Management

The amount of residue cover left on the field is greatly affected by the type of operation and the implements that have been used (Table 1). Each implement's design, adjustments, and depth of soil disturbance, and to a lesser extent, its speed and the condition of the residue, will have an effect on the percentage of both fragile and non-fragile residue remaining on the soil surface. Other factors that affect residue cover are type of residue, chopping versus leaving residue unchopped, carryover of residue, degree of grazing after harvest, type of field operations, soil moisture and weather conditions, and timing of field operations. The effect of each of these factors varies considerably. The fragility of the residue is important and will determine the amount of residue that will remain on the soil surface as it interacts with the other factors. Typical amounts of residue coverage left after harvest of various crops are listed in Table (Table 2) (Al-Kaisi, 2009). Valzano et al., (2005) defined three crop residue management practices:

- **Residue retention** involves leaving stubbles on the soil surface, treated or untreated (Fig 5). The untreated stubble is considered standard harvesting by cutting high or low with no modification of stubble levels. The treated stubble is considered to have levels reduced by cutting low or by windrowing, baling or removal (chaff carts). This method of stubble management protects the soil surface from wind and water erosion, while retaining carbon at the soil surface.

Fig. 5: Residue retention under No-tillage

- **Residue incorporation** involves the use of tillage implements to incorporate remnant plant residue into the soil following harvesting (Fig 6). Traditionally this practice was considered useful in returning organic matter to the soil and protecting the soil from erosion. However, it can contribute to the transference of plant pathogens from one crop to another, offers less surface protection than other stubble retention practices, and can adversely affect soil structure and porosity.

Fig. 6: Residue incorporation under reduced tillage

- **Residue burning** involves the burning of residues. Ideally, stubbles should be burnt just prior to sowing so as to minimize the time in which a soil is exposed to potential erosion. India produced a record of 257.4 million tons of foodgrains in 2012-13. It also produces a large amount of crop residues (500-550 million tons) annually (NAAS, 2012; MNRE, 2009). These crop residues are used as animal feed, soil mulch, manure, thatching for rural homes and fuel for domestic and industrial purposes. However, a large portion of these crop residues (90-140 Mt) is burnt on-farm primarily to clear fields to facilitate timely planting/seeding of succeeding crops (Fig. 7). In comparison to burning, residue retention increases soil carbon and nitrogen stocks, provides organic matter necessary for soil macro-aggregate formation and fosters cellulose–decomposing fungi and thereby carbon cycling. It is estimated that additional amount of about 1.6 t/ha of crop residue is being added in to the field compared to farmers practice, if one feet height of residue is left on the field under no-tillage (NT)/reduced tillage (RT) (Table 1)

Fig. 7: Widespread burning of wheat residue in Central India

iii) Weed and Input Management

Weed control is the other main bottleneck. Thus, increased use of herbicides is pre-requisite for adopting conservation agriculture. Countries that use relatively higher amounts of herbicides are already facing such problems of pollution and environmental hazards. Therefore, use of appropriate herbicide at right dose and right time is very much required to manage weeds under CA. Difficulty in use of inputs like application of nutrients, pesticides and irrigations under higher levels of crop residues. Nutrient management may become complex because of higher residue levels in surface layers and reduced options for application of nutrients, particularly through manure. Application of fertilizers, especially N entirely as basal dose at the time of seeding may result in a loss in its efficiency and environmental pollution. Sometimes, increased application of specific nutrients may be necessary and specialized equipments are required for differential placement of seeds and fertilizers, which contributes to higher costs.

Lack of appropriate machinery: Sowing of a crop in the presence of residues of preceding crop is a greater problem. Different designs of direct drilling machines *viz* zero till drill, no till plant drill, strip till drill, and rotary slit no till drill are available. Spread of conservation agriculture is evident in the region of IGP region under Rice-Wheat cropping system. This has been achieved by the joint effort of Rice-Wheat consortium- NARS-CGIAR (Gupta *et al.*, 2005). However, the spread of these practices into other regions of India is very little. This was mainly due to i) mindset/ attitudinal change of farmer, ii) CA strategies are different from those we have adopted over past decades and iii) non-availability of suitable farm equipment/ Farmers' choice.

Fig. 8: Soybean sown by No-till seed drill

On other hand, new variants of zero-till seed-cum-fertilizer drill/planters such as Happy Seeder, Turbo Seeder and Rotary-disc drill have been developed for direct drilling of seeds even in the presence of surface residues (loose and anchored up to 10 t ha^{-1}) in the IGP region. These machines are found to be very useful for managing crop residues for conserving moisture and nutrients as well as controlling weeds (IARI, 2012). Agricultural machinery or tools, which support conservation agriculture generally refer to the cultivation systems with minimum or zero tillage and *in-situ* management of crop residues (Fig. 8; Table 3).

Different designs of direct drilling machines *viz.*, zero till seed drill & no till plant drill (commercially known as *Happy seeder* & *Turbo Seeder*), strip till seed drill, roto till drill and rotary slit no till drill have been developed with controlled traffic measures or energy efficient and cost-effective seeding of crops with less soil disturbances.

Table 3: Residue addition in conservation agriculture in Soybean-Wheat system

Wheat Stubble Retention	Addition of residue (Air-dry weight kg/ha)	
Farmers Practice (10-15cm)	676	
Reduced Tillage/No-tillage (1 foot)	2283 (+1607)	

Water and energy saving: Zero tillage farming on 5 m ha. Indo-Gangetic plains reportedly saved 360 million m^3 water. It also reduces the number of operating hours of the pumps, thus reducing CO_2 emission and consumption of electrical energy. Approximately 2500 MJ energy is being saved through no-tillage practices compared to conventional tillage practices.

Farmers' perception: Limiting factor in adoption of residue incorporation systems in conservation agriculture by farmers include additional management skills, apprehension of lower crop yields and/or economic returns, negative attitudes or perceptions, and institutional constraints. In addition, farmers have strong preferences for clean and good looking tilled fields vis-à-vis untilled shabby looking fields. Furthermore, pest infestation particularly termite infestation under residue retained field conditions under conservation agriculture.

Prospects of Conservation Agriculture in Rainfed Regions

CA has been proposed as widely adapted set of management principles that can assure more sustainable agricultural production.

Impact of CA on Soil Properties

Due to poor irrigation facilities, limited water resource and uncertain rainfall pattern during monsoon results in low and stagnating productivity particularly for the soybean-wheat system in Vertisols of central India. Besides this, soil related constraints like low infiltration, high incidence of inundation, accelerated runoff and soil erosion in sloping land, low soil organic carbon (SOC) and low nutrient input by resource poor farmers have also affect crop yields in the region. Thus, conservation tillage practices like, no tillage and reduced tillage has emerged as an alternative to conventional tillage practices that help in restoring soil health and enhancing crop productivity.

Experiences at Indian Institute of Soil Science (IISS) showed that reduced tillage is a suitable option for growing soybean–wheat crops in Vertisols with saving of energy and labour (Subba Rao *et al.*, 2009). Soil organic carbon (SOC) is vital for sustainable soil quality and food production systems, which can be improved under conservation agriculture practices compared to conventional system. The soil properties like, organic carbon and pools, physical and biological properties were also improved under conservation agricultural practices.

Effect of Conservation Tillage on Soil Properties

✓ Long term study clearly indicated that conservation tillage promoted formation of macro-aggregates from micro-aggregates in vertisols of Central India.

✓ Increased SOC content and stabilized soil carbon by capturing them inside aggregates and consequently decreasing their decomposition rate by physically occluding them from microbial attack helping in sequestration of carbon in soil profile.

✓ Soil moisture retention of undisturbed soil core samples were higher in treatments where residues were retained compared to CT where residue were removed, however nitrogen level did not influence the soil water retention at Field capacity (Table 4).

Table 4: Soil water retention (%, v/v) at field capacity of the top 5 cm soil layer

	Soil water retention at 0.33 bar (%, v/v)			
Treatment	$N_{50\%}$	$N_{100\%}$	$N_{150\%}$	Average
NT	35.2	35.8	35.4	35.5
RT	35.6	35.5	35.8	35.6
MB	34.4	34.8	35.0	34.7
CT	32.8	33.4	33.1	33.1
Average	34.5	34.9	34.8	

NT-No-tillage, RT-reduced Tillage, MB-Moulboard, CT-Conventional tillage

Source: IISS, Annual Report (2011-12)

Residue Burning under conventional tillage practices

Crops residues left on the field were burnt under conventionally tilled plots, whereas, it was retained under conservation tillage system (RT/NT). Soil temperature was recorded before and after burning (5 min after fire was subsided) and it varied from 39 to 40 °C and 45 to 55 °C, respectively (Fig. 9 & 10). It was inferred that there was increase in soil temperature to the tune of 15°C due to residue burning (Table 4).

Table 4: Soil temperature during residue burning

Particulars	Soil temperature
Before burning	34.8-35.7 °C
During burning	51-53 °C
5 min after burning	40.0-41.3 °C
	~15°C increase in temperature due to burning

Fig. 9: Residue burning under Conventional Tillage

Fig. 10: Residue retention under Conservation agriculture

Infiltration Rate

Infiltration rate was measured using double ring infiltrometer immediately after harvest of *Kharif* crops during 2013 (Fig. 11). During initial year there was no clear cut effect of tillage treatments on steady state infiltration rate and values were on the higher side under different tillage treatments. However, during second crop cycles we could record representative steady state value under different tillage systems. Data indicated that higher steady state infiltration was recorded under no-tillage and reduced tillage as compared to conventional tillage (Table 5). This was mainly due to minimum soil disturbances as well as addition of crop residues which favoured better soil physical conditions.

Fig. 11: Measurement of Infiltration Rate

Table 5: Steady state infiltration rate (cm/hr) under different tillage systems

Year	Conventional Tillage (CT)	Reduced Tillage (RT)	No-Tillage (NT)
2013*	0.77	0.92	0.99

*Averaged over four cropping system

Effect of Conservation Tillage on Mean Weight Diameter

The soil moisture content and mean weight diameter (MWD) at surface layer (0-15 cm) was significantly affected by the tillage treatments after three crop cycles. Under reduced tillage (RT) practices, soil moisture content was higher than conventional tillage (CT) with an increase of 3-13%. Similarly, mean weight diameter (MWD) was significantly higher under reduced tillage (RT) compared to conventional tillage (CT) fields in surface layer, however, in subsurface layer there was no significant difference (Fig. 12) (Somasundaram et al., 2013 unpublished data).

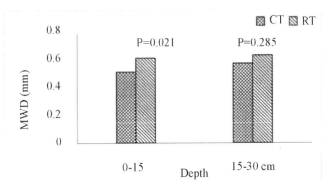

Fig. 12: Effect of reduced tillage on MWD

Effect of Conservation Agricultural Practices on Soil Organic Carbon

The SOC is vital for sustainable soil quality and food production systems. It can be better stored/sequestered through conservation tillage practices in crop land soils in comparison with conventional tillage practices (Lal, 2004). Crop residues are precursors of the SOC pool, and returning more crop residues to the soil associated with an increase in SOC concentration/sequestration. To quantify the impact of conservation agricultural practices on various fractions of soil organic carbon (SOC), soil samples were analyzed for various fractions of SOC namely labile fraction, oxidizable carbon and total organic carbon.

i) Active Carbon (Labile fraction)

Active carbon is a better indicator of the field management practices. This portion of carbon is also known as labile fractions. Small change in labile fractions of soil organic carbon (SOC) may give an early indication of soil degradation or agradation in response to management practices. The active carbon content of the soil was determined after the completion of first crop cycle. The results showed that soils under pigeon pea (Pp) based cropping system registered relatively higher active carbon (372 mg/kg) in the top 0-5cm soil layer as compared to other cropping systems. Among the tillage systems, higher active carbon was recorded in no-tillage (383mg/kg) and reduced tillage (360 mg/kg) as compared to that in conventional tillage (335 mg/kg) in the 0-5 cm soil layer. Similar trend was also observed for the 5-15 cm soil layer. However, both tillage and cropping system effect on active carbon content was not significant after first crop cycle (Fig.13). However, it is expected that in the long run say after 5-6 crop cycles the effect of tillage and cropping systems on active carbon content would be clear due to the additive effect of surface crop residue addition and less soil disturbances under conservation tillage/ agriculture system.

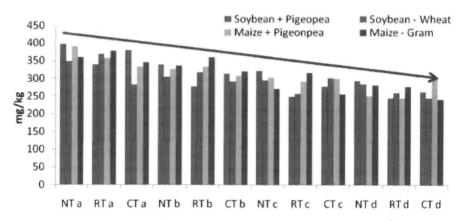

a: 0-5cm, b:5-15, c:15-30, d:30-45; *CT: Conventional Tillage, RT- Reduced Tillage, NT- No-till*
Fig. 13: Active carbon under conservation Agricultur

ii) Oxidizable Carbon

Crop residues retained on the soil surface in conservation agriculture, in general, serve a number of beneficial functions, including soil surface protection from erosion, enhancing infiltration and cutting run-off rate, decreasing surface evaporation losses of water, moderating soil temperature and providing substrate for the activity of soil micro-organisms, and a source of soil organic carbon (SOC). It is estimated that wide dissemination of conservation agriculture (which leaves at least 30% of plant residue cover on the surface of the soil after planting) could offset as much as 16% of worldwide fossil fuel emissions (CTIC, 1996). A study conducted at IISS, Bhopal also reveals effect of tillage systems on SOC was found to be significant only at surface layer (0-15cm) and higher SOC value observed under reduced tillage (RT) as compared to conventional tillage (CT) after three years of crop cycles (Fig.14). Further, reduction in tillage operation coupled with residue retention helps in maintaining the soil organic carbon.

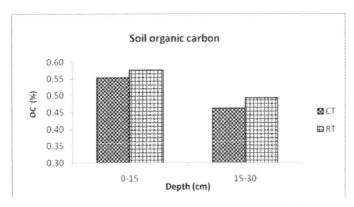

Fig. 14: Organic carbon (%) status under conservation agriculture

iii) Total organic carbon (TOC)

TOC of the soil samples collected after the first crop cycle was determined by dry combustion method using CHN analyzer. However, significant difference could not be observed from TOC data. Carbon data revealed that about 6.1-10.9% increase of total organic carbon under RT and NT over CT in 0-5 cm soil layer. However, tillage system did not have significant effect on total organic carbon as compared to initial TOC. However, conspicuous change in carbon content is expected to occur only after 5-6 years under conservation agriculture.

Soil samples collected after completion of two crop cycles were analyzed for total organic carbon (TOC) content. The data showed that irrespective of tillage systems top layers (0-5cm and 5-15cm) registered relatively higher TOC than sub-surface layers. Among the tillage systems, no-tillage was significantly different (based P≤ 0.01) from the conventional tillage. However, cropping systems had inconspicuous effect on the TOC even after two crop cycles (Table 6), presumably poor crop performance due to heavy rains during *kharif* season.

Table 6: Weighted average of total organic carbon (%) in 0-45 cm under conservation agriculture

	2013-14		
Cropping System	NT	RT	CT
Soybean - Wheat	1.00	0.90	0.80
Soybean + Pigeonpea (2:1)	0.90	0.99	0.89
Maize- Gram	0.98	0.90	0.86
Maize + Pigeonpea (1:1).	0.94	0.84	0.87
	0.96[a]	0.91[ab]	0.86[b]
Tillage	P value 0.003		
Cropping system	0.561		
Tillage X Cropping system	0.067		

NT-No-tillage, RT-Reduced tillage, CT-Conventional Tillage; Initial TOC Value was 0.88 % in 0-45cm soil depth (i.e weighted average).

Effect of Conservation Agriculture on Crop Yields

One of the primary challenges of our time is to feed a growing and more demanding world population with reduced external inputs and minimal environmental impacts, all under more variable and extreme climate conditions in the future (Pittelkow *et al.*, 2014; Lobel *et al.*, 2008; Godfray and Garnett, 2011). Conservation agriculture represents a set of three crop management principles that has received strong international support to help address this challenges. However, conservation agriculture is highly controversial/ debatable, with respect to both its effects on crop yield and its applicability in different farming contexts (Stevenson *et al.*, 2014). Pittelkow *et al.* (2014) reported from global meta analysis using 5463 paired yield observations from 610 studies (48 crops and 63 countries) that no-till system reduces yields, yet this response is variable under certain conditions no-till can produce equivalent or greater results than conventional tillage. Interestingly, when no-till is combined with the other two conservation agriculture principles of residue retention and crop rotation, its negative impacts are minimized. Moreover, no-till in combination with the other two other principles significantly increases rainfed crop productivity in dry climates.

In a similar fashion, results of Long term tillage experiments (2000-2010) conducted at rainfed vertisols (IISS, Bhopal) revealed that yield levels of conservation tillage (i.e no-tillage and reduced tillage) was on par with conventional tillage, besides, greater saving of energy and labour under conservation tillage (Fig. 15).

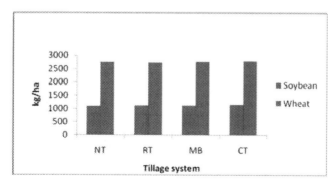

NT-No-tillage, RT-reduced Tillage, MB-Mouldboard, CT-Conventional tillage
Fig. 15: Average yield of soybean and wheat under different tillage systems

Another study conducted at IISS, Bhopal (2010-2014) revealed that tillage did not have significant effect on crop yields after completion of three crop cycles. Yield data indicated that tillage had no effect on soybean grain equivalent (SGE) after three crop cycles. Among the cropping systems studied, maize-gram recorded higher soybean grain equivalent yield followed by soybean+ pigeon pea (2:1) and soybean-wheat cropping system. Barring soybean-wheat system and soybean-fallow under CT, all other crop yields were higher under RT after third crop cycle. This was possibly attributed to improvement in soil structure three years of continuous reduced tillage (Fig. 16).

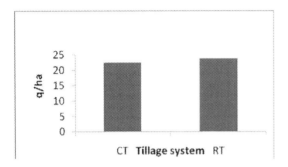

Fig. 16: Effect of tillage on soybean grain equivalent yield (q/ha) (average yield of six cropping systems)

The success stories of CA in the IGP region is evident, however, the percolation of this technology is still very little under rainfed region. Therefore, the following selected cropping systems provide opportunities to introduce, test, evaluate and improve productivity under conservation agriculture.

1. Soybean based cropping system in the medium and deep black soil regions in states of Madhya Pradesh, Maharashtra and Rajasthan (3.2mha).

2. Rice-fallow and rice legume dominated region in the hot-sub-humid eastern plateau region in the states of Chhattisgarh, Jharkhand, Bihar, Odisha, West Bengal and Assam (12 mha).

3. Maize-wheat cropping system dominated sub mountain hill and foot hills regions in the state of Himachal Pradesh, Jammu & Kashmir, Uttarakhand, Jharkhand, and Chandigah (1.86 mha).

4. Kharif-fallow-Rabi cropped areas in semi-arid alluvial soil regions.

5. Pearlmillet/Sorghum, clusterbean-wheat/mustard/ chickpea in the state of Rajasthan, Maharashtra, Western UP and Haryana (5.7 mha).

6. Cotton based cropping systems of the shallow and medium black soil region of Maharastra, Madhya Pradesh and Gujarat semi-arid regions of North-western IGP, Maharastra, Madhya Pradesh and Karnataka (2.55 mha).

Conclusion

CA is definitely a sustainable production system which not only sequesters carbon but also enhances productivity and soil quality as well as promoting ecologically and economically sustainable production system in a particular region. Moreover, CA practices benefit farmers in the short term through reduced cultivation costs. It has been observed that no tillage practice is most time, energy and cost-effective system as compared to the conventional tillage/practices. Agricultural tools and equipments for such practices are now available in our country at a reasonable price, in spite of that adoption of CA practices in other parts especially in rainfed regions of the country is very meager. Therefore, location specific CA technology/machinery generation–dissemination–adoption of CA practices has to be looked into for broader perspective and it has to go hand in hand, that is, close partnership with farmers. However, the success of CA lies with the integration of other techniques, such as integrated pest management, plant nutrient management, and weed and water management.

References

Abrol, I.P and Sunita, S. (2005). Conservation agriculture for transition to sustainable agriculture, Meeting report. *Current Scienc,* 88(5): 686-687..

Baker, C.J., K.E. Saxton, and W.R. Ritchie. (2002). No-tillage Seeding: Science and Practice. 2nd Edition. Oxford, UK: CAB International.

Conservation Technology Information Centre. (2002). Tillage type definitions. Website: http://www.ctic.purdue.edu/ Core4/CT/Defintions.html.

Godfray, H.C.J. and Garnett, T. (2014). Food security and sustainable intensification. *Phil. Trans. R. Soc. B,* 369: 20120273.

FAO (2011). What is Conservation Agriculture? FAO CA website (http://www.fao.org/ag/ca/1a.html), FAO, Rome.

Faulkner, E.H. (1945). Ploughman's Folly. Michael Joseph, London. p. 142.

Friedrich, T., Derpsch, R. and Kassam, A (2012). Overview of the Global Spread of Conservation Agriculture, http://factsreports.revues.org/1941 Published 12 September, 2012.

Friedrich, T., Kassam, A.H., Shaxson, F. (2009). Conservation Agriculture, In: Agriculture for Developing Countries, Science and Technology Options Assessment (STOA) Project, European Technology Assessment Group, Karlsruhe, Germany.

Fukuoka, M. (1975). One Straw Revolution, Rodale Press, English translation of shizen noho wara ippeon no kakumei, Hakujusha Co., Tokyo. p. 138..

Hobbs, P. R. (2007). Conservation Agriculture: What Is It and Why Is It Important for Future Sustainable Food Production?. *The Journal of Agricultural Science*, 145 (2): 127-137.

IARI, (2012). Crop residues management with conservation agriculture: Potential, constraints and policy needs. Indian Agricultural Research Institute, New Delhi, vii+32 p.

Kassam, T. Friedrich & R. Derpsch (2018): Global spread of Conservation Agriculture, International Journal of Environmental Studies, https://doi.org/10.1080/00207233.2018.1494927

Kumar, S., Sharma, K.L., Kareemulla, K., Ravindra Chary, G., Ramarao C.A., Srinivasa Rao, Ch. and Venkateswarlu B. (2011). Techno-economic feasibility of conservation agriculture in rainfed regions of India. *Current Science*, 101(9): 1171-1181.

Lal, R. (2004). Soil carbon sequestration impacts on global climate change and food security. *Science,* 304: 1623-1627.

Lobell, D.B., Burke, M.B., Tebaldi, C., Mastrandrea, M.D., Falcon, W.P. and Naylor, R.L. (2008). Prioritizing climate change adaptation needs for food security in 2030. *Science,* 319: 607–610.

NAAS (2012). Management of crop residues in the context of Conservation Agriculture, Policy paper No. 58, National Academy of Agricultural Sciences, New Delhi. p. 12.

MNRE (Ministry of New and Renewable Energy Resources) (2009) Govt. of India, New Delhi. (www.mnre.gov.in/ biomassrsources).

Sayre, K.D., and P.R. Hobbs. (2004). The Raised-Bed System of Cultivation for Irrigated Production Conditions. In R. Columbus, Ohio, and New York, USA: Ohio State University and Marcel Dekker, Inc. pp. 337-355

Stevenson, J. R., Serraj, R. and Cassman, K. G. (2014). Evaluating conservation agriculture for small-scale farmers in sub-Saharan Africa and South Asia. *Agriculture Ecosystem and Environment* 187:1–10.

Subba Rao, A., Biswas, A.K., Sammi Reddy, K., Hati, K.M. and Ramana, S. (Eds) (2009). IISS:Two Decades of Soil Research, Indian Institute of Soil Science, Bhopal, 1-132.

TIFAC (2018) Estimation of Surplus Crop Residues in India for Biofuel Production (Joint Report of TIFAC & IARI October, 2018)

Wall, P.C. (2007). Tailoring conservation agriculture to the needs of small farmers in developing countries: An analysis of issues. *Journal of Crop Improvement,* 19:137-155.

WCCA, (2009). 4th World Congress on Conservation Agriculture Report, Congress held at New Delhi, 4-7 February, 2009.

5

Conservation Agriculture *vis-à-vis* Resource Conservation Technologies (RCTs): Why and How?

R. Saha, J. Mitra and Alka Paswan

ICAR-Central Research Institute for Jute & Allied Fibres, Barrackpore West Bengal – 700 121, India

Introduction

The Indian agriculture is very complex and carrying out multi-functionalities of providing food, nutrition and ecological security besides employment and livelihood for over 700 million people. Indeed, India has made a marvelous achievement in attaining self-sufficiency in food grain production after the induction of Green Revolution which eventually resulted in maintaining all-time high buffer stock in warehouses of our country. Such rosy picture in production trends turned to be bleak in the past few years. There are various reasons behind this. Shrinking resources of prime lands, deforestation and accelerated erosion, deterioration of soil physical environment, increasing waterlogging and salinity in canal irrigated areas, declining water table in well-irrigated areas, poor management of rainwater, lower efficiency of inputs such as water, fertilizers and agrochemicals, rapid industrialization coupled with pollution and environmental degradation and hazards have aggravated the problem. At the same time, increase in atmospheric concentrations of greenhouse gasses (GHG), of which the most common is carbon dioxide (CO_2), is the primary cause of global warming and the Intergovernmental Panel on Climate change (IPCC) estimates that the current greenhouse gases (GHGs) concentrations are 30% more than the pre-industrial level. C is accumulating in the atmosphere at a rate of 3.5 Pg (Pg = 1015 g or billion tons) per annum, the largest proportion of which resulting from the burning of fossil fuels and the conversion of tropical forests to agricultural production. Under such circumstance, there is an imperative need to produce more from less arable land and water through meticulous management of basic agricultural resources such as soil, water and biological inputs.

The conservation agriculture (CA)-based RCTs, practiced over an estimated 100 M ha area worldwide and across a variety of climatic, soil and geographic zones (Derpsch and Friedrich, 2009), have proved to be energy and input efficient, besides addressing the emerging environment and soil health problems (Saharawat *et al.*, 2010). The CA technologies involving no- or minimum tillage with direct seeding and bed planting, residue management (mainly residue retention) and crop diversification have potential for improving productivity and soil quality, mainly by soil organic matter (SOM) build-up (Bhattacharyya *et al.*, 2013). The RCTs bring many possible benefits including reduced water and energy use (fossil fuels and electricity), reduced greenhouse gas (GHG) emissions, soil erosion and degradation of the natural resource base, increased yields and farm incomes, and reduced labor shortages (Pandey *et al.*, 2012). The efficiency and sustainability of any production system depends on system-based management optimization of crop yields, economic benefits, and environmental impacts.

CA/ RCTs for Indo-Gangetic Plains

The Indo-Gangetic Plains (IGP) is the breadbasket of south Asia. The region witnessed higher growth rates for food grain production compared with other regions of the world. Most of this area is under rice-wheat cropping systems covering a total of 13.5 million hectares in India, Nepal, Bangladesh and Pakistan. Extensive irrigation infrastructure, mechanization and easy access to production inputs, marketing and grain procurement services have contributed to these increases, especially in the western parts of the IGP. However, growth rates have decreased even as there is a wider recognition of environmental issues arising from the intensive and sometimes excessive use of inputs. The long-term sustainability of these systems is now a subject of attention. There is a general consensus that quality of natural resource base needs to be improved for enhanced productivity in the eastern IGP. Also, it is believed that future productivity growth would come about through better risk management strategies in the drought and flood prone marginal tracts of the eastern Gangetic plains. Targeting the resource conserving technologies offers newer opportunities to provide for better livelihood for the resource poor, densely populated, small and marginal farmers of eastern Gangetic plains. Various approaches of resource conservation technologies are as follows:

No tillage or Minimum Tillage

Tillage practice under conventional crop production delays planting, escalate costs, reduce profits, needs more water for crop production. No tillage (NT) or minimum tillage involves the use of a tillage implement that creates a narrow slot for the seed and does not disturb or turn over the soil in the process of planting the crop. The traditional approaches of ploughing which include 3-4 tillage operations are completely skipped. Hence, cost of production is reduced and timely planting of crop (wheat) is ensured. Another benefit of earlier sowing under NT is that *Phalaris minor*, a herbicide-resistant weed in wheat, is less competitive than when wheat is sown late under conventional tillage (CT), (Malik *et al.*, 2002). Evidence on yield effects of zero tillage is highly variable. Where NT is combined with mulching, a commonly described pattern is for yields to fall initially (Baudron *et al.*, 2011), and then to increase over the subsequent decade or so, eventually exceeding yields in conventional tillage-based agriculture. Wider adoption of NT in wheat was due to a combination of both increased yields and reduced production costs. NT has effectively reduced the demand for water in rice-wheat cropping systems in more than 1 million ha of area in the Indo-Gangetic Plains. Gupta (2003) found that 13–33% less irrigation water was used under NT compared with CT in wheat. According to the Vincent and Quirke (2002), the Indian economy would gain about US$1400 million over the next 30 years from adoption of NT in rice–wheat system of Northwestern India due to savings in tillage and herbicide costs and higher wheat yields. Further, it is estimated that if we could bring about 5 million ha of area under conservation tillage systems in IGP, the CO_2 emission could be reduced to the extent of 70 Mg per year. Other advantages NT provides over CT are: improved soil health, fuel savings (75%) and improved level of organic carbon (Malik *et al.*, 2002). Minimum-tillage or NT as a CA component may increase soil C compared with CT but these increases are often confined to near-surface layers (<10 cm). However, the potential of CA for storing C depends on antecedent soil C concentration, cropping system, management duration, soil texture, slope and climate. In addition, the soil properties like bulk density, penetration resistance, aggregation and infiltration were improved remarkably under NT as compared to conventional tillage systems. NT has a direct mitigation effect as it converts the green house gases like CO_2 into O_2 in the atmosphere and enriches soil organic matter (Venkateswarlu and Shanker 2009).

Fig. 1: Sowing with Happy seeder

Management of Crop Residues

Crop residues are the parts or portion of a plant or crop left in the field after harvest, or that part of the crop which is not used domestically or sold commercially or discarded during processing. A vast potential is available to efficiently recycle crop residues, especially in rice-wheat belt of Punjab, Haryana and western Uttar Pradesh, where it is burnt in situ. Burning crop residues due to lack of efficient and user-friendly technologies for in-situ recycling not only leads to loss of considerable amount of N,P,K and S but also contributes to global NO_2 and CO_2 budget. More than 340 Mt of crop residues from various crops are produced annually of which major quantity is contributed by rice and wheat (nearly 240 Mt). This accounts for nearly 6 Mt of major nutrients of which at least one third is trappable for recycling. CA practices require a critical level of crop residues with objective, to protect the soil against weather aggressions and water erosion, to maintain soil moisture (Lal, 1997), to suppress weed growth and to provide shelter and food for the soil biota. Crop residues are also important source of nutrients and maintain or enhance soil chemical, physical and biological properties and prevent land degradation. The importance of crop residue cover as part of the CA system has been emphasized by several researchers. Crop residue recycling in rice-wheat was found to increase rice as well as wheat yields by 13 and 8%, decrease cost effectiveness by 5 and 3% and energy efficiency by 13 and 6%, respectively, compared to residue retrieval, whereas yield advantage was to the tune of 9 and 3% compared to residue burning. However, decomposition rate and release of N from residues depends on soil, climatic conditions and the C:N ratio of plant residues (Prasad and Power, 1997).

Fig. 2: Crop residues as cover in vegetable crop production

Crop Rotation and Cropping System

Use of crop rotations or intercropping is considered vital in CA systems, as it offers an option for higher diversity in plant production and thus in human and livestock nutrition, and pest/weed management that are no longer realized through soil tillage. Legumes grown in rotation can provide a range of benefits to the agro-ecosystem, including increases in subsequent crop yields and reductions in input costs due to nitrogen fixation and reductions in crop diseases; greater distribution of channels or biopores created by diverse roots (various forms, sizes and depths); better distribution of water and nutrients through the soil profile and increased humus formation. High-residue producing crops may sequester C than crops with low residue input. Intensification of cropping systems such as increased number of crops per year, double cropping, and addition of cover crops can increase soil C storage under NT (Luo *et al.*, 2010). Furthermore, a diversity of crops in rotation leads to a diverse soil flora and fauna, as the roots excrete different organic substances that attract different types of bacteria and fungi, which in turn, play an important role in the transformation of these substances into plant available nutrients. An effective crop rotation in a cropping system not only helps to increase the crop productivity and soil fertility, but also improve the water use efficiency by reducing weeds, providing conducive micro-climate for plant growth and development, reduction in soil thermal regime and improving physical properties of the soil.

Diversification/Intensification: A shift from sole cropping to a diversified/intensified farming system is highly warranted. The increased cropping intensity/diversification is intended to minimize risk, improve biodiversity and diversify income sources and enhance resource sustainability. It will be a key strategy for future gains in crop production. Short duration pulses, oilseeds and other high value crops will find their definite niche as sequential or intercrops, rather than replacing the major cereal crops having higher yield stability (IIFSR 2015). Hence, an increased cropping intensity will contribute substantially to additional demands of food and cash crops. Pigeonpea, the most important wet season grain legume crop in south Asia has shown potential for rice crop diversification in Indo-Gangetic Plain (IGP). Similarly, bio-intensive diversified cropping systems would enable small and marginal farmers to utilize limited land and water resources in more efficient manner.

Integrated Farming Systems: Integrated Farming Systems hold a special position in conservational agriculture as in this system nothing is wasted, the byproduct of one system becomes the input for other. For example, crop residues from the field can be used for animal feed, while manure from livestock can enhance agricultural productivity by improving soil fertility as well as reducing the use of chemical fertilizers. Moreover, the system helps poor small farmers, who have very small land holding and a few heads of livestock to diversify farm production, increase cash income, improve quality and quantity of food produced and exploit unutilized resources. Animals play key and multiple roles in the functioning of the farm. These not only provide meat, milk, eggs, wool, and hides; but can be converted into prompt cash in times of need. Animals transform plant energy into useful work: animal power is used for ploughing, transport, marketing and water lifting for irrigation.

Cover Crops

Keeping the soil covered is a fundamental principle of CA. Crop residues are left on the soil surface, but cover crops may be needed, if the gap is too long between harvesting one crop and establishing the next. Cover crops may fulfill additional agronomic, ecological or economical functions in CA systems that can supplement those performed by the main commercial crops (Hartwig and Ammon 2002). CA systems with one commercial crop such as rice or maize, relayed by or intercropped with a cover crop (*Brachiaria sp., Stylosanthes sp., Crotalaria sp.* and *Cajanus sp.)* produces extra biomass at the end of the rainyseason which can be used either as green manure or grazed on the field. Cover crops increase total biomass production that allows mulching the soil permanently (Calegari 2006), even under humid tropical conditions where residues decompose rapidly. They also improve the stability of the CA system, not only on the improvement of soil properties but also for their capacity to promote an increased biodiversity in the agro-ecosystem.

Cover crops, when properly intercropped with commercial ones, can significantly increase radiation interception by a plant canopy, during before or after the commercial crop cycle (Picard *et al.*, 2010). Vegetative cover also protects the soil against the impacts of raindrops; keep soil-shaded; and maintain the highest possible moisture content. They may also contribute to the mineral nutrition of the main crop(s) through nitrogen fixation in the case of legumes, mulch mineralization or manure returns from animals that feed on them. Part of the biomass they produce may contribute to farm incomes, e.g. additional grain production for human food or as extra fodder resources. Beside their above-ground functions, cover crops fulfill important functions below the ground. Their root systems contribute to preventing or remediating soil compaction, tapping soil moisture from deeper horizons below the root zone of the main crops or recycling nutrients such as nitrates, K, Ca and Mg that are easily leached to deeper soil horizons. There are various crop alternatives to be used as vegetative cover, such as grains, legumes, root crops and oil crops.

Fig. 3: Cover crop as resource conservation technology

Site-specific Nutrient Management (SSNM)

Site-specific nutrient management (SSNM) provides an approach for need based feeding of crops with nutrients while recognizing the inherent spatial variability. It involves monitoring of all pathways of plant nutrient flows / supply, and calls for judicious combination of fertilizers, bio fertilizers, organic manures, crop residues and nutrient efficient genotypes to sustain agricultural productivity. It avoids indiscriminate use of fertilizers and enables the farmer to dynamically adjust the fertilizer use to fill the deficit optimally between nutrient needs of the variety and nutrient supply from natural resources, organic sources, irrigation water etc. It aims at nutrient supply at optimal rates and times to achieve high yield and efficiency of nutrient use by the crop. It also helped in increase of organic carbon by 55.9%. It is, therefore, pertinent to further disseminate this technology, which has potential to enhance the productivity in the range of 3-4 t ha^{-1}. SSNM approach involves three steps – establishing attainable yield targets, effectively use existing nutrient sources and application of fertilizers to fill the deficit between demand and supply of nutrients.

Leaf Colour Chart

The user friendly real time N management practices based on the soil supply capacity and crop demand (Leaf Colour Chart; LCC, green seeker) are the potential approaches to improve the N use efficiency in rice farming systems. A LCC developed in Japan, is used to measure green color intensity of rice leaves to assess the nitrogen requirements by non-destructive method (Nachimuthu *et al.*, 2007), and is being standardized with chlorophyll meter. In hybrid as well as inbred rice, N management through LCC proved superior to locally recommended N application in three splits. It was found possible to curtail 20-30 kg of fertilizer N/ha without sacrificing rice yield, when N is applied as per LCC values. N application at LCC<3 in Basmati and at LCC<4 in coarse and hybrid rice was found optimum. Moreover, in LCC-based N management, basal application of N can be skipped without any disadvantage in terms of grain yield, and agronomic, physiological or recovery efficiency of fertilizer N.

Fig. 4: Leaf colour chart and green seeker for N management in crop field

Laser land leveler

Undulated soil surface influences the farming operations, energy use, aeration, crop stand and yield mainly through nutrient-water interactions. The general practices of land levelling used by the farmers in India is either through use of plankers drawn by draft animals or by small tractors. These levelling practices are not so perfect even after best effort for levelling which results in less input efficiencies and low yield at the cost of more water. Laser leveling is a process of smoothening the land surface (± 2 cm) from its average elevation using laser equipped drag buckets to achieve precision in land leveling. Precision land leveling involves altering the fields in such a way as to create a constant slope of 0 to 0.2%. This practice makes use of large horsepower tractors and soil movers that are equipped with global positioning systems (GPS) and/or laser-guided instrumentation so that the soil can be moved either by cutting or filling to create the desired slope/level. (Walker *et al.*, 2003). Similar increased yield and reduced irrigation water application in the case of no tillage wheat and laser land leveling were reported in India and China (Jat *et al.*, 2009). For instance, Kahlown *et al.*, (2006) showed that the use of RCTs, including no tillage, laser leveling and bed and furrow planting, reduced irrigation water applications between 23 and 45% while increasing yield. It has been noticed that 31.16 % of water could be saved due laser land leveling compared to traditional land levelling.

Fig. 5: Laser land levelling before sowing of crop

Bed Planting

During last decade, practice of raised bed planding has been emerging with a greater pace in Indo-Gangetic plains. The major concern of this system is to enhance the productivity and save the irrigation water. Bed-planting has the potential to conserve significant quantities of water (30–50%) (Kukal *et al.*, 2005). Other benefits of bed-planting include, reduced seed rates, conserved rainwater, facilitated mechanical weed control, minimized lodging in the wheat crop (Gupta *et al.*, 2003); cost reduction and conservation of resources. Fertilization application practices are also easily performed by trafficking in the furrow bottoms and the fertilizers can be banded through the surface residues, reducing thereby potential nutrient losses under permanent raised bed planting. The raised bed planting technique also provides an opportunity for crop diversity through inclusion of different crops as well as feasibility of inter or relay cropping, thereby opening avenues for generating alternate sources of productivity growth through efficient use of resource base. Bed-planting is widely adopted in the Indo-Gangetic Plains, proved to be a successful conservation technology. However, the use of direct dry seeding on flat and raised beds while resulting in considerable water savings generally had negative impacts on rice yield (Choudhury *et al.*, 2007). Raised bed planting system provides additional options in northwest Indo-Gangetic plains to generate alternate source of productivity.

Fig 6: Raised bed preparation before sowing

Direct Seeded Rice (DSR) Technology

Traditionally rice is transplanted after puddling and wheat is sown after pulverizing the soil. This reflects an edaphic conflict in traditional soil management for rice and its consequent deleterious effects on the soil environment for the succeeding wheat and other upland crops. The shortages of labor and water, and soil fertility issues are causing increasing interest in shifting from puddling and transplanting to DSR. It has been found that low wages and adequate availability of water favour transplanting, whereas, high wages and low water availability favour DSR. DSR can reduce the labor requirement by 50% compared with transplanting (Santhi et al., 1998). The DSR system provides incentives for saving water. In Northwest India, about 35–57% water savings have been reported in research experiments in DSR sown into unpuddled soils. Direct-seeded and transplanted rice grown on raised beds decreased water use by 12–60% when compared with flooded, transplanted rice in the IGP (Gupta et al., 2003).

System of Rice Intensification (SRI)

At present, SRI methods have been adopted in almost 50 countries, including major rice-producing nations such as India, China, Vietnam and the Philippines. SRI is usually understood as a package of possible practices, which have to be adapted to local conditions. SRI produce higher yields, contour farming and strip cropping. Farming on the contour creates small ridges that slow runoff water, and it increases the rate of water infiltration, reduces the hazard of erosion and redirects runoff from a path directly down slope to a path around the hill slope. Farming on the contour rather than up and down the slope reduces fuel consumption and is easier. Contour farming can reduce soil erosion by as much as 50% compared to farming up and down hills. Where sheet and rill erosion is a concern on sloping land, the strips are laid out on the contour or across the general slope. Where wind erosion is a concern, the strips are laid out as closely perpendicular to the prevailing erosive wind direction as possible. Thus strip cropping reduce sheet and rill and wind erosion; increased infiltration and available soil moisture; reduced dust emissions into the air; improved water quality; improved visual quality of the landscape; improved wildlife habitat; improved crop growth and soil quality.

Fig. 7: SRI method of rice cultivation

CA/RCTs for North East Hilly Eco-systems

The North Eastern Hilly (NEH) region of India is characterized by diverse agro climatic and geographical conditions. The region receives 122.7 per cent of the country's rainfall against 8.7 per cent of the geographical area. A large fraction of the rain water received is lost as runoff and *in-situ* retention is very low. About 63.1 per cent of the total geographical area is under forests, 16.6 per cent under crops and rest either under non-agricultural uses or uncultivated land. NEH region has remained economically backward, though there is ample potential for development due to the presence of abundant natural resources. There is also problem of growing crops during *rabi*, although the region falls in moisture index zone of 6-8, yet there is acute scarcity of moisture from November to March. Further, irrigation resources are meager and more than 90 per cent of the cropped area is rainfed. Thus, resource conserving techniques (RCTs) using locally available resources encompass practices that enhance resources or input-use efficiency and provide immediate, identifiable and demonstrable economic benefits such as reduction in production costs, saving in water, fuel, labour requirements and timely establishment of crops resulting in improved yields (Ghosh *et al.*, 2010).

Incorporation of Weed Biomass

The Northeastern hill region of India, being a hot spot region of biodiversity and "a low" fertilizer consumption region, recycling of organic materials is gaining importance particularly in rice-based systems. The farm yard manure (FYM) is the common source of organic materials in this region. However, the availability of FYM is limited, and its per capita availability is less as compared with other parts of India. Under such conditions, finding an alternative source of organic materials is a high priority. The nutrients contents of some of the popular weed species available in this area are given in Table 1. Study conducted with locally available grasses *(Jungle grass)* and weeds *(Ambrosia sp.)* showed that incorporation of FYM or jungle grass or *Ambrosia sp.*, continuously for 5 years in puddled rice soil improved soil organic carbon (SOC) by 21.1%; the stability of micro-aggregates, moisture retention capacity, and infiltration rate of the soil by 82.5, 10, and 31.3%, respectively; and soil bulk density decreased by 12.6% (Saha and Mishra, 2009). Coupled with this improvement in soil quality, the incorporation of organic residues resulted in significant increase in rice yield. The study revealed that FYM is the best source of organic application. However, the locally available jungle grasses and *Ambrosia sp.* are equally good as an organic amendment, which would also ease the problem of disposal of these grasses during peak monsoon.

Table 1: Nutrient contents of various weed biomass

Scientific names of the weeds	N content (%)	P Content (%)	K content (%)
Eupatorium odoratum	3.36	0.10	0.82
Eichhornia crassipes	3.01	0.90	0.15
Ipomea sp.	2.01	0.33	0.40
Ambrossia sp.	3.15	0.11	0.79
Lanta camara	2.41	0.08	1.37
Milkania micrantha	2.94	0.18	1.71
Azolla caroliniana	2.38	0.51	2.75

Permanent Organic Soil Cover

Mulching is the practice of covering soil around plants with an organic material to make conditions more favourable for plant growth, development and crop production. Mulches accelerates plant growth by increasing soil temperature, stabilizing soil moisture and reducing weed growth and hence increases crop yields. They also help in preserve vegetables from decay by preventing contact with the soil. Live mulching consists of growing crops having dense foliage in between two rows of other crops to control evaporation as well as weeds. Generally, fast growing leguminous crops such as sunnhemp are grown before or simultaneously with seasonal grain crops such as rice/maize and incorporated into the soil to act as mulch, which after decomposition supplies nutrients to the main crop as well.

In rice based cropping systems (Rice-Potato, Rice-tomato, Rice-radish, Rice-Pea, Rice-toria, Rice-lentil, Rice-gram and Rice coriander), the mean response of crop to mulching (rice straw @ 7.5 t ha^{-1}) was significant with 16 per cent increase in yield and 33 per cent in net returns (Rautray, 2005). In rice-linseed cropping system, normal tillage with rice straw mulching (@ 3.5 t ha^{-1}) recorded significantly higher seed yield over without mulching (Kalita *et al.*, 2005).

Fig. 8: Mustard crop with grass mulch

Zero Tillage

The practice of ploughing the field (tillage) for seed bed preparation has been in vogue since time immemorial. Farmers have perceived that tillage or soil loosening would improve soil fertility, increase its ability to absorb rainwater, and help in controlling unwanted weed flora. In this manner, tillage practiced over the years has led to accelerated oxidation, resulting in reduction in soil's organic matter content, finally leading to reduced overall productivity. By definition, zero tillage also known as conservation tillage consists of a one pass operation which places seed and fertilizer into an undisturbed seedbed, packs the furrow and retains adequate surface residues to prevent soil erosion. It involves planting seeds into soil that hasn't been tilled after the harvest of the previous crop. The crop germinates in residual water left by the previous crop, saving up to a million litres of water per hectare. Zero tillage seeding offers the benefits of retaining surface residues and reduced soil water loss.

Fig. 9: Rice crop under zero tillage

In-situ Residue Management for Carry Over Soil Moisture

An innovative, simple and very low-cost technique of *in–situ* moisture conservation has been developed for *rabi* crop (mustard) using residue of preceding maize crop grown during rainy season. Maize was sown in June with recommended agronomic practices. Biomass of a local weed *Ambrossia* spp, the only external input, has been applied between rows of standing maize at 20 days before its harvest in the month of September to recharge *in-situ* soil moisture profile by preventing run off from the field at the later part of rainy season. Immediately after harvest of maize, its stalk is spread all over the field just above the applied *Ambrossia* and kept as such till sowing of mustard. This way *Ambrossia* spp and maize stalk act as "double mulch" not only to provide optimum soil moisture for sowing of mustard in October but also to recharge the soil profile for support growth and development of mustard during the whole growing season. Mustard was sown in October between maize rows by removing maize stalk and put again between mustard rows on the same day immediately after sowing and kept till harvest of mustard.

Fig. 10: Ambrossia biomass application before maize harvest (Sept.)

Fig. 11: Mustard under maize stalk (Early Nov.)

Green Manuring

Green manuring can be defined as a practice of ploughing or turning down tender and fresh green biomass into the soil, for the purpose of improving fertility and physical condition of the soil. Ideal green manure crop must of faster growing in nature, tolerant to adverse climatic condition, and should be effective nitrogen fixer with adequate *Rhizobium* nodulation potential and quickly decomposable. Incorporation and decomposition of green manure has solubilizing effect on N,P,K and micronutrients like Zn, Fe, Mn and Cu in the soil (Saraf and Patil, 1995) and deficiency of different nutrient element can be mitigated by the way of recycling through green manure. Crops like rice bean, soybean, sunnhemp, dhaincha, mung bean, black gram etc. are commonly used for green manuring in our country. The green manure crop should be allowed to grow up to 45-60 days and should be incorporated into the soil around flowering time. Growing of sunnhemp (*Crotolaria* sp.) as green manure crop along with upland rice and incorporating at one month after germination provides nutrients and conserves soil and water in this region. On an average, 8-10 q (on dry wt. Basis) of biomass is added to the soil along with nitrogen fixed through the root nodules of sunnhemp. The nitrogen added through sunnhemp (150-210 kg N ha^{-1}) dhaincha (140-195 kg N ha^{-1}) and mung bean/ black gram (80-140 kg N ha^{-1}) is highly beneficial for the succeeding crop.

Fig. 12: Incorporation of Sunnhemp biomass at One month after germination

Thrust Areas for Future

- Long-term effects of mechanization of CA and RCTs on crop, soil, biodiversity and climate in various production systems and agro-ecologies for the profitability and sustainability of different cropping system should be the future agenda of research under natural resource management program.
- Designing long-term experiments to study the impact of conservation agriculture on soil health, water and nutrient use efficiency, C sequestration, GHGs emission and ecosystem services.
- Region-specific interventions for crop diversification through substitution/ intensification and matching production technologies.
- Development of low power requiring tools fortillage and crop establishment practices with least disturbances to soil/ soil cover (*in situ* crop residues) especially under conditions of small farms.

- For alleviation of subsoil compaction due to indiscriminate movement of heavy traffic in the field, systematic research work on controlled traffic ZT-drills needs to be initiated.

- Identification and standardization of new cropping, intercropping and novel farming system combinations including livestock and fisheries, which can be economically viable.

- Developing complete package of practices for CA for prominent cropping systems in each agro-ecological region, particularly in rainfed and dryland eco-systems.

- Site-specific nutrient management/ balanced nutrient supply systems and precision input management for intensive major cropping systems to optimize resource use and enhance efficiencies.

- Development of cost-effective technologies for *in-situ* crop residue management and efficient application of different kinds of fertilizers and herbicides.

- Promoting conservation agriculture practices especially in water harvesting, nutrient, pest and disease management.

- Evolving efficient water and soil management practices in addition to identification of crops and varieties with high water use efficiency.

- Development of economically viable strategies to prevent and manage herbicide resistance under such situation should be a major research area.

- Farmer's involvement in participatory research and demonstration trials can accelerate adoption of CA.

Conclusion

Agriculture, the dynamic system governed by several biotic and abiotic factors, needs to be sustained. Residue management along with conservation agriculture is the best direction of farming community rather than a destination. The long term fertility of the soil can be maintained only if output of plant nutrient is compensated with the addition of either crop residues and through green manuring, crop mulching etc. This will improve the farmer's income, put in use on-farm resources and ultimately ensure food security in marginal areas in the hilly ecosystem. In the light of today's urgent need, there should be an allout effort for the development of our society through adoption of such novel strategies, which would trigger an ever-green revolution in this remote part of India.

References

Baudron, F., Tittonell, P., Corbeels, M., Letourmya, P. and Giller, K. E. (2011). Comparative performance of conservation agriculture and current smallholder farming practices in semi-arid Zimbabwe. *Field Crops Research,* 132: 117-128.

Bhattacharyya, R., Tuti, M.D., Kundu, S., Bisht, J.K. and Bhatt, J.C. (2012). Conservation tillage impacts on soil aggregation and carbon pools in a sandy clay loam soil of the Indian Himalayas. *Soil Science Society of America Journal,* 76: 617–627.

Choudhury, B.U., Bouman, B.A.M. and Singh, A.K. (2007). Yield and water productivity of rice–wheat on raised beds at New Delhi, India. *Field Crops Research,* 100: 229-239.

Derpsch, R. and Friedrich, T. (2009). Global overview of Conservation Agriculture adoption. Invited Paper, 4th World Congress on Conservation Agriculture: Innovations for Improving Efficiency, Equity and Environment. 4-7 February 2009, New Delhi, ICAR. (www.fao.org/ag/ca).

Ghosh, P.K., Das, Anup, Saha, R., Kharkrang, Enboklang., Tripathi, A.K., Munda, G.C. and Ngachan S.V. (2010). Conservation Agriculture towards Achieving Food Security in North East India. *Current Science,* 99(7): 915-921.

Gupta, R. (2003). Agenda notes 12th Regional Co-ordination Committee Meeting of Rice–wheat Consortium for Indo-Gangetic Plains. February 7-9, Islamabad, Pakistan, pp. 39-50.

Gupta, R.K., Naresh, R.K., Hobbs, P.R., Jiaguo, J. and Ladha, J. K. (2003). Sustainability of post-green revolution agriculture: the rice–wheat cropping systems of the Indo-Gangetic Plains and China. In: [Ladha, J.K., Hill, J., Gupta, R.K., Duxbury, J., Buresh, R.J. (Eds.)], Improving the Productivity and Sustainability of Rice–wheat Systems: Issues and Impact. ASA Special Publication 65. ASA, Madison, WI, USA Chapter 1, pp. 1-25.

Hartwig, N.L. and Ammon, H.U. (2002). Cover crops and living mulches. Weed Science 50: 688-699. Hobbs, P.R., Sayre, K. and Gupta, R. (2008). The role of conservation agriculture in sustainable agriculture. *Philosophical Transactions of the Royal Society, London, Biological Science,.* 363: 543-555.

IIFSR. 2015. Vision 2050. Indian Institute of Farming Systems Research. www.iifsr.res.in

Jat, M.L., Gathala, M.K., Ladha, J.K., Saharawat, Y.S., Jat, A.S., Kumar, V., Sharma, S.K., Kumar, V. and Gupta, R.K. (2009). Evaluation of precision land levelling and double zero tillage systems in the rice–wheat rotation: water use, productivity, profitability and soil physical properties. *Soil and Tillage Research,* 105: 112-121.

Kahlown, M.A., Azam, M. and Kemper, W.D. (2006). Soil management strategies for rice-wheat rotations in Pakistan's Punjab. *Journal of Soil and Water Conservation.* 61: 40-44.

Kalita, H., Bora, P.C. and Debnath, M.C. (2005). Effect of sowing date and tillage on soil properties, nutrient uptake and yield of linseed (*Linum usitatissimum*) grown in winter rice (*Oryza sativa*)- fallows. *Indian Journal of Agronomy,* 50(1): 70-72.

Kukal, S.S., Humphreys, E., Singh, Y., Timsina, J. and Thaman, S. (2005). Performance of raised beds in rice–wheat systems of northwestern India. In: [Roth, C.H., Fischer, R.A., Meisner, C.A. (Eds.)], Evaluation and Performance of Permanent Raised Bed Cropping Systems in Asia, Australia and Mexico. *ACIAR Proceedings,* 121: 26-40.

Lal, R. (1997). Residue management, conservation tillage and soil restoration for mitigating greenhouse effect by CO2 enrichment. *Soil and Tillage Research,* 43: 81-107.

Luo, Z., Wang, E. and Sun, O. J. (2010). Can no-tillage stimulate carbon sequestration in agricultural soils? A metaanalysis of paired experiments. *Agriculture, Ecosystem and Environment,* 39: 224-231.

Malik, R. K., Yadav, A., Singh, S., Malik, R. S., Balyan, R.S., Banga, R.S., Sardana, P.K., Jaipal, S., Hobbs, P.R., Gill, G., Singh, S., Gupta, R.K. and Bellinder, R. (2002). Herbicide Resistance Management and Evolution of Zero-tillage - A Success Story. Research Bulletin, CCS Haryana Agricultural University, Hisar, India, p. 43.

Nchimuthu, G., Velu, V., Malarvizhi, P., Ramasamy, S. and Gurusamy, L. (2007). Standardisation of leaf colour chart based nitrogen management in direct wet seededrice (*Oryza sativa* L.). *Journal of Agronomy*, 6: 338-343.

Pandey, D., Agrawal, M. and Bohra, J.S. (2012). Greenhouse gas emissions from rice crop with different tillage permutations in rice–wheat system. *Agriculture, Ecosystem and Environment*, 159: 133-144.

Picard, D., Ghiloufi, M., Saulas, P. and de Tourdonnet, S. (2010). Does under-sowing winter wheat with a cover crop increase competition for resources and is it compatible with high yield? *Field Crops Research*, 115: 9-18.

Prasad, R. and Power, J.F. (1997). Soil fertility management for sustainable agriculture. New York, USA, Lewis Publishers. p. 356.

Rautray, S.K. (2005). Effect of mulching on yield and economies of rainfed rice (*Oryza sativa*) –based cropping sequences in lower Assam. *Indian Journal of Agronomy*, 50(1): 13-15.

Saha, R. and Mishra, V.K. (2009). Effect of organic residue management on soil hydro-physical characteristics and rice yield in eastern Himalayan region, India. *Journal of Sustainable Agriculture*, 33(2): 161-176.

Saharawat, Y.S., Singh, B., Malik, R.K., Ladha, J.K., Gathala, M., Jat, M.L. and Kumar, V. (2010). Evaluation of alternative tillage and crop establishment methods in a rice–wheat rotation in North Western IGP. *Field Crops Research*. 116: 260-267.

Santhi, P.K., Ponnuswamy, K. and Kempuchetty, N. (1998). A labour saving technique in direct-sown and transplanted rice. *Field Crops Research*, 23: 35-36.

Saraf, C.S. and Patil, R.R. (1995). Fertilizer use in pulse based cropping systems. *Fertilizer News*, 40(5): 55-65.

Venkateswarlu, B. and Shanker, A.K. (2009). Climate change and agriculture: Adaptation and mitigation strategies. *Indian Journal of Agronomy*. 54: 226-230.

Vincent, D. and Quirke, D. (2002). Controlling Phalaris minor—in the Indian rice–wheat belt. In: ACIAR Impact Assessment Series No. 18. ACIAR, Canberra, Australia, p. 35.

Walker, T.W., Kingery, W. L., Street, J.E., Michael, S., Cox, J., Larry, O., Patrick, D., Gerard, and Xiang, H.F. (2003). Rice Yield and Soil Chemical Properties as Affected by Precision Land Leveling in Alluvial Soils. *Agronomy Journal*, 95: 1483-1488.

6

Conservation Agriculture and Resource Conservation Technologies in Indo-Gangetic Plains: Status and Challenges Ahead

D. K. Kundu

ICAR-Central Research Institute for Jute & Allied Fibres, Barrackpore West Bengal-700 121, India

Introduction

The Indo-Gangetic Plain is one of the world's major foodgrain producing regions. The Indian states falling under this region, viz. Punjab, Haryana, Uttar Pradesh, Himachal Pradesh, Bihar and West Bengal, are also the major rice-wheat growing states spread over 10.5 million hectares in the country. During the past 30 years, agricultural production growth in this region has been able to keep pace with population demand for food in the country mainly due to adoption of green revolution technologies inducing yield growth, followed by area expansion. But this opportunity is ceasing very fast due to limited scope for increasing the availability of arable land and natural resources. The other issue is the conservation of the basic resources of land and water for sustainability of agriculture in the Indo-Gangetic Plain. It is generally believed that the rice-wheat system has strained the natural resources in this region and more inputs are required to attain the same yield levels (Swarup and Singh, 1989; Kumar and Yadav, 1993; Lal *et al.*, 2004). It is, therefore, imperative now to promote alternative technologies that would help conserve the much needed but gradually depleting natural resources while boosting productivity growth in the long-run by maintaining soil health and production environment. As a part of this strategy, conservation agriculture and resource conserving technologies play a major role in sustaining and enhancing the productivity of the rice-wheat system at a lower cost of production.

Conservation Agriculture

Conservation Agriculture (CA) is an approach developed to manage farmland for sustainable crop production, while simultaneously preserving soil and water resources (Erenstein, 2011). CA is described by FAO as a concept for resource saving agricultural crop production which is based on enhancing the natural and biological processes above and below the ground. According to Dumanski *et al.*, (2006), CA is not "business as usual", based on maximizing yields while exploiting the soil and agro-ecosystem resources. Rather, CA is based on optimizing yields and profits, to achieve a balance of agricultural, economic and environmental benefits. It promotes maintenance of a permanent soil cover, minimum soil disturbance, and diversification of plant species. It enhances biodiversity and natural biological processes above and below the ground surface, which contribute to increased water and nutrient use efficiency and to improved and sustained crop production.

It advocates combined social and economic benefits gained from combining production and protecting the environment, including reduced input and labor costs. With CA, farming communities become providers of more healthy living environments for the wider community through reduced use of fossil fuels, pesticides, and other pollutants, and through conservation of environmental integrity and services. It aims at reversing the process of degradation inherent to the conventional agricultural practices like intensive agriculture, burning/removal of crop residues. Hence, it aims to conserve, improve and make more efficient use of natural resources through integrated management of available soil, water and biological resources combined with external inputs. It can also be referred to as resource efficient agriculture. CA practices perused in many parts of the world are built on ecological principles making land use more sustainable (Behera *et al.*, 2010, Lal, 2013). CA basically relies on 3 principles, which are interlinked and must be considered together for appropriate design, planning and implementation processes. These are: (i). minimal mechanical soil disturbance, (ii). permanent organic soil cover, and (iii). diversified crop rotation.

CA principles are universally applicable to all agricultural landscapes and land uses with locally adapted practices. Soil interventions such as mechanical soil disturbance are reduced to an absolute minimum or avoided, and external inputs such as agrochemicals and plant nutrients of mineral or organic origin are applied optimally and in ways and quantities that do not interfere with, or disrupt, the biological processes. CA facilitates good agronomy, such as timely operations, and improves overall land husbandry for rainfed and irrigated production. Complemented by other known good practices, including the use of quality seeds, and integrated pest, nutrient, weed and water management, etc., CA is a base for sustainable agricultural production intensification. It opens increased options for integration of production sectors, such as crop-livestock integration and the integration of trees and pastures into agricultural landscapes.

Resource Conservation Technologies

The resource conservation technologies (RCTs) primarily focus on resource savings through minimal tillage, ensuring soil nutrients and moisture conservation through crop residues and growth of cover crops, and adoption of spatial and temporal crop sequencing. These pro-sustainable technologies and the practices therein have long been practised by the farmers in the Indo-Gangetic Plains but got eroded in recent times. With squeezing net returns and increasing threats of sustainability, the viability of rice-wheat farming was looming large. These issues were discussed and concerns were raised by the planners and policy makers. Considerable efforts are being made to popularize and increase the adoption of RCTs in the region. In the absence of specific information, however, the policies and efforts are inadequate and ineffective to achieve the desired results in this direction.

Some of the RCTs that are being promoted in the rice-wheat belt of the Indo-Gangetic Plains are: zero tillage, laser land levelling, bed planting, surface seeding, rotary tillage, use of leaf colour chart, mechanical rice transplanter, etc. The adoption of RCTs is expected to yield benefits to the farmers in terms of reduced losses due to soil erosion, saving of energy and irrigation costs, savings on labour input, increased productivity and water-use efficiency, reduced pumping of groundwater, increased nutrient-use efficiency and adoption of new crop rotations. Although considerable debate is currently going on these issues, in the absence of specific information, the planners and policymakers find it difficult to make suitable policies for popularization/increased adoption of RCTs.

Some RCTs, while attractive in the short-term, may be unsustainable in the longer-term. An example of this is the use of zero tillage without residue retention and without suitable rotations which, under some circumstances, can he more harmful to agroecosystem productivity and resource quality than a continuation of conventional practices (Sayre, 2000).

Are Conservation Agriculture and Resource Conserving Technologies synonymous?

In literature, the phrases "conservation agriculture" (CA) and "resource-conserving technologies" (RCTs) have been frequently used as if their meanings are similar. However, there is a clear difference between the two.

The "resource conserving technologies" refer to those practices that enhance resource- or input-use efficiency. This covers a lot of ground. New varieties that use nitrogen more efficiently may be considered RCTs. Zero or reduced tillage practices that save fuel and improve plot-level water productivity may also be considered RCTs, as may land-leveling practices that help save water. There are many more.

In contrast, "conservation agriculture" practices refer to only those RCTs with the following characteristics:

i)   Soil cover, particularly through the retention of crop residues on the soil surface,

ii) Sensible, profitable crop rotations, and

iii) A minimum level of soil movement, e.g., reduced or zero tillage.

The purpose of this chapter is to provide an Indian perspective on conservation agriculture and the use of resource conserving technologies. Recent advances in the development and use of conservation agriculture in Indo-Gangetic Plains are emphasised. This information will be helpful to the researchers, development workers and farming communities of the region.

Resource Conservation Technologies for Conservation Agriculture

The conservation agriculture-based resource conservation technologies (RCTs) like those listed below have proved to be energy and input efficient, besides addressing the emerging environment and soil health problems:

(i). zero or minimum tillage, (ii). crop residue cover, (iii). cover crops, (iv). precision farming, (v). use of GPS and GIS systems, (vi). site specific nutrient management, (vii). leaf color chart, (viii). laser land leveller, (ix). crop rotation and cropping system, (x). diversification, (xi). integrated farming systems, (xii). rainwater harvesting, (xiii). off season/summer ploughing, (xiv). bed planting, (xv). direct seeded rice, (xvi). system of rice intensification, and (xvii). contour farming and strip cropping.

The RCTs involving no- or minimum tillage with direct seeding and bed planting, residue management (mainly residue retention) and crop diversification have potential for improving productivity and soil quality, mainly by soil organic matter (SOM) build-up. The RCTs bring many possible benefits including reduced water and energy use (fossil fuels and electricity), reduced greenhouse gas (GHG) emissions, soil erosion and degradation of the natural resource base, increased yields and farm incomes, and reduced labor shortages.

Status of Conservation Agriculture in India and Abroad

Globally, CA is being practiced on about 125 M ha. The major CA practising countries are USA (26.5 M ha), Brazil (25.5 M ha), Argentina (25.5 M ha), Canada (13.5 M ha) and Australia (17.0 M ha). In India, CA adoption is still in the initial phases. Over the past few years, adoption of zero tillage and CA has expanded to cover about 1.5 million hectares (Jat *et al.*, 2012). The major CA based technologies being adopted is zero-till (ZT) wheat in the rice-wheat (RW) system of the Indo-Gangetic Plains (IGP). In other crops and cropping systems, the conventional agriculture based crop management systems are gradually undergoing a paradigm shift from intensive tillage to reduced/zero-tillage operations. In addition to ZT, other concept of CA need to be infused in the system to further enhance and sustain the productivity as well as to tap new sources of growth in agricultural productivity. The CA adoption also offers avenues for much needed diversification through crop intensification, relay cropping of sugarcane, pulses, vegetables etc. as intercrop with wheat and maize and to intensify and diversify the RW system. The CA based resource conservation technologies also help in integrating crop, livestock, land and water management research in both low- and high-potential environments.

In India, efforts to adopt and promote conservation agriculture technologies have been underway for nearly a decade but it is only in the last 8 to 10 years that the technologies are finding rapid acceptance by farmers. Efforts to develop and spread CA have been made through the combined efforts of several State Agricultural Universities, ICAR institutes and the Rice-Wheat Consortium for the Indo-Gangetic Plains. The spread of technologies is taking place in India in the irrigated regions in the Indo-Gangetic Plains where rice-wheat cropping systems dominate. CA systems have not been promoted in other major agro-ecoregions like rainfed semi-arid tropics and the arid regions of the mountain agro-ecosystems.

Spread of these technologies is taking place in the irrigated regions of the Indo-Gangetic Plains where the rice-wheat cropping system dominates. The focus of developing and promoting resource conservation technologies has been on zero-till seed-cum fertilizer drill for sowing of wheat in rice-wheat system. Other interventions include raised-bed planting systems, laser equipment aided land leveling, residue management practices, alternatives to the rice-wheat system etc. It has been reported that the area planted with wheat adopting the zero-till drill has been increasing rapidly (Sangar *et al.*, 2005), and presently 25% – 30% of wheat is zero-tilled in rice-wheat growing areas of the Indo-Gangetic Plains of India. In addition, raised-bed planting and laser land leveling are also being increasingly adopted by the farmers of the north-western region.

Potential Benefits of Conservation Agriculture

Adoption and spread of zero-tilled wheat has been a success story in north-western parts of India due to: (1) reduction in cost of production by Rs 2,000 to 3,000 ha^{-1} (RWC-CIMMYT, 2005); (2) enhancement of soil quality, i.e. soil physical, chemical and biological conditions (Jat *et al.*, 2009, Gathala *et al.*, 2011); (3) enhancement in the long term C sequestration and build-up in soil organic matter constitute a practical strategy to mitigate Green House Gas emissions and impart greater resilience to production systems to climate change related aberrations (Saharawat *et al.*, 2012); (4) reduction of the incidence of weeds, such as *Phalaris minor* in wheat; (5) enhancement of water and nutrient use efficiency (Saharawat *et al.*, 2012); (6) enhancement of production and productivity (4 – 10%) (Gathala *et al.*, 2011); (7) advanced sowing date; (8) reduction in greenhouse gas emission and improved environmental sustainability (Pathak *et al.*, 2011); (9) avoiding crop residue burning reduces loss of nutrients, and environmental pollution, which reduces a serious health hazard; (10) providing opportunities for crop diversification and intensification-for example in sugarcane based systems, mustard, chickpea, pigeonpea etc. (Jat *et al.*, 2005); (11) improvement of resource use efficiency through residue decomposition, soil structural improvement, increased recycling and availability of plant nutrients (Jat *et al.*, 2009); and (12) use surface residues as mulch to control weeds, moderate soil temperature, reduce evaporation, and improve biological activity (Jat *et al.*, 2009, Gathala *et al.*, 2011).

Because of the zero-tilled wheat benefits, the CA based crop management technologies have been tried in other cropping systems in India, but there are large knowledge gaps in CA based technologies which indicates there is a need to develop, refine, popularize and disseminate these technologies on a large scale.

Prospects of Conservation Agriculture

The promotion of CA under Indian context has the following prospects.

(i) Reduction in cost of production: This is a key factor contributing to rapid adoption of zero-till technology. Cost reduction is attributed to savings on account of diesel, labour and input costs.

(ii) Reduced incidence of weeds: Most studies tend to indicate reduced incidence of *Phalaris minor*, a major weed in wheat, when zero-tillage is adopted resulting in reduction in use of herbicides.

(iii) Saving in water and nutrients: Limited experimental results and farmers experience indicate that considerable saving in water (up to 20% – 30%) and nutrients are achieved with zero-till planting and particularly in laser leveled and bed planted crops. De Vita *et al.*, (2007) stated that higher soil water content under no-tillage was due to reduced water evaporation. They also found that across growing seasons, soil water content under no-till was about 20% greater than under conventional tillage.

(iv) Increased yields: In properly managed zero-till planted wheat, yields were invariably higher compared to traditionally prepared fields for comparable planting dates. CA has been reported to enhance the yield level of crops due to associated effects like prevention of soil degradation, improved soil fertility, improved soil moisture regime (due to increased rain water infiltration, water holding capacity and reduced evaporation loss) and crop rotational benefits. Yield increases as high as $200 - 500$ kg ha^{-1} are found with no-till wheat compared to conventional wheat under a rice-wheat system in the Indo-Gangetic plains (Hobbs and Gupta, 2004). While some studies claim that CA results in higher and more stable crop yields, on the other hand there are also numerous examples of no yield benefits and even yield reductions particularly during the initial years of CA adoption.

(v) Environmental benefits: Conservation agriculture involving zero-till and surface managed crop residue systems are an excellent opportunity to eliminate burning of crop residue which contribute to large amounts of greenhouse gases like CO_2, CH_4 and N_2O. Burning of crop residues, also contribute to considerable loss of plant nutrients, which could be recycled when properly managed.

(vi) Crop diversification opportunities: Cropping sequences/rotations and agroforestry systems when adopted in appropriate spatial and temporal patterns can further enhance natural ecological processes. Limited studies indicate that a variety of crops like mustard, chickpea, pigeonpea, sugarcane, etc., could be well adapted to the new systems.

(vii) Resource improvement: No tillage when combined with surface management of crop residues begins the processes whereby slow decomposition of residues results in soil structural improvement and increased recycling and availability of plant nutrients. Surface residues acting as mulch, moderate soil temperatures, reduce evaporation, and improve biological activity.

Constraints for Adoption of Conservation Agriculture

The following are a few important constraints which impede broad scale adoption of CA:

- Lack of appropriate seeders especially for small and medium scale farmers
- The widespread use of crop residues for livestock feed and fuel
- Burning of crop residues
- Lack of knowledge about the potential of CA to agriculture leaders, extension agents and farmers
- Shortage of skilled and scientific manpower

Challenges in Conservation Agriculture

Conservation agriculture as an upcoming paradigm for raising crops will require an innovative system perspective to deal with diverse, flexible and context specific needs of technologies and their management. Research and development on conservation agriculture (CA), thus will call for several innovative features to address the challenge. Some of these are briefly mentioned below.

(a) Understanding the system: Conservation agriculture systems are much more complex than conventional systems. Site specific knowledge has been the main limitation to the spread of CA system (Derpsch *et al.*, 2010). Managing these systems efficiently will be highly demanding in terms of understanding of basic processes and component interactions, which determine the whole system performance. For example, surface maintained crop residues act as mulch and therefore reduce soil water losses through evaporation and maintain a moderate soil temperature regime. However, at the same time crop residues offer an easily decomposable source of organic matter and could harbour undesirable pest populations or alter the system ecology in, some other way. No-tillage systems will influence depth of penetration and distribution of the root system which, in turn, will influence water and nutrient uptake and mineral cycling. Thus the need is to recognize conservation agriculture as a system and develop management strategies.

(b) Building a system and farming system perspective: A system perspective is built working in partnership with farmers. A core group of scientists, farmers, extension workers and other stakeholders working in partnership mode will therefore be critical in developing and promoting new technologies.

(c)   Technological challenges: While the basic principles which form the foundation of conservation agriculture practices, that is, no tillage and surface managed crop residues are well understood, adoption of these practices under varying farming situations is the key challenge. These challenges relate to development, standardization and adoption of farm machinery for seeding with minimum soil disturbance, developing crop harvesting and management systems.

(d)   Site specificity: Adapting strategies for conservation agriculture systems will be highly site specific, yet learning across the sites will be a powerful way in understanding why certain technologies or practices are effective in a set of situations and not effective in another set.

(e)   Long-term research perspective: Conservation agriculture practices, e.g. no-tillage and surface-maintained crop residues result in resource improvement only gradually, and benefits come about only with time. In many situations, benefits in terms of yield increase may not come in the early years of practicing conservation agriculture. Understanding the dynamics of changes and interactions among physical, chemical and biological processes is basic to developing improved soil-water and nutrient management strategies (Abrol and Sangar, 2006). Therefore, research in conservation agriculture must have longer term perspectives.

Policy Issues

Conservation agriculture implies a radical change from conventional agriculture. There is need for policy analysis to understand how CA technologies integrate with other technologies, and how policy instruments and institutional arrangements promote or deter CA (Raina *et al.*, 2005). While R&D efforts over the past decade have contributed to increasing farmer acceptance of zero tillage for wheat in rice-wheat cropping systems, this has raised a number of institutional, technological, and policy related issues which must be addressed if CA practices are to be adopted in large scale in the region on a sustained basis. Following are some of the important policy considerations for promotion of CA.

- *Scaling up conservation agriculture practices*: One of the reasons for poor percolation of the technology to the farmers was the past bias or mindset about tillage by the majority of farmers. Under such situations, farmers participatory on-farm research to evaluate/refine the technology in initial years followed by large scale demonstration in subsequent years is needed.

- *Shift in focus from food security to livelihood security:* Food security policy based on cereal production must now replace a well-articulated policy goal for livelihood security. This will help the diversification of dominant rice-wheat cropping systems (occupying about 13.5 million ha) in the Indo-Gangetic Plains, the cultivation of which in conventional tillage practice has overexploited the natural resources in the region. The nature of cropping patterns and the extent of crop diversification are influenced by policy interventions.

- *Developing, improving, standardizing equipment for seeding, fertilizer placement and harvesting ensuring minimum soil disturbance in residue management for different edaphic conditions* will be key to success of CA. For example, in hilly tracts, for small land holders bullock drawn equipment will have greater relevance. Ensuring quality and availability of equipment through appropriate incentives will be important. In these situations, the *subsidy support from national or local government to firms for developing low cost machines* will help in the promotion of CA technologies.

- CA technologies bring about significant changes in the plant growing micro-environment. These include changes in moisture regimes, root environment, emergence of new pathogens and shift in insect-pest scenario etc. There is a need to develop complementary crop improvement programmes, aimed at developing cultivars which are better suitable to new systems. Farmers-participatory research would appear promising for *identifying and developing crop varieties suiting to a particular environment or locations.*

- *Besides resources, systematic monitoring of the socio-economic, environmental and institutional changes should become an integral part of the major projects on CA.*

- Policy support for *capacity building by organizing training on CA is needed.* Availability of trained human resources at ground level is one of the major limiting factors in adoption of CA. In the long term, CA should be included in curricula from primary school to university levels, including agricultural colleges.

- *Institutionalize CA*: CA has to be mainstreamed in relevant ministries, departments or institutions and supported by adequate provision of material, human and financial resources to ensure that farmers receive effective and timely support from well trained and motivated extension staff.

- *Support for the adaptation and validation of CA technologies in local environments*: Adaptive research in collaboration with local communities and other stakeholders is required to tailor CA principles and practices to local conditions. Issues that should be addressed include crop species, selection and management of crop and cover crop and rotations, maintenance of soil cover and CA equipment. The resource poor and small holder farmers in India do not have economic access to new seeds, herbicides and seeding machineries etc. (Sharma *et al.*, 2012). This calls for policy frame work to make easily available critical inputs.

- *Support the development of CA equipment and ensure its availability*: While some countries produce CA equipment, most of the available implements and equipment are imported. In the short term, consideration could be made on removing or reducing tariffs on imported CA equipment and implements to encourage and promote their availability. In the medium to long run, local manufacture of these will increase availability, ensure that equipment is adapted to local conditions, increase employment opportunities and reduce costs. The larger and more complex equipment is expensive and users may have to hire it. There is an opportunity to develop a local hire service industry by providing equipment, and training on machine maintenance and business skills.

- *Promote payments for environmental services and fines for faulty practices:* Adopters of CA improve the environment through carbon sequestration, prevention of soil erosion or the encouragement of groundwater recharge. It provides ecosystem services, thus, farmers could be rewarded for such services, which have a great impact on the quality of life for all. On the other hand, burning of huge quantities of crop residues has adverse environmental impacts. In a prosperous state like Punjab in India, 81% of the rice straw (crop residue) is burnt, leading to the loss of a huge quantity of nutrients and pollution of the environment. There is a need for a strong policy intervention for prohibiting such an unscientific practice by imposing a fine.

- *Building partnership:* CA systems are very complex and their efficient management needs understanding of basic processes and component interactions which determine the system performance. A system perspective is the best to build working in partnership with farmers, who are at the core of farming systems and best understand this system. Scientists, farmers, extension agents, policy makers and other stakeholders in the private sector working in partnership mode will be important in developing and promoting new technologies.

- *Credit and subsidy:* The other important thing for successful adoption of CA is the need to provide credit to farmers to buy the equipment, machinery, and inputs through banks and credit agencies at reasonable interest rates. At the same time government need to provide a subsidy for the purchase of such equipment by farmers.

References

Abrol, I.P., Sangar, S. (2006). Sustaining Indian agriculture - Conservation agriculture the way forward. *Current Science,* 91 (8): 1020-2015.

Behera, U.K., Amgain, L.P., Sharma, A.R. (2010). Conservation agriculture: principles, practices and environmental benefits.*in* U.K. Behera, T.K. Das, A.R. Sharma (Eds.), *Conservation Agriculture,* Division of Agronomy, Indian Agricultural Research Institute, New Delhi – 110012. pp. 28-41.

De, P., Vita, E. Di Paolo, Fecondo, G., Fonzo, N., Pisante, Di M. (2007). No-tillage and conventional tillage effects on durum wheat yield, grain quality and soil moisture content in Southern Italy. *Soil & Tillage Research,* 92: 69-78.

Derpsch, R., Friedrich, T., Kassam, A., Li, H.W. (2010). Current status of adoption of no-till farming in the world and some of its main benefits. *International Journal of Agricultural and Biological Engineering,* 3: 1-25.

Dumanski, J., Peiretti, R., Benetis, J., McGarry, D. and Pieri C. (2006). The paradigm of conservation tillage. *Proceedings of World Association of Soil and Water Conservation,* P1. pp. 58-64.

Erenstein, O. (2011). Cropping systems and crop residue management in the Trans-Gangetic Plains: Issues and challenges for conservation agriculture from village surveys. *Agricultural Systems,* 104(1): 54-62.

Gathala, M.K., Ladha, J.K., Kumar, V., Saharawat, Y.S., Kumar, V., Sharma, P.K., Sharma, S. and Pathak, H. (2011). Tillage and Crop Establishment Affects Sustainability of South Asian Rice-Wheat System. *Agronomy Journal,* 103: 961-971.

Hobbs, P.R. and Gupta, R.K. (2004). Problems and challenges of no-till farming for the rice-wheat systems of the Indo-Gangetic Plains in South Asia. *in* R. Lal, P. Hobbs, N. Uphoff, D.O. Hansen (Eds.), *Sustainable Agriculture and the Rice-Wheat System,* Ohio State University and Marcel Dekker, Inc, Columbus, Ohio, and New York, USA. pp. 101-119.

Jat, M.L. Singh, S., Rai, H.K., Chhokar, R.S., Sharma, S.K. and Gupta, R.K. (2005). Furrow Irrigated Raised Bed Planting Technique for Diversification of Rice-Wheat System of Indo-Gangetic Plains. *Journal of Japan Association for International Cooperation for Agriculture and Forestry,* 28(1): 25-42.

Jat, M.L., Gathala, M.K., Ladha, J.K., Saharawat, Y.S., Jat, A.S., Kumar Vipin, Sharma, S.K.V. Kumar and Gupta, R. (2009). Evaluation of Precision Land Leveling and Double Zero-Till Systems in Rice-Wheat Rotation: Water use, Productivity, Profitability and Soil Physical Properties. *Soil and Tillage Research,* 105:112-121.

Jat, M.L., Saharawat, Y.S. and Gupta, R. (2011). Conservation agriculture in cereal systems of south Asia: Nutrient management perspectives. *Karnataka Journal of Agricultural Science,* 24: 100-105.

Jat, M.L., Malik, R.K., Saharawat, Y.S., Gupta, R., Bhag, M. and R. Paroda (2012). Proceedings of Regional Dialogue on Conservation Agricultural in South Asia. New Delhi, APAARI, CIMMYT, ICAR. New Delhi, India, p. 31

Kumar, A. and Yadav, D.S. (1993). Effect of long-term fertilization on soil fertility and yield of rice–wheat cropping system. *Journal of the Indian Society of Soil Science,* 41: 178-180.

Lal, R. (2013). Climate-resilient agriculture and soil Organic Carbon. *Indian Journal of Agronomy,* 58 (4): 440-450.

Lal, R., Hansen, D.O., Hobbs, P.R. and Uphoff, N. (2004). Sustainable Agriculture and the International Rice-Wheat System (mimeo). pp. 495-512.

Pathak, H., Saharawat, Y.S., Gathala, M. and J.K. Ladha (2011). Impact of resource-conserving technologies on productivity and greenhouse gas emission in rice-wheat system. *Greenhouse Gases: Science and Technology*, 1: 261-277.

Raina, R.S., Sulaiman, R., Hall, A.J., Sangar, S. (2005). Policy and institutional requirements for transition to conservation agriculture: An innovation systems perspective. *In* I.P. Abrol, R.K. Gupta, R.K. Mallik (Eds.), *Conservation Agriculture – Status and Prospects*, Centre for Advancement of Sustainable Agriculture, New Delhi, pp. 224-232.

RWC-CIMMYT (2005). Agenda Notes. 13th Regional Technical Coordination Committee Meeting. RWC-CIMMYT, Dhaka, Bangladesh.

Saharawat, Y.S., Singh, B., Malik, R.K., Ladha, J.K., Gathala, M., Jat, M.L. and V. Kumar (2010). Evaluation of alternative tillage and crop establishment methods in a rice–wheat rotation in North Western IGP. *Field Crops Research*, 116: 260-267.

Saharawat, Y.S., Ladha, J.K., Pathak, H., Gathala, M., Chaudhary, N. and M.L. Jat (2012). Simulation of resource-conserving technologies on productivity, income and greenhouse gas emission in rice-wheat system. *Journal of Soil Science and Environmental Management*, 3 (1): 9-22.

Sayre, K. (2000). "Effects of Tillage. Crop Residue Retention and Nitrogen Management on the Performance of Bed-Planted. Furrow Irrigated Spring Wheat in Northwest Mexico". Paper presented at the Conference of the International Soil Tillage Research Organization, 15; Fort Worth, Texas, USA: 2-7 July, 2000.

Sangar, S., Abrol, J.P., Gupta, R.K. (2005). *Conservation Agriculture: Conserving Resources Enhancing Productivity*. CASA, NASC Complex, New Delhi, p. 19

Sharma, A.R., Jat, M.L., Saharawat, Y.S., Singh, V.P. and R. Singh (2012). Conservation agriculture for improving productivity and resource-use efficiency: prospects and research needs in Indian context. *Indian Journal of Agronomy* 57 (IAC Special Issue), pp. 131-140.

Swarup, A. and Singh, K.N. (1989). Effect of 12 years' rice– wheat cropping and fertilizer use on soil properties and crop yields in a sodic soil. *Field Crops Research*, 21: 277-287.

7

Impact of Climate Change on Insect Pest Dynamics and Its Mitigation

S. Satpathy, B.S. Gotyal and V. Ramesh Babu

ICAR-Central Research Institute for Jute and Allied Fibres, Barrackpore West Bengal - 700 121, India

Introduction

Climate change refers to change of weather variables over a long period of time. The Inter-Governmental Panel on Climate Change (IPCC) defined climate change as a "change in the state of the climate that can be identified by changes in the mean and/or the variability of its properties and that persists for an extended period, typically decades or longer". This may occur due to natural variability or as a result of human activity. The change may be in the form of magnitude or variability of a single or multiple weather variables or weather phenomena. Global shift in temperature and precipitation pattern as well as increasing frequency of extreme weather events are consequence of climate change which is perceived over several decades. Climate change is indeed increasingly recognized as a considerable risk to agriculture particularly with respect to direct impact on crop production and yield stability (Barzman, 2015).

Effect of Climate Change in Global Warming and Greenhouse Gases

Global warming as the resultant of climate change has become an issue of serious concern worldwide for existence of life on the earth. Over past hundred years, the global temperature has increased by $0.8^{0}C$ and is expected to reach $1.1-5.4^{0}C$ by the end of next century. Major factors affecting the global warming and the climate change is the concentration of greenhouse gases like carbon dioxide, methane and nitrous oxide. On the other hand, CO_2 concentration in the atmosphere has increased drastically from 280 ppm to 370 ppm and is likely to be doubled in 2100 (IPCC, 2007).

Since 1958, the rate of increase in atmospheric carbon dioxide concentration has accelerated from decade to decade. Safety limit for atmospheric CO_2 is 350 ppm. The annual CO_2 concentration growth rate was larger during the last 10 years (1995-2005 average: 1.9 ppm per year). Atmospheric CO_2 levels remained higher than 350 ppm since early 1988. The accumulation of greenhouse gases in the atmosphere has warmed the planet and caused changes in the global climate. No driver other than GHGs provides a scientifically sound explanation of most of the warming observed both globally and nationally over the past few decades. The IPCC (2007) report corroborated previous scenarios that by 2100, the precipitation will decrease in the sub-tropics, and extreme events will become more frequent. However, changes in climate are already being observed—the last 60 years were the warmest in the last 1000 years and changes in precipitation patterns have brought greater incidence of floods or drought globally.

Impact of Climate Change in Agriculture

The impact of climate change in agriculture is inevitable and very complex as its effect is observed in multiple ways. The assessment of all the major climatic factors, GHG concentration and the consequent effect of changes in these factors is very crucial. The climate change can interfere and influence the normal plant physiological activities, magnitude of biotic stresses and ultimately the crop production. Global warming and climate change will have major implications for ecosystem services, water availability, crop production, and food security. Increased temperatures have drastically affected rice (*Oryza sativa* L.) production because of reduced crop duration in Philippines (10% reduction in rice yield per 1^0C rise in temperature) (Peng *et al.*, 2004). An increase of 6^0C in temperature and a precipitation deficit of 300 mm reduced maize (*Zea mays* L.) yield by 36% in the European Union (Ciais *et al.*, 2005).

As in the global scale, the Indian climate has shown gradual shift from the normalcy by manifestation of increasing trends in annual temperature with an average of 0.56°C rise over last 100 years (IPCC, 2007; Rao *et al.*, 2009). The post monsoon period is experiencing more temperature and number of hotter days is more frequent in winter. Even though, there was slight increase in total rainfall received, number of rainy days decreased. The declining total rainfall in the rainfed zone of the country has gradually become a regular phenomenon (De and Mukhopadhyay, 1998, Rao *et al.*, 2009). The semi-arid regions of the country had maximum probability of prevalence of droughts of varying magnitudes (20-30%), leading to sharp decline in water tables and crop failures (Lal, 2003; Rao *et al.*, 2009; Samra, 2003). By the end of next century (2100), the temperature in India is likely to increase by 1-5°C (De and Mukhopadhyay, 1998; Lal, 2003; IPCC, 2007). There will be 15-40% increase in rainfall with high degree of variation in its distribution. Apart from this, the country is likely to experience frequently occurring extreme climatic events (NATCOM, 2004; IPCC, 2007).

Being a tropical country, India is more challenged with impacts of looming climate change. Already, the productivity of Indian agriculture is limited by its high dependency on monsoon rainfall which is most often erratic and inadequate in its distribution (Chand and Raju, 2009). Erratic climate driven changes will pose new problems and that is going to be a big challenge to attain the desired agricultural production.

Effect of Climate Change on Insect Pests of Crops

A major impact of climate change is the risk from new and emerging invasive species of insect pests and potential adaptations in the existing indigenous pests with greater pest status challenging the conventional practices and approaches adopted for their management (Barzman, 2015).

In India, the average annual losses caused by insect pests have been estimated to be 17.5% in eight major field crops. Losses attributable to insect damage are likely to increase as a result of decreased crop diversity and increased incidence of insect pests resulting from global warming. The seasonal and long term changes in climatic condition affect the insect pests of crops in multiple ways. Insects have short generation times and high reproductive rate, and hence they are more likely to be most affected by climate change because environmental factors have a strong influence on the development, reproduction, and survival of insect pests and their natural enemies (Bale *et al.*, 2002). Potential responses will include changes in phenological patterns, habitat selection and expansion, and/or contraction of geographic distribution. Species responses depend on the flexibility of life history characteristics, and different growth rates and diapause requirements may influence species distribution and population increase (Bale *et al.*, 2002). Climatic change is also directly or indirectly influences the distribution and severity of crop pests (Macfadyen *et al.*, 2018).

Global warming will lead to faster development of immature stages, and adults will emerge much earlier than before.In totality these changes in climate will influence the insect pests by affecting the diversity, abundance geographical distribution, development, population dynamics, host-plant resistance to insects and effectiveness of management interventions. Predicting the effect of climate change on pests is very complex due to the complicated interacting effect of increasing atmospheric CO_2 concentration, changing climatic regimes and altered frequency /intensity of extreme events (Bebber *et al.*, 2013)

Reduced Faunal Diversity

In general South Asia and India in particular due to the existence of diverse agro-ecological conditions, climate and soils enjoys rich faunal and floral biodiversity. The richness of biodiversity in the greatest natural wealth favorably influences the agriculture of any region. Due to the change in the climate pattern in recent decades owing to increasing industrialization and over-exploitation of natural resources many species of plants, animals and insects are decreasing at an alarming rate.

The loss of biological diversity is still accelerating which may reduce the ecosystem's resilience to the climatic changes. The relative abundance of different insect species may change rapidly because of climate change, and the species unable to withstand the stresses may be lost in the near future. Climate change will become a major factor for the extinction of arthropod species. Mountain species and those restricted to high latitudes are most likely to become extinct as a result of climate change. The species adapted to cold conditions will be forced to move uphill to higher latitudes as a result of global warming (Thomas *et al.*, 2004). The insect diversity in a habitat indicates the stability of an ecosystem as they are very good indicators of environmental change (Gregory *et al.*, 2009), plays an important role in food chains, are excellent pollinators for many of the economically important crops. About 6.83% of world insect species are inhabitant in India (Alfred, 1998). The negative effects of climate change are accelerating the rate of biodiversity loss, worldwide. The loss of biodiversity may impact negatively the structure, composition and functioning of ecosystems and wildlife habitat leading to outbreaks of destructive insect-pests and diseases (Timoney, 2003; IPCC, 2007).

Variable Insect Abundance

Climate change will have a major effect on geographic distribution of insect pests and low temperatures are often more important than high temperatures in determining the distribution. The distribution and abundance of plants and animals in nature heavily depend on species specific climate requirements essential for their growth, survival and reproduction. The differential rates of range adjustments between annual and perennial plant species along with local extinctions will definitely affect distribution and survival of insect fauna associated with them (Thomas *et al.*, 2004). Earlier researches have shown that altitude wise shifts in insect distributions along with their host plants in response to changing climate are already in progress. With rise in temperature, the insect-pests are expected to extend their geographic range from tropics and subtropics to temperate regions at higher altitudes.

Global warming-resultant altitude-wise range expansion and increased overwintering survival of corn earworms *Heliothis zea* (Boddie) and *Helicoverpa armigera* (Hubner) may cause heavy yield loss and put forth major challenge for pest management in maize, a staple food crop of USA (Diffenbaugh *et al.*, 2008). Range extension in migratory species like *Helicoverpa armigera* (Hubner), a major pest of cotton, pulses and vegetables in North India is predicted with global climate warming (Sharma *et al.*, 2005). The legume pod borers *H. armigera* and *Maruca vitrata* (Geyer) presently confined to tropical climates in Asia, Africa, and Latin America, are most likely to move to northern Europe and North America over the next 50 years as a result of global warming and climate change (Sharma, 2014).

Temperature does not act in isolation to influence pest status, and therefore it is important to consider interactions with other variables, such as rainfall, humidity, radiation and CO_2 concentrations (Harrington *et al.*, 2001). Spatial shifts in the distribution of crops under changing climatic conditions will also influence the distribution of insect pests in a geographical region (Parry and Carter, 1989). However, whether or not an insect pest would move with a crop into a new habitat will depend on other environmental conditions, such as the presence of overwintering sites, soil type, and moisture.

Biology, Population Dynamics and Pest Out Breaks

Higher temperatures within the maximum developmental threshold will result in rapid development and increases in pest populations as the time to reproductive maturity will be reduced considerably. In addition to the direct effects of temperature changes on development rates, improvement in food quality because of abiotic stresses may result in dramatic increase in the growth rates of some insect species (White, 1984), whereas the growth of certain insect pests may be affected adversely (Maffei *et al.*, 2007). Overwintering of insect pests will increase as a result of climate change, producing larger spring populations as a base for a build-up in numbers in the following season. These may be vulnerable to parasitoids and predators if the latter also overwinter more readily. Temperature has a strong influence on the viability and incubation period of *H. armigera* eggs, which can be predicted on the basis of degree-days required for egg hatching (Dhillon and Sharma, 2007). Insect populations vary across seasons and locations. Both the onset of insect infestation and population build up are influenced by the weather conditions during the crop growing season and/or during the preceding period. Because no two seasons have similar weather conditions, the insect population dynamics also varies across seasons and locations, for example, there is a large variation in the peak population of Oriental armyworm, in Andhra Pradesh, across seasons. Maximum moth catches in light traps were recorded during the 31st standard week in 1980 and the 40th standard week during 1976. (Sharma *et al.*, 2002). Many insects such as *H. armigera, M. separata,* and *Spodoptera litura* (F.), which are migratory, may be well able to exploit new opportunities by moving rapidly into new areas as a result of climate change (Sharma *et al.*, 2002; Sharma 2005, 2014).

Changes in climatic variables have led to increased frequency and intensity of outbreaks of insect-pests. Outbreak of sugarcane woolly aphid, *Ceratovacuna lanigera* Zehntner in sugarcane belt of Karnataka and Maharashtra states during 2002-03 resulted in 30% yield losses. Persistence of abnormal weather condition associated with misuse of insecticides resulted in outbreak of rice plant hoppers, *Nilaparvata lugens* (Stal) and *Sogatella furcifera* (Horvath) in North India inflicting crop failure in more than 33,000 ha paddy area. In 2006, mealybug, *Phenacoccus solenopsis* Tinsley outbreak happened in the cotton, vegetables and ornamental plants in the cotton growing belts of the country, largely due to prevalence of higher temperature and changed cropping environment (introduction of Bt cotton) causing yield loss to the extent of 30-40% (Dhawan *et al.*, 2007).

Recently, *Phenacoccus solenopsis* has emerged as a new threat to jute crop. The infestation of *P. solenopsis* on jute was to the extent of 60-80% during summer 2009 in North 24-Parganas district of West Bengal. An insight into the weather parameters persisting during the past six years indicated that there was increase in the maximum and minimum temperature compared to the previous years, besides low rainfall and less number of rainy days occured between January and May. The warm and dry condition during the summer months might be the pre-disposing factors for mealybug outbreak. In subsequent years due to the prevalence of similar climatic conditions the outbreak was repeated again in jute (Satpathy *et al.*, 2016).

Alteration in Host Plant Resistance and Insect-plant Interaction

The climatic variables invariably bring changes in the host plant resistance against insect pests and diseases. Quicker disease cycle by more virulent pathogens and modified physiologies of insect pests can overcome the plant resistance (Shrestha, 2019).

Climate change may alter the interactions between insect pests and their host plants (Bale *et al.*, 2002; Sharma *et al.*, 2005, 2014). Under stressful environment, plant becomes more susceptible to attack by insect-pests because of weakening of their own defensive system resulting in pest outbreaks and more crop damage (Rhoades, 1985). Resultant weakened host defences due to the stress caused by lack of adaptation to suboptimal climatic conditions will be exploited by the insects. Climate change will affect the availability of water and, thus, indirectly affect plant growth and insect–host plant interactions. Water-stressed plants of sorghum suffer greater damage by the spotted stem borer, *C. partellus*, and the sugarcane aphid, *Melanaphis sacchari* (Zehnt.), than the plants grown under irrigated conditions (Soman *et al.*, 1994). Unusually severe drought increases the damage by insect species, such as spotted stem borer, *C. partellus*, in sorghum (Sharma *et al.*, 2005). Sorghum genotypes with moderate levels of resistance to *C. partellus* exhibit a susceptible reaction under drought stress. Insect–host plant interactions will also undergo a change in response to the effects of CO_2 on nutritional quality and secondary metabolites of the host plants. Increased levels of CO_2 will enhance plant growth, but they may also increase the damage caused by some phytophagous insects (Gregory *et al.*, 2009). In the CO_2-enriched atmosphere, many species of herbivorous insects will confront less nutritious host plants that could induce both lengthened larval developmental times and greater mortality. Increased CO_2 may also cause a slight decrease in nitrogen-based defences (e.g., alkaloids) and a slight increase in carbon-based defences (e.g., tannins). The lower foliar nitrogen content due to higher CO_2 causes an increase in food consumption by herbivores up to 40%. Endophytes, which play an important role in conferring tolerance to both abiotic and biotic stresses in plants, may also undergo a change caused by climate change (Newton *et al.*, 2009). The environmental factors like high temperature have been found affecting transgene expression in *Bt* cotton resulting in reduced production of Bt toxins. This led to enhanced susceptibility of the crops to insect-pests like bollworms *viz., Heliothis virescens* (F.) (Kaiser, 1996).

Effectiveness of Pest Management Strategies

Higher temperatures will make dry seasons drier and, conversely, may increase the amount and intensity of rainfall, making wet seasons wetter than at present. Such changes may alter the efficacy of pest control strategies (biological control and synthetic insecticides) (Barzman, 2015). Rapid dissipation of insecticide residues because of increases in temperature and precipitation will require more frequent application of insecticides. A hot and humid climate will increase crop protection costs through more frequent use of pesticides for crop protection (Chen and McCarl, 2001). Temperature has shown a positive effect on organochlorines, organophosphates, and carbamates (although there are a few exceptions) but has shown a negative effect on synthetic pyrethroids (Wang and Shen, 2007). Increase in temperature reduced the toxicity of two pyrethroids, lambda cyhalothrin and bifenthrin and spinosin against maize stem borer (Musser and Shelton, 2005). Rainfall reduces insecticide toxicity; however, the effects vary with the intensity and amount of rainfall, formulation, and the adjuvants used.

As there will be an array of climate effects on host plant and the pests, the cumulative impact on the tritrophic relation of host plant-pest-natural enemies will further magnify. Hosts may pass through vulnerable life stages more quickly at higher temperatures, reducing the chances of parasitism which may adversely affect the survival and multiplication of parasitoids. Differences in thermal preferences of crop pests and their natural enemies can lead to a loss in synchronization between the two and an increased risk of pest outbreaks (Furlong and Zalucki, 2017).

With changing climate, incidence of entomopathogenic fungi might be favoured by prolonged humidity conditions and obstinately be reduced by drier conditions (Newton *et al.*, 2011). There is a need to take a critical look at the effectiveness of various insecticides, activity of natural enemies and entomopathogens under global warming and climate change for sustainable crop protection.

Mitigation Strategies to Reduce the Adverse Effect of Increasing Pest Status Due to Climate Change

So far the consequence of climate change and increasing pest pressure on the food security at global level have drawn lesser importance than the human or animal health affected by climate change. Such impacts may be extended beyond the farm, to local, national and international food security, as well as environmental, economic and social sustainability. Immediate action is therefore needed on multiple levels and geographical scales to counter the effects through well planned pest management approach as the mitigation measure. The existing integrated pest management (IPM) strategies are to be revisited to emphasise on locally adapted preventive strategies in the background of diversified agroecosystems resilient enough to tolerate extreme weather fluctuations (Barzman *et al.*, 2015). IPM recommendations and tools recommended for diverse agroecosystems and climatic conditions are to be analysed in a focused manner with the objective of adapting to climate-induced change and recognising the potential of pest management for climate change mitigation.

In line with 'Climate-smart agriculture' (CSA), as promoted by FAO (2010), that advocates the reorientation of entire agricultural systems in order to support development and ensure food security in a changing climate; the new approach of climate-smart pest management (CSPM) has been developed (Heeb *et al.*, 2019).

CSPM provides recommendations across multiple stakeholder levels and geographical scales to guide producers, extension agents, researchers, policy makers and the wider public and private sector in the development of proactive and reactive strategies against climate-induced global changes in crop pests. It also explains how CSPM supports the achievement of national, regional and global climate change adaptation and mitigation goals.

Climate Smart Pest Panagement (CSMP)

The concept of CSPM is comparatively new which is based on a set of inter disciplinary approaches and strategies needed for primary production to adapt, with its supporting functions (e.g. extension and research) and enabling environment (e.g. policies, infrastructure and social/human capital), to the changing climatic environment, which, together with land use characteristics, sets and influences the boundaries for geographical distributions of species (crops, pests and natural enemies). Implementation of CSPM concept ensures holistic and more effective solution to the increased threat posed by new and existing crop pests to agricultural production and ecosystem services that increases resilience of farmer livelihoods and overall local and national food security to climate change.

CSPM also considers the priority on dip in yield losses posed by pest damage and GHG emissions intensity per unit of food produced. The benefits of CSPM can be greater realized when it is implemented in the concept of CSA. CSPM will provide farmers with the information and tools in hand to immediately and proactively put into action pest prevention practices (e.g. crop diversification, establishment of natural habitats and careful water and nutrient management) that will enhance the health of their farms and the surrounding landscape, and reduce susceptibility to pest-induced disturbance.

With the backup of sufficient and accurate information on weather parameters that upsurge the pest population, the farmers can prevent the build-up of pest population threshold by identifying and adopting specific pest prevention practices. In cases where pest populations reach economic injury levels, CSPM enables farmers to make rapid, informed decisions regarding the most appropriate reactive pest control strategy. CSPM is not the individual farmer's approach rather co-ordinated research and extension effort which ensures relevant, locally adapted services to all the farmers in the community. The research based information on impacts of climate change on crop-pest-natural enemy dynamics, and quantifying the consequences of these impacts, will facilitate the development of targeted adaptive responses that are currently lacking. In addition, the analysis of historical weather and climate data, as well as the development of niche models to determine pest species' potential distribution under varying climatic scenarios, will allow pest risk forecasting to become a viable tool to guide proactive pest prevention/management strategies (Yonow et al., 2018). Besides, CSPM also places a significant

116 Conservation Agriculture and Climate Change

emphasis on investment, infrastructure, and policies to catalyse adoption of CSPM approaches.

Conclusion

Keeping in pace with the growing world population there will be need to increase the food production by 60% globally to meet the additional demand by 2050. Climate change and global warming will influence the insect pests by affecting the diversity, abundance geographical distribution, development, population dynamics, host-plant resistance to insects and effectiveness of management interventions. Distribution of insect pests and their natural enemies will also be influenced by changes in the cropping patterns triggered by climate change. However, climate change is already having impacts on agriculture, including through its effects on the biology, distribution and outbreak potential of pests across all land uses and landscapes. In recent years the Indian agriculture has faced many insect outbreaks, enhancement in pest status and report of new hosts for the existing pests resulting in greater damage to cereals, grain legumes, vegetables, fruit crops, and forest trees. There is variation in response of insect species to atmospheric temperature and carbon dioxide in different regions and hosts. Enhancing the productivity with reduced pesticide application and decreased GHG emission in the backdrop of changing climate and dynamics of insect pests is to be ensured which necessitates immediate implementation of adaptation strategies at farm and landscape levels to decrease vulnerabilities of individual farmers to the adverse effects of climate change. In this context, the coordinated approach through CSPM with the support of farmers, extension workers, scientists and public and private sector stakeholders to reorient pest management approaches and develop an appropriate enabling environment to manage evolving climate change-induced pest threats and invasions more effectively. Orientation of research in understanding the complexity of climate change effect on various crop pests along with implementation of associated long term policies will help to subside the menace of anticipated climate change on agricultural productivity.

References

Alfred, J.R.B. (1998). Faunal Diversity in India: An Overview. In: *Faunal Diversity in India* (Eds.: Alfred, J.R.B. *et al.*,). ENVIS Centre, Zoological Survey of India, Calcutta, pp. 1-495

Bale, J.S., Masters, G.J., Hodkinson, L.D., Awmack, C., Bezemer, T.M., Brown, V.K., Butterfield J., *et al.*, (2002). Herbivory in global climate change research: direct effects of rising temperature on insect herbivores. *Global Change Biology,* 8:1–16.

Barzman, M., Lamichhane, J.R., Booij, K., Boonekamp, P., Desneux, N., Huber, L., Kudsk, P., Langrell, S.R.H., Ratnadass, A., Ricci, P., Sarah, J-L. and Messean, A. (2015) Research and development priorities in the face of climate change and rapidly evolving pests. In: Lichtfouse E (ed) Sustainable Agriculture Reviews, Vol 17. Springer, Cham, pp 1–27. https ://doi.org/10.1007/978-3-319- 16742 -8_1

Impact of Climate Change on Insect Pest Dynamics and Its Mitigation   117

Bebber, D.P., Ramotowski, M.A. and Gurr, S.J. (2013). Crop pests and pathogens move polewards in a warming world. *Nature Climate Change*, 3:985–988. https ://doi.org/10.1038

Chand, R. and Raju, S.S. (2009). Instability in Indian Agriculture during different phases of technology and policy. *Indian Journal of Agricultural Economic*, 64: 187-207.

Chen, C.C. and McCarl, B.A. (2001). An investigation of the relationship between pesticide usage and climate change. *Climatic Change*, 50:475–487.

Ciais, P., Reichstein, M., Viovy, N., Granier, A., Ogee, J., Allard, V., Aubinet, M., *et al.*, (2005). Europe-wide reduction in primary productivity caused by the heat and drought in 2003. *Nature*, 437: 529–533.

De, U.S. and Mukhopadhyay, R.K. (1998). Severe heat wave over the Indian subcontinent in 1998 in perspective of global climate, *Current Science*, 75: 1308-1315.

Dhawan, A.K., Singh, K., Saini, S., Mohindru, B., Kaur, A., Singh, G. and Singh, S. (2007). Incidence and damage potential of mealybug, *Phenacoccus solenopsis.* Tinsley, on cotton in Punjab.*Indian Journal of Ecology*, 34: 110-116.

Dhillon, M.K. and Sharma, H.C. (2007). Effect of storage temperature and durationon viability of eggs of *Helicoverpa armigera* (Lepidoptera: Noctuidae). *Bulletin of Entomological Research*, 97:55–59.

Diffenbaugh, N.S, Krupke, C.H., White, M.A. and Alexander, C.E. (2008). Global warming presents new challenges for maize pest management. *Environmental Research Letters*, 3:1-9. doi.org/10.1002/ps.998

FAO (2010) "Climate-Smart" agriculture: policies, practice and fnancing for food security, adaptation and mitigation. FAO, Rome, Italy. http://www.fao.org/docrep/013/i1881e/i1881e00.pdf

Furlong, M.J. and Zalucki, M.P. (2017). Climate change and biological control: the consequences of increasing temperatures on host–parasitoid interactions. *Current Opinion in Insect Science*, 20:39–44. https ://doi. org/10.1016/j.cois.2017.03.006

Gregory, P.J., Johnson, S.N., Newton, A.C. and Ingram, J.S.I. (2009). Integrating pests and pathogens into the climate change/food security debate. *Journal of Experimental Botany*, 60: 2827-2838

Harrington, R., Fleming, R.A. and Woiwod, P. (2001). Climate change impacts on insect management and conservation in temperate regions: can they be predicted? *Agricultural and Forest Entomology*, 3:233–240.

Heeb, L., Jenner, W., Romney, D. (2016). Promising innovative extension approaches for climate-smart agriculture: the plantwise example. In: Sala S, Rossi F, David S (eds) Supporting agricultural extension towards climate smart agriculture: an overview of existing tools. Global Alliance for Climate Smart Agriculture, pp. 50–57. http://www.fao.org/3/a-bl361e.pdf.

IPCC. (2007). Climate Change- Impacts, Adaptation and Vulnerability. In: (Eds.: Parry, M.L., Canziani, O.F., Palutikof, J.P., van der Linden, P.J., Hanson, C.E.) Cambridge University Press, Cambridge, UK, pp. 976

Kaiser, J. (1996). Pests overwhelm Bt cotton crop, *Nature*, 273: 423

Lal, M. (2003). Global climate change: India's monsoon and its variability. *Journal of Environmental Studies and Policy*, 6: 1-34.

Macfadyen, S., McDonald, G. and Hill, M.P. (2018). From species distributions to climate change adaptation: knowledge gaps in managing invertebrate pests in broad-acre grain crops. *Agriculture, Ecosystems & Environment*, 253: 208–219. https ://doi.org/10.1016/j.agee.2016.08.029

Maffei, M.E., Mithofer, A. and Boland, W. (2007). Insects feeding on plants: Rapid signals and responses proceeding induction of phytochemical release. *Phytochemistry*, 68: 2946–2959.

Musser, F.R. and Shelton, A.M. (2005). The influence of post-exposure temperature on the toxicity of insecticides to *Ostrinia nubilalis*(Lepidoptera: Crambidae). *Pest Management Science*, 61:508–510.

NATCOM (2004). India's initial national communication to the United Nations framework-convention on climate change. Ministry of Environment and Forests, p. 268.

Newton, A.C., Begg, G. and Swanston, J.S. (2009). Deployment of diversity for enhanced crop function. *Annals of Applied Biology*, 154:309–322.

Newton, A.C., Johnson, S.N. and Gregory, P.J. (2011). Implications of climate change for diseases, crop yields and food security. *Euphytica*, 179: 3-18

Parry, M.L., and Carter, T.R. (1989). An assessment of the effects of climatic change on agriculture. *Climatic Change*, 15:95–116.

Peng, S., Huang, J., Sheehy, J.E., Laza, R.C., Visperas, R.M., Zhong, X., Centeno, G.S., Khush, G.S., and Cassman, K.G. (2004). Rice yields decline with higher night temperature from global warming. *Proceedings of the National Academy of Sciences of the United States of America*, 101:9971–9975.

Rao, G.G.S.N., Rao, A.V.M.S. and Rao, V.U.M. (2009). Trends in rainfall and temperature in rainfed India in previous century. In: *Global climate change and Indian Agriculture case studies from ICAR network project*, (Ed.: PK Aggarwal), ICAR Publication, New Delhi. pp.71-73

Rhoades, D.F. (1985). Offensive-defensive interactions between herbivores and plants: their relevance in herbivore population dynamics and ecological theory. *American Naturalist*, 125: 205-238

Samra, J.S. (2003). Impact of Climate and Weather on Indian Agriculture. *Indian Society of Soil Science*, 51: 418-430

Satpathy, S., Gotyal, B.S. and Selvaraj, K. (2016). First report of cotton mealybug, *Phenacoccus solenopsis* Tinsley on cultivated jute (*Corchorus olitorius* L.) in India. *Entomologia Generalis*, 36(1): 55-61

Sharma, H.C., M.K. Dhillon, J. Kibuka, and S.Z. Mukuru. (2005). Plant defence responses to sorghum spotted stem borer, *Chilo partellus* under irrigated and drought conditions. *International Sorghum and Millets Newsletter*, 46:49–52.

Sharma, H.C., Sullivan, D.J. and Bhatnagar, V.S. (2002). Population dynamics of the Oriental armyworm, *Mythimna separate* (Walker) (Lepidoptera: Noctuidae) in South-Central India. *Crop Protection*, 21:721–732.

Sharma, H.C. (2014). Climate change effects on insects: implications for crop protection and food security. *Journal of Crop Improvement*, 28: 229-259.

Shrestha, S. (2019). Effects of Climate Change in Agricultural Insect Pest. *Acta Scientific Agriculture*, 3(12): 74-80.

Soman, P., Nwanze, K.F., Butler, D.R. and Reddy, Y.V.R. (1994). Leaf surface wetness in sorghum and resistance to shoot fly, *Atherigona soccata*: role of soil and plantwater potentials. *Annals of Applied Biology*, 124:97–108.

Thomas, C.D., Cameron, A., Green, R.E. *et al.*, (2004). Extinction risk from climate change. *Nature*, 427: 145-148.

Timoney, K.P. (2003). The changing disturbance regime of the boreal forest of the Canadian Prairie Provinces. *Forest Chronicle*, 79: 502-516.

Wang, X.Y. and Shen, Z.R. (2007). Potency of some novel insecticides at various environmental temperatures on *Myzus persicae*. *Phytoparasitica*, 35: 414–422.

White, T.C.R. (1984). The abundance of invertebrate herbivores in relation to the availability of nitrogen in stressed food plants. *Oecologia*, 63:90–105.

Yonow, T., Kriticos, D.J., Kirichenko, N. and Ota, N. (2018). Considering biology when inferring range-limiting stress mechanisms for agricultural pests: a case study of the beet armyworm. *Journal of Pest Science*, 91:523–528. https://doi.org/10.1007/s10340-017-0938-9

Approaches of Conservation Agriculture as Climate Change Adaptation Strategies

8

Crop Diversification in CA /RCTs under Climate Change Perspective: Special Reference to Jute & Allied Fibres

A.K. Ghorai and Debarati Datta

Crop Production Division, ICAR-Central Research Institute for Jute and Allied Fibres, Barrackpore, Kolkata – 700 121, West Bengal

Introduction

Conservation Agriculture (CA) is a set of soil management practices that minimize the disruption of the soil's structure, composition and natural biodiversity. These include: i) maintenance of permanent or semi-permanent soil cover (using either a previous crop residue or specifically growing a cover crop for this purpose); ii) minimum soil disturbance through tillage (just enough to get the seed into the ground); iii) regular crop rotations to help combat the various biotic constraints. CA also uses or promotes where possible or needed various management practices listed below: i) utilization of green manures/ cover crops (GMCC's) to produce the residue cover; ii) no burning of crop residues; iii) integrated disease and pest management; iv) controlled/limited human and mechanical traffic over agricultural soils. When these CA practices are used by farmers one of the major environmental benefits is reduction in fossil fuel use and greenhouse gas (GHG) emissions.

In India, efforts to develop, refine and disseminate conservation-based agricultural technologies have been underway for nearly two decades and made significant progress since then even though there are several constraints that affect adoption of CA. Particularly, tremendous efforts have been made on no-till in wheat under a rice-wheat rotation in the Indo-Gangetic plains. The technologies of CA provide opportunities to reduce the cost of production, save water and nutrients, increase yields, increase crop diversification, improve efficient use of resources, and benefit the environment. However, there are still constraints for promotion of CA technologies, such as lack of appropriate seeders especially for small and medium scale farmers, competition of crop residues between CA use and livestock feeding, burning of crop residues, availability of skilled and scientific manpower and overcoming the bias or mindset about tillage. The need to develop the policy frame and strategies is urgent to promote CA in the region. (Bhan and Behera, 2014).

Crop Diversification

Recognition that climate change could have negative consequences for agricultural production has generated a desire to build resilience into agricultural systems. One rational and cost-effective method may be the implementation of increased agricultural crop diversification. Crop diversification can improve resilience in a variety of ways: by engendering a greater ability to suppress pest outbreaks and dampen pathogen transmission, which may worsen under future climate scenarios, as well as by buffering crop production from the effects of greater climate variability and extreme events. Such benefits point toward the obvious value of adopting crop diversification to improve resilience, yet adoption has been slow. However, crop diversification can be implemented in a variety of forms and at a variety of scales, allowing farmers to choose a strategy that both increases resilience and provides economic benefits, (Lin, 2011).

Makate *et al.* (2016) indicated that the importance of crop diversification as a viable climate smart agriculture practice that significantly enhances crop productivity and consequently resilience in rural smallholder farming systems. They, therefore, recommend wider adoption of diversified cropping systems notably those currently less diversified for greater adaptation to the ever-changing climate. They also found that crop diversification depends on the land size, farming experience, asset wealth, location, access to agricultural extension services, information on output prices, low transportation costs and general information access.

Crop Diversification in Conservation Agriculture in Jute

Fluctuating market price of farm outputs and changing climatic condition have pushed back the farmers to a risky environment where farm outputs and net profits from farming enterprizes are often unreliable. Crop diversification in resource conservation technologies reduces the risk of crop failure, generates more income, increases soil fertility and its moisture content depending on situations and practices adopted by the farming communities. Jute-rice-pulses/oilseeds/vegetable based cropping systems under varying edapho-climatic conditions to make jute farming more reliable and profitable mitigating different climatic stresses using different components of conservation agriculture technologies including water harvesting *in situ* (Ghorai, 2019) will helps in life saving irrigation and promotes ground water recharges too.

Jute is primarily a rainfed crop and its national average yield is around 23 q/ha. Under optimum moisture and nutrition supply the realizable fibre yield potential is around 35-40 q/ha and thus the yield gap is around 12-17 q/ha. For mesta, the national average fibre yield around 11-14 q/ha whereas its fibre yield potential is around 30-32- q/ha with an untapped potential of 18-19 q/ha. This poor fibre yield recovery is mainly due to unassured moisture supply during its growth period combined with its poor nutrition management. The rainfall deficit during early growth stages have been found to vary from 25-50 per cent in various years from (Ghoari *et al.*, 2014). Improved Agronomic management practices and soil water conservation technics were developed to improve jute and mesta fibre productivity under drought situation (Ghorai *et al.*, 2008-2014). Fibre retting and fibre quality are also very often hampered due to insufficient retting water availability during its harvest. Experimental results to combat drought in jute and mesta are shown below.

Drought Management Strategies for Jute and Mesta using RCTs

Under rain fed situation and 40-50% deficit rainfall condition, open furrow sowing of jute at RDF (N:P:K ::60:30:30) helps in collection of rainwater in furrows, increases in infiltration in root zone which in turn helps in proper germination of its tiny seeds (2mg/seed) and maintains better moisture regime during its growth period and produced 27-31 q jute fibre/ha where national average was 23 q/ha (Ghorai *et al.*, 2008) Fig. 1. Recommended fertilizer dose (N:P:K::60:30:30), one flood irrigation in flat bed sowing, could produce 31.24 q jute fibre /ha. At RDF, open furrow sowing of jute and one irrigation on furrows, reduced the irrigation requirement by 40 percent and produced 33.81 q jute jute fibre /ha which was 2.57 q higher over traditional flatbed method of sowing. At RDF and one flood irrigation followed by soil mulching (at field capacity) by CRIJAF nail weeder produced 34.78 q jute fibre /ha and is 3.54 q/ha over traditional flood irrigation system. In large scale FLDS' (55 ha), in farmers' field on drought management of jute, soil mulching by CRIJAF nail weeder could save 1-2 irrigations in different locations of Hoogly, Murshidabad and 24 PGS (N) districts of WB and produced 30-35 q jute fibre/ha (CRJAF Ann. Rep. 2015-16 & 2016-17).

Drought Management Strategies Developed for Roselle using RCTs

At RDF (N:P:K ::60:30:30), rainfed open furrow sowing produced 27.85 q jute roselle fibre/ha which is 1.52 q over traditional flat bed sowing. At RDF (N:P:K ::60:30:30) rainfed flat bed sowing produced 26.33 q jute roselle /ha. Rainfed roselle, under water conservation by soil mulching using CRIJAF nail weeder, produced 30.36 q roselle fibre /ha and is 4.03 q/ha over traditional rainfed system at RDF. In rainfed system flat bed sowing, N:P:K ::60:30:30 +

30 kg, elemental sulphur yielded 29.90 q roselle fibre /ha which was 3.11 q higher than RDF and N:P:K ::80:40:40 yielded 27.85 q roselle fibre /ha which 1.06 q higher than RDF, Ghorai *et al.*, 2013) .

Fig. 1: Open furrow jute sowing in jute to combat drought stress

Rice straw mulch @ 5-10 tonnes/ha increased jute fibre yield from 25-33 per cent over bare soil (Ghorai *et al.*, 2006 and 2008). In pointed gourd cultivation, rice straw mulch @ 8-20 tonnes/ha increased fruit yield (77-323% in different years), controlled composite weeds effectively, increased electrical conductivity, available phosphorus, potassium and organic matter content of soil (Ghorai, 2004). Soil health improvement by residue mulching and crop residue in jute has been reported by Mukesh Kumar *et al.*, (2108) in jute. Straw mulch was found effective for irrigation management in wheat, mustard and chickpea (Mandal *et al.*, 1991). Surface mulching in well drained jute fields with 10 to 15 t rice or wheat straw per ha reduced weed dry-matter production by 68 to 82% and gave fibre yields upto 40 q/ha. Mulched plots maintained better hydrothermal regime of soil, were relatively cooler (6-8°C) and maintained more soil moisture (18 to 23.4%) than bare soil (12.6 to 18%) cultivation (Ghorai *et al.*, 2008).

Further mixed cropping practiced in this system produced 6.40 to 11.5 q red amaranth, 6.40 to 6.72 q white amaranth and 9.38 to 15.00 q summer radish. It fetched a net return of Rs 18,472 to Rs 20,949 compared to Rs 16,147 per ha under manual weeding. Raising smother crops of red amaranth (Fig. 2, seed @ 10-30 kg/ha) suppressed weeds from 22-54% over two manual weedings, it produced 35 to 37 quintal jute fibre along with 19 to 59 quintal red amaranth (21 DAE) per hectare and gave a net return upto Rs 20,949/ha (Ghorai, 2008).

Fig. 2. Straw mulching in jute for moisture conservation and weed smothering

Crop Diversification of Jute Field by Mixed Cropping

Summer spinach (Local cv. Haldi Bari), summer radish (cv. Paus mula) and red amaranth (cv. Jabakusum) while grown with jute as intercrops (with butachlor 5G @1.5 to 2 kg /ha during sowing plus one hand weeding) produced 89 q palak, 42 q summer radish, 72 q red amaranth and 32 to 38 q jute fibre/ha (Fig. 3). These interim vegetable harvests will strengthen famers economy by early realizing the cost of manual weeding. This will create early solvency, minimize loss of input through weeds and more remunerative than manual weeding twice. (Ghorai *et al*, 2003). This system fetched a gross return of Rs. 1.63 to1.68 from each one rupee invested as total variable cost. The gross return/TVC ratio was 1.60 for manual weeding twice. Sorghum bicolor (seed yield) 7 q/ha and Basella were found suitable mixed crop with jute.

Crop Diversification with Legumes in Jute System as Smother Weed

Inclusion of legumes in the cropping system has been known since times immemorial. Legume is a natural mini-nitrogen manufacturing factory in the field and the farmers by growing these crops can play a vital role in increasing indigenous nitrogen production. Legume help in solubilizing insoluble P in soil, improving the soil physical environment, increasing soil microbial activity, and restoring organic matter, and also has smothering effect on weed. The carryover of N derived from legume grown, either in crop sequence or in intercropping system for succeeding crops, is also important. In a country like India, where the average consumption of plant nutrients from chemical fertilizers on national basis is very low, the scope for exploiting direct and residual fertility due to legumes has obviously a great potential (Ghosh *et al.*, 2007).

Fig. 3: Crop diversification of jute field for better resource utilization, weed control by smothering and generate more income per unit land for resource poor farmers

Intercropping/Strip Cropping Pulses with Jute

Intercrop sowing time: 3rd March to 10th March. Intercropping green gram (Pant mung 5) with jute suppressed *Cyperus rotundus* upto 56 % by smothering affect. This system produced 10q pulse grain, fibre yield 28 q/ha and 2 tonnes pulse waste/ha. The fibre equivalent yield upto 49 q/ha over 38 q/ha under two manual weeding (Ghorai *et al.*, 2015 and 2016). Weed smothering in jute by intercropping green gram with jute (1:1), Fig. 4a. The sunlight entry in soil was cut off by mung (cv. TMB 37, Sukumar, Birat etc) canopy upto 95%. Weed biomass was 76 per cent lower than manual weeding twice AT 25 DAS. Green gram yield (TMB-37, matures in 52-55 DAS) in intercrop was 9.0 /ha. mixed crop 7.2 q/ha and relay crop 6.2 q/ha yield and corresponding jute fibre yields were 30.5, 30.6 and 24.0 q respectively. The pulse grain yield of cv. Sukumar, (small grain variety, matures in 65 DAS) was 7.0 q/ha and the fibre yield was 31 q/ha (CRIJAF Ann Rep 2015 & 2016).

Fig. 4: Pulse intercropping with jute for weed smothering and sustainable jute farming

In farmers [Nadia, Murshidabad, 24 PGS (N), Hoogly. Howrah, Jalpaiguri Paschim Midnapur, etc] and normal rainfall situation in 24 pgs (N), the mean green gram yield was 5-5 q/ha (max 9 q) and mean fibre yield was 27-30 q/ha. Under very high rainfall (180 mm from 19th April to 30th May) the mean green gram yield was 3.20 q/ha and fibre yield was 29.4 q/ha. In mixed cropping green gram yield ranged from 2.68-3.80 q/ha and mean jute fibre yield was 32 q/ha [in Murshidabad and 24 PGS (N)]. In farmer's field it eliminated 2nd weeding which about 30 man days/ha. The green gram intercropping acted as an insurance crop as it produced nearly 2.6 q pulse grain grains along with 22-25 q jute fibre, even after the young jute crops were topped by hail storm at its initial stages. Rice yield was better in intercrop field over control.

Crop Diversification in Jute and Roselle by Strip Cropping with Maize, Okra and Groundnut

Jute + green gram (4:4, cv. RMG 62) strip cropping (Fig. 5a) produced 25.30 q jute and 5.52 q green gram/ ha. It added 17-20 q nitrogen rich (2.35% N) green biomass/ha. Jute + Maize (3:1) strip cropping + red amaranth (Fig. 5b) produced 28.63 jute fibre, 41667 maize cobs and 58.35 q red amaranth/ha. The fibre equivalent yield of this system was 55.74 q/ha.

Fig. 5a: Pulses + jute /mesta strip cropping in conservation agriculture

Fig. 5b: Risk management in jute and roselle by crop diversification

Crop Rotation in Jute based Cropping System

This was primarily done for residual moisture use of rice field in jute-rice-pulse cropping pattern, residue recycling, soil enrichment and providing protein security to resource jute farmers in zero till and minimal tillage system. At ICAR- CRIJAF, jute and rice was strip cropped to diversify jute farming. After harvest of rice in strips (sown dry), Pant mung 5, RMG-62, RMO-40, Pant U-31 (utera crop, Zero tillage) were sown (30th September with N:P:K:: 26:52:52) in between rice stubbles in minimal tillage system which yielded 16, 15, 11.82 and 5.5 q pulse grains/ha in rainfed condition and were synchronous in maturity (90 days). Scented rice cv. Tulai panji (WB) yielded maximum up to 4.5 t/ha after jute harvest. Lentil (Fig. 6), pea, Khesari and mustard were sown at double seed rates at 10-15 days ahead of rice harvest on muddy soil with N:P:K::26:52:52. For mustard the total nitrogen was 80 kg/ha. Rice was harvested keeping stubbles of 6-8 inches height to prevent seed lose by bird eating and one post sowing irrigation was given having good drainage facility in the field. Immediately after rice harvest (10 DAS), Preatilachlor 50 EC @ 0.9 kg *ai*/ha was applied to control weeds in paira crop. In this paira cropping system the yield of lentil, pea, Khesari and mustard were 8.8 to 14.85 q; 17.3 to 27 q/ha; 15.6 q/ha and 10.6 to 15.9 q/ha (Ghorai, 2008; CRIJAF Ann. Rep. 2017-18 & 2108-19).

Fig. 6: Paira crop of lentil and sequences crop of mung (after jute) in jute

Pulse, Jute, Rice, Pulses Relay Crop

Mung (Pant mung 5), moth bean (RMO 40) and black gram (Pant U 31) were sown before jute in relay (Fig. 7) system to add nutrient rich pulse wastes (1.5 to 2 t/ha) and rainwater conservation for easy germination of jute under rain fed farming. Addition of 2 tonnes pulse waste in sequence produced 37.4 jute fibre/ha with higher water productivity 1035 lit/kg fibre) over 30.3 q/ha under control (low water productivity 1277 lit/kg fibre), [Ann. Rep. CRIJAF 2016-17].

Fig. 7: Relay crop of green gram and moth bean in jute based cropping system

Crop Diversification in Jute and Mesta by Mixed Cropping with *Crotalaria* spp

Green manuring in *situ* is viable option to add nutrition in soil, increase water holding capacity of soil and thus will helps the jute crop under drought stress. *Crotalaria quinquefolia* seed @ 20 to 40 kg/ha mixed with jute at 0 DAS produced raw jute fibre to the tune of 35 to 38.67q/ha along with 108 to 147 q green biomass as mixed green manure. The sole crop jute yield was 38.74 q/ha. As sole green manure at 45 DAE, *Crotalaria quinquefolia* produced 250q biomass/ha and with 50 kg chemical nitrogen it produced 32 q raw rice/ha in sequence. Nitrogen content of *Crotalaria quinquefolia* at 45 and 60 DAE are 2.41 and 1.95% respectively. For sunnhemp nitogen content at 30 and 60 DAS are 2.30 and 1.90 % respectively. This combination increased rice and lentil yield in sequence. Second weeding is not required due to canopy pressure (Ghorai *et al.*, 2006).

Zero Tillage Jute after Summer Rice/ Onion to Minimize Cost of Cultivation

Savings in the cost of ploughing, irrigation (minimizes fossil fuel use), fertilizer and drudgery in sowing stimulating many farmers to adopt zero tillage jute after summer rice or onion harvest. Sowing of jute after onion saves the cost of fertilizer due to high residual fertility left after its harvest. In conventional jute cultivation at least two plough and one planking are required for sowing of jute which requires about Rs. 7500/ha. Farmers simply broadcast jute on unploughed even rice soils full of hair cracks sufficient enough to retain tiny jute seeds in it which does not disperse during flood irrigation. In no till soil, soil being even in nature, the irrigation requirement and time of irrigation are also lesser than ploughed soil. The water advancing front is much higher in zero till soil than ploughed and uneven surface soil which usually forces the water to go deeper than required and decreases the irrigation efficiency both in terms of quantity of water requirement and time of application. Pilot studies were made at Kalapunja, Paschim Midnapur in Zero till jute in 2018 & 2109.

Fig. 8a: Zero till jute after summer rice (cv. NJ 7010) at Paschim Midnapur, WB in clay soil in 2018

Fig. 8b: Zero till jute (NJ 7010) after onion harvest at ICAR-CRIJAF, in alluvial soil WB

Crop Diversification in Waterlogged Rice Soil using Gunny Bags Based Soil Columns

Submerged rice field limits cultivation of dicot vegetable in it because of its anaerobic nature. Use of biodegradable jute gunny bag/jute fabrics reinforced soil column (Fig. 9) have opened up a new vista for growing dicotyledonous crops within rice field avoiding anoxia and providing adequate oxygen to these crops, facilitating proper drainage by gravitational and lateral flow through meshy gunny bags/jute hessians (Ghorai *et al.*, 2013). Crop diversification in rice field using gunny bags/ jute fabrics will make effective utilization of resources applied in rice field, reduce the irrigation requirement of long duration vegetables due to its long association (2-3 months) in dwarf and semi dwarf rice filed (Annul Report DARE, 2012-13). This crop diversification in rice field increases the cropping intensity (upto five hundred percent under irrigated system), generates additional return and creates more employment opportunities from unit area and time. Extensive adoption of jute fabrics in subaquatic rice field for crop diversification will thus increase the marketing opportunities of raw jute fibre in nontraditional areas and improve the livelihood security of resource poor raw jute farmers of South east Asia.

Fig. 9: Gunny bag reinforced soil columns in waterlogged summer rice field for vegetable cultivation

In kharif, 3 to 5.5 tonnes of fine rice (cv. Banskati and Ksitish) and 15-50 tonnes/ha of vegetables (different cucurbits, solanaceous vegetables and field beans) were harvested from this rice-vegetable relay cropping system (Ghorai *et al.*, 2014) using gunny bag reinforced soil columns (30 cm ht. x 45 cm dia.) avoiding anoxia. In summer rice, different cucurbit yield varied from 55-150 q/ha, ginger 600 q/ha, amorphaphalus 120 q/ha, and colocasia 20 to 25 tonnes tuber/ha (Ghorai *et al.*, 2013) were obtained in waterlogged rice field without affecting rice yield (5.4 t/ha). The hydrograph of ponding ranged between 0-30 cm during kharif and 0-10 cm in Boro season. Water productivity of rice field is also thus increased as remunerative vegetables are grown in rice field. The columns duration was almost eight months in rice field. The oxygen diffusion rate (ODR) of gunny bag/hessian reinforced soil columns (upto 15 cm depth) was two and half time higher (280 $\mu g/m^2/sec$) over conventional ridges (115 $\mu g/m^2/sec$) in. The moisture content of this gunny bag/hessian reinforced soil columns (30 cm depth) varied from 18.5-23 % over 23-25.5 % in conventional ridges.

This process consumed about 1166 gunny bags (weighing 7-8 q jute fibre, 50 kg capacity) /ha. These gunny bags decays in 4 months leaving the columns bare. After summer rice harvest, the left over bare soil columns were further reinforced using fresh old gunny bags treated with prescribed pesticides for Kharif season. In same field in kharif season, super early vegetable (1st week of June, cauliflower, cabbage, coriander, brassica, late cucurbits, early field beans, brinjal, tomato etc) – scented rice relay has been successful avoiding anoxia. In medium land (0 to 30 cm ponding depth), as high 4000-6000 jute bags can be consumed for producing early cole crops, ginger, dioscorea, peas, French beans etc. avoiding waterlogging stress (column height ranged from 11-30 cm).

Table 1: Vegetables with in kharif rice in dual culture on medium land Using gunny bag based soil columns and its economics

Crops	Jute bags consumes/ha	Rice yield (t/ha)	Inter/relay crop yield (q/ha)	Gross return from relay/inter crops (Rs./ha)	Cost of cultivation of relay/inter crops (Rs./ha)	Net return from relay/inter crops (Rs./ha)	Return per unit cost (Rs/Rs)
Cabbage	12405	4.5	51	51133	39818	11315	1.28
Brinjal	12405	4.5	152	150000	66316	83684	2.26
Carrot	12405	4.5	28.3	42500	36816	5684	2.80
Tomato	12405	4.5	69	103000	37316	65584	0.30
Onion	12405	4.5	12.3	11333	33316	-21983	0.34
Chilli	12405	4.5	11.8	35400	36316	-916	0.97
Cauliflower	12405	4.5	29	29000	36316	-7316	0.80
Coriander	12405	4.5	9.3	46500	36316	10184	1.28
Radish	12405	4.5	33	50000	36311	13689	1.38
C.D (5%)	NS	NS	5.60	7343	7495	--	0.31
CV (%)	--	--	7.37	9.74	10.86	--	14.21

Source: Ghorai et al. (2016); CRIJAF Annual Report (2016)

Fig. 10: Crop diversification in waterlogged Kharif and Rabi rice field using high value vegetable crops

Fig. 11: Crop diversification in waterlogged Kharif and summer rice land using gunny bag based soil columns

Conclusion

In the face of shrinking natural resources and ever increasing demand for larger food and agricultural production arising due to high population and income growths, agricultural diversification is the main course of future growth of agriculture. Increased return of crop residue from greater annualized crop production can lead to higher soil organic matter levels and a greater potential for nutrient cycling, an effect that will increase with time. Sensible crop diversification should be adopted with ultimate goal to employ economically viable, diversified crop combinations, to help moderate possible weeds, disease and pest problems, to enhance soil bio-diversity, to reduce labour peak and to provide farmers with new risk management opportunities.

References

DARE (2012-2013). Crop Management: Crop diversification in waterlogged rice field using jute reinforced soil column. p. 44.

Ghorai, A.K., Saha, S. and Chakraborty, A.K. (2014). Concentrated jute and mesta leaf manures: Its role on summer radish production and its comparative performance with mustard oil cake. *Indo-American Journal of Agriculture & Veterinary Science*, 2(1): 26-30.

Ghorai, A.K (2019). Ground water recharge and lifesaving irrigation potential of dugout ponds for *in situ* jute retting in rice paddies. DOI: 10.13140/RG.2.2.16733.15848.

Ghorai, A.K. (2004). Analysis of pointed gourd (Tricosanthes dioica L.) cultivation with and without rice straw mulch: A case study. *SAARC Journal of Agriculture*, (2): 73-87.

Ghorai, A.K. (2008). Integrated weed management of jute (*Corchorus olitoriùs*). *Indian Journal of Agronomy*, 53(2): 149-151.

Ghorai, A.K. (2016). Irrigation methods and soil water conservation practices for improving water productivity in jute. *Annual Report CRIJAF*. pp 27-28.

Ghorai, A.K. (2019). Zero tillage jute after summer rice/onion to minimize cost of cultivation. Jaf News, 17(1): 21.

Ghorai, A.K. and Kundu, D.K (2016). Avoid burning rice and wheat straw, compost it using sunhemp and minimise green house gas emission. 'Resource Based Inclusive Agriculture and Rural Development: Challenges and opportunities" under climate resilient agriculture and stress management", Ramkrishna Mission Vivekananda University (RKMVU). p. 40.

Ghorai, A.K. and Saha, S. (2006). Crotalaria quinquefolia a highly potent green manure crop in jute-rice-lentil cropping system. *Jaf News*, 4(2): 101.

Ghorai, A.K., Chakraborty, A. K., Pandit, N.C. and Mandal,R.K. (2006). Integrated weed management of jute (*Corchorus* spp.L). *Indian Journal of Weed Science*, 38 (1&2): 163-164.

Ghorai, A.K., Chakraborty, A. K., Pandit, N.C. and Mondal, R.K. (2004). Jute – sorghum mixed cropping under rainfed condition for fibre, grain and fodder production. *Jaf News*, 2(1): 9-10.

Ghorai, A.K., Chakraborty, A.K., Pandit, N.C., Mondal, R.K. and Biswas, C.R. (2004.) Grass weed control in jute by Targa super (quizalofop-ethyl 5% EC). *Pestology*, 28: 31-34.

Ghorai, A.K., Choudhury, H, Kundu, D.K. and Kumar, S. (2015). Use of gunny bags andjute fabrics in Agricultural field, Kheti, ICAR, New Delhi.

Ghorai, A.K., Chowdhury, H and Kundu, D.K. (2011). Use of jute fabrics in agricultural field Annual Report CRIJAF (2011-12, 2012-13, 2013-14).

Ghorai, A.K., Chowdhury, H., Kundu, D. K., Kumar, M., and Mahapatra, B.S. (2013). In situ jute retting (Corchorus olitorius L.) using native culture in low volume water in polyethylene lined micro pond and poly culture in and around it. Technical bulletin, ICAR-CRIJAF, Barrackpore. No.6/2013, p. 28.

Ghorai, A.K., Hembram, P., Chowdhury M., Kumar, H., and Kundu, D.K. (2012). Dry transplanting of rice in kharif season in jute-rice-pulse sequence under deficit rainfall. *Jaf News*, 10(1): 1415.

Ghorai, A.K., Kumar, M. and Kar, C.S. (2016). Weed smothering in jute with greengram intercropping. *Indian Journal of Weed Science*, 48(3): 343–344.

Ghorai, A.K., Kumar, M. and Roy, S. (2017). Development of low cost and eco-friendly technologies for jute. *Annual Report CRIJAF* (2017-19), pp. 45-46.

Ghorai, A.K., Kundu, D.K, Kumar, S. and Shamna, A (2016) Use of gunny bags/jute fabrics in agricultural field for sustainable family farming for food, nutrition and livelihood security (pp 107-110). In book Editerd by Bitan Mondal, Debasish Sarkar, Siddhartha Dev Mukhopadhyay, Souvik Ghosh, Bidhan Chandra Roy and Sarthak Choudhury. Renu Publishers, New Delhi-16.

Ghorai, A.K., Kundu, D.K., Kumar, S., Shamna, A. and Datta, D. (2020). Gunny bag-based soil columns for crop diversification in rice field to enhance livelihood security of land scarce farmers. *Current Science*, 119(7): 1190-1195.

Ghorai, A.K., Kundu, D.K., Satpathy, S. and Ghosh, R. (2014). Crop diversification in anaerobic rice field using gunny bag reinforced soil columns. *SAARC Agri News*, 8(2): 7-8.

Ghorai, A.K., Saha, S and Hembram, HP.K. (2010). Drought management of jute indeficit rainfall in Jute and allied fibres, Production, utilization and marketing. Ed. Palit et al., Indian Fibre Society Eastern Region, pp. 187-193.

Ghorai, A.K., Saha, S., Saren, B.K, Hembram, P.K., Mandal, B.K., Thokle, J.G.,More, S.R., Srilata, T., Jagannadham, G.,Tripathi, M.K., Kumar, S., Kundu, D.K. and Mahapatra, B.S. (2013). Drought Management of jute and mesta crop under deficit rainfall. Technical bulletin No.5/2013. p. 67.

Ghorai,A.K. and Kundu, D.K. (2014). Do not burn straw in rice field, convert it to organic manure (Mather kharpordabennaa, jaibsaarbanan). *Feere Asuk Sobu*j, 1(5): 3.

Lin, Brenda, B. (2011): Resilience in Agriculture through Crop Diversification: Adaptive Management for Environmental Change. *Bio Science*, 61(3): 183–193.

Mallick, M. and Patnaik, U. (2017). Crop Diversification and Sustainable Agriculture in India. 136.

Mandal, B.K.; Saha, Ashok, T.K. Kundu and Ghorai, A.K.. (1991). Wheat (Triticum aestivum) based intercropping and the effct of mulch on growth and yield Indian J. Agron. 36:23-29

Mandal, B.K.; Saha, Avhijit; Ghorai, A.K. and Saha, Ashok. (1989). Effect of preceding crop and mulch on succeeding rice. *Indian J. Agron.* 28(1) 93-96

Mukesh Kumar, D.K. Kundu, A.K. Ghorai, S. Mitra, S.R. Singh (2018): Carbon and nitrogen mineralization kinetics as influenced by diversified cropping systems and residue incorporation in Inceptisols of eastern *Indo-Gangetic Plain.Soil & Tillage Research.*178(108-117)

Raicy Mani Christy and Elango Lakshmanan (2017). Percolation pond as a method of managed aquifer recharge in a coastal saline aquifer: A case study on the criteria for site selection and its impacts. *J. Earth Syst. Sci.* (2017) 126:66 c_ Indian Academy of Sciences DOI 10.1007/s12040-017-0845- 8

9

Resource Conservation through Crop Residue Management for Sustainable Agriculture

M.S. Behera, Laxmi Sharma and Pradipta Samanta

ICAR-Central Research Institute for Jute and Allied Fibres, Barrackpore West Bengal – 700 121, India

Introduction

Resource Conserving Technology refers to any management approach or technology that increases factor productivity including land, labour, capital and inputs. The resource conserving technologies (RCTs) involving no or minimum tillage, direct seeding, bed planting and crop diversification with innovations in residues management are the possible alternatives to the conventional energy and input-intensive agriculture. The RCTs with innovations in residue management avoid straw burning, improve soil organic C, enhance input efficiency and have the potential to reduce GHGs emissions (Pathak *et al.*, 2011). Conservation agriculture is one the effective Resource Conservation Technologies, which is based on three principles which are minimum soil disturbance, maintenance of permanent soil covers by the use of crop residues and cover crops and diverse plant associations which included crop rotations. Globally, Conservation Agriculture is being practiced in about 125 M ha area. USA with 26.5 M ha contributes to the major CA-practicing countries followed by Brazil (25.5 M ha), Argentina (25.5 M ha), Canada (13.5 M ha) and Australia (17.0 M ha). However, in the context of climate change to increase the production and productivity, India's CA adoption is still in preliminary phases in area of 1.5 million ha with adoption of Zero Tillage (Jat *et al.*, 2012, www.fao.org/ag/ca/6c.html). The major CA-based technologies being adopted is zero-tillage (ZT) wheat in the rice-wheat system of Indo-Gangetic Plains.

Conservational Tillage: A Step Towards RCTs

According to CTIC (1992), any tillage and planting system that maintains at least 30% of the soil surface covered by residue after planting to reduce water erosion, or where soil erosion by wind is a primary concern, maintain atleast 1000 pounds of flat, soil grain residue equivalent on the surface during the critical wind erosion period. Conservation agriculture comprises of various conservational tillage methods like Zero Tillage (ZT) involving nominal mechanical seedbed preparation and dependence on herbicides or cover crops or both to control weeds.

Zero tillage is adopted where straw residue for previous crops are on the soil surface without any form of incorporation by planting in loose and anchored residues using Turbo seeder or a Combo Happy Seeder and partially bailed out for use as animal feed. For the success of a permanent zero till system, some crop Residue must be left in the fields. This method is prevalent in no till or conservation tillage practice where atleast 30% of the soil surface is covered with crop residue. When crop residue of 20 to 30 cm height is left on the field, an estimated 3-4 tonnes of organic are returned to the soil. (Singh *et al.*, 2010) found that incorporation of crop residues decreased BD and increased infiltration rate, WHC, microbial population, soil fertility as compared to no residue treatment.

Other approaches consist of seedbed preparation on ridges along with leftover residue between ridges on the surface, known as ridge tillage and other popular method like mulch tillage. All the methods retain protective amounts of residue on the surface creating a permanent or semi-permanent organic soil cover, which protects the soil against the harmful effects of exposure to rain and sun to provide the micro and macro organisms in the soil with a constant supply of food and alter the microclimate in the soil for optimal growth and development of soil organisms, including plant roots. It can implement either through mulching or cover crops or combination of both. Mulching helps in prevention of splash erosion and surface run-off. Multi-purpose cover crops like *Crotalaria juncea L.* can also be planted to improve soil aggregation, thus ensuring soil biological activity. Cover crops, such as Sesbania can produce a green biomass of up to 30 ton per hectare within 60 days and control most of the weeds, leaving fields almost weed-free. It consists of minimum mechanical soil disturbance, soil cover with plant biomass/cover crops and diversified associations. The level of soil cover should be ideally 100% of the soil surface, but never less than 30% and should always contain sufficient organic carbon to maintain and enhance soil organic matter levels.

Crop Residues: A Missed Opportunity

Crop residue is defined as the vegetative crop material left on the field after a crop is harvested, pruned or processed. With increased industrial demand for crop residues, conservation agriculture (CA) offers a good opportunity managing the residues in a productive and profitable manner. Their Felder and Wall, 2008 described the retention of crop residue as the key drive in the positive realization of the benefits of the CA. Utilization of crop residues can be through the brown route or the green route at the farm level. Agricultural crops generate considerable amounts of leftover residues, which increases with increase in food production. About more than 686 MT of crop residues are generated by India annually from 26 Field Crops. The major contribution of 545 MT residue is from Cereals, oilseeds, pulses and sugarcane, 61 MT from horticultural crops (Coconut, Banana and Arica nut) and 80 MT from other crops. Straw, a low-density crop residue is the dominant residue. Unlike the cereal crops, crops like pigeon pea, cotton, rapeseed, mustard, mulberry and plantation crops produce woody residues. Considering the state wise data, the generation of crop residues is highest in Uttar Pradesh (60 Mt) followed by Punjab (51 Mt) and Maharashtra (46 Mt). Incorporation of residue requires tractor energy and also lead to temporary immobilization of nutrients specially nitrogen. When it is intended to incorporate residues of wheat before rice transplanting, it is best to incorporate them using disk plough, irrigate and apply a bag or urea to facilitate early decomposition Reduced tillage and in situ incorporation of crop residues at 5 Mg ha^{-1} along with 150 kg N ha^{-1} were optimum to achieve higher yield of wheat after rice in sandy loam soils of Indo-Gangetic plains of India (Gangwar *et al.*, 2006). Residue retention has now become a viable option due to availability of a new machines like Turbo Seeder. For example, the Combo Happy Seeder are used for direct sowing of wheat crop in the residue of preceding rice-crop.

The dominant uses of crop residues in India are as fodder for cattle, fuel for cooking, thatch materials for housing and as organic resources. It can be used as a dry bed on the floor of domestic animals during winter. It serves as a source of energy for the fauna, which facilitate biological tillage. The dominant residues are those of rice, wheat, sugarcane and cotton accounting for 72.9% of the total residue production.

Burning of Crop Residues: No Choice Left

Burning causes harm to physical, chemical and biological process in the soil resulting in adversely affecting the soil properties. Lack of labour, increased mechanization particularly the use of combine harvesters, declining numbers of livestock, higher cost for removal of crop residue by conventional method, longer period required for composting and unavailability of alternative economically viable solutions along with financial assistance from government often compel the farers to burn the residues without being aware of the adverse consequences of on-farm burning of crop residues. It is recorded that reduction in yield and nutrient loss due to burning monetary loss may be of Rs 500 crore per year (Bimbraw, 2019). Crop residue burning significantly increases the quantity of air pollutants such as CO_2, CO, NH_3, NO_X, SO_X, Non-methane hydrocarbon. It is estimated that 87 Mt of surplus crop residues is burnt in different croplands (TERI 2019). It is more practiced in irrigated agriculture and mechanized rice, wheat system of North-West India. Trace gas and aerosols emissions due to open field burning of such large quantity of residue leads to adverse implications of the local and regional environment, which also has linkages to the global climate change. Presently, more than 80 percent of total rice straw produced annually is being burnt by the farmers in 3-4 weeks during October-November. Burning of 10 quintal of rice straw causes loss of 400 kg organic carbon, 5.5 kg of nitrogen, 2.2 kg phosphorus, 25 kg potassium and 1.2 kg sulphur. The heat generated due to burning elevates 7 cm top soil temperature up to 40°C, thereby declining of microbial population in the soil.

Management of Crop Residues in Conservation Agriculture

Crop residues varies from crop to crop with pulses having greater RPR Ratio (residue to Final Economic Produce Ratio) followed by coffee, tea, and cereals crops like maize, jowar, bajra, wheat and rice. Out of the total nutrient uptake, residue content in paddy straw is 25% Nitrogen (N), Phosphorus (P), 50% Sulphur (S) and 75% Potassium (K). The role of crop residues on carbon sequestration in soils would be an added advantage in relation to climate change and GHGs mitigation. Hence, efficient management of crop residues would help in enhancing soil physical and chemical properties along with checking emergence of weeds. The following are some of the management practices of crop residues:

Improved Economics and System Productivity: Residue Recycling through Sequential cropping and Intercropping helps in improving both productivity and profitability. Legume residues when added to intercropped system and sole cropping help in nitrogen fixation and reduce evaporation losses. In-situ management of crop residue can also be facilitated by happy seeder which can increase 29% nitrogen uptake and 37% enhanced yield of 37%, compared to burning. In addition to this, residues when levelled by rotavator and chopped through chopper enhances the content of soil nutrients and less weed infestations. Supplying machineries for conservation agriculture on subsidized rates, custom hiring systems and soft loans for purchase of implements are already in place by many State Governments to encourage use of crop residue. About 7-8 acres of rice straw residue can be managed for sowing of wheat seed per day by 1 happy seeder (Jat, Kapil, Kamboj, *et al.*, 2013). Dual purpose legumes and leuceana leaf mulching were found to be beneficial to increase the systems productivity and soil fertility on long term in wheat-maize and green gram-mustard-cowpea cropping system. In-situ management of rice straw could improve the wheat crop yield by 2-10% (Kumar *et al.*, 2015 and NAAS 2017). Cover crops, when taken into crop rotation also helps in nutrient recycling. However, development of region-specific crop residues inventories including total production from different crops, their quality, utilization and amount burnt on-farm, for evolving management strategies should be taken up.

Stability of Soil Temperature: Residue retention on soil buffers diurnal fluctuations in soil temperature- keep soils cool in summer and warm in winter season. As there is reduced direct heating of soil surface, extreme fluctuations in soil temperature and volatility in nitrogen is minimized, thus favorable micro-climate is created. This would be added advantage for the small farmer, who have constraint of using high level of synthetic fibre. Introducing C-credit schemes for the farmers can be introduced, who follow conservation agriculture for carbon sequestration and GHGs mitigation by use of crop residue.

Moisture Conservation: Retention of crop residues help in reducing the irrigation requirement of the crop like elimination of pre-sowing irrigation and lesser irrigation requirement due to reduced evaporation losses. Experiments recorded maximum consumptive use of water (426.9 and 302.4 m) was recorded with FYM @ 5t/ha +dust mulch + straw mulch closely followed by Kaolin + straw mulch and dust mulch + straw mulch and minimum water consumptive use with no mulch in pearl millet. Chakraborty *et al.*, (2008) reported water savings of 3-11% and improved water use efficiency of 25% in wheat crop due retention of rice straw as mulch. Crop Conservation tillage with residue incorporated showed aggregate stability (MWD rang: 0.51-0.83 mm, WSMA range: 1.2-62.8 %) as compared to other tillage treatments.

Livestock Feed & Mushroom Cultivation: Crop residues are low-density fibrous materials. With high silica content, rice straw is termed as poor feed for animals. To meet the nutritional requirements of animals, the residues need processing and enriching with urea and molasses, and supplementing with green fodders (leguminous/non-leguminous) and legume (sunhemp, horse gram, cowpea, gram) straws. In order to make rice straw more nutritive for feed, the crop should be harvested/cut close to the ground as possible as stems with less silica content are more digestible than leaves. The inedible crop residues can further serve as good excellent substrates for the cultivation of both whit button mushroom and straw mushroom). These lignocellulosic agricultural crop residues like wheat, rice straw, corn cobs, cotton stalks, coconut and banana residues, corn husks, maize and sorghum stover can enhance the additional farmers income for mushroom cultivation. Composting can also be derived from crop residue by pit composting where rice straw, cow dung and water is combined and left buried in the pit for 4-6 months. Temperature is however an important factor during composting because higher temperatures results in destruction of the pathogens and disinfection the organic matter.

Production of Bio-oil: Bio-oil can be produced from crop residues by the process of fast pyrolysis, which requires temperature of biomass to be raised to 400-500^{0}C within a few seconds, resulting in a remarkable change in the thermal disintegration process. The bio-oil produced from corn sticks contain high present of heavy constituents C20-C38 and high percent of volatile bio-oil T C6-C9. Biomass gasifier-based systems converts solid biomass into the more user-friendly gaseous form, which can be used directly in IC engines to generate power. For producing 1 kWh of electricity (using 100% producer gas engine), about 1.2 to 1.4 kg of biomass is required. If operated on high capacities and treated well, these producer gas can also run cold storages, which is the basic requirement for famers at village level.

Packaging: Crop residue being natural material can be used in packaging as a cushion like packing of crockery, glass, electronic goods. Companies like Bio-lutions India uses a variety of fibrous plant residue from sugarcane, pineapple leaves, mulberry and areca to make biodegradable packaging products on large scale at reduced cost. (NewsMinute.com).

Constraints of using Crop Residues with Conservation Agriculture

i. Initial time and budget involvement for developing crop residues management policy for each state defining clearly various competing uses.

ii. Difficulties in sowing and application of fertilizer, pesticides and problem of pest infestation with increased pest and weed problems during the transition period.

iii. Lack of Subsidies for buying machineries for residue management

iv. Requires more attention on timings and placement of nutrients, pesticides and irrigation

v. Nutrient management becomes complex because of higher residues levels and lack of adequate know-how on reduce to product ratio of different ratios.

vi. Lack of destinations for supply of raw material and slow-paced transportation in rural areas where transportation cost of residue as raw material is more than profit earned.

vii. Residue retention on soil surface is the integral component of CA systems, which may not be feasible under most Indian conditions. In some areas, where the soils are susceptible to compaction, absence of soil cover under no-till may result in increased weed growth and depressed yields

viii. Farmers have strong preferences for clean and good-looking fields as against untilled shabby looking fields and apprehensive of lower crop yields and /or economic returns

Conclusion

Lack of proper management of crop residue not only results in loss of resources but also an untrodden opportunity to augment farmer's income. The high production levels have to be not only sustained but also achieved with emphasis on energy savings and low emission technologies considering climate change impacts. Development of proper low-cost crop residue management strategies should be based on the crops taken along with enhancing factors that will accelerate their decomposition. Conservation Agriculture is a new paradigm by achieving sustained agricultural production and is a major step in the transition to sustainable agriculture (Farooq and Siddique, 2015). For ensuring the country's food security both in short and long-term perspectives and making agriculture sustainable, the soil resource base must be strong and healthy. The prospect for crop residue utilization in nonconventional ways is limited. Through many policies like National Policy for Management of Crop Residue (NPMCR) are introduce to prevent agricultural residue burning and promote technologies for optimum utilization and in situ management of crop residues, still a hyper-local focused approach should be supplemented swiftly as one size doesn't fit all. Not only agricultural sector but sectoral linkages with environment, energy and economy can be approached for a wide-awake initiative. The practice of residue burning, or removal doesn't find suitable for environment as well as soil fertility and productivity thus, there is a need to fine tune nutrient management for continuous incorporation along with providing impetus and providing prioritization to the potential of individual industries.

References

Bimbraw, A.S. (2019). Generation and Impact of Crop Residue and its Management. *Current Agricultural Research,* 7(3). doi : http://dx.doi.org/10.12944/CARJ.7.3.05

Chakraborty, D., Nagarajan, S., Aggarwal, P., Gupta, V.K., Tomar, R.K., Garg, R.N., Sahoo, R.N., Sarkar, A., Chopra, U.K., Sarma, K.S.S. and Kalra, N. (2008). Effect of mulching on soil and plant water status, and the growth and yield of wheat (*Triticum aestivum* L.) in a semi-arid environment. *Agricultural Water Management,* 95: 1323–1334

Conservation Technology Information Center (CTIC) (1992).1992 National Survey of Conservational Tillage Practices. CTIC, West Lafayette, IN.

FAO. (2012). Food and Agriculture Organization of the United Nations, 2012. Available online at http://www.fao.org/ag/ca/6c.html.

Gangwar, K.S., Singh, K.K., Sharma, S.K. and Tomar, O.K. (2006). Alternative tillage and crop residue management in wheat after rice in sandy loam soils of Indo-Gangetic plains. *Soil and Tillage Research,* 88(1-2): 242-252.

Jat, M. L., Malik, R.K., Saharawat, Y.S., Gupta, R., Bhag, M., & Raj Paroda. (2012). Proceedings of Regional Dialogue on Conservation Agricultural in South Asia, New Delhi, India, APAARI, CIMMYT, ICAR, p. 32.

Kumar *et al.,* (2015). Socioeconomic and Environmental Implications of Agricultural Residue Burning: A Case Study of Punjab, India. Springer Briefs in Environmental Science. ISBN 978-81-322-2146-3. DOI 10.1007/978-81-322-2014-5. 2015.

NAAS (2017). National Academy of Agricultural Sciences (NAAS). Innovative Viable Solution to Rice Residue Burning in Rice-Wheat Cropping System through Concurrent Use of Super Straw Management System-fitted Combines and Turbo Happy Seeder.

NewMinute: https://www.thenewsminute.com/article/no-plastics-bengaluru-company-makes-packaging-products-crop-residue-80664.

Pathak, H, Saharawat, Y.S., Gathala, M., and Ladha, J.K. (2011). Impact of resource-conserving technologies in the rice-wheat system. *Greenhouse Gas Science & Technology,* 1:261–277.

Singh, S.K., Kumar, D., Lal, S.S. (2010) Integrated use of crop residues and fertilizers for sustainability of potato (Solanum tuberosum) based cropping systems in Bihar. *Indian Journal of Agronomy,* 55 (3): 203-208.

TERI (2019) Development of Spatially Resolved Air Pollution Emission Inventory of India. New Delhi: The Energy and Resources Institute

10

Agroecological Dynamics and Strategic Cultural Management of Weeds in Conservation Agriculture

S. Sarkar and B. Majumdar

ICAR-Central Research Institute for Jute and Allied Fibres, Barrackpore West Bengal - 700 121, India

Introduction

Generally speaking, Conservation Agriculture (CA) includes minimum soil disturbance, permanent soil cover, and crop rotation (Hobbs, 2007). CA has been recognized as an efficient way for sustainably increasing crop yields in different countries (Hobbs *et al.*, 2008; Pittelkow *et al.*, 2015). Farmers following CA will face several managerial problems, and weed management is the most challenging (Wall, 2007; Giller *et al.*, 2009; Farooq *et al.*, 2011). Weed growth in an agroecological system and crop yields are inversely related; so, knowledge on weed dynamics and weed management are vital for achieving the possible yield gains in CA systems. There are few studies that examined both the direct and interactive effects of the 3 basic CA principles on weed dynamics (Chauhan *et al.*, 2012; Giller *et al.*, 2009; Farooq *et al.*, 2011).

For comprehensive understanding the current topic, a diagrammatic representation of an annual weed's life cycle (Fig. 1) will be useful and the same corresponds to four likely areas managing the weeds:

(i) Possibility of seed dormancy and loss of viable weed seeds in and on the soil.

(ii) Reducing/ altering weed seedling establishment.

(iii) Diminishing seed production by weeds.

(iv) Preventing weed seed dispersal.

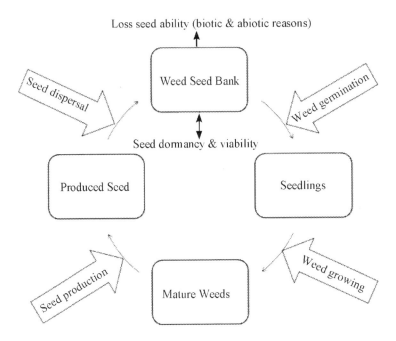

Fig. 1: Diagrammatic representation of transitions in annual weeds lifecycle

Lifecycle of biennial and perennial weeds differ from the annual weeds' lifecycle. But the basic principle still works. It may be noted that the effect of weeds is not always bad, in some situation weeds are useful for contributing ground cover, nutrient stabilization, food sources, and sometimes (<threshold population) do not affect crop yield.

Factors Affecting Weed Dynamics in Cnservation Agriculture

Tillage practice

In this article it was referred to any tillage practice where the top soil is ploughed/ harrowed at least once during the crop growing season as conventional tillage (CT). Distinction is noted between zero soil disturbance (no-till; NT) and minimum or reduced tillage. The tillage effect on the quantum of the weed seedbank depends on several factors (Mohler, 1993). So, studies suggested contradicting results, such as tillage has no effect (Barberi *et al.*, 2001), reduces (Murphy *et al.*, 2006), or increases (Sosnoskie *et al.*, 2006) on weed seedbank. Several studies showed that the response to tillage on weed seedbank depends on the individual weed species (Farooq *et al.*, 2011), and Mohler (1993) noted that weed response to tillage involves a complex interaction between weather, duration of study, and long-term field history factors. Weed species of competition with economic crop depends on the vertical distribution of seeds in the seedbank. In NT system, weed seeds penetrate the soil through cracks, slowly through freeze-dry cycles, resulting in an accumulation of weed seeds (60-90%) in the top 5 cm of the soil (Hoffman *et al.*, 1998). Whereas, conventional tillage encourages generalized patterns of seed distributions (Fig. 2). Therefore, tillage-induced changes in seed distribution affect germination and seedling establishment indirectly in weed species. Better germination is observed in tilled soils for most weed species. Tillage itself provides germination stimulus by breaking dormancy for weeds requiring light flashes, scarification, fluctuating temperatures, ambient CO_2 concentrations, and/or higher -NO_3concentrations (Benech-Arnold *et al.*, 2000). In normal situation, weed germination is generally higher at the top soil layer and declines with increasing soil depth. So, a higher proportion of NT soil seedbanks germinate as compared to the CT seedbanks because NT seedbanks are placed in the top soil layer (Fig. 2a) (Gallandt *et al.*, 2004). Soil living and other field insects, rodents, and birds consume a significant amount of weed seed (along with fallen grains) in fields, and therefore, act as valuable means by reducing the chance of entering soil seedbank (Jacob *et al.*, 2006; Chauhan *et al.*, 2010). Generally surface accumulation of seeds in NT system (Fig. 2a) improves access to the predators for weed seeds and therefore, increase the seed removal rates. Weed seed viability is affected by the changes in seed distribution altered by tillage conditions (Fig. 2). Seeds on the soil surface sometimes lose viability due to desiccation and harsh weather (Anderson, 2005). Therefore, based on the severity of the environmental condition, the accumulation of weed seeds on un-tilled soil surfaces might increase the non-viable weed seed count in the soil seedbank. Tillage produces lower resistance to weed seedling penetration, which affects the growth of germinated weed seeds (Verhulst *et al.*, 2010). Properly tilled soils also permit seedlings to appear from deeper soil layer compared to un-tilled soils (Chhokar *et al.*, 2007; Franke *et al.*, 2007). Seedling establishment is also affected by tillage practice. If seeds are located directly on the surface of un-tilled soils, the radicle of germinated weed seeds have difficultly penetrating NT soil surfaces, resulting in fatal sprouting (Mohler, 2001[b]).

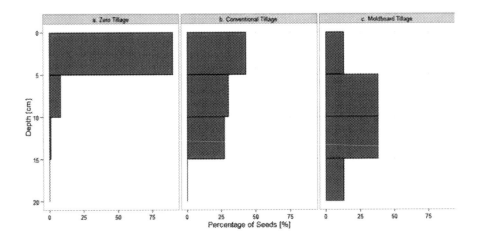

Fig. 2: Weed Seed distribution in different tillage in controlled model experiments

Mohler (1993) through model shown how seed characteristics influenced seedling emergence under different tillage systems. It was found that the number of emerged seedlings in the first year of NT was higher than other tillage systems assuming that a seed sprinkling in year zero followed by tillage. However, with the assumption of no new seed input in the second year, in following years tilled soils would have more emerged seedlings compared to NT soils starting in year two. Additionally, again assuming no seedbank replenishment, over time the NT systems exhausted seedbanks more rapidly (Fig. 3). It is considered that conventional tillage system offers mechanical weed control and is a useful tool for controlling established weeds. Whereas, in general, once a weed is established in NT fields, the only options for termination before seed-set are herbicides and hand weeding. Seed dispersal are also affected by different tillage practices. Field machinery operations such as tilling provide opportunities to introduce or spread weed seeds (Buhler *et al.*, 1997). One study showed cultivation (after ploughing) following harvest significantly increased weed seed dispersal (Heijting *et al.*, 2009), and another found that the weed seeds (*Avena* sp.) travelled 2-3 m in the direction of tillage, while in un-tilled soils the distance was negligible (Barroso *et al.*, 2006). Therefore, reducing tillage can decrease the spread of weed seeds.

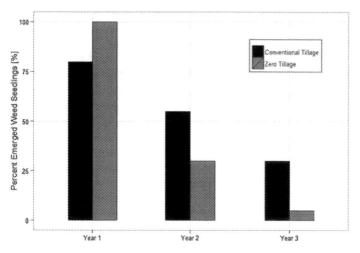

Fig. 3: Effect of tillage on number of emerged weed seedlings over time having no weed seed addition

Because NT fields are similar to undisturbed (not ploughed) roadside conditions and perennial grasslands, weeds may spread more easily from such environments (Tuesca *et al.*, 2001). Often volunteer crops are problematic in NT systems, although this depends on the crop rotation followed (Derksen *et al.*, 1994). Higher weed species diversity has been observed in NT seedbanks as compared to the weed communities in tilled soils (Sosnoskie *et al.*, 2006). The commonly considered itea that NT systems supports perennial weeds might work for some cases, but not always true. Ecological succession theory suggests perennial weeds dominate in the undisturbed/ untilled systems. Indeed, high disturbance environments such as CT systems have been shown to favour annual broadleaved weeds, while NT systems favour perennial weeds and those annual grasses which easily germinate on the soil surface (Tuesca *et al.*, 2001; Taa *et al.*, 2004). Dorado *et al.*, (1999) reported that in reduced tillage, such weeds grow early whose lifecycles and resource requirement is similar to the crops grown, and not according to the annual or perennial classification. Blackshaw *et al.*, (2001) reported that changing to NT in rotations with two or more crops (winter wheat cropping system) not increased the perennial weeds. Whereas, in jute-based cropping system under CA, the higher density of broad-leaved weeds was recorded in CT compared to NT, however, it was at par with no tillage with additional crop residue (NT+R). *Cynodon dactylon*, a perennial grass density was significantly higher in NT system. *Cyperus rotundus* density did not vary significantly among different tillage systems (Saha *et al.*, 2018).

Crop residues

In either CT or NT systems, crop residue (CR) may be kept in the field as CT + CR and NT + CR. In CT + CR the residue is incorporated into the soil. Although incorporated crop residue may affect weeds through modifications of nutrient dynamics; the effects will be highly dependent on the type of tillage, the C:N ratio of the incorporated residue, the type of soil, and the environment (Liebman and Mohler, 2001). Crop residues placed on the soil surface affects weed seed germination through changes in the seed physical and chemical environment. Reduction in light and soil surface insulation are the two main physical effects. Soil surface insulation affects on both soil temperature and soil moisture. It was reported that even under heavy crop residue loads, most weed seeds lying on the soil surface receive sufficient light to trigger seed germination (Teasdale and Mohler, 1993). In residue retention condition, less weed seed germination may not be due to insufficient availability of light. Surface residue decreases the daily maximum soil temperature, but has little effect on the daily minimum temperature (Teasdale and Mohler, 1993). Most agronomic crops and many weeds require soil temperatures above a certain threshold in order to germinate—lower average soil temperatures would therefore, delay germination of both. If soil moisture is in sufficient quantity, the weed seed germination is not affected due to crop residue, but in water-limited environments crop residue may promote weed seed germination (Wicks et al., 1994; Vidal and Bauman, 1996). It was reported that residue was not effective in suppressing weeds in drier field condition or in relatively drier crop year (Buhler et al., 1996; Mashingaidze et al., 2012; Ngwira et al., 2014). The bio-chemical environment of the weed seed is changed through allelopathy, when surface residue is allowed to retain. Allelopathic effects from crop residue tend to have more pronounced effects on small weed seeds (Liebman and Davis, 2000). Wheat and rice residue showed allelopathy which may be used for weed management (Wu et al., 2001; Khanh et al., 2007). Crutchfield et al. (1986) reported that at least 3.4 t/ha wheat straw was needed in order to significantly reduce weed biomass in a wheat-maize rotation study, whereas, in maize monocrop system in Zimbabwe, 6 t/ha of maize stover was required (Ngwira et al., 2014). In CA systems it is advised to keep at least 30% of the ground covered; while the same quantum of residue will give soil quality benefits, but not reduce weed pressure (Liebman and Mohler, 2001). In some specific situation, crop residues can indirectly affect weed growth (by reducing light, physical hindrance and allelopathy) and reduce weed seed production (Franke et al., 2007).

Crop rotation

Possibly the most effective method to manage weeds in CA system is crop rotations. Every crop puts some unique biotic and abiotic constraints on the weed germination and growth. Some crops promote the growth of some specific weeds, while preventing growth of other type of weeds. Rotating crops in CA will rotate the selection pressures, preventing one specific weed species from being repeatedly successful, and thus restricting establishment of the same. Growing crops in rotations, with different planting dates is the key factor in reducing the size of the weed seedbank, because it changes the timing of field activities. Studies showed decreased weed seed densities in crop rotations having different planting dated, while rotations with identical planting dates do not (Barberi *et al.*, 2001). Inclusion of crops with allelopathic effects in a rotation also decrease weed seed germination. The maize in the 3-year rotation had lower weed densities compared to the maize-soybean rotation, resulting less seedbank germination (Schreiber, 1992). Prevention of weed seed-set is the most powerful mechanisms of weed control offered by rotations. Weeds can be successful in reproducing when they mimic the crop life cycle (Liebman and Dyck, 1993; Moyer *et al.*, 1994). Weed growth that is poorly synchronized with the crop grown is easy to control. Weeds growing in the temporary fallow situation and without a crop may be controlled using non-selective herbicide, and weeds in their vegetative stage during crop harvest are automatically removed before they could set seed.

Cultural Practices for Weed Management in CA

In CA system (without tillage), weed control wholly depends on herbicide and other agronomical/ cultural manipulations. In cultural practices for weed management under CA include - (i) adjusting the crop planting date, (ii) modifications of planting density, and/or spatial arrangement, (iii) resource management, (iv) cultivar selection, (v) preventing weed seed introduction, and (vi) planned and purposeful crop rotations.

Adjusting the sowing/ planting date of crops

Many weed seeds germinate during specific seasons only because of inherent seed dormancy. Knowing the approximate date of germination of weeds, crop planting dates can be adjusted accordingly, so that either (i) the crop can be emerged before the weed germination for a competitive advantage or (ii) weeds are allowed to germinate and are killed before crop planting. Planting earlier by even a few days can give the crop a significant competitive advantage over weeds (Mohler, 2001[a]). Weed suppression in case of early crop planting was demonstrated in case of *Phalaris minor* in rice-wheat systems in India. Application of NT technology allowed wheat crops to be sown 1-2 weeks earlier, favoured the wheat to establish before emergence of *Phalaris minor* (Chhokar and Malik, 1999; Chauhan *et al.*, 2012).

Adjusting the crop density and spatial arrangement of the crop

Crop density increase hastens resource use by the crop compared to weeds, and may be useful in such condition. Increased crop density produces more biomass, which often may not significantly produce higher yields. However, in suffice soil moisture conditions, utilizing crop densities higher than 150 plants/m^2 in wheat (Lemerle *et al.*, 2004), 4 plants/m^2 in maize (Tollenaar *et al.*, 1994), and 100 plants/m^2 in rice (Zhao *et al.*, 2007) had proved to lower weed densities and increase crop yields. It was reported that reducing crop row spacing reduces weeds, although it does not always increase yields (Mohler, 2001[a]). However, other reports showed that increased planting density significantly reduced weed biomass and raised yields in number of crops (Weiner *et al.*, 2001; Olsen *et al.*, 2012; Marin and Weiner, 2014). It was found that directional orientation of crops can provide more sunlight to the crop, and thus restricting the quantum of light available for weed growth. In general, north-south orientation of crops get such benefit for most latitudes (Mohler, 2001[a]).

Resource management

In soils with low native levels of fertility, band placement of fertilizers reduced weed biomass compared to broadcasting of fertilizer; however, sub-surface band placement was more effective than surface banding (Liebman and Mohler, 2001; Derksen *et al.*, 2002). In irrigated environments, spatial and temporal variation of soil moisture offers opportunities for weed control. When the top layer of soil is dry, planting large-seeded crops into deep can provide crops with an initial advantage over weeds (Liebman and Mohler, 2001). Another option under these conditions is to apply irrigation for quick germination of weeds, then killing them using non-selective herbicides, afterwards crops can be raised on non-weedy situation (Stale seedbed) (Shaw, 1996; Chauhan *et al.*, 2012; Mulvaney *et al.*, 2014).

Cultivar selection

Specific varieties of crops showed more competitiveness with weeds than other varieties of rice (Zhao *et al.*, 2006) and maize (Marin and Weiner, 2014). Scientists opined that development of CA- specific cultivars for higher weed competitiveness in such condition is a desirable area of research (Mahajan and Chauhan, 2013). Breeding programs to find and use competitive ability characteristics under CA is challenging because of complexity of characteristics and greater spatial and temporal variations. It was emphasized that development of such varieties would be highly beneficial not only for weed control, but for other CA-specific characteristics also (Herrera *et al.*, 2013).

Restriction of weed seed introduction

Often weed seeds are introduced to agricultural fields through applied manure, crop seed, and even irrigation water (Dastgheib, 1989; Kelley and Bruns, 1975). Use of clean crop seed, discarding and disposing contaminated crop seed, and filtering irrigation water are effective ways for reducing such types of weed seed introduction in CA as well as conventional crop production system.

Planned and purposeful crop rotations

Planned crop rotations for 4 years or more reduced herbicide application use in both tilled and un-tilled systems (Anderson, 2015; Liebman *et al.*, 2008). Perennial alfalfa in a rotation had contributed weed control for 3 years, and was effective in NT systems (Ominski *et al.*, 1999; Ominski and Entz, 2001).

General considerations

It is widely accepted that many weed control methods are ineffective if used solely, but when used in amalgamated form could able to reduce weeds. One negative point in CA system, as often mentioned, that it increases dependency on herbicides as compared to conventional tilled agro-systems. In CA system, glyphosate (non-selective herbicide) may be heavily used, especially to manage perennial weeds (Moyer *et al.*, 1994). It was reported from Canada that adoption of NT has not increased herbicide use significantly (Derksen *et al.*, 1996), and in USA, wheat growing under NT systems showed weeds under control using cultural strategies and reduced herbicide usage by 50% compared to CT (Anderson, 2005).

Conclusion

Crop rotation and retaining surface residue in crop fields, are two important factors for weed management in conservation agriculture system. All 3 principles of CA together provide more advantages, but if only one CA principle is used, it may be more problematic, as far as weed management is concerned. With special reference to weed control, NT should not be used in monoculture systems. More options for weed management in CA systems are - selecting new varieties with more weed competitiveness; modifying planting dates of crop, planting densities, row-spacing, fertilizer placement etc (Nichols *et al.*, 2015). Moreover, research aiming practical field solutions are needed where interactions between different components (especially tillage and residue retention) of CA practices should get emphasis for better weed management under CA systems.

152 Conservation Agriculture and Climate Change

References

Anderson, R.L. (2005). A multi-tactic approach to manage weed population dynamics in crop rotations. *Agronomy Journal*, 97: 1579-1583.

Anderson, R.L. (2015). Integrating a complex rotation with no-till improves weed management in organic farming. A review. *Agronomy for Sustainable Development*, 35(3): 967-974.

Barberi, P., Bonari, E., Mazzoncini, M., Garda-Torres, L., Benites, J., Martinez-Vilela, A. (2001). Weed density and composition in winter wheat as influenced by tillage systems. Conservation agriculture, a worldwide challenge. In: *Proceedings of the First World Congress on Conservation Agriculture*, Madrid, Spain, 1-5 October, pp. 451-455.

Barroso, J., Navarrete, L., Sanchez del Arco, M., Fernandez-Qunitanilla, C., Lutman, P., Perry, N., Hull, R. (2006). Dispersal of *Avena fatua* and *Avena sterilis* patches by natural dissemination, soil tillage and combine harvesters. *Weed Resesearch*, 46: 118-128.

Benech-Arnold, R.L., Sanchez, R.A., Forcella, F., Kruk, B.C., Ghersa, C.M. (2000). Environmental control of dormancy in weed seed banks in soil. *Field Crops Research*, 67: 105-122.

Blackshaw, R., Larney, F., Lindwall, C., Watson, P., Derksen, D. (2001). Tillage intensity and crop rotation affect weed community dynamics in a winter wheat cropping system. *Canadian Journal of Plant Science*, 81: 805-813.

Buhler, D., Mester, T. and Kohler, K. (1996). The effect of maize residue and tillage on emergence of *Setariafaberi*, *Abutilon theophrasti*, *Amaranthus retroflexus* and *Chenopodium album*. *Weed Research*, 36: 153-165.

Buhler, D.D., Hartzler, R.G., Forcella, F. (1997). Implications of weed seedbank dynamics to weed management. *Weed Science*, 45: 329-336.

Chauhan, B., Migo, T., Westerman, P., Johnson, D. (2010). Post-dispersal predation of weed seeds in rice fields. *Weed Research*, 50: 553-560.

Chauhan, B.S., Singh, R.G., Mahajan, G. (2012). Ecology and management of weeds under conservation agriculture: a review. *Crop Protection*, 38: 57-65.

Chhokar, R., Malik, R., Balyan, R. (1999). Effect of moisture stress and seeding depth on germination of little seed canary grass (*Phalaris minor* Retz). *Indian Journal of Weed Science*, 31: 78-79.

Chhokar, R., Sharma, R., Jat, G., Pundir, A., Gathala, M. (2007). Effect of tillage and herbicides on weeds and productivity of wheat under rice-wheat growing system. *Crop Protection*, 26: 1689-1696.

Chhokar, R.S., Malik, R.K. (1999). Effect of temperature on germination of *Phalaris minor* Retz. *Indian Journal of Weed Science*, 31: 73-74.

Crutchfield, D.A., Wicks, G.A., Burnside, O.C. (1986). Effect of winter wheat (*Triticum aestivum*) straw mulch level on weed control. *Weed Science*, 34: 110-114.

Dastgheib, F. (1989). Relative importance of crop seed, manure and irrigation water as sources of weed infestation. *Weed Research*, 29: 113-116.

Dekersen, D.A., Thomas, G.P., Lafond, G.P., Loeppky, H.A. and Swanton, C.J. (1994). Impact of agronomic practices on weed communities fallow within tillage systems. *Weed Science*, 42: 184-194.

Derksen, D., Blackshaw, R., Boyetchko, S. (1996). Sustainability, conservation tillage and weeds in Canada. *Canadian Journal of Plant Science*, 76: 651-659.

Derksen, D.A., Anderson, R.L., Blackshaw, R.E., Maxwell, B. (2002). Weed dynamics and management strategies for cropping systems in the northern Great Plains. *Agronomy Journal*, 94:174-185.

Dorado,J., Del Monte,J., Lopez-Fando, C. (1999). Weed seedbank response to crop rotation and tillage in semi arid agroecosystems. *Weed Science*, 47: 67-73.

Farooq, M., Flower, K., Jabran, K., Wahid, A., Siddique, K.H. (2011). Crop yield and weed management in rainfed conservation agriculture. *Soil and Tillage Research,* 117: 172-183.

Franke, A., Singh, S., McRoberts, N., Nehra, A., Godara, S., Malik, R., Marshall, G. (2007). *Phalaris minor* seedbank studies: longevity, seedling emergence and seed production as affected by tillage regime. *Weed Research,* 47: 73-83.

Gallandt, E.R., Fuerst, E.P., Kennedy, A.C. (2004). Effect of tillage, fungicide seed treatment, and soil fumigation on seed bank dynamics of wild oat (*Avena fatua*). *Weed Science,* 52: 597-604.

Giller, K.E., Witter, E., Corbeels, M., Tittonell, P. (2009). Conservation agriculture and smallholder farming in Africa: the heretics' view. *Field Crops Research,* 114: 23-34.

Heijting, S., Van der Werf, W., Kropff, M. (2009). Seed dispersal by forage harvester and rigid-tine cultivator in maize. *Weed Research,* 49: 153-163.

Herrera,J., Verhulst, N., Trethowan, R., Stamp, P., Govaerts, B. (2013). Insights into genotype x tillage interaction effects on the grain yield of wheat and maize. *Crop Science,* 53: 1845-1859.

Hobbs, P.R. (2007). Conservation agriculture: what is it and why is it important for future sustainable food production? *Journal of Agricultural Science,* 145: 127.

Hobbs, P.R., Sayre, K., Gupta, R. (2008). The role of conservation agriculture in sustainable agriculture. *Philosophical Transactions of the Royal Society B: Biological Sciences,* 363: 543-555.

Hoffman, M.L., Owen, M.D., Buhler, D.D. (1998). Effects of crop and weed management on density and vertical distribution of weed seeds in soil. *Agronomy Journal,* 90: 793-799.

Jacob, H.S., Minkey, D.M., Gallagher, R.S., Borger, C.P. (2006). Variation in Post dispersal weed seed predation in a crop field. *Weed Science,* 54: 148-155.

Kelley, A., Bruns, V. (1975). Dissemination of weed seeds by irrigation water. *Weed Science,* 23: 486-493.

Khanh, T., Xuan, T., Chung, I. (2007). Rice allelopathy and the possibility for weed management. *Annals of Applied Biology,* 151: 325-339.

Lemerle, D., Cousens, R., Gill, G.S., Peltzer, S., Moerkerk, M., Murphy, C., Collins, D., Cullis, B.R. (2004). Reliability of higher seeding rates of wheat for increased competitiveness with weeds in low rainfall environments. *Journal of Agricultural Science,* 142: 395-409.

Liebman, M., Davis, A., (2000). Integration of soil, crop and weed management in low-external-input farming systems. *Weed Research,* 40: 27-48.

Liebman, M., Dyck, E. (1993). Crop rotation and intercropping strategies for weed management. *Ecological Applications,* 3 (1): 92-122.

Liebman, M., Gibson, L.R., Sundberg, D.N., Heggenstaller, A.H., Westerman, P.R., Chase, C.A., Hartzler, R.G., Menalled, F.D., Davis, A.S., Dixon, P.M. (2008). Agronomic and economic performance characteristics of conventional and low-external-input cropping systems in the central corn belt. *Agronomy Journal,* 100: 600-610.

Liebman, M., Mohler, C.L. (2001). Weeds and the soil environment. *In* Liberman, M., Mohler, C.L. and Staver, C.P. (eds.). Ecological Management of Agricultural Weeds. Cambridge University Press, UK, pp. 210-268.

Mahajan, G., Chauhan, B.S. (2013). The role of cultivars in managing weed in dry-seeded rice production systems. *Crop Protection,* 49: 52-57.

Marin, C., Weiner, J., (2014). Effects of density and sowing pattern on weed suppression and grain yield in three varieties of maize under high weed pressure. *Weed Research,* 54: 467-474.

Mashigaidze, N., Madakadze, C., Twomlow, S., Nyamangara, J. and Hove, L. (2012). Crop yield and weed growth under conservation agriculture in semi-arid Zimbabwe. *Soil and Tillage Research,* 124: 102-110.

Mohler, C.L. (2001[a]). Enhancing the competitive ability of crops. *In* Liberman, M., Mohler, C.L. and Staver, C.P. (eds.). Ecological Management of Agricultural Weeds. Cambridge University Press, UK, pp. 269-321.

Mohler, C.L. (2001[b]). Mechanical management of weeds. *In* Liberman, M., Mohler, C.L. and Staver, C.P. (eds.). Ecological Management of Agricultural Weeds. Cambridge University Press, UK, pp. 139-209.

Mohler, C.L., (1993). A model of the effects of tillage on emergence of weed seedlings. *Ecological Applications*, 3(1): 53-73.

Moyer, J., Roman, E., Lindwall, C., Blackshaw, R. (1994). Weed management in conservation tillage systems for wheat production in North and South America. *Crop Protection*, 13: 243-259.

Mulvaney, M., Verhulst, N., Herrera,J., Mezzalama, M., Govaerts, B. (2014). Improved wheat performance with seed treatments under dry sowing on permanent raised beds. *Field Crops Research*, 164:189-198.

Murphy, S.D., Clements, D.R., Belaoussoff, S., Kevan, P.G., Swanton, C.J. (2006). Promotion of weed species diversity and reduction of weed seedbanks with conservation tillage and crop rotation. *Weed Science*, 54: 69-77.

Ngwira, A., Aune,J. B., Thierfelder, C., (2014). On-farm evaluation of the effects of the principles and components of conservation agriculture on maize yield and weed biomass in Malawi. *Experimental Agriculture*, 50: 591-610.

Nichols, V., Verhulst., Cox, R. and Govaerts. (2015). Weed dynamics and conservation agriculture principles. *Field Crops Research*, 183: 56-68.

Olsen, J.M., Griepentrog, H.-W., Nielsen, J., Weiner, J. (2012). How important are crop spatial pattern and density for weed suppression by spring wheat? *Weed Science*, 60: 501-509.

Ominski, P., Entz, M., (2001). Eliminating soil disturbance reduces post-alfalfa summer annual weed populations. *Canadian Journal of Plant Science*, 81: 881-884.

Ominski, P., Entz, M., Kenkel, N. (1999). Weed suppression by *Medicago sativa* in subsequent cereal crops: a comparative survey. *Weed Science*, 47: 282-290.

Pittelkow, C.M., Liang, X., Linquist, B.A., Van Groenigen, K.J., Lee, J., Lundy, M.E., van Gestel, N., Six,J., Venterea, R.T., van Kessel, C. (2015). Productivity limits and potentials of the principles of conservation agriculture. *Nature*, 517: 365-368.

Saha, R., Kumar, M., Behera, M.S., Sharma, L., Majumdar, B., Saha, A.R., Barman, D., Mazumdar, S.P., Naik, R.K. and Kundu, D.K. (2018). Conservation agricultural practices of jute-based cropping systems under climate change scenario. *Annual Report 2018-19*. ICAR- CRIJAF, Barrackpore, Kolkata. P. 124.

Schreiber, M.M. (1992). Influence of tillage, crop rotation, and weed management on giant foxtail (*Setaria faberi*) population dynamics and corn yield. *Weed Science*, 40: 645-653.

Shaw, D.R, (1996). Development of stale seedbed weed control programme for southern row crop. *Weed Science*, 44: 413-416.

Sosnoskie, L.M., Herms, C.P., Cardina, J. (2006). Weed seedbank community composition in a 35-yr-old tillage and rotation experiment. *Weed Science*, 54: 263-273.

Taa, A., Tanner, D., Bennie, A.T. (2004). Effects of stubble management, tillage and cropping sequence on wheat production in the south-eastern highlands of Ethiopia. *Soil and Tillage Research*, 76: 69-82.

Teasdale, J.R., Mohler, C. (1993). Light transmittance, soil temperature, and soil moisture under residue of hairy vetch and rye. *Agronomy Journal*, 85: 673-680.

Tollenaar, M., Dibo, A., Aguilara, A., Weise, S., Swanton, C. (1994). Effect of crop density on weed interference in maize. *Agronomy Journal*, 86: 591-595.

Tuesca, D., Puricelli, E., Papa, J. (2001). A long-term study of weed flora shifts in different tillage systems. *Weed Research*, 41: 369-382.

Verhulst, N., Govaerts, B., Verachtert, E., Castellanos-Navarrete, A., Mezzalama, M., Wall, P., Chocobar, A., Deckers,J., Sayre, K. (2010). Conservation agriculture, improving soil quality for sustainable production systems. In: Lal, R., Stewart, B.A. (Eds.), *Advances in Soil Science: Food Security and Soil Quality*., pp. 137-208.

Vidal, R.A., Bauman, T.T. (1996). Surface wheat (*Triticum aestivum*) residues, giant foxtail (*Setaria faberi*), and soybean (*Glycine max*) yield. *Weed Science*, 44: 939-943.

Wall, P.C. (2007). Tailoring conservation agriculture to the needs of small farmers in developing countries: an analysis of issues. *Journal of Crop Improvement*, 19: 137-155.

Weiner,J., Griepentrog, H.W., Kristensen, L. (2001). Suppression of weeds by spring wheat *Triticum aestivum* increases with crop density and spatial uniformity. *Journal of Applied Ecology*, 38: 784-790.

Wicks, G.A., Crutchfield, D.A., Burnside, O.C. (1994). Influence of wheat (*Triticum aestivum*) straw mulch and metolachlor on corn (*Zea mays*) growth and yield. *Weed Science*, 42: 141-147.

Wu, H., Pratley,J., Lemerle, D., Haig, T. (2001). Allelopathy in wheat (*Triticum aestivum*). *Annals of Applied Biology*, 139: 1-9.

Zhao, D., Atlin, G., Bastiaans, L., Spiertz,J. (2006). Cultivar weed-competitiveness in aerobic rice: heritability, correlated traits, and the potential for indirect selection in weed-free environments. *Crop Science*, 46: 372-380.

Zhao, D., Bastiaans, L., Atlin, G., Spiertz, J. (2007). Interaction of genotype x management on vegetative growth and weed suppression of aerobic rice. *Field Crops Research*, 100: 327-340.

11

Prospects of Organic Farming as Resource Conservation Technology

Brij Lal Lakaria, Satish Bhagwatrao Aher[1], Pramod Jha
A.B. Singh, B.P. Meena, S. Ramana and J. K. Thakur

ICAR-Indian Institute of Soil Science, Bhopal
Madhya Pradesh- 462 038 India
[1]ICMR- National Institute for Research in Environmental Health
Bhauri, Bhopal- 462 030, Madhya Pradesh

Introduction

Natural resources, especially those of soil, water, plant and animal diversity, vegetation cover, renewable energy resources, climate, and ecosystem services are fundamental for the structure and function of agricultural systems and for social and environmental sustainability, in support of life on earth. Historically the path of global agricultural development has been narrowly focused on increased productivity rather than on a more holistic integration of natural resource management with food and nutritional security. A holistic, or system-oriented approach, is preferable because it can address the issues associated with the complexity of food and other production systems in different ecologies, locations and cultures. Modern agriculture largely depends on the use of fossil fuel-based inputs, such as chemical fertilizers, pesticides, herbicides and energy intensive farm machinery. While the applications of such high input technologies have undoubtedly increased production and labour efficiency. There is a growing concern over their adverse effect on soil productivity and environmental quality which is emerging to recognize that the farmer has a great social responsibility as a land owner than merely agribusiness considerations. Therefore, there is need to develop agricultural techniques that are ecologically sound, economically viable, and socially responsible. Sustainable agriculture in the context of development helps to achieve production efficiency, protect ecosystem functions, enhance resilience to climate change, ensure healthy communities, satisfies basic needs and ensures optimum use/conservation of natural resources. Organic/biodynamic farming is among the various types of sustainable agriculture. Traditionally these methods has been playing very vital role in the conservation of natural resources in terms of soil health sustenance, environmental protection, food safety, waste recycling, biodiversity conservation with an economically viable approach. The chapter discusses the role of organic/biodynamic agriculture in natural resource conservation.

Agriculture and Sustainability

The agricultural production can be sustainable if it promotes practices that improve soil quality, while reducing erosion, salinization and other forms of degradation to achieve greater resilience to drought, better fertilizer efficiency, and reduced greenhouse gas emissions. There is minimum use of pesticides and herbicides by applying integrated pest management, crop rotation and crop diversification. It employs environmental management systems to ensure proper treatment of solid waste, manure and waste water. It also ensure the safe storage, application and disposal of agricultural chemicals to maintain congenial habitats to support wildlife and conserve biodiversity.

Therefore, there is need to develop agricultural techniques that are ecologically sound, economically viable, and socially responsible. These activities should focus on environmental sustainability across agricultural supply chains and multi-use landscapes. Sustainable agriculture in the context of development helps to achieve production efficiency, protect ecosystem functions, enhance resilience to climate change, ensure healthy communities, and satisfy basic needs (Aher *et al.*, 2012).

Resource Exploitation/ Unsustainable Utilization in Conventional Farming

Increasing population and to some extent capitalism in the second half of the 20th century forced to develop shortcut and fast output system. Accordingly modern agricultural production system has developed in a direction contrary to the requirements of productive and sustainable system for cohesiveness and functional diversity. The trend of modern agricultural system is toward an open flow of material rather than being cyclic use of resources. The growing use of off-farm production inputs such as mineral fertilizers, plant protection agents, growth promoters, farm machineries etc. indicates that closed systems are being increasingly developed and thus, losing their cohesiveness. Instead of being complex and diverse, modern farming systems tend to be one sided (specialized) in response to pressure toward economic gains. Production is becoming increasingly confined to only a few lines of activity, limited crop rotations and a small number of species and varieties, generally with a greatly restricted genetic base (e.g. hybrid varieties) (Sharma, 2001).

Fig. 1: Issues with modern agriculture

Modern agriculture largely depends on the use of fossil fuel-based inputs, while the applications of such high input technologies have undoubtedly increased production and labour efficiency. Besides high crop productivity, modern agriculture has created drastic impacts on natural/agricultural resources (Fig. 1).

Organic Farming

The organic agriculture has many definitions and explanations but all converge to state that it is a system that relies on ecosystem management rather than external agricultural inputs. According to the National Organic Standards Board of the US Department of Agriculture (USDA) the word 'Organic' has the following official definition: "An ecological production management system that promotes and enhances biodiversity, biological cycles and soil biological activity. It is based on the minimal use of off-farm inputs and on management practices that restore, maintain and enhance ecological harmony." (Lieberhardt, 2003). According to the Codex Alimentarius Commission (FAO, 2001), "organic agriculture is a holistic production management system that avoids use of synthetic fertilizers, pesticides and genetically modified organisms, minimizes pollution of air, soil and water, and optimizes the health and productivity of interdependent communities of plants, animals and people".

It is a system that begins to consider potential environmental and social impacts by eliminating the use of synthetic inputs, such as synthetic fertilizers and pesticides, veterinary drugs, genetically modified seeds and breeds, preservatives, additives and irradiation. These are replaced with site-specific management practices that maintain and increase long-term soil fertility and prevent pest and diseases.

To meet these objectives, organic agriculture farmers need to implement a series of practices that optimize nutrient and energy flows and minimize risk, such as: crop rotations and enhanced crop diversity; different combinations of livestock and plants; symbiotic nitrogen fixation with legumes; application of organic manure; and biological pest control (Scialabba and Hattam, 2002). All these strategies seek to make the best use of local resources. Organic farming is distinguished from conventional agriculture by exercising particular respect for human values, the environment, nature, and animal welfare, etc. This regard is incorporated in the basic principles of organic farming, as formulated by the International Federation of Organic Agriculture Movements (IFOAM). (IFOAM, 2002). Organic farming is an agricultural system which originated early in the 20th century in reaction to rapidly changing farming practices. Certified organic agriculture accounts for 70 million hectares globally, with over half of that in Australia.

Principles of Organic Farming

- Production of food of high quality in sufficient quantities
- Operation within natural cycles and closed systems as far as possible, using local resources
- The maintenance and long term improvement of the fertility and productivity of soils
- Creation of a harmonious balance between crop production and animal husbandry
- Securing of high levels of animal welfare

- Fostering of local and regional production and supply chains, and

- The provision of support for the establishment of an entire production, processing and distribution chain that is both socially and ecologically justifiable

These basic principles provide organic farming with a platform for ensuring the health of environment for sustainable development, even though the sustainable development of mankind is not directly specified in the principles.

Driving Forces of Organic Agriculture

Consumer or market-driven organic agriculture

Products are clearly identified through certification and labeling. Consumers take a conscious decision on how their food is produced, processed, handled and marketed. The consumer therefore has a strong influence over organic production.

Service-driven organic agriculture

In countries such as in the European Union (EU), subsidies for organic agriculture are available to generate environmental goods and services, such as reducing groundwater pollution or creating a more biologically diverse landscape.

Farmer-driven organic agriculture

Some farmers believe that conventional agriculture is unsustainable and have developed alternative modes of production to improve their family health, farm economies and/or self-reliance. In many developing countries, organic agriculture is adopted as a method to improve household food security or to achieve a reduction of input costs. Produce is not necessarily sold on the market or is sold without a price distinction as it is not certified. In developed countries, small farmers are increasingly developing direct channels to deliver non-certified organic produce to consumers. In the United States of America (USA), farmers marketing small quantities of organic products are formally exempt from certification.

Resource Conservation Through Organic Farming

The organic or biodynamic farming ensures the conservation and optimum use of natural resources as shown in Fig. 2. The explicit goal of organic/biodynamic agriculture is to contribute to sustainability enhancement. The soil and water protection and conservation techniques of sustainable agriculture used to combat erosion, compaction, salinization and other forms of degradation are evident in organic/biodynamic farming. The use of crop rotations, organic manure and mulches improves soil structure and encourages the microbial diversity. Mixed and relay cropping provide a more continuous soil cover and thus a shorter period when the soil is fully exposed to the erosive power of the rain, wind and sun. Terracing is practiced to conserve moisture, and soils are used in appropriate situations and particular attention is paid in irrigated areas to on-farm water management. Organic/biodynamic farming reduces or eliminates water pollution and helps conserve water and soil on the farm. A few developed countries compel or subsidise farmers to use organic techniques as a solution to water pollution problems (e.g. Germany, France).Organic farmers rely on natural pest controls (e.g. biological control, plants with pest control properties) rather than synthetic pesticides which, when misused, are known to kill beneficial organisms (e.g. natural parasites of pests, bees, earthworms), cause pest resistance, and often pollute water and land.

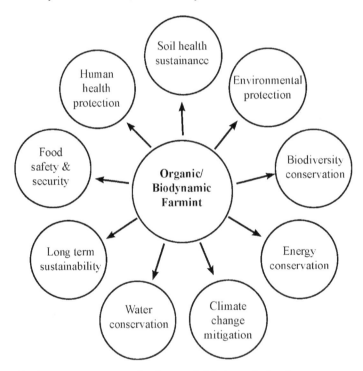

Fig. 2: Resource conservation in organic/biodynamic farming

Reduction in the use of toxic synthetic pesticides, which the World Health Organization (WHO) estimates to poison three million people each year, should lead to improved health of farm families. Organic farmers aim to make the maximum use of the recyclable on-farm crop residues (straws, stovers and other non-edible parts) either directly as compost and mulch or through livestock as farmyard manure. It eliminates the use of synthetic nitrogenous fertilizers that greatly lower the risks of nitrogen contamination of water. Crop rotation is a widely used method of fertility maintenance and pest and disease control, which is used in large and small scale farming in both developed and developing countries, especially under intensification. In addition to it, fodder legumes are well-known fertility-building crops and are grown on vast areas in sub-tropical Asia and in semi-arid regions for the dual purpose of feeding livestock and adding nitrogen to the farm fertility cycle. Grain legumes may also produce a reasonable crop without nitrogenous fertilizer. Leguminous crops in rotations add various amounts of nitrogen to the overall farm system through biological fixation; other nitrogen-fixing plants such as *Azolla* may also be used. Biological nitrogen fixation is a powerful technique but it often requires some addition of minerals to the soil, especially phosphorus. Natural and organic fertilizers from outside the farm are also used (e.g. rock phosphate, potash, guano, seaweed, slaughterhouse by-products, ground limestone, seaweed, wood-ash). Following proper crop rotations encourage a diversity of food crops, fodder crops and under-utilized plants. This helps in conservation of plant genetic resources besides improving overall farm production and soil fertility. Integrating livestock into the system adds income through organic meat, eggs and dairy products, as well as draught animal power. Tree crops and on-farm forestry integrated into the system provide shade and windbreaks while providing food, income, fuel and wood. Integrated agri-aquaculture may also be found within diverse organic/biodynamic agricultural systems. Economic objectives are not the only motivation of organic farmers; their intent is often to optimize land, animal, and plant interactions, preserve natural nutrient and energy flows, and enhance biodiversity, all of which contribute to the overall objective of sustainable agriculture to preserve natural resources and ecosystems for future generations.

Crops under Organic Farming in India

Among different crops currently cultivated under organic farming in our country, the cereals include wheat, paddy, jowar, bajra, maize; while pigeonpea, greengram, blackgram, and chickpea are the predominant pulse crops. Groundnut, castor, mustard, sesame are main oilseeds while cotton, sugarcane, particularly for sugarcandy (gur) are the other commodity crops. Not only these, there are certain spice crops (ginger, turmeric, chillies, cumin); plantation crops (tea, coffee, cardamom); fruit crops (banana, sapota, custard apple and papaya) and vegetables such as tomato, brinjal, cucurbits, cole crops, leafy vegetables grown under orgnic farming in India.

Acreage and Production

At present a total area under organic certification process (registered under National Programme for Organic Production) is 3.67 million hectare (2019-20).This includes 2.299 million hectare cultivable area and another 1.37 million hectare for wild harvest collection. Among all the states, Madhya Pradesh has covered largest area under organic certification followed by Rajasthan, Maharashtra, Gujarat, Karnataka, Odhisa, Sikkim and Uttar Pradesh. Sikkim has achieved a remarkable distinction of converting its entire cultivable land (more than 75000 ha) under organic certification. India produced around 2.75 million MT (2019-20) of certified organic products which includes all varieties of food products namely Oil Seeds, Sugar cane, Cereals & Millets, Cotton, Pulses, Aromatic & Medicinal Plants, Tea, Coffee, Fruits, Spices, Dry Fruits, Vegetables, Processed foods etc. The production is not limited to the edible sector but also produces organic cotton fiber, functional food products etc. Among different states Madhya Pradesh is the largest producer followed by Maharashtra, Karnataka, Uttar Pradesh and Rajasthan. In terms of commodities oilseeds are the single largest category followed by sugar crops, cereals and millets, tea & coffee, fiber crops, fodder, pulses, medicinal/ herbal and aromatic plants and spices & condiments. The total volume of export during 2019-20 was 6.389 lakh MT. The organic food export realization was around INR 4,686 crore (689 million USD). Organic products are exported to USA, European Union, Canada, Switzerland, Australia, Japan, Israel, UAE, New Zealand, Vietnam etc. In terms of export value realization processed foods including soya meal (45.87%) lead among the products followed by oilseeds (13.25%), plantation crop products such as tea and coffee (9.61%), cereals and millets (8.19%), spices and condiments (5.20%), dry fruits (4.98%), sugar (3.91%), medicinal plants (3.84%) and others (apeda.gov.in/apedawebsite/organic/Organic_Products.htm.)

Conclusion

The modern agriculture depends on the use of external inputs which are energy intensive. Their continuous use leads to resource exploitation or unsustainable utilization resulting in environmental damage, loss of biodiversity, soil health issues, food contamination, human and animal health issues etc. Organic/ biodynamic farming is an ecological production management system that promotes biodiversity, and is based on the minimal use of off-farm inputs. Organic/biodynamic agriculture is a traditional method of conservation of natural resources. Thus, adopting organic farming ensures optimum use of natural resources that address many local as well as global issues.

References

Aher, S.B., Bhaveshananda, S. and Sengupta, B., (2012). Organic agriculture: Way towards sustainable development. *International Journal of Environmental Sciences*, 3(1): 209-216.

Food and Agriculture Organization of the United Nations (FAO) (2001). Codex Alimentarius – Organically Produced Foods, FAO, Rome. Fertilizer statistics 2003-04, The Fertilizer Association of India, New Delhi, p 77.

International Federation of Organic Agriculture Movements (IFOAM). (2002). Basic Standards for organic farming and processing, 2nd draft 2001.

Lieberhardt B. (2003). What is organic agriculture? What I learned from my transition. 'Organic Agriculture: Sustainability, Markets and Policies', Organization for Economic Cooperation and Development (OECD) and CAB 1, Wallingford, UK. p. 31-44.

Scialabba E.N., Hattam C. (2002). Organic Agriculture Environment and Food Security. Food and Agriculture Organization of the United Nations (FAO), Rome p. 252.

Sharma, A.K. (2001). A handbook of organic farming. Agrobios India, Jodhpur, Rajasthan, India, p. 627.

12

Cover Crops: Potential and Prospects in Conservation Agriculture

Debarati Datta, Sourav Ghosh[1], R. Saha and C. P. Nath[2]

ICAR-Central Research Institute for Jute and Allied Fibres, Barrackpore West Bengal -700 121, India
[1]ICAR- Directorate of Onion and Garlic Research
Maharashtra- 410 505, India
[2]ICAR–Indian Institute of Pulses Research, Kanpur, U.P - 208 024, India

Introduction

Modern intensive agriculture systems based on the principles of yield maximization have been plagued with multiple constraints such as declining factor productivity, decreasing resource use efficiency (fertilizers, water, labour etc.), soil organic matter decline, salinization, soil structure degradation, water and wind erosion, reduced water infiltration rates, soil compaction due to surface sealing and crusting, declining ground water table, pest and disease outbreak, weed resistance and weed shifts (Paustian *et al.*, 2016). To address these overwhelming issue, conservation agriculture (CA) has been defined by FAO as a concept of resource saving agricultural crop production which is based on enriching the natural and biological processes above and below ground. The major purpose for its advancement is to reduce the production cost, save resources, enhance yields and utilize resources like nutrient and water efficiently. Conservation tillage denotes soil management system that follows the principle of at least 30% of the soil surface coverage with crop residues after seeding of the crop (Jarecki and Lal, 2003). CA, a resource-efficient agriculture helps to conserve, improve and utilize natural resources efficiently through integrated management of available soil, water and biological resourwces combined with external inputs and thus contributes to environmental conservation as well as enhanced and sustained agricultural production in the long term. The practice and wider extension of conservation agriculture thus requires a deeper understanding of its ecological underpinnings for managing its various elements for sustainable intensification.

Cover crops are close-growing plants having dense foliage and root system, provide soil and seeding protection, and improve soil properties between periods of normal crop production. Cover cropping is a widespread novel technique and is seen as a green culture due to its multifarious considerable benefits. Having entered into the realm of green culture, practice of cover cropping stands to gain status as a bonafide environmental ethic. Cover crops, a fundamental component of the stability of the conservation agriculture provides numerous ecosystem services (Schipanski *et al.*, 2014) which includes improvement in soil properties, biological diversity, protecting the soil surface and dissipating raindrop energy, weed management, nitrogen fixation, pest and disease control, increasing infiltration, anchoring soil and adding carbon deep in the soil profile via roots, sequestering nutrients and supplying nutrients to the following crop (Cordeau *et al.*, 2015). Small-scale farmers prefer a cover crop which has multiple purposes such as food, fodder, fuelwood which fits into their normal cropping system. This chapter provides a comprehensive overview of practices and paybacks using cover crops necessary for augmenting the productivity, profitability and feasibility of CA systems.

Methods of Cover Cropping

Live mulches

Living mulches are cover crops that co-exist with the cash crop during the growing season and continue its growth even after the cash crop's harvest (Hartwig and Ammon, 2002). They can be chosen and managed for lessening competition with the main cash crop yet take full benefit after competing with weeds. Living mulch systems can be viable for orchards, agronomic crops, such as cowpeas, beans, soybeans, mucuna, corn, soybean and vegetables and are dependent on adequate moisture for the cash crop. Living mulches especially grasses can serve as a sink to tie up excess nitrogen and hold it until the next growing season. In conservation tillage systems, living mulches can improve nitrogen budgets, provide weed and erosion control, and may contribute to pest management and help mitigate environmental problems. The system requires close monitoring and careful control of competition between the living mulch and grain crop to maintain crop yields.

Companion cropping

Companion cropping can be a good weed suppressing technique but research helps in determining how suitable it may be in specific circumstances. Yields are generally not affected since they are complementary and not competitive. Evidence supports that it may decrease weed competition, build soil organic matter, reduce soil erosion, and improve water penetration. Forage legumes have higher potential as a companion crop as they obtain over 90% of the nitrogen from atmospheric fixation in comparison to 50% by grain legumes (Fisher, 1996). Under certain climatic conditions when spring soil moisture is limiting, cover or companion crops can exhaust moisture and be detrimental to crops regardless of weed control advantages.

Intercropping and mixed cropping

Intercropping involves growing of two or more crops simultaneously in alternate rows having proper row arrangement, whereas in mixed cropping row arrangements are not maintained and different crop seeds are broadcasted randomly. Intercropping with legumes is an admirable practice for controlling soil erosion and sustaining crop production. Legumes enrich soil by fixing the atmospheric nitrogen, converting it from an inorganic form to forms that are readily taken up by plants and thus provides ecological stability. Advantages includes yield enhancement over different seasons, proper use of growth resources, excellent weed, pests and diseases control, physical support and shelter to the other crop. Intercropping of low canopy legumes like groundnut, blackgram, soybean, cowpea in wider spaces of maize, sorghum, castor provides ground cover besides reducing erosional hazards and biological insurance for boosting arable land productivity. Yield of sorghum+pigeonpea intercropping was reported to be increased by 30% under contour furrow system due to in situ moisture conservation.

Mixed cropping of low canopy legumes with wide space crops provide continuous cover to the ground, protect against beating action of rain drops and ensures at least one crop under adverse condition against complete failure of crop. Cover crops like dhaincha, black gram, groundnut, soybean, sunnhemp restore fertility, controls weeds, conserve rainwater, reduce energy and costs besides improving morphological characters and utilise moisture from different depths of feeding zone. For instance, maize can be mixed with pigeon pea and *Crotalaria juncea* both planted when the maize plants are 30 cm height. After the maize harvest these species accelerate their development, completely overgrowing the maize residues which can be followed by subsequent crops like beans, sorghum or sunflower.

Strip cropping

Strip cropping is a well-accepted conservation measure in controlling soil erosion worldwide. It works by reducing slope length, reducing run off velocity, arresting erosion by acting as a biological barrier and increasing water movement in the soil profile. Various strip cropping includes contour strip cropping, field strip cropping, wind strip or permanent strip cropping which protects nearby crops from soil erosion and harmful winds. Tall growing crops like jowar, bajra, maize and low growing crops like legumes are planted in alternate straight and long but narrow, parallel strips across the direction of prevailing wind. In buffer strip cropping, strips are established to take care of critical or steep sloping fields and are planted with perennial legumes, grasses or shrubs.

Alley cropping

A system in which arable crops are grown in alleys formed by trees. The hedges serve as windbreaks and increases soil fertility restoration. Alley cropping protects fragile soils through a network of roots produced by the trees and supplemental ground-cover resulting from fallen leaves and the companion crop. The alley cropping with leguminous trees *viz.*, subabul (*Leucaena leucocephala*) has been most widely used on field bunds for producing mulch material for moisture conservation, nutrient recycling and erosion control on sloping lands up to 30%. The contour-paired rows of, Leuceana hedge, *Glyricidia maculate* and Eucalyptus trees, and wide grass barrier of 0.75 m at 1.0 m interval in maize brought down runoff from 40 to 30% of rainfall and soil loss from 21 to 8 t/ha on 4% sloping land (Narain *et al.*, 1992).

Choice of Cover Crops

Cover crop options and niches are as diverse as the farming systems in the region (Table 1&2). A good cover crop has the following characteristics:

- Quick growing, completely covering the ground to prevent erosion and sunlight entry and having multiple uses
- Produces heavy leaf biomass and aggressive enough to compete with weeds.
- Be rustic and require low management costs, avoid competition for resources with the cash or subsistence crop

The principal motive to start the first years of conservation agriculture with cover crops should be to leave a lot of residues on the surface, which decomposes slowly because of the high C/N ratio. Grasses and cereals, due to their abundant rooting system may condition the soil within a shorter time span. Following this, legumes can be incorporated which can enrich the soil with nitrogen and decompose rapidly because of low C/N ratio. The basis for cover crops selection should be determined by the presence of lignin and phenolic acids, which gives higher resistance for residues decomposition. The grain species (oats and wheat) show more resistance to decomposition than common vetch (legume). Cellulose-rich plants or plant parts degrade far more rapidly than ligneous ones like those of mature grasses. Cereal and grasses by virtue of the sheer volume and mass-distribution of their fibrous root systems bind the soil in a manner that gives it excellent structural attributes and thus increases resistance to entrainment by overland flow whereas broadleaf legumes due to their overlapping arrangement of leaves intercepts rainfall and minimizes splash.

Table 1: Choice of cover crops based on season, family and benefits

Types	Cover crops	Benefits	References
Winter cover crops	Rye, brassica, hairy vetch, red clover, oats	Erosion control, nitrogen fixation, Improves soil structure, organic matter addition	Moncada and Sheaffer, 2010
Summer cover crops	Cowpea, Sunnhemp, Sudan grass, buckwheat	Improves organic matter, reduce erosion, weed suppression, enhance soil microbes	Creamer and Baldwin, 2000
Legumes	Pea, bean (summer), Hairy vetch, field pea, crimson clover (winter)	Nitrogen fixation, erosion prevention, attracts beneficial insects	Clark, 2008
Non legumes	Barley, oat, wheat, brassica, mustard	Weed suppressor, nutrient scavenging, reduced soil erosion	Clark, 2008

Source: Sharma *et al.* (2018)

Table 2: Agro-ecological adaptation of cover crops

Humid lowland	Butterfly pea, black gram
Cold condition	Lucern, clover, butterfly pea
Flooded areas	Rice bean
Drought tolerant	Pigeon pea, butterfly pea, cowpea, mucuna, lucern
Low fertility soil	Pigeon pea, mucuna, hairy vetch, cowpea, rye

Legumes

Legumes are excellent cover crops and are mainly selected due to their nitrogen fixing ability from the atmosphere and enriching the soil. Besides, legumes benefits by attracting beneficial insects, helping control erosion, and adding organic matter to soils (Fig. 1). Biennials and perennials include red clover, white clover, sweet clover, and alfalfa.

- **Crimson clover (*Trifolium incarcatum*)** - cool season annual legume, more cold hardy than most other clovers, grows great on all soil types in 1-2 months, supplies lots of foliage and roots, attracts lots of beneficial insects.

- **Hairy vetch (*Vicia villosa*)** - cool season annual climbing legume, most cold hardy of the popular legumes, able to grow in acid soil, can grow in temp. below -5 °F, weed suppressor, grows great with a non-legume like rye as a nurse crop, supply over 100 pounds of nitrogen per acre.

Fig. 1: Some widely used cover crops-crimson clover, hairy vetch, sunnhemp, dhaincha, rye, sudan grass

- **Beans (*Dolichos lablab*) and pea (*Pisum sativum*)** - Lablab is a biennial herbaceous legume, outstanding under drought, nematode susceptible. Most peas are cold hardy and usually supply 30 to 80 pounds of available nitrogen. Forage peanut displays excellent growing capacity on sandy and clay loam soil with production of 8t/ha/yr dry matter.

- **Sunnhemp (*Crotalaria juncea*)** - vigorous growing green manure crop, which can be incorporated 10 weeks after sowing and does not withstand waterlogging. The green matter yield is 15-20 t/ha and quantity of nitrogen fixed is 75-80 kg/ha.

- **Dhaincha (*Sesbania aculeata* and *Sesbania rostrata*)** - *Sesbania aculeata* is a quick growing succulent green manure crop, which can be incorporated at 8 to 10 weeks after sowing. The green matter yield is 10-20 tonnes per ha and quantity of nitrogen fixed is 75 to 80 kg/ha. *Sesbania rostrata* is a green manure crop, which is stem and root nodulating and thrives well under waterlogged condition. A green matter yield of 15 to 20 t/ha equivalent to 150-180 kg N/ha is obtained.

Brassicas

Brassica crops, such as mustard, rapeseed, radishes or turnips, can help scavenge for nutrients deep in the soil profile. They are increasingly used as winter or rotational cover crops in vegetable.

- **Radish (*Raphanus sativus*), canola (*Brassica napus*)** - fast growing cool season annual non-legumes with lots of foliage, weed suppressor. Radishes are great for repelling many bug pests, has a large taproot that can break through compacted layers, resulting in friable soil condition and improvement in rainfall infiltration and storage. Among different winter species, higher dry matter and nutrient content was recorded in radish (Calegari *et al.*, 1993) (Table 3). Canola grows well under moist and cool conditions and may function as biofumigants, suppressing root pathogens and plant-parasitic nematodes (Creamer *et al.*, 2000).

Grass cover

Grass cover crops include annual cereals (rye, wheat, barley, oats), annual or perennial forage grasses (ryegrass) and warm-season grasses (sudan grass). Grass species are very useful for scavenging nutrients left over from a previous crop but if grown for maturity, the amount of available nitrogen for the succeeding crop may get reduced due to high C/N ratio. Thus, way out is to supply extra nitrogen by seeding a legume-grass mix.

- **Rye (*Secale cereale*), winter wheat (*Triticum aestivum*), and oats (*Avena sativa*), sudan grass (*Sorghum sudanense*)** - cool season fast growing grass non-legumes. Rye has wide adaptability to soil and climatic condition, the most cold hardy of all green manures, drought tolerant, fast growing, ability to grow on acid soil, reduces N leaching, has rapid growth, and produces a large quantity of biomass. Rye can grow in temps below -20°F. Rye must be terminated at least 4-6 weeks before planting spring crops to prevent seed germination. Oat stems help trap snow and conserves moisture. Black oat (*Avena strigosa*) provides excellent soil cover better than any other cover crop and suppresses weed. Sudan grass, a forage cover crop suppresses plant-parasitic nematodes as they produce highly toxic substances during decomposition and also helpful for loosening compacted soil (Blanco-Canqui *et al.*, 2012).

Mixtures

Mixtures are ideal for producers who have multiple goals. For example, a mixture of rye, brassicas and legumes can be planted when producers are trying to achieve ground cover early. Legumes aid in nitrogen fixing whereas grasses and brassicas are better at reducing N leaching. White lupine in combination with black oat provides excellent weed control and reduces anthracnose attack notably.

- **Buckwheat (*Fagopyrum esculentum*) and sunflower (*Helianthus annuus*)**- fast growing warm season annual non-legumes that can break up hardpan, kill weeds, and pull up insoluble subsoil phosphorus up to the topsoil. Buckwheat can grow fast on any soil type as it requires 4-6 weeks to mature.

- **Marigold (*Tagetes erecta*) and other herbs** - warm season annual non-legumes. Thick planted marigold is highly effective for nematode control. Garlic, onions and hot peppers control many soil pests.

Table 3: Dry matter, nutrient and C/N ratio of summer and winter cover crop

Cover crop	Dry matter (kg/ha)	Organic carbon (kg/ha)	N (kg/ha)	P (kg/ha)	K (kg/ha)	C/N
Summer species*						
Pigeon pea	9153	5153	240	13	240	22
Mucuna	7500	3911	192	10	108	21
Winter species**						
Radish	4771	2219	86	11	156	19
White lupin	4012	1925	75	5	55	26
Black oat	3680	1818	48	7	84	39
Hairy vetch	2942	1395	81	7	81	17

*Sown in association with corn, and dry matter determined at approximately 7 months.
** Sown after cotton, and dry matter determined at approximately 4 months.

Source: Calegari *et al.* (1993)

Role of Cover Crops in CA

Cover crops can fit into diverse niches particularly fruit and vegetable systems. Here are some of the ways cover crops can benefit the soil and ultimately bottom line:

Making the transition to no-till

Farmers who have used conservation tillage alongside cover crops have been transitioning to no-till or strip-till. Changing tillage systems may pose a menace to yield and profit, but the use of cover crops while making the transition can offer a buffer to the expected yield. No-till alone provides reduced labour and machinery cost, but pairing no-till with cover crops substantially buffer the shift. Reports suggest that cover crops start paying off in the second year of no-till for corn, and will break even in the first year for soybeans.

Improving resource use efficiency

Cover crops have the potent to increase nitrogen and water-use efficiencies, resulting from higher return of vegetative residues to the soil (Teasdale and Mohler, 2000). The fixed nitrogen from addition of leguminous cover crops can offset a portion of fertilizer N requirement for subsequent cash crops, reduce purchase of nitrogen fertilizer that is produced using fossil fuels and therefore enhance nitrogen use efficiency of the whole crop system. Studies show that the living roots can scavenge and recycle from 170 to 340 kg of mineral N/ha/yr that would otherwise be lost due to leaching (Delgado *et al.*, 2007), as well as avoiding off-site N_2O emissions. One well-characterized repayment of cover crop that includes year-long soil cover is the ability to capture and retain residual as well as mineralized nitrogen (Gardner and Drinkwater, 2009). Furthermore, leguminous cover crops also fix atmospheric N into plant-usable forms allowing nitrogen fertilizer savings. Thus, agricultural management of cover crops should be prudently monitored for the synchronization of N release with subsequent crop need. Cover crops can cut the need for insecticide application by attracting beneficial insects for predating economically damaging insects. Cover crops of *Vicia villosa* and *Secale cereale* in pecan orchards sustained increased densities of predatory insects which feed on aphids (Bugg *et al.*, 1993).

Soil biological diversity

Cover crops perform a dual function as they benefit the soil in two specific ways, through soil physical conditioning and fertility building. Under permanent soil cover, the crop residues produce a layer of mulch which besides, protecting the soil from the physical impact of rain and wind, also stabilizes the soil moisture and temperature in the surface layers. Thus this zone becomes a habitat for larger insects down to soil borne fungi and bacteria. It has been affirmed that quality of life in the soil is what determines its potential productivity. The roots and microbial organisms macerate the mulch, incorporate, mix it with the soil and decompose it which generally adds to the SOM reservoir and contributes to the physical stabilization of the soil structure. Larger components of the soil fauna, such as earthworms, provide a soil structuring effect producing very stable soil aggregates as well as uninterrupted macropores i.e. earthworm burrows leading from the soil surface down to the subsoil and allowing rapid water infiltration. This process carried out by the soil biota, the edaphon, can be regarded as "biological tillage". Cover crops such as pea and oat showed evidences for higher earthworm populations compared to bare fallow plots in the cover crop-spring barley crop rotation whereas mustard recorded less earthworm population (Roarty *et al.*, 2017). Ryegrass maintained a higher microbial population in the upper 2 cm layer of conservation (CT) and no tillage (NT) compared to no-cover plots; while CT had greater bacterial and fungi colony-forming units (CFUs) in 2 to 10 cm depth. Cover crops under NT registered hundred fold higher CFUs than other treatments (Zablotowicz *et al.*, 2007). Results showed that oats and cereal rye cover crop favoured Arbuscular mycorrhizal (AM) fungi whereas hairy vetch was associated with non AM fungi (Finney *et al.*, 2017). Long-term cover cropping has been shown to enhance organic C availability, which strongly stimulated the abundance and the diversity of microbial communities (Schmidt *et al.*, 2018). Furthermore, Chavarria *et al.* (2016) reported an increment in soil bacterial phospholipid fatty acid and soil enzyme activities to 6.8 and 20%, respectively, under cover crop mixtures.

Improved soil physical properties

Cover crop diversification strategies can directly alter soil water dynamics through changes in evaporation, transpiration, and loss pathways. Cover crops positively impact soil water content by:

- reducing runoff losses and protecting the soil from raindrop impact, lessening slaking of surface aggregates, preventing pore sealing and crust formation and thus increasing the opportunity time for water infiltration.

- decreasing evaporative losses through mulching effect during cover crop growth and termination.

In areas that lack moisture, cover crops increase resiliency by increasing the efficiency of rain infiltration and existing irrigation systems by creation of macropores as compared to bare soil (Fig. 2). This network encourages rainwater to enter the root zone, keep the soil cooler due to decreased evaporation and reduce crop heat stress even during drought. Cover crops can also reduce the evaporation rate by isolating the soil from sun heating and ambient air temperature and increasing resistance to water vapour flux by reducing wind speed. Especially effective at covering the soil surface are grass type cover crops such as rye, wheat, and sorghum-sudangrass hybrid and water-efficient legumes such as lentils in dryland areas which conserve more moisture than conventional bare fallow. Similarly, study conducted for 7 years by successive use of cereal rye as cover crop in maize-soybean cropping system was found to be effective in improving soil water table and decreasing bulk density by 3.5% (Haruna and Nkongolo, 2015). Cover crops increased water retention in soil at water potentials related to field capacity and plant available water by 10% - 11% and 21% - 22%, respectively (Basche *et al.*, 2016). Another study revealed that cover crops including brome grass, vegetation and strawberry clover lessesned surface soil strength by 38 to 41 %, increased infiltration rate by 37 to 41 % and cumulative water uptake by 20 to 101 % (Folorunso *et al.*, 1992).

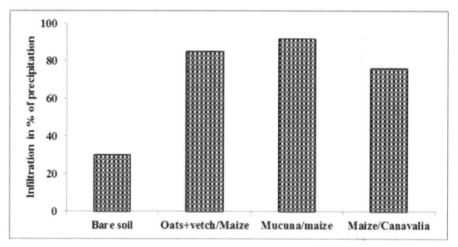

Fig. 2: Effect of cover crops on rainwater infiltration in a maize production system compared to bare soil.
Source: Debarba and Amado (1997)

Cover crops have the potential to produce long term tangible benefits by breaking up compacted layers in the soil, reducing the need for overly-aggressive mechanical tillage resulting in better traffic conditions. For no-till farmers, cover crops provide a non-mechanical method for repairing and maintaining an ideal soil structure. Recent research with forage radish suggests that its large taproot acts as biological plough and can alleviate compaction (Williams and Weil, 2004). Deep rooting cover crops create macropores that spurs the formation of soil aggregates improving nutrient and gaseous exchange.

Cover crop residues reduce daily fluctuations of soil temperature and reduce soil temperature maximums and minimums. The cooler soil temperatures, which benefit the cash crop throughout the summer, can delay spring planting compared to a system without a cover crop. Spring soil temperature is particularly important in cover crop-conservation tillage systems. The harmful effects of planting when the soil temperature is too low were demonstrated in Colorado for conservation tillage with continuous corn but no cover crops. Low soil temperatures contributed to reduced corn yields over 5 years (Halvorson and Reule, 2006).

Soil nutrient and organic matter enrichment

From the standpoint of soil nutrient enrichment, cover crops contribute to the nutrient pool through mineralization of decaying biomass under the influence of soil biota, which release the nutrients in absorbable form. Roots and microbial processes provide labile C for aggregate formation and stabilization. Increased SOM means greater paybacks to soil biodiversity, which is the heart of all SOM dynamics and aggregate stability which, in turn, reduces tendency to erode. Hairy vetch and sunnhemp cover crop in winter wheat-sorghum main crop increased soil organic carbon by 30% (Blanco-Canqui *et al.*, 2011). Where the decaying material is leguminous, nitrogen enrichment is predominant due to the capacity of their roots to fix atmospheric nitrogen in symbiotic association with rhizobacteria. *Sesbania aculeata* and green leaf manuring crops like *Argemone mexicana* and *Tamarindus indicus,* when applied continuously for 4-5 seasons has excellent buffering effect in sodic soils.

Soil and water conservation

Cover crops are becoming a part of intensified cropping systems under conservation tillage in semihumid or humid regions. Cover crops with fine and extensive roots such as rye and oats may reduce concentrated flow erosion during winter by strengthening the aggregates whereas white mustard and fodder radish are less effective in preventing soil erosion. The protective canopy formed by a cover crop slows the velocity of runoff from rainfall and snowmelt, reducing crusting and soil loss due to sheet and rill erosion (Table 4 & 5). Reduced erosion protects and improves water quality by keeping agricultural chemicals in the ground and out of drainage water. In Missouri, on a silt loam soil, inclusion of rye or wheat cover crop reduced soil loss in no-tillage silage corn from 9.8 to 0.4 tons/a/yr (Wendt and Burwell, 1985). Cover crop residue reduced runoff and increased infiltration by 50 to 800% compared to its removal at Alabama (Truman *et al.*, 2005). Infiltration rates were 100% greater for a sandy loam soil when grain sorghum was no-till planted into crimson clover in comparison to a tilled seedbed without a cover crop. In Maryland, pure and mixed stands of hairy vetch and rye conserved soil moisture and contributed to greater corn yield (Clark *et al.*, 2007). Again, surface evaporation was five times less under no-till with surface mulch than with conventional tillage during May to September (Corak *et al.*, 1991).

Table 4: Percentage of soil cover during vegetative period of winter cover crop (3 years average)

Species	% soil cover (April sown)			
	15DAE	45 DAE	75 DAE	105 DAE
Black oat+hairy vetch	15	35	93	100
Black oat+white lupine	20	36	96	100
Black oat	17	35	84	100
Hairy vetch	14	30	82	98
Radish	12	43	88	99
Pea	17	34	69	45

Source: Florentin (1999)

Table 5: Percentage of soil cover during vegetative period of summer cover crop (3 years average)

Species	% soil cover (October sown)			
	45DAE	90 DAE	120 DAE	160 DAE
Sunnhemp	20	45	40	35
Lablab	75	100	90	90
Cowpea	63	45	40	-
Mucuna	50	89	95	90
Pigeon pea	35	90	95	80
Tephrosia	10	15	25	40

Source: Derpsch and Florentin (1992)

Resilience to climate variability

Cover crops as a part of conservation agriculture can improve the resilience of the crop system against weather variability and minimize soil nutrient losses. In agriculture system, various strategies to mitigate biogeochemical include reduction in greenhouse gas emissions, reduce nitrogen fertilizer production, increasing sink for greenhouse gases in soil. Cover crops management help in climate change adaption through reduction of erosion from rains, retention of mineralized nitrogen due to warming, increases soil water management options during soil saturation period or in droughts, break compacted layers and hard pans, permit a rotation in a monoculture.

Soil Carbon Sequestration

Intensive crop cultivation leads to SOC losses of 30 to 40% in comparison to natural vegetation (Don *et al.*, 2011). Sequestration of SOC in conventional tillage and no-till soils can be influenced due to the difference in plant carbon inputs and rate of mineralization. Carbon storage not only improves soil productivity but also abets global warming. Research results highlight that SOC at 0-10 cm was greater from the cover crops (hairy vetch, rye, mixture of hairy vetch and rye) as compared to no cover crops in cotton and sorghum.

Cover crops in rotation was linearly correlated in soil depth of 22 cm with annual SOC change at the rate of 0.32 ± 0.08 Mg/ ha/yr ($R^2 = 0.19$) as evidenced by 54 years of rotation. The newly predicted data after using cover crops for 155 years would have SOC accumulation of 16.7 ± 1.5 Mg/ ha/yr (Popelau *et al.*, 2015). The average soil C sequestration rate is 1.3 t CO_2 e/ha/yr and the net GHG mitigation potential for winter cover crops is 1.9 t CO_2e/ha/yr. Cover crops enable warming mitigation through greenhouse gas flux by 100 - 150 g CO_2 e/m²/yr. A case study estimation to calculate the change of surface albedo due to cover crops showed mitigation around 12 - 46 g CO_2 e/m²/yr over 100 year period.

Reduced nitrate leaching

Leaching of nitrate is a subject of major concern due to its direct upshot on drinking water, eutrophication of sea water and atmospheric pollution with ammonia. Losses through nitrate leaching under conventional system ranges between 10 to 30% of applied nitrogen (Sharma *et al.*, 2012). Cover crops have an ability of capturing nutrients in its roots, prohibiting nitrate leaching into groundwater. Nitrate leaching was 40% lower in legume cover crop treatment and 70% lower in non-legume cover crop treatment than fallow system (Tonitto *et al.*, 2006). Growing of perennial ryegrass (*Lolium perenne* L.) as a cover crop in barley at Sweden reduced nitrate leaching by less than 5 mg/ L as compared to 10-18 mg/ L without cover crop (Bergstrom and Jokela, 2001). Oats reduced nitrate concentration by 26% whereas reduction in rye was 48% (Kaspar *et al.*, 2012).

Weed management

Weeds are likely to show greater resilience in competition with crops due to changes in CO_2 concentrations and increasing temperature owing to their greater physiological plasticity and diverse gene pool. Weed management effects of cover crop under conservation tillage include:

- Suppress weeds as a smother crop that outcompetes weeds for water and nutrients. Good choices for weed suppression include lablab, cowpea, lucerne and buckwheat. In some cases, green cover has been shown to reduce weed density by 90 percent in corn crops.

- Acting as a weed-suppressive mulch after termination of the cover crop

- Allelopathy effects both from the living cover crop and after termination (Jabran *et al.*, 2015). Buckwheat, oat and sunflower are good types of allelopathic plants. Cereal rye is known to release phenolic and benzoic acids that can inhibit weed seed germination and development.

Cover crop residue or leaf canopy can alter the weed's microclimate as it can block available light, change soil temperature, deprive the weeds of nutrients and thus ensure that weed seeds are not added to the soil seed bank. Brown manuring, a no-till version of green manuring involves desiccating *Sesbania* before flowering with 2, 4-D for smothering the persistent weeds and improving the nutrient status of the soil (Maitra and Zaman, 2017). In Alabama, a conservation tillage system using rye or black oat cover crops eliminated the application for post-emergence herbicides in soybean and cotton (Price *et al.*, 2006). Even if cover crops cannot completely eliminate the need for herbicide application, they can curtail the total number of applications to a great extent. While the potential savings in herbicide costs will not necessarily pay for the full cost of cover crop seeding, when combined with possible yield advantages or even yield losses at harvest, cover cropping may provide a positive return in the first year or two of use.

Insect and disease management

Host beneficial diverse microbes that daunt disease form an inhospitable soil environment for many soil borne diseases, encourage beneficial insect predators and parasitoids that can diminish insect damage below economic thresholds, release compounds that reduce nematode pest populations. Some cover crops like marigolds and crimson clover can attract beneficial insects, provide nectar to pollinators and repel damaging ones. Certain cover crops like pongamia and neem leaves are reported to control insects. Some special cover crops like garlic, onions, hot peppers, basil, marigolds, thyme and other herbs can control soil borne diseases in the soil or on the nearby foliage for instance tomato. Annual ryegrass and cereal rye cover crops are found significant in reducing soybean cyst nematodes. Brassica cover crop substituted profitably for chemical nematicides when grown after malting barley or silage corn (Koch, 1995). Cover crops used as trap crop in corn are effective in controlling corn earworm and tarnish bugs (Hoorman, 2009). An experiment conducted using barley, canola, rapeseed and ryegrass as crop rotation with potato resulted in reduction of *Rhizoctonia* by 15% to 50%. Conservation tillage with rye cover crop encouraged natural control of *Helicoverpa armigera* in non-Bt cotton by protecting the habitat of fire ants which are predators of *Helicoverpa armigera* (Reddy, 2016).

Agronomic Management of Cover Crops under CA Systems

Proper seeding time and seedling protection

Most growers sow cover crops after the harvest of economic crop specially under milder climate. The seeds can be no-till drilled instead of broadcast, for better cover crop stands. In warmer regions, it is possible to establish cover crops with some early-maturing vegetable in late spring or early summer. Cover crops fit well into an early vegetable–winter grain rotation sequence. Another strategy is to interseed cover crops with short growing season during the main crop growth in winter grain cropping systems. Winter rye is able to establish well without seed covering, as long as sufficient moisture is present. Dry seeding of sorghum is done under limited moisture condition. In Bangalore regions, gap filling may be done in the fields of groundnut and finger millet with seeds of cowpea, horse gram, niger or sunflower in the event of gaps more than 25% of area during aberrant weather situation. One potential risk of a growing spring cover crop is that its residue or biomass could interfere with the economic crop. For example, if growers plant too quickly after terminating a cover crop, the planter could have difficulty navigating through the stringy, tough residue resulting in hair-pinning (poor seed soil contact where air around the sown seed prevents it from absorbing moisture from the soil) and release of toxin substances as a result of decomposition.

Judicious nutrient management

Nutrient tie-up can be an issue with crop following a grassy cover crop like rye or wheat cover that has reached the boot stage. Mixtures of grass and legume cover crops often mitigate the nitrogen immobilization of pure grass cover crops. The grass component scavenges residual nitrogen effectively, while the legume adds fixed nitrogen that is a boon to the cash crop. Besides, extra nitrogenous fertilizers may be banded in addition to a starter fertilizer dose to combat risk from grass cover. Terminating the rye cover crop earlier, when it is under 6 inches tall, can also stave off this phenomenon.

Herbicide for knock down

Conservation tillage and the allelopathic effects of cover crop residue can both contribute to the suppression of weeds and reduce the need for herbicide application to the main crop. For grassy covers like cereal rye or barley, glyphosate should work just fine, when temperatures are warm enough for plants to be actively growing. When legumes and other types of cover crops are added into the mix, dicamba and 2,4-D may be necessary to knock them down.

Scout for pests

Cover crops can be used in conservation tillage systems to attract beneficial insects. Use of marigold in tomato may reduce nematode problem. Besides, use of thrips deterring cover crop may be used in onion. A live strip of cover crops between crop rows may serve as habitat and a food source until the main crop is established (Schomberg, 2005). Armyworm and cutworm moths like to lay their eggs in grassy cover crop fields, which provide a good food supply for the emerging caterpillars.

Meeting moisture requirement

In drier areas late killing of a winter cover crop may result in moisture deficiency for the main summer crop. However, in warm, humid climates where no-till methods are practiced, the cover crop if grown longer will aid in more residue and water conservation for the main crop. Cover crop may more than compensate for the extra water removed from the soil during the later period of growth. In addition, in very humid regions or on wet soils, the ability of an actively growing cover crop to pump water out of the soil by transpiration may be an advantage.

Termination time

There are a number of ways to terminate a cover crop: tillage, crimping, spraying the cover with herbicides before planting the spring crop. It is important to choose the precise moment at which the vegetative cover crop is controlled, because most of the species used can germinate if the plants are allowed to mature, as may happen with oats, rye, chickpea, vetches and forage radish. For quick decomposition it is recommendable to knock it down with herbicides. In case of direct seeding over the cover crop it is recommended to seed 8 to 12 days after managing the cover crop for cover crops with low to medium C/N ratio (12 to 22) and 12 to 20 days for cover crops with high C/N ratio (>24). The best moment to control the majority of cover crop species is at the full flowering stage after they have accumulated maximum biomass. In legumes, the pods from the first flowering should be already formed but not yet mature. Vetches should have some mature pods. Oats and rye can be best managed at the milk stage. Both sunnhemp and pigeon pea need to be controlled before flowering due to a high regrowth rate and excessive wood development in the stems.

Economics of Cover Crop Establishment and Use

Cover crop seed costs vary considerably from year to year and from region to region, but historically, legume cover crops cost about twice as much to establish as small grain covers which can be offset by the value of N that legumes can replace. Factors affecting the economics of cover crop use include: the cover crop selected, time and method of establishment, the cash value applied to the environment and soil productivity. Literature having an economic assessment in relation to effects on returns or fertilizer inputs is available, but limited information is available on the benefits to soil (Blanco-Canqui *et al.*, 2015). The experiment conducted using cattle grazed and non-grazed winter rye as a cover crop in cotton had average returns of $81/ha (Schomberg *et al.*, 2014). Moreover, it was found that cover crops in a no-tillage system and minimum tillage would be $25.60/acre and $15.10/acre, respectively. Categories accounting benefits for nutrient credits include the reduction in fertilizer purchase, credit in erosion reduction, reduction in the purchase of herbicides, insecticides and increase in yield of cash crop after cover crop. Overall, total net benefits found to be $1354/acre/yr for adoption of long-term cover crops (Adusumilli *et al.*, 2016).

Future Prospects and Challenges

Establishment and management of cover crops can be a great challenge in some soils due to nitrogen immobilisation (e.g., "nitrogen robbery" due to high carbon residues causing soil microbes to use up available soil nitrogen). Improper residue management may result in accumulation of residue in the planter, provoke an unequal drying of the soil and thus delay the warming-up of the seedbed, interfere with sowing and fertilizing activities and eventually hinder the emergence of seedlings. Besides, there may be problem of hairpinning and allelopathic effect on main crop under residue laden condition. Cover crops adoption in semiarid or water-limited regions is limited because cover crops may reduce plant-available water for the main crop. Proper management including early termination and species selection are strategies to grow cover crops in water-limited regions. Cover crops could act as an alternate host for insects and pathogens in the offseason. Cover crops may act as a refugee for cutworms and other insects that may later attack the cash crop. The requirement of labour, timeline of cover crop, limited machinery for the cover crops further make it hard for farmers to grow cover crops.

Conclusion

With an escalation in the growth of global population which is expected to reach 9 billion by 2050, agriculture faces a major predicament of moderating its threats on the natural resource degradation, environmental pollution, soaring production costs and climate change while meeting the future food demand. The need for its correction is a pressing urgency and much attention has been given for restoring soil health and to address other agricultural side-effects like nutrient leaching, water pollution, and soil erosion, by exploring alternative practices like reduced tillage or crop rotations. Direct sowing, reduced tillage, permanent soil cover in the form of residues and cover crops are diverse modalities of conservation agriculture, which represents dramatic change in soil management for modern agriculture. Cover cropping is appreciated as a viable sustainable agricultural practice as it can provide multifarious benefits like preventing soil erosion, weed suppression, providing soil aggregate stability, improving soil hydraulic properties, enhancing the SOC, carbon sequestration, reducing the nitrate leaching and hence reduction of groundwater pollution. These benefits largely develop from the physically, chemically and biologically distinct agroecosystem that cover crops shape compared to that under bare fallow. Although the principal crop affords adequate ground coverage at full canopy, interim cover cropping is a necessity under physical erosion. Cover cropping may impact soil microbial functionality responsible for important soil ecosystem services. Thus, cover crops as a part of conservation agriculture can help to improve the resilience of the crop system against climate hazards and minimize soil nutrient losses. This in the long run will benefit the soil quality and hence, all the aspects of farmer's livelihoods related to better crop yields.

References

Adusumilli, N., Davis, S. and Fromme, D. (2016). Economic Evaluation of Using Surge Valves in Furrow Irrigation of Row Crops in Louisiana: A Net Present Value Approach. *Agricultural Water Management,* 174: 61-65.

Basche, A.D., Kaspar, T.C., Archontoulis, S. V., Jaynes, D. B., Sauer, T. J., Parkin, T. B. and Miguez, F. E. (2016). Soil Water Improvements with the Long-Term Use of a Winter Rye Cover Crop. *Agricultural Water Management,* 172:40-50.

Bergström, L.F. and Jokela, W.E. (2001). Ryegrass cover crop effects on nitrate leaching in spring barley fertilized with $^{15}NH_4$$^{15}NO_3$. *Journal of Environmental Quality,* 30(5):1659-1667.

Blanco-Canqui, H., Claassen, M. and Presley, D. (2012). Summer Cover Crops Fix Nitrogen, Increase Crop Yield, and Improve Soil-Crop Relationships. *Agronomy Journal,* 104:137-147.

Blanco-Canqui, H., Mikha, M.M., Presley, D.R. and Claassen, M.M. (2011). Addition of Cover Crops Enhances No-Till Potential for Improving Soil Physical Properties. *Soil Science Society of America Journal,* 75:1471-1482.

Blanco-Canqui, H., Shaver, T.M., Lindquist, J.L., Shapiro, C.A., Elmore, R.W., Francis, C.A. and Hergert, G.W., (2015). Cover crops and ecosystem services: Insights from studies in temperate soils. *Agronomy Journal,* 107(6): 2449-2474.

Bugg, R. L. and Dutcher, J. D. (1993). *Sesbania exaltata* (Rafinesque-Schmaltz) Cory (Fabaceae) as a warm-season cover crop in pecan orchards: effects on aphidophagous Coccinellidae and pecan aphids. *Biological Agriculture & Horticulture,* 9(3):215-229.

Calegari, A., Mondardo, A., Bulisani, E.A., Wildner, L.D.P., Costa, M.D., Alcantara, P.B., Miyasaka, S. and Amado, T.J.C., (1992). Adubação verde no sul do Brasil. *Rio de Janeiro: AS-PTA,* p.346.

Chavarria D. N., Verdenelli R. A., Serri D. L., Restovich S. B., Andriulo A. E. and Meriles J. M., (2016). Effect of cover crops on microbial community structure and related enzyme activities and macronutrient availability. *European Journal of Soil Biology,* 76: 74–82.

Clark, A. (2008). Managing Cover Crops Profitably. Diane Publishing, Collingdale, PA

Clark, A..J., Meisinger, J..J., Decker, A. M. and Mulford, F. R. (2007). Effects of a Grass-Selective Herbicide in a Vetch–Rye Cover Crop System on Corn Grain Yield and Soil Moisture. *Agronomy Journal,* 99(1):43-48.

Corak, S. J., W. W. Frye and M. S. Smith. (1991). Legume mulch and nitrogen fertilizer effects on soil water and corn production. *Soil Science Society of America Journal,* 55(5):1395-1400.

186 Conservation Agriculture and Climate Change

Cordeau S., Guillemin J. P., Reibel C., Chauvel, B. (2015). Weed species differ in their ability to emerge in no-till systems that include cover crops. *Annals of Applied Biology*, 166:444–455.

Creamer, N. G. and Baldwin, K. R. (2000). An Evaluation of Summer Cover Crops for Use in Vegetable Production Systems in North Carolina. *Horticultural Science*, 35: 600-603.

Debarba, L. and Amado, T. J. C., (1997). Desenvolvimento de sistemas de produção de milho no sul do Brasil com características de sustentabilidade. *Revista Brasileira de Ciência do Solo*, 21(3): 473-480.

Delgado, J. A., Dillon, M. A., Sparks, R. T., and Essah, S. Y. C. (2007). A decade of advances in cover crops. *Journal of Soil and Water Conservation*, 62(5):110A–117A.

Derpsch, R. and Florentin, M. A., (1992). La mucuna y otras plantas de abono verde para pequenas propiedades. *Ministerio de Agricultura y Ganadería. Asunción, Paraguay*.

Don, A., Schumacher, J. and Freibauer, A. (2011). Impact of Tropical Land-Use Change on Soil Organic Carbon Stocks—A Meta-Analysis. *Global Change Biology*, 17:1658-1670.

Finney, D. M., Buyer, J. S. and Kaye, J. P. (2017). Living cover crops have immediate impacts on soil microbial community structure and function. *Journal of Soil and Water Conservation*, 72(4):361-373.

Fisher, N. M. (1996). The potential of grain and forage legumes in mixed farming systems. Legumes in Sustainable Farming Systems, Occasional Symposium No. 30 British Grassland Society, 290-299.

Florentin, M. (1999). Uso de abonos verdes en los sistemas de producción de lospequeños productores de San Pedro, Paraguay. In: VIEDMA, L., Coord. Curso de Siembra Directa. mag/dia, procisur, gtz. Encarnación, Paraguay.pp. 44-57

Folorunso, O., Rolston, D., Prichard, T. and Loui, D. (1992). Soil Surface Strength and Infiltration Rate as Affected by Winter Cover Crops. *Soil Technology*, 5:189-197.

Gardner, J. B. and Drinkwater, L. E. (2009). The fate of nitrogen in grain cropping systems: a meta-analysis of 15N field experiments. *Ecological Applications*, 19: 2167-2184.

Halvorson, A. D. and C. A. Reule. (2006). Irrigated corn and soybean response to nitrogen under no-till in northern Colorado. *Agronomy Journal*, 98:1367–1374.

Haruna, S. I. and Nkongolo, N. V. (2015). Cover Crop Management Effects on Soil Physical and Biological Properties. *Procedia Environmental Sciences*, 29: 13-14.

Hoorman, J. J. (2009). Using cover crops to improve soil and water quality. *Agriculture and Natural Resources. The Ohio State University Extension Press, Lima, Ohio*, pp.1-4.

Jabran, K., Mahajan, G., Sardana, V. and Chauhan, B. S. (2015). Allelopathy for weed control in agricultural systems. *Crop Protection*, 72:57-65.

Jarecki, M. K. and Lal, R. (2003). Crop management for soil carbon sequestration. *Critical Reviews in Plant Sciences*, 22(6):471-502.

Kaspar, T., Jaynes, D., Parkin, T., Moorman, T. and Singer, J. (2012). Effectiveness of Oat and Rye Cover Crops in Reducing Nitrate Losses in Drainage Water. *Agricultural Water Management*, 110: 25-33.

Koch, D. W. (1995). Brassica utilization in sugarbeet rotations for biological control of cyst nematode. SARE Project Report #LW91-022. Western Region SARE. Logan, Utah.

Maitra, S. and Zaman, A. (2017). Brown manuring, an effective technique for yield sustainability and weed management of cereal crops: a review. *International Journal of Bioresource Science*, 4(1):1-5.

Narain, P., Singh, G. and Joshie, P. (1992). Technological needs of vegetative land protection measures. In: Proceedings 7th ISCO Conference, 28-30 September,1992. Sydney, Austrialia. pp. 638-643.

Paustian, K., Lehmann, J., Ogle, S., Reay, D., Robertson, G.P. and Smith, P. (2016). Climate-smart soils. *Nature*, 532(7597):49-57.

Poeplau, C. and Don, A. (2015). Carbon Sequestration in Agricultural Soils via Cultivation Of Cover Crops-A Meta-Analysis. *Agriculture, Ecosystems & Environment*, 200:33-41.

Price, A. J., Reeves, D. W. and Patterson, M. G. (2006). Evaluation of weed control provided by three winter cover cereals in conservation-tillage soybean. *Renewable Agriculture and Food Systems*, 21(3):159-164.

Reddy, P. P. (2016). Cover/Green Manure Crops. In: Sustainable Intensification of Crop Production, Springer, Singapore, 55-67.

Roarty, S., Hackett, R. A. and Schmidt, O. (2017). Earthworm Populations in Twelve Cover Crop and Weed Management Combinations. *Applied Soil Ecology*, 114:142-151.

Schipanski, M. E., Barbercheck, M., Douglas, M. R., Finney D. M., Haider, K., Kaye J. P. (2014). A framework for evaluating ecosystem services provided by cover crops in agroecosystems. *Agricultural Systems*, 125: 12–22.

Schmidt, R., Gravuer, K., Bossange, A. V., Mitchell, J. and Scow, K. (2018). Long-term use of cover crops and no-till shift soil microbial community life strategies in agricultural soil. *PLoS One*.13:e0192953.

Schomberg, H. H. (2005). Enhancing sustainability in cotton production through reduced chemical inputs, cover crops and conservation tillage. SARE Project Report #LS01-121. Southern Region SARE. Griffin, GA.

Schomberg, H. H., Fisher, D. S., Reeves, D. W., Endale, D. M., Raper, R. L., Jayaratne, K.S.U., Gamble, G. R. and Jenkins, M. B. (2014). Grazing winter rye cover crop in a cotton no-till system: Yield and economics. *Agronomy Journal*, 106(3):1041-1050.

Sharma, P., Shukla, M. K., Sammis, T.W., Steiner, R. L. and Mexal, J. G. (2012). Nitrate-Nitrogen Leaching from Three Specialty Crops of New Mexico under Furrow Irrigation System. *Agricultural Water Management*, 109:71-80.

Sharma, P., Singh, A., Kahlon, C. S., Brar, A. S., Grover, K. K., Dia, M. and Steiner, R. L. (2018). The Role of Cover Crops towards Sustainable Soil Health and Agriculture—A Review Paper. *American Journal of Plant Sciences*, 9:1935-1951.

Teasdale, J. R. and Mohler, C. L. (2000). The quantitative relationship between weed emergence and the physical properties of mulches. *Weed Science*, 48:385–392.

Tonitto, C., David, M. and Drinkwater, L. (2006). Replacing Bare Fallows with Cover Crops in Fertilizer-Intensive Cropping Systems: A Meta-Analysis of Crop Yield and N Dynamics. *Agricultural Water Management*, 112:58-72.

Truman, C. C., Shaw, J. N. and Reeves, D. W. (2005). Tillage effects on rainfall partitioning and sediment yield from an Ultisol in central Alabama. *Journal of Soil and Water Conservation*, 60:89-98.

Wendt, R. C. and R. E. Burwell. (1985). Runoff and soil losses for conventional, reduced, and no-till corn. *Journal of Soil and Water Conservation,*. 40:450-454.

Williams, S. M. and Weil, R. R. (2004). Crop cover root channels may alleviate soil compaction effects on soybean crop. *Soil Science Society of America Journal*, 68:1403-1409

Zablotowicz, R.M., Locke, M.A. and Gaston, L.A. (2007). Tillage and Cover Effects on Soil Microbial Properties and Fluometuron Degradation. *Biology and Fertility of Soils*, 44:27-35.

13

Biodynamic Farming and Organic Farming: Traditional Approach for Resource Conservation

Mahua Banerjee and R. Saha[1]

Department of Agronomy, Institute of Agriculture, Visva-Bharati Sriniketan, Birbhum, West Bengal, 731 236, India
[1]ICAR-Central Research Institute for Jute and Allied Fibres Barrackpore, West Bengal -700 121, India

Introduction

Organic farming is basically a holistic management system, which promotes and improves the health of the agro-ecosystem related to biodiversity, nutrient biocycles, soil microbial and biochemical activities. Organic and bio-dynamic farming emphasises management practices involving substantial use of organic manures, green manuring, organic pest management pratices and so on. It has also come to mean that it is a system of farming that prohibits the use of artificial fertilisers and synthetic pesticides.

Biodynamic agriculture is an advanced form of organic farming system that is gaining increasing attention because of its added advantage of soil health and food quality. It is an alternative variant where the chemical fertilisers are totally replaced by microbial (biological) nutrient givers such as bacteria, algae, fungi, mycorhiza, actinomycetes. Biological Pest management of crops is undertaken by employing predators, parasites and other plethora of natural enemies of pests, in addition to all the rest of option that help to avoid resorting to chemical pesticides. These agents could be augmented into farms or promoted through such activities that favour their flourished activities. Composting, Green manuring, crop rotations. Intercropping, mixed cropping etc. as well as bird perches, trap crops promote such biological activities.

Origin

Biodynamic agriculture was initially developed out of eight lectures in 1924 given by Rudolf Steiner (1861–1925), an Australian scientist and Philosopher, to a group of farmers near Breaslau which was then located in eastern part of Germany and is now called Wroclaw in Poland (Paull, 2013a). In response to the interesting observations from farmers after the introduction of chemical fertilizers, Steiner gave these lectures to the group of farmers. The farmers observed that soils were becoming depleted, soil health, quality of crops and livestocks drastically deteriorated. The lectures were published in November 1924; the first English translation appeared in 1928 as *The Agriculture Course* (Diver, 1999a). Thus, biodynamic agriculture was the first ecological farming system developed as a grassroot alternative to chemical agriculture. It was the first of the organic agriculture movements that treats soil fertility, plant growth, and livestock care as ecologically interrelated tasks, emphasizing spiritual and supernatural perspectives.

Steiner emphasized that the methods the proposed should be tested experimentally. For this purpose, Steiner established a research group, the "Agricultural Experimental Circle of Anthroposophical Farmers and Gardeners of the General Anthroposophical Society"(Paull, 2013b, c). Between 1924 and 1939, this research group attracted about 800 members from around the world, including Europe, the Americas and Australasia. Another group, the "Association for Research in Anthroposophical Agriculture" (*Versuchsring anthroposophischer Landwirte*), directed by the German agronomist Erhard Bartsch, was formed to test the effects of biodynamic methods on the life and health of soil, plants and animals; the group published a monthly journal, *Demeter* (Paull, 2011a, c). Bartsch was also instrumental in developing a sales organisation for biodynamic products, Demeter, which still exists today. The Research Association was renamed the Imperial Association for Biodynamic Agriculture (*Reichsverband für biologisch-dynamische Wirtschaftsweise*) in 1933. It was dissolved by the National Socialist regime in 1941. In 1931 the association had 250 members in Germany, 109 in Switzerland, 104 in other European countries and 24 outside Europe. The oldest biodynamic farms are the Wurzerhof in Austria and Marienhöhe in Germany (Koepf and Plato, 2001)

In 1938, Ehrenfried Pfeiffer's text, *Bio-Dynamic Farming and Gardening*, was published in five languages – English, Dutch, Italian, French, and German; this became the standard work in the field for several decades. In July 1939, at the invitation of Walter James, 4th Baron Northbourne, Pfeiffer travelled to the UK and presented the Betteshanger Summer School and Conference on Biodynamic Farming at Northbourne's farm in Kent. The conference has been described as the 'missing link' between biodynamic agriculture and organic farming because, in the year after Betteshanger, Northbourne published his manifesto of organic farming, *Look to the Land*, in which he coined the term 'organic farming' and praised the methods of Rudolf Steiner (Paull, 2011a, b). In the 1950s, Hans Mueller was encouraged by Steiner's work to create the organic-biological farming method in Switzerland; this later developed to become the largest certifier of organic products in Europe, *Bioland* (Kristiansen and Mainsfield, 2006).

Method of Biodynamic Farming

In common with other forms of organic agriculture, biodynamic agriculture uses management practices that are intended to "restore, maintain and enhance ecological harmony". Central features include crop diversification, the avoidance of chemical soil treatments and off-farm inputs generally, decentralized production and distribution, and the consideration of celestial and terrestrial influences on biological organisms (Lotter, 2003). The Demeter Association recommends that "minimum of ten percent of the total farm acreage be set aside as a biodiversity preserve. That may include but is not limited to forests, wetlands, riparian corridors, and intentionally planted insectaries. Diversity in crop rotation and perennial planting is required: no annual crop can be planted in the same field for more than two years in succession. Bare tillage year round is prohibited so land needs to maintain adequate green cover".

The Demeter Association also recommends that the individual design of the land "by the farmer, as determined by site conditions, is one of the basic tenets of biodynamic agriculture. This principle emphasizes that humans have a responsibility for the development of their ecological and social environment which goes beyond economic aims and the principles of descriptive ecology." (Leiber *et al.*, 2006) Crops, livestock, and farmer, and "the entire socioeconomic environment" form a unique interaction, which biodynamic farming tries to "actively shape through a variety of management practices. "The farmer seeks to enhance and support the forces of nature that lead to healthy crops, and rejects farm management practices that damage the environment, soil plant, animal or human health....the farm is conceived of as an organism, a self-contained entity with its own individuality," holistically conceived and self-sustaining (Alsos *et al.*, 2011). "Disease and insect control are addressed through botanical species diversity, predator habitat, balanced crop nutrition, and attention to light penetration and airflow. Weed control emphasizes prevention, including timing of planting, mulching, and identifying and avoiding the spread of invasive weed species".

Principles of Biodynamic Farming/Organic Farming

Key principles are: mixed farming, crop rotation, organic cycle optimization:

i) Organize the production of crops and livestock and the management of farm resources so that they harmonize rather than conflict with natural system.

ii) Use and development of appropriate technologies based upon an understanding of biological systems.

iii) Achieve and maintain soil fertility for optimum production by relying primarily on renewable resources.

iv) Use diversification to pursue optimum production.

v) Aim for optimum nutritional value of staple food.

vi) Use decentralized structures for processing, distributing and marketing of products.

vii) Strive for equitable relationships between those who work and live on the land.

viii) Create a system, which is aesthetically pleasing for those working in this system, and for those viewing it from outside, e.g., it should enhance rather than scare the landscape of which it forms a part.

ix) Maintain and preserve wild life and their habitats.

Several practices in organic agriculture are aimed at minimizing the downward movement of nutrients in the soil profile:

i) The rotation with deep-rooted crops.

ii) The avoidance of high solubility nutrient sources.

iii) The avoidance of mould board ploughing in favour of chisel ploughing.

iv) The insertion of nutrients into the rotation onto a sod crop, if possible, to maximize uptake.

v) The seasonal use of cover crops in and around the major cash crops.

Biodynamic Preparations

Steiner prescribed nine different preparations to aid fertilization, and described how these were to be prepared. The main purposes of these preparations were to enhance soil quality and stimulate plant life. They consist of mineral, plant or animal manure extracts, usually fermented and applied in small proportions to compost, manures, the soil, or directly onto plants, after dilution and stirring procedures called dynamizations. Steiner believed that these preparations mediated terrestrial and cosmic forces into the soil (Kirchmann *et al.*, 2008). The original biodynamically prepared substances are numbered 500 - 508, where the first two are used for preparing fields, and the other seven are used for making compost. A long term trial (DOK experiment) evaluating the biodynamic farming system in comparison with organic and conventional farming systems, found that both organic farming and biodynamic farming resulted in enhanced soil properties, but had lower yields than conventional farming. Regarding compost development beyond accelerating the initial phase of composting, some positive effects have been noted (Reeve *et al.*, 2005).

Field Preparations

Field preparations, for stimulating humus formation:

- **500**: A humus mixture prepared by filling a cow's horn with cow manure and burying it in the ground (40–60 cm below the surface) in the autumn. It is left to decompose during the winter and recovered for use as fertilizer the following spring.

- **501**: Crushed powdered quartz stuffed into a cow's horn and buried in the ground in springtime and taken out in autumn. It can be mixed with 500 but is usually prepared on its own. The mixture is sprayed under very low pressure over the crop during the wet season, as a supposed antifungal (Linda, 2013)

Compost Preparations

The compost preparations recommended by Steiner employ herbs which are frequently used in alternative medical remedies. Many of the same herbs Steiner referenced are used in organic practices to make foliar fertilizers, green manure, or in composting. The preparations Steiner discussed were:

- **502**: Yarrow blossoms (*Achillea millefolium*) stuffed into the urinary bladders from red deer (*Cervus elaphus*), placed in the sun during summer, buried in the ground during winter, and retrieved in the spring.

- **503**: Chamomile blossoms (*Matricaria recutita*) stuffed into the small intestines of cattle, buried in humus-rich earth in the autumn, and retrieved in the spring.

- **504**: Stinging nettle *(Urtica dioica)* plants in full bloom stuffed together underground surrounded on all sides by peat for a year.

- **505**: Oak bark (*Quercus robur*) chopped in small pieces, placed inside the skull of a domesticated animal, surrounded by peat, and buried in the ground in a place near rain runoff.

- **506**: Dandelion flowers *(Taraxacum officinale)* stuffed into the mesentery of a cow, buried in the ground during winter, and retrieved in the spring.

- **507**: Valerian flowers *(Valeriana officinalis)* extracted into water.

- **508**: Silica-rich horsetail plant *(Equisetum arvense)* and used as a foliar spray to suppress fungal diseases in plants Horsetail *(Equisetum)*.

- Biodynamic preparations are intended to help moderate and regulate biological processes as well as enhance and strengthen the life (etheric) forces on the farm. The preparations are used in homeopathic quantities, meaning they produce an effect in extremely diluted amounts. As an example, just 1/16th ounce ¾ a level teaspoon of each compost preparation is added to seven- to ten-ton piles of compost (Linda, 2013).

Recent biodynamic research supports the static pile approach as a viable compost option. In the July-August 1997 issue of *Biodynamics*, Dr. William Brinton of Woods End Agricultural Research Institute published "Sustainability of Modern Composting: Intensification Versus Costs and Quality." Brinton argues that low- tech composting methods are just as efficient in stabilizing nutrients and managing humus as the management and capital intensive compost systems that employ compost turners and daily monitoring. These findings are particularly encouraging to farmers choosing the low-input approach to this age-old practice of transforming organic matter into valuable humus (Brinton, 1997).

Cover Crops and Green Manures

Biodynamic farmers make use of cover crops for dynamic accumulation of soil nutrients, nematode control, soil loosening, and soil building in addition to the commonly recognized benefits of cover crops like soil protection and nitrogen fixation. Biodynamic farmers also make special use of plants like rapeseed, mustard, and oilseed radish in addition to common cover crops like rye, legumes. Cover crop strategies include under sowing and catch cropping as well as cover crops and summer green manures.

Green manuring is a biological farming practice that receives special attention on the biodynamic farm. Green manuring involves the soil incorporation of any field or forage crop while green, or soon after flowering, for the purpose of soil improvement. The decomposition of green manures in soils parallels the composting process in that distinct phases of organic matter breakdown and humus buildup are moderated by microbes. Many biodynamic farmers, especially those who follow the guidelines established by Dr. Ehrenfried Pfeiffer, spray the green residue with a microbial inoculant (BD Field Spray®)

prior to plowdown. The inoculant contains a mixed culture of microorganisms that help speed decomposition, thereby reducing the time until planting. In addition, the inoculant enhances formation of the clay-humus crumb which provides numerous exchange sites for nutrients and improves soil structure (Diver, 1999c).

Crop Rotations & Companion Planting

Crop rotation - the sequential planting of crops is honed to a fine level in biodynamic farming. A fundamental concept of crop rotation is the effect of different crops on the land. Generally, crops should be rotated considering whether they are "humus-depleting" and "humus- restoring" crops; "soil-exhausting" and "soil- restoring" crops; and "organic matter exhausting" and "organic matter restoring" (Koepf, *et al.*, 1976).

Companion planting, a specialized form of crop rotation commonly used in biodynamic gardening, entails the planned association of two or more plant species in close proximity so that some cultural benefit (pest control, higher yield) is derived. In addition to beneficial associations, companion planting increases biodiversity on the farm which leads to a more stable agroecosystem (Pieters, 1938).

Liquid Manures and Herbal Teas

Herbal teas, also called liquid manures or garden teas, are an old practice in organic farming and gardening ¾ especially in biodynamic farming. Herbal teas usually consist of one fermented plant extract, while liquid manures are made by fermenting a mixture of herb plants in combination with fish or seaweed extracts. The purpose of herbal teas and liquid manures are manifold; here again, they perform dual roles by supporting *biological* as well as *dynamic* processes on the farm; i.e., source of soluble plant nutrients; stimulation of plant growth; disease-suppression; carrier of cosmic and earthly forces. To reflect their multi-purpose use, they are sometimes referred to as immune-building plant extracts, plant tonics, biotic substances, and bio stimulants. Herbal teas and liquid manures aim to influence the phyllosphere; composts, tillage, and green manures influence the rhizosphere. Foliar-applied biotic extracts can sometimes initiate a systemic whole plant response known as induced resistance. It also induces physical modification of the leaf surface to inhibit pathogen spore germination or promote antagonistic (beneficial) microbes to compete against disease-causing organisms (pathogens) (Diver, 1999c).

Planting Calendar

The approach considers that there are lunar and astrological influences on soil and plant development—for example, choosing to plant, cultivate or harvest various crops based on both the phase of the moon and the zodiacal constellation the moon is passing through, and also depending on whether the crop is the root, leaf, flower, or fruit of the plant (Desai, 2007). Lunar and astrological cycles also play a key role in the timing of biodynamic practices, such as the making of BD preparations This aspect of biodynamics has been termed "astrological" and "pseudoscientific" in nature (Novella, 2017).

Seed Production

Biodynamic agriculture has focused on the open pollination of seeds (with farmers thereby generally growing their own seed) and the development of locally adapted varieties (Nemoto, 2007).

Community Supported Agriculture

Biodynamics considers farm as a self-contained entity or farm organism. It completes the circle with suitable marketing schemes to support the economic viability of farms. The Demeter label for certified bio dynamically grown foods is one possibility. A second consequence of this view is the Community Supported Agriculture movement (Groh and Steven, 1997).

Community Supported Agriculture, or CSA, is a direct marketing substitute for small-scale growers. In a CSA, the farmer grows food for a group of shareholders (or subscribers) who pledge to buy a portion of the farm's crop that season. This arrangement gives growers up- front cash to finance their operation and higher prices for produce, since the middleman has been removed. Besides receiving fresh, high-quality produce, shareholders also know that they're directly supporting a local farm which is very much important to ensure rural self-reliance (Diver, 1999 a, b).

Advantages

- Produce optimal soil conditions for high yields and good quality crops.
- Balanced nutrient supply
- Improve plant growth and physiological activities of plants
- Improve soil physical conditions: granulation and good tilth, good aeration, easy root penetration, improved water holding capacity, better soil aggregation, carbon as source of energy.

- Improve soil chemical properties such as supply and retention of soil nutrients and promote favorable chemical reactions

- Reduced need of purchased inputs

- Minimum pollution by utilization of wastes and by products

- Organically grown crops are more healthy and nutritionally superior

- Organically grown crops are more resistant to disease and insects and hence few chemical sprays are required

- Increasing consumer demand for produces free of toxic chemical residues and greater prices for organics

- Prevent environmental degradation and can be used to regenerate degraded lands

- Crop diversification secure more incomes

Limitations

Factors Constraining Greening of Indian Agriculture

Although there are positive signs for green agriculture in India it is not growing at a swiftness to enhance its market attractiveness so as to motivate larger section of farming community to opt for organic agriculture. Major problems that delay the growth of organic agriculture in India can be listed as follows:

Factors limiting Bio-inputs Market

In exploring the factors limiting bio-input market we attempted to analyse them from multiple stakeholder's perspectives presented briefly in the following:

Producers'/Distributors'/Traders' point of view

- Lack of proper infrastructure for allocation and conservation of bio-inputs is a major constraint hinders the access of these inputs to farmers.

- Existence of poor quality bio-inputs in market reduces the reliability of input providers. Lack of quality control mechanisms for bio-inputs furthers the mistrust among farmers.

- Given the low penetration of bio-inputs market and the limited shelf-life it is disincentivizing the traders to store and sell bio-inputs

From Users' (farmers') point of view

- Bio-fertilizers and bio-pesticides are perceived as less yielding.

- Some climatic regions and soil conditions are not suitable for specific strains of organic production.

- For some strains limited shelf life is also constrain as most of the bio-input last only for about 4-6 months.

- Given the mandated gestation period of around three years for a conventional farm to become an organic farm the benefits perceived by farmers in general and small and marginal farmers in particular tend to be limited as they have short term orientation. As a result, even if they are aware they are hesitant to switch over to organic (green) agricultural practices.

From Promoters' (Government's) point of view

- Agricultural departments, research institutions and extension services have for long been oriented towards chemical input agriculture as a result there is a requirement for reorienting these officials towards organic (green) agriculture

- Changing the cropping and cultivation patterns is slow and time-consuming process given the high levels of illiteracy and large number of small and marginal farmers it makes the change process difficult.

- Subsidies on chemical fertilizers and pesticide obstruct the growth of organic agriculture.

Limiting Factors Organic Produce Market

- Lack of market information in general and organic market information in particular is biggest problem for Indian agriculture.

- Quality of Indian food industry is always a constraint for growth, low consistency of quality and contamination in food products is a impediment in capturing the available market especially the international market.

- Given the high levels of transaction costs for getting farms certified as organic it is a major deterrent for enhancing organic production in the country.

- Government has shown limited interest for organic agriculture, though the activities from government side are increasing but till date there is no direct support from government side in terms of subsidy or market support towards organic agriculture.

- Lack of proper infrastructure in terms of roads from remote villages, cold storage facilities and slow transportation infrastructure affects the cost, quality and reach of producers and

- Indian organic agriculture is very fragmented and there are no organizations for managing the entire value chain of organic products.

Present Status

Today biodynamic agriculture is practiced on farms around the world, on various scales, and in a variety of climates and cultures. As of 2019, biodynamic techniques were used on 202,045 hectares in 55 countries. However, most biodynamic farms are located in Europe, the United States, Australia, and New Zealand. Germany accounts for 41.8% of the global total; the remainder average 1750 ha per country (Anonymous, 2019). Biodynamic methods of cultivating grapevines have been taken up by several notable vineyards (Reeve *et al.*, 2005). Now a day, biodynamics is practiced in a variety of circumstances, ranging from temperate arable farming, viticulture in France, cotton production in Egypt, to silkworm breeding in China. Demeter International is the primary certification agency for farms and gardens using the methods (Leiber *et al.*, 2006).

In the United States, biodynamic farming dates from 1926. From 1926 through to 1938, 39 farmers and gardeners in USA pursued biodynamic practices. The Biodynamic Farming & Gardening Association was founded in 1938 as a New York state corporation (Paull, 2019c).

In Great Britain, biodynamic farming dates from 1927. In 1928 the *Anthroposophical Agricultural Foundation* was founded in England; this is now called the *Biodynamic Agriculture Association*. In 1939, Britain's first biodynamic agriculture conference, the Betteshanger Summer School and Conference on Biodynamic Agriculture, was held at Lord Northbourne's farm in Kent; Ehrenfried Pfeiffer was the lead presenter (Paull, 2019b).

In Australia, the first biodynamic farmer was Ernesto Genoni who in 1928 joined the Experimental Circle of Anthroposophical Farmers and Gardeners, followed soon after by his brother Emilio Genoni. Ernesto Genoni's first biodynamic farm was at Dalmore, in Gippsland, Victoria, in 1933. The following year, Ileen Macpherson and Ernesto Genoni founded Demeter Biological Farm at Dandenong, Victoria, in 1934 and it was farmed using biodynamic principles for over two decades. Bob Williams presented the first public lecture in Australia on biodynamic agriculture on 26 June 1938 at the home of the architects Walter Burley Griffin and Marion Mahony Griffin at Castlecrag, Sydney. Since the 1950s research work has continued at the Biodynamic Research Institute (BDRI) in Powelltown, near Melbourne under the direction of Alex Podolinsky. In 1989 Biodynamic Agriculture Australia was established, as a not for profit association (Paull, 2019a).

In France, the International Federation of Organic Agriculture Movements (IFOAM) was formed in 1972 with five founding members, one of which was the Swedish Biodynamic Association. The University of Kassel had a Department of Biodynamic Agriculture from 2006 to March 2011.

Emerson College (UK) was founded in 1962 and named after Ralph Waldo Emerson, American poet and transcendentalist. Since then it has held courses inspired by the philosophy and teachings of Rudolf Steiner, including on biodynamic agriculture (Paull, 2010).

Research Evidences

- A study on "Soil Quality and Financial Performance of Biodynamic and Conventional Farms in New Zealand" was published by Dr. John Reganold in the April 16, 1993 issue of *Science*. It reported contrasted soil quality factors and the financial performance of paired biodynamic and conventional farms in New Zealand (Koepf, 1993). In a comparison of 16 adjacent farms, the biodynamic farms exhibited superior soil physical, biological, and chemical properties and were just as financially viable as their conventional counterparts.

- *Agriculture of Tomorrow* contains research reports from 16 years of field and laboratory work conducted by the German researchers Eugen and Lilly Kolisko. They emphasized the esoteric nature of biodynamics: the moon and plant growth; the forces of crystallization in nature; planetary influences on plants; homeopathy in agriculture; experiments with animals to study the influence of homeopathic quantities; capillary dynamolysis; research on the biodynamic preparations (Reganold, *et al.*, 1993).

- Panchgavya contains many useful microorganisms such as fungi, bacteria, actinomycetes and various micronutrients. The formulation act as tonic to enrich the soil, induce plant vigour with quality production. Physico-chemical studies have revealed that panchgavya possess almost all macro and micronutrients and growth hormones (IAA, GA) required for plant growth. Predominance of fermentative microorganisms such as yeasts and Lactobacillus helps improve the soil biological activity and promote the growth of other microorganisms. For foliar spray 3-4% panchgavya solution is quite effective. Four to five sprays ensure optimum growth and productivity: (a) two sprays before flowering at 15 days' interval, (b) two sprays during flowering and pod setting at 10 days' interval and (c) one spray during fruit/pod maturation. Application of panchgavya has been found to be very effective in many horticultural crops such as mango, guava, acid lime, banana, spice turmeric, flower-jasmine, medicinal plants like Coleus, Ashwagandha, vegetable like cucumber, spinach, okra, radish and grain crops such as maize, green gram and sunflower. Panchgavya has also been found to be reducing nematode problem in terms of gall index and soil nematode population. As due to application of panchgavya a thin oily film is formed on the leaves and stem, it reduces evaporation losses and ensures better utilization of applied water (Yadav., 2010).

Biodynamic Farming and its Resource Conservation Potential

The dimension and intensity of problems faced with the use of chemical fertilizers are quite big and may become massive in future if no corrective measures are taken soon in spirit and action. By all means, this is a good example of disturbing the balance of nature and inviting troubles with excessive human involvement in natural processes; that is, the process of regenerative agriculture.

The best option is to encourage the use of organic resources to meet plant nutrition as stand alone and as a supplementary to chemical fertilizers so that we balance the use of resources to attain high crop production without creating further more problems of hazards to plant, human and animal and to the safety of environment.

A number of diverse organic resources are available for use in agriculture. These are mostly of plant origin and are utilized either directly or after having been used as food for animals and humans. They are mostly the wastes generated in agriculture, forests and in animal and human habitations.

	Crop residues after grain removal (straw/stover etc.)
	Oilcakes after oil extraction
	Biomass of weeds or uncultivated plants (water hyacinth)
	Wastes from fruit/vegetable processing factories
	Fodder/feeds fed to animals ---- animal wastes
Plant	Fodder/feed fed to animals and their products consumed (meat, milk, eggs etc.) by man ---- animal wastes
	Farm produce consumed by man (grains, fruits, vegetables etc.) --- human wastes
	Crop wastes due to post harvest losses
	Fruit/vegetable waste after consumption from the main produce
	Forest waste - litter

Further, they have been the time-tested materials for improving soil fertility and crop productivity from the time immemorial. Organic resources are low cost materials on a bulk basis, but on the basis of nutrients their value is high.

The striking features of organic resources are that they are generated mainly close to the point of end use, are renewable, familiar to the farmer and his family since ages, and need little equipment support. In addition, these are expected to be available in increasing amounts in the future as the volume of agricultural production rises and as the animal and human population increases.

Unlike fertilizers, organic resources are "the complete plant food" rather than components of it. Nevertheless, these are not exceptions to causing environmental problems, but these are a part of natural systems and under the best regulatory processes of nature. Recycling of organic resources is undoubtedly the best for more value addition and is recommended the world over.

In India, over 700 million tons of biodegradable waste is produced annually. The nutrient potential of various wastes is estimated at 19 million tons in terms of NPK and economic value at Rs.86 billion. The potential is huge and if this happens, we can as well meet the entire nutrient and health requirements of crops through organic resources. However, 30% utilization of this potential is said to be normally possible, that is around 6 million tons of the so called NPK (if this is taken as the only yardstick) can be contributed to agricultural crop production by proper recycling of the biological wastes.

The saddest thing is that the potential of biological resources is estimated only on its NPK content than on its content of microbial population, organic carbon and other plant nutrients that contribute more in mobilizing nutrients for plants from the soil and atmosphere and in improving physical, chemical and biological properties of the soil. The biological fixation of atmospheric nitrogen on an annual basis the world over is estimated at 2×10^8 tons. Typical levels of nitrogen fixed by soil microbes are also high on a hectare basis.

Organic resources of nutrients as indicated earlier require proper recycling so that they retain maximum nutrients they contain till they are used for feeding the plants. For crop residues, the best way is to plough them in and get them well decomposed in the soil by using bio-cultures that are now available for their fast decomposition.

The food, human and animal wastes of all kinds need to be properly composted at rural and urban levels so that their physical, chemical and biological values are increased and they become more profitable for crop production. The composting of all human, animal and industrial wastes at urban level is more important as the large amounts of farm produce get transported to urban areas and the waste generated from these do not go back to the farms unless they are specially transported to, rather these are often burnt or lie rotting on the roads or in dumping yards creating all sorts of health and hygiene problems for urban dwellers. Many proficient technologies are now available and being used increasingly for composting and recycling of these wastes. A good compost made from an organic wastes including plant and animal wastes, largely generated in rural areas, can have a potential of more than 100 kg/ton in terms of all nutrients including a potential for 60 kg/ton of NPK (60 kg of NPK /ton).

The potential of organic resources including manures and composts are not often measured in terms of microbial populations in these materials, which help in mobilizing nutrients from the soil and atmosphere. Most bio-manures contain 10^7-10^{12} populations (viable cells) of many microbes including bacteria, fungi, actinomycetes etc. Certainly, if this is taken into consideration, these materials will be seen more important than any encapsulated product such as a chemical fertilizer.

The most important thing at this juncture is to value the efforts of composting all sorts of wastes and making these composts available to farmers as we make chemical fertilizers available to them providing all sorts of subsidies. Soil and plant health insurance should be more important than ailment insurance that is achieved through chemicals.

Having created a disparity in nutritional status of the soil, we now require each nutrient to be supplied through a product to keep crops growing. What is and what will be the cost of doing this? Farmers have become and will become bankrupt, and symptoms of this are already seen with some committing suicide, some leaving agriculture, some shifting to organic agriculture, and still many more standing with a begging bowl in front of the government. In this pandemic threatened era of agriculture, where large amount of labours are returning to agriculture sector due to lack of job opportunity in the other sectors, it is the right time to give a boost to this form of agriculture and resource management which will ensure both environmental security and livelihood security in a sustainable manner.

Some Important Formulations Used in India

Preparation of liquid manures for soil enrichment

Many variants of liquid manures are being used by farmers of different states. Few important and widely used formulations are given below:

Sanjivak – Mix 100 kg cow dung, 100 lit cow urine and 500 gm jaggary in 300 lit of water in a 500-lit closed drum. Ferment for 10 days. Dilute with 20 times water and sprinkle in one acre either as soil spray or along with irrigation water.

Jivamrut – Mix cow dung 10 kg, cow urine 10 lit, Jaggary 2 kg, any pulse grain flour 2 kg and Live forest soil 1 kg in 200 lit water. Ferment for 5 to 7 days. Stir the solution regularly three times a day. Use in one acre with irrigation water.

Amritpani – Mix 10 kg cow dung with 500 gm honey and blend thoroughly to form a creamy paste. Add 250 gm of cow desi ghee and mix at high speed. Dilute with 200 lit water. Sprinkle this suspension in one acre over soil or with irrigation water. After 30 days apply second dose in between the row of plants or through irrigation water.

Panchgavya – Mix fresh cow dung 5 kg, cow urine 3 lit, cow milk 2 lit, curd 2 lit, cow butter oil 1 kg and ferment for 7 days with twice stirring per day. Dilute 3 lit of Panchgavya in 100 lit water and spray over soil. 20 lit panchgavya is required per acre for soil application along with irrigation water.

Enriched Panchgavya (or Dashagavya) – Ingredients - cow dung 5 kg, cow urine 3 lit, cow milk 2 lit, curd 2 lit, cow deshi ghee 1 kg, sugarcane juice 3 lit, tender coconut water 3 lit, banana paste of 12 fruits and toddy or grape juice 2 lit. Mix cow dung and ghee in a container and ferment for 3 days with intermittent stirring. Add rest of the ingredients on the fourth day and ferment for 15 days with stirring twice daily. The formulation will be ready in 18 days. Sugarcane juice can be replaced with 500 g jaggery in 3 lits water. In case of non-availability of toddy or grape juice 100g yeast powder mixed with 100 g jaggery and 2 lit of warm water can also be used. For foliar spray 3-4 lit panchgavya is diluted with 100lit water. For soil application 50 lit panchagavya is adequate for one ha. It can also be used for seed treatment.

Botanical pesticides

Many plants are known to have pesticidal properties and the extract of such plants or its refined forms can be used in the management of pests. Among various plants identified for the purpose, neem has been found to be most effective.

Neem (*Azadirachta indica*) – Neem has been found to be effective in the management of approximately 200 insects, pests and nematodes. Neem is very effective against grasshoppers, leaf hoppers, plant hoppers, aphids, jassids, and moth caterpillars. Neem extracts, are also very effective against beetle larvae, butterfly, moth and caterpillars such as Mexican bean beetle, Colorado potato beetle and diamondback moth. Neem is very effective against grasshoppers, leaf minor and leaf hoppers such as variegated grasshoppers, green rice leaf hopper and cotton jassids. Neem is fairly good in managing beetles, aphids and white flies, mealy bug, scale insects, adult bugs, fruit maggots and spider mites.

Some other pest control formulations

Many organic farmers and NGOs have developed large number of pioneering formulations which are effectively used for control of various pests. Although none of these formulations have been subjected to scientific justification but their wide acceptance by farmers speak of their usefulness. Farmers can try these formulations, as they can be prepared on their own farm without the need of any purchases. Some of the popular formulations are listed below:

Cow urine – Cow urine diluted with water in ratio of 1: 20 and used as foliar spray is not only effective in the management of pathogens & insects, but also acts as effective growth promoter for the crop.

Fermented curd water – In some parts of central India fermented curd water (butter milk or *Chaach*) is also being used for the management of white fly, jassids aphids etc.

Dashparni extract – Crush neem leaves 5 kg, *Vitex negundo* leaves 2 kg, Aristolochia leaves 2 kg, papaya (*Carica Papaya*) 2 kg, *Tinospora cordifolia* leaves 2 kg, *Annona squamosa* (Custard apple) leaves 2 kg, *Pongamia pinnata* (Karanja) leaves 2 kg, *Ricinus communis* (Castor) leaves 2 kg, *Nerium indicum* 2 kg, *Calotropis procera* leaves 2 kg, Green chilly paste 2 kg, Garlic paste 250 gm, Cow dung 3 kg and Cow Urine 5 lit in 200 lit water ferment for one month. Shake regularly three times a day. Extract after crushing and filtering. The extract can be stored up to 6 months and is sufficient for one acre.

Neem-Cow urine extract - Crush 5 kg neem leaves in water, add 5lit cow urine and 2 kg cow dung, ferment for 24 hrs with intermittent stirring, filter squeeze the extract and dilute to 100 lit, use as foliar spray over one acre. Useful against sucking pests and mealy bugs.

Mixed leaves extract - Crush 3 kg neem leaves in 10 lit cow urine. Crush 2 kg custard apple leaf, 2 kg papaya leaf, 2kg pomegranate leaves, 2 kg guava leaves in water. Mix the two and boil 5 times at some interval till it becomes half. Keep for 24 hrs, then filter squeeze the extract. This can be stored in bottles for 6 months. Dilute 2-2.5 lit of this extract to 100 lit for 1 acre. Useful against sucking pests, pod/fruit borers.

Chilli-garlic extract - Crush 1 kg *Ipomea* (besharam) leaves, 500 gm hot chilli, 500 gm garlic and 5 kg neem leaves in 10 lit cow urine. Boil the suspension 5 times till it becomes half. Filter squeeze the extract. Store in glass or plastic bottles. 2-3 lit extract diluted to 100 lit is used for one acre. Useful against leaf roller, stem/fruit/pod borer.

Broad spectrum formulation - 1 - In a copper container mix 3 kg fresh crushed neem leaves and 1 kg neem seed kernel powder with 10 lit of cow urine. Seal the container and allow the suspension to ferment for 10 days. After 10 days boil the suspension, till the volume is reduced to half. Ground 500 gm green chillies in 1 lit of water and keep overnight. In another container crush 250gm of garlic in water and keep overnight. Next day mix the boiled extract, chilli extract and garlic extract. Mix thoroughly and filter. This is a broad spectrum pesticide and can be used on all crops against wide variety of insects. Use 250 ml of this concentrate in 15 lit of water for spray.

Broad spectrum formulation - 2 Suspend 5 kg neem seed kernel powder, 1kg Karanj seed powder, 5 kg chopped leaves of besharam (*Ipomea* sp.) and 5kg chopped neem leaves in a 20lit drum. Add 10-12 lit of cow urine and fill the drum with water to make 150 lit. Seal the drum and allow it to ferment for 8-10 days. After 8 days mix the contents and distil in a distiller. Distillate will act as a good pesticide and growth promoter. Distillate obtained from 150lit liquid will be sufficient for one acre. Dilute in appropriate quantity and use as foliar spray. Distillate can be kept for few months without any loss in characteristics.

Cow-pat pit (CPP)

Prepare a brick lined pit measuring 90 x 60 cm and 30 cm deep without any lining in the bottom. Mix 60 kg fresh cow dung with 200gm crushed and powdered egg shells and 300 gm basalt dust (or blue granite dust or bore well soil). Mix thoroughly to obtain smooth paste. Fill the mixture in to pit up to 12 cm height. Dog 5 holes in the paste and put one teaspoon full (3 gm each) of preparation 502 to 506 in each hole. Preparation 507 is mixed with water and half is poured in one hole and half sprinkled over the entire surface. Cover the surface with wet gunny bag.

After four weeks, aerate the dung by turning it with the help of a fork. Smooth out again and cover. Thereafter turn every week. CPP compost will be ready in 12 weeks time.

CPP can be used in various ways depending upon the requirement and crop/ plants. Use 100 gm CPP/acre, mix with BD 500 or 501 and use as spray.

CPP can be used as soil inoculant (@ 2 kg/acre) mixed with composts. CPP can also be used as foliar spray (@ 5kg/acre) right from the beginning of crop to up to fruit/pod formation stage with an interval of 7 to 15 days. CPP can also be used as paste on stem of fruit trees. CPP can also be used as inoculant to biodynamic composts in place of 502 to 507.

Rishi Krishi

Drawn from Vedas, the Rishi Krishi method of natural farming has been mastered by farmers of Maharashtra and Madhya Pradesh. In this method, all on-farm sources of nutrients including composts, cattle dung manure, green leaf manure and crop biomass for mulching are exploited to their best potential with continuous soil enrichment through the use of Rishi Krishi formulation known as "*Amritpani*" and virgin soil. 15 kg of virgin rhizosperic soil collected from beneath of Banyan tree (*Ficus bengalensis*) is spread over one acre and the soil is enriched with 200 lit Amritpani. It is prepared by mixing 250 g ghee into 10 kg of cow dung followed by 500 g honey and diluted with 200 lit of water. This formulation is utilized for seed treatment (*beej sanskar*), enrichment of soil (*bhumi sanskar*) and foliar spray on plants (*padap sanskar*). For soil treatment it need to be applied through irrigation water as fertigation. The system has been demonstrated on a wide range of crops i.e. fruits, vegetables, cereals, pulses, oilseeds, sugarcane and cotton.

Panchgavya Krishi

Panchgavya is a special bioenhencer prepared from five products obtained from cow; dung, uine, milk, curd and ghee. Dr Natrajan, a Medical practitioner and scientist from Tamilnadu Agricultural University, has further refined the formulation suiting to the requirement of various horticultural and agricultural crops. Ingredients and methods of preparation of Panchgavya and enriched Panchgavya (Dashgavya) has already been described in preceding pages. The cost of production of panchgavya is about RS. 25-35 per litre (Yadav, 2010).

Business Ideas for Future Generation

India has traditionally been a country of organic agriculture, but the growth of modern scientific, input intensive agriculture has pushed it to wall. But with the increasing awareness about the safety and quality of foods, long term sustainability of the system and accumulating evidences of being equally productive, the organic farming has emerged as an alternative system of farming which not only address the quality and sustainability concerns, but also ensures a debt free, profitable livelihood option.

Organic vegetable farming, Organic vegetable seed, Organic fish farm, Organic grocery wholeseller, Organic poultry products, Organic food exports, Organic food blogging, Organic livestock breeding, Organic feed for livestock, Organic dairy and dairy products, Organic desert shops, Organic mushroom farm, Organic cosmetics products, Organic baby food, Organic hand pressed oils, Organic food canning, Organic jams and pickles, Natural cleaning products, Organic food flavours and colouring, Organic restaurants, Organic chocolate and candy, Organic juice bar, Organic gardening service, Organic spice, Organic compost sales, Organic food boutique, Organic rooftop gardening, Organic food bakery, Organic farm consultant, Organic liquor, Organic food nutritionist, Organic market.

Biodynamic and Organic Farming: A Viable Support for Women in Agriculture

The health and environmental advantages of organic goods provides a platform for women engaged in this field and encourages the development of sustainable and easily accessible sales outlets for women producers from the remotest corners of India. The Women of India Organic Festival give a successful platform to empower women farmers and entrepreneurs in a fruitful manner. This festival not only promotes the women and farmers engaged in organic farming but also encourage others to follow suit and encourage organic farming culture. Besides promoting organic farming, it also ensures a healthier lifestyle and livelihood security in the rural families and ensure social and economic empowerment of the rural women.

Conclusion

Biodynamics uses scientifically sound organic farming practices that build and sustain soil productivity as well as plant and animal health. The philosophical tenets of biodynamics ¾ especially those that emphasize energetic forces and astrological influences ¾ are harder to grasp, yet they are part and parcel of the biodynamic experience.

That mainstream agriculture does not accept the subtle energy tenets of biodynamic agriculture is a natural result of conflicting paradigms. In mainstream agriculture the focus is on physical- chemical-biological reality. Biodynamic agriculture, on the other hand, recognizes the existence of subtle energy forces in nature and promotes their expression through specialized "dynamic" practices.

A third view, expressed by a local farmer, accepts the premise that subtle energy forces exist and may affect biological systems, but holds there is not enough information to evaluate these influences nor make practical agronomic use of them.

The fact remains that biodynamic farming is practiced on a commercial scale in many countries and is gaining wider recognition for its contributions to organic farming, food quality, community supported agriculture, and qualitative tests for soils and composts. From a practical viewpoint, biodynamics has proved to be productive and to yield nutritious, high-quality foods.

It is apt to quote the statement of the founder of the Organic Farming Association and an Assistant Dean at University of Havana. "Many people think that farming is a simple and mundane act, but they are wrong. It is the soul of any great culture, because it requires not only a great deal of accumulated knowledge, but also putting this knowledge to use every single day. Knowledge of the weather, the soil, plants, animals, the cycles of the nature: all of this is used every day by a farmer to make the decisions that have to be made in order to produce the food that we eat. To use it may seem like food comes from a factory, but in reality it comes from a culture that, generation after generation, has been created to produce that food".

References

Alsos, G.A., Carter, S., and Ljunggren, E. (2011), *The Handbook of Research on Entrepreneurship in Agriculture and Rural Development* Cheltenham, GB:Edward Elgar Publishing.

Anonymous, (2019). Demeter Monitor 2018/2019 (page 6)" (PDF) (in Dutch). Retrieved 2019-03-15.

Brinton, William F. (1997). Sustainability of modern composting: intensification versus costs & quality. *Biodynamics*. July-August. pp. 13-18.

Desai, B. K. (2007). Sustainable agriculture: a vision for the future. New Delhi: B T Pujari/New India Pub. Agency. pp. 228–29. ISBN 978-81-89422-63-9.

Diver (1999), Introduction Archived 2011-05-26 at the Wayback Machine.

Diver (1999), Planetary Influences Archived 2011-05-26 at the Wayback Machine.

Diver, S. (1999). Biodynamic farming and compost preparation (available at: http://www.attra.org/attra-pub/biodynamic.html).

Groh, T. and Steven McFadden (1997). Farms of Tomorrow Revisited: Community Supported Farms, Farm Supported Commu- nities. Biodynamic Farming and Gardening Association, Kimberton, PA. p. 294.

Kirchmann, H., Thorvaldsson, G., Bergstrom, L., Gerzabek, M., Andren, O., Eriksson, L.O., Winninge, M. (2008). Fundamentals of organic agriculture (PDF). Organic Crop Production-Ambitions and Limitations. pp. 13-18.

Koepf, Herbert and Plato, Bodo von, (2001). Die biologisch-dynamische Wirtschaftsweise im 20. Jahrhundert, D., 2001

Koepf, Herbert H. (1993). Research in Biodynamic Agriculture: Methods and Results. Bio-Dynamic Farming and Gardening Association, Kimberton, PA.p. 78.

Koepf, Herbert H., Bo D. Pettersson and Wolfgang S. (1976). Biodynamic Agriculture: An Introduction. Anthro-posophic Press, Hudson, New York. p. 430.

Kristiansen, P. and Mansfield, C. (2006) Overview of organic agriculture, in Paul Kristiansen, Acram Taji, and John Reganold (2006), Organic Agriculture: A global perspective, Collingwood, AU: CSIRO Publishing.

Leiber, F., Fuchs, N. and Spieß, H. (2006). Biodynamic agriculture today, in Paul Kristiansen, Acram Taji, and John Reganold *Organic Agriculture: A global perspective*, Collingwood, AU: CSIRO Publishing.

Linda, Chalker-Scott (2013). The Science Behind Biodynamic Preparations: A Literature Review. *HortTechnology*. 23 (6): 814–819.

Lotter, Donald W. (2003). Organic Agriculture. *Journal of Sustainable Agriculture*. 21 (4): 59-128.

Nemoto, K., Nishikawa, Y. (2007). Seed supply system for alternative agriculture: Case study of biodynamic agriculture in Germany. *Journal of the Faculty of Agriculture*. Shinshu University, Japan. 43 (1–2): 73–81.

Novella, S. (2017). Biodynamic Farming and Other Nonsense. NeuroLogica Blog. The New England Skeptical Society.

Paull, J. (2010) From France to the World: The International Federation of Organic Agriculture Movements (IFOAM), *Journal of Social Research & Policy*, 1(2):93–102.

Paull, J. (2011a) The Betteshanger Summer School: Missing link between biodynamic agriculture and organic farming, *Journal of Organic Systems*, 6(2):13–26.

Paull, J. (2011b). Biodynamic Agriculture: The Journey from Koberwitz to the World, 1924–1938. *Journal of Organic Systems*. 6 (1): 27–41.

Paull, J. (2011c). The secrets of Koberwitz: the diffusion of Rudolf Steiner's agriculture course and the founding of biodynamic agriculture. *Journal of Social Research & Policy*. 2 (1): 19–29.

Paull, J. (2013a) Breslau (Wrocław): In the footsteps of Rudolf Steiner, *Journal of Bio-Dynamics Tasmania*, 110: 10–15.

Paull, J. (2013b) Koberwitz (Kobierzyce); In the footseps of Rudolf Steiner, *Journal of Bio-Dynamics Tasmania*, 109 (Autumn), pp. 7–11.

Paull, J. (2013c) A history of the organic agriculture movement in Australia. In: Bruno Mascitelli, and Antonio Lobo (Eds.) Organics in the Global Food Chain. Connor Court Publishing, Ballarat, ch.3, pp. 37–61.

Paull, J. (2019a) Dalmore Farm: Victoria's first biodynamic farming venture (1933–1934), *Journal of Bio-Dynamics Tasmania*, 131: 26–31.

Paull, J. (2019b) The Pioneers of Biodynamics in Great Britain: From Anthroposophic Farming to Organic Agriculture (1924-1940), *Journal of Environment Protection and Sustainable Development*, 5(4): 138-145.

Paull, J. (2019c) The Pioneers of Biodynamics in USA: The Early Milestones of Organic Agriculture in the United States, *American Journal of Environment and Sustainable Development*, 6(2):89-94.

Pieters, A.J. (1938). Soil-Depleting, Soil- Conserving, and Soil-Building Crops. USDA Leaflet No. 165. p. 7.

Reeve, Jennifer R., Carpenter-Boggs, Lynne, Reganold, John P., York, Alan L., McGourty, Glenn, McCloskey, Leo P. (2005). Soil and Winegrape Quality in Biodynamically and Organically Managed Vineyards. *American Journal of Enology and Viticulture*. 56 (4): 367-376.

Reganold, J.P., Palmer, A.S., Lockhart, J.C., Macgregor, A.N. (1993). Soil quality and financial performance of biodynamic and conventional farms in New Zealand. Science. April 16. pp. 344-349.

Yadav, A.K. (2010). Organic Agriculture (Concept, Scenario, Principals and Practices), National Project on Organic farming, Deptt of Agriculture and Cooperation, Govt. of India.

14

Environmental Friendly Insect Pest Management under Resource Conservation Technologies

B.S. Gotyal, S. Satpathy and V. Ramesh Babu

ICAR- Central Research Institute for Jute and Allied Fibres, Barrackpore West Bengal -700 121, India

Introduction

Environmental friendly Insect Pest Management (IPM) relies on preventive rather than reactive strategies. The cropping program should focus primarily on preventive practices above and below ground such as crop management and soil management to build the farm's natural defenses. Reactive management *viz.*, reactive inputs for pest management and reactive inputs to reduce plant stress are reserved for problems not solved by the preventive or planned strategiesfor supplemental pest management practices and planned supplemental soil practices to reduce crop stress and/or optimize yield and crop quality. Earlier IPM models are designed from the scientific perspective with a focus on ecological, environmental friendly and evolutionary aspects of pest management to reduce or prevent economic losses.There are four major components in the new environmentally pest management model that address various pest management options, the knowledge, and resources the grower has to address the pest issue, planning and organization of information to take appropriate management actions, and maintaining good communication to acquire and disseminate knowledge about pests and their management. These models mainly address the human, environmental, social and economic factors that influence food production.

IPM for Resource Conservation

IPM is a comprehensive approach to insect management to protect against yield loss. It provides many benefits for farmers, society and the environment.

IPM practices offer farmers flexibility, good use of resources, opportunities to increase yields and profits, new technology and reduced potential for insect resistance. The conservation of natural resources such as natural enemies of pests and their augmentation is of prime importance. Besides, the intrinsic property of renewability, reversibility and resilience of botanicals and bio pesticides make them most dependable tools for sustainable IPM. Hence, to maintain ecological balance and to manage the pests, the use of bio-agents and bio pesticides/botanicals must receive priority attention. Sustainable farming practices commonly include:

- Crop rotations that mitigate insect and other pest problems; provide alternative sources of soil nitrogen; reduce soil erosion; and reduce risk of water contamination by agricultural chemicals.

- Pest control strategies include integrated pest management techniques that reduce the need for pesticides by practices such as scouting/ monitoring, use of resistant cultivars, timing of planting, and biological pest controls.

- Increased mechanical/biological weed control; more soil and water conservation practices; and strategic use of green manures.

- Use of natural or synthetic inputs in a way that poses no significant hazard to humans or the environment.

Environmental Friendly Insect Pest Management

The concept of pest control has changed to pest management over the years knowing that a balanced approach to managing pest populations to levels that do not cause economic losses is better than eliminating or eradicating (except for newly introduced invasive pests), for environmental and economic reasons. Although the term control is frequently used in literature and conversations, it generally refers to management. A thorough knowledge of IPM principles and various ecofriendly management options for all possible pest problems is important as some are preventive and others are curative. Some of the recommended practices may not be practical in all situations and the grower or the pest control professional has to choose the option(s) appropriate for their situation. It is also essential to understand the inherent and potential interactions among these management options to achieve desired results. The following are common ecological and environmental friendly control options that can be used at different stages of crop production to prevent, reduce, or treat pest infestations. Each of them may provide a certain level of control, but their additive effect can be significant in preventing yield losses.

Host Plant Resistance

Host plant resistance among crop plants is a major part of integrated pest management (IPM). The need for a sustainable food and fibre supply, the importance of arthropod pests as constraints on crop yield and quality, and the shortcomings of a pesticide- dominated approach to pest management highlight the need for novel approaches to managing pests. As has often been noted, plant resistance is in many ways an ideal pest management tactic: easy to use, inexpensive to the producer, cumulative in its effects, and mostly compatible with other tactics (Wiseman, 1994). The applied research on host-plant resistance is mostly guided by a framework developed (Painter, 1951). Host plant resistance has been used as a control measures for various agricultural pests for many years (Smith, 1989). It is relatively constant, cheap, and non-polluting and is compatible with other methods of pest control. A strategy that involves the use of pest-resistant and pest-tolerant cultivars developed through traditional breeding or genetic engineering (Nelson *et al.*, 2018). These cultivars possess physical, morphological, or biochemical characters that reduce the plant's attractiveness or suitability for the pest to feed, develop, or reproduce successfully. In jute till date there is no source of resistance insect pests which are regular pests of jute crop. Occasionally the infestation of jute hairy caterpillar, *Spilosoma obliqua* infestation becomes alarming due to outbreak in large area causing extensive damage to jute crop. Few accessions of wild species germplasm were evaluated to find out the source of resistance against this pest and the associated mechanism of resistance. Among the wild jute species, *C. aestuans* accession namely, WCIN-179 was found to be highly resistant to jute hairy caterpillar, *S.obliqua* (Gotyal *et al.*, 2013). This accession is confirmed to be resistant as evidenced through the biology, feeding preference and oviposition behavior of *S. obliqua* recorded on *C. aestuans* (Accession: WCIN-179) as compared to other *Corchorus* species. The wild species (*C. aestuans* accession number, WCIN-179) showed resistance to hairy caterpillar and had recorded 71% less oviposition and 52% less preference for feeding by *S. obliqua* compared to cultivated variety, *C. olitorius* (cv. JRO-204) (Gotyal *et al.*, 2015). These kind of cultivars/ germplasms resist or tolerate pest damage and thus reduce the yield losses. This option is the first line of defense in IPM.

Fig. 1: Jute hairy caterpillar resistant wild jute, *C. aestuans* (WCIN 179)

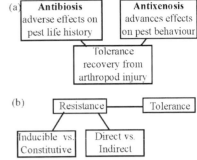

Fig. 2: Mechanism of host plant resistance

Cultural Control

Adopting good agronomic practices that avoid or reduce insect pest infestations and damage refers to cultural control. Choosing clean seed or plant material is critical to avoid the chances of introducing pest's right from the beginning of the crop production. Adjusting planting dates can help escape pest occurrence or avoid most vulnerable stages. In case of jute stem weevil infestation, removal and destruction of stubbles and self-sown plants avoid the carry of the pest and reduce the infestation. Sowing of the crop, both *tossa* and *white* jute varieties in the late April remarkably reduce the incidence of stem weevil as compared to those sown in late March or early April (Gotyal *et al.*, 2019). Line sown crop harbors less mite population and suffers less from mite damage in jute crop. Destroying crop residue and thorough cultivation will eliminate breeding sites and control soil-inhabiting stages of the pest. Sanitation practices to remove infected/infested plant material, regular cleaning field equipment, avoiding accidental contamination of healthy fields through human activity are also important to prevent the pest spread. Plowing is also an important control option to destroy the crop residue and expose the soil-inhabiting stages of several vegetable pests (Kunjwal and Srivastava 2018). Intercropping of non-host plants or those that deter pests or using trap crops to divert pests away from the main crop are some of the other cultural control strategies in IPM (Nielsen *et al.*, 2016). Line sowing with CRIJAF seed drill and maintenance of optimum plant population of 5-6 lakhs/ discourage population buildup of hairy caterpillar in the jute crop.

Fig. 3: Gregarious hairy caterpillar on jute plant

Physical or Mechanical Control

This approach refers to the use of a variety of physical or mechanical techniques for pest exclusion, trapping (in some cases similar to the behavioral control), removal, or destruction (Dara *et al.*, 2018). In jute crop regular monitoring to spot early oviposition and egg masses in the early stage, when the caterpillars remain gregarious on leaf, it is easy to destroy them after plucking such infested leaves and then dipping them in insecticidal solution. The hand picking and destruction of egg masses / gregarious larvae of hairy caterpillar in the early hours when they are active on plant parts can reduce the damage to great extent (Gotyal, *et al.*, 2019).

Pest exclusion with netting or row covers, handpicking or vacuuming to remove pests, mechanical tools for weed control, traps for rodent pests, modifying environmental conditions such as heat or humidity in greenhouses, steam sterilization or solarization, visual or physical bird deterrents such as reflective material or sonic devices are some examples of physical or mechanical control.

Biological Control

Natural enemies such as predatory arthropods and parasitic wasps can be very effective in causing significant reductions in pest populations in certain circumstances (Hajek and Eilenberg 2018). Natural enemies which help to subside pest population are an important component of sustainablea gricultural production that aims to decrease the application of chemicals. The activity of natural enemies mainly depends on the stability of the agro

ecosystem, its habitat management, which is a form of conservation biological control. Such ecological approach favors natural enemies and enhances biological control in agricultural systems. Natural enemies also require colonizing ability which allows it to keep pace with changes to the habitat in space and time. Application of selective and biorational insecticides are compatible with the activity of natural enemies. In jute also many natural enemies and entomopathogens have been reported to parasitize, predate and infect the insect pests causing damage to the crop. In jute ecosyste *Protapanteles obliquae* (Wilkinson) is one of the most important larval parasitoid of jute causing extensive parasitisation. It is a density dependent larval parasitoid and first time reported in jute ecosystem (Selvaraj, *et. al.,* 2015). The full grown *P. obliquae* emerged out through the ventro-lateral body region of the jute hairy caterpillar larva (mostly 2nd-3rd instars). The activity of this parasitoid was noticed from mid-May to mid-July during the cropping season. The parasitoid, *P. obliquae* is a gregarious, endoparasitoid specific to *S. obliqua* and parasitize to the extent of 38% up to third instar of larvae.

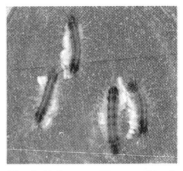

Fig. 5: Hairy caterpillar parasitized by *Protapanteles obliquae*

Fig. 6: Adult *Protapanteles obliquae*

Periodical releases of commercially available natural enemies or conserving natural enemy populations by providing refuges or avoiding practices that harm them are some of the common practices to control endemic pests. Biological control has been successfully used in greenhouses and specialty crops such as strawberries grown in the field (Zalom *et al.,* 2018). To address invasive pest issues, classical biological control approach is typically used where natural enemies from the native region of the invasive pest are imported, multiplied, and released in the new habitat of the pest (Cock 2018). The release of irradiated, sterile insects is another biological control technique that has been effectively used against a number of pests (Klassen and Curtis 2005). The parasitoid, *Meteorus spilosomae* identified and the report confirms the parasitisation by

Fig. 7: *Meteorus spilosomae*

M. spilosomae, a larval parasitioid of *S. obliqua* in jute ecosystem for the first time in 2012 (Selvaraj *et al.*, 2013). This parasitoid is reported to cause up to 77% parasitisation under field condition, indicated the possibility of these parasitoids to be used as potential natural enemy of *S. obliqua* of jute through conservation, augmentation and mass multiplication.

The parasitoid which emerged from the mummified mealybug in jute was identified as *Aenasius bambawalei*. The extent of average parasitisation on mealybugs in these plants ranged from 15-32% with peak activity during the month of second fortnight of June to late August (Satpathy, *et al.*, 2014). The parasitisation is dependent upon the density of mealybug population. The widespread occurrence of *P. solenopsis* on diverse host plants, natural enemies also proliferated in many agro-ecosystems. This was the first report of parasitoids from the jute and mesta based ecosystem.

Fig. 8: Adult *Aenasius bambawalei*
Courtsey: Moazzem Khan

Fig. 9: Melaybug parasitized by *Aenasius bambawalei*

Microbial Control

Using entomopathogenic bacteria, fungi, microsporidia, nematodes, or viruses, and fermentation byproducts of some microbes against arthropod pests, plant parasitic nematodes, and plant pathogens generally come under microbial control (Lacey, 2017).

Nuclear Polyhedrosis Virus (NPV) for insect pest control is an entomopathogenic virus which can destroy larval stage of insects of the order Lepidoptera. The infected larva will die within 3-7 days. In jute ecosystem it occurs naturally and produces about 93% of disease in the hairy caterpillar larvae. The natural epizootic on hairy caterpillar caused by infection of this NPV has been evidenced in farmers' field. This NPV has the potential for its use as microbial biocontrol agent of hairy caterpillar (Gotyal *et al.*, 2019).

Fig. 10: NPV infected hairy caterpillar larva

Behavioral Control

Protection of a resource from a pest is usually achieved by poisoning the pest with a toxic pesticide, but it can also be achieved by manipulating a behavior of the pest. The manipulation of a pest's behavior to protect a resource is not a new concept. The practice of trap cropping, i.e. using a sacrificial resource for the pest to attack, in order to protect a valued resource, has been known for centuries. However, in the last 30 years or so, largely due to improvements in analytical techniques and an increased desire to reduce the reliance on broad-spectrum insecticides, there has been increased interest in behavioral manipulation for pest management. Behavioral control utilizes some chemicals to modify insect pest behavior, and control pest without the use of toxins, thereby playing an important role in area-wide control system. At present, behavioral modification method (e.g. pheromones) have been used to confuse or trap the male population. The behavior of the pest can be exploited for its monitoring and control through baits, traps, and mating disruption techniques (Morrison *et al.*, 2016). Baits containing poisonous material will attract and kill the pests when distributed in the field or placed in traps. Pests are attracted to certain colors, lights, odors of attractants or pheromones. Devices that use one or more of these can be used to attract, trap or kill pests.

A pheromone trap is a type of insect trap that uses pheromones to lure insects. Pheromones are chemicals used by insects and other animals to communicate with each other. Pheromone lures confuse adult insects and disrupt their mating potential, and thus reduce their offspring. Insects send these chemical signals to help attract mates, warn others of predators, or find food. Using specific pheromones, traps can be used to monitor target pests in agriculture or in residential areas. By constantly monitoring for insects, it may be possible to detect an infestation before it occurs. Early detection of pest insects using pheromone traps can also lessen damage to agriculture and other plants.

Tips to Remember if using Pheromone Traps

- Each pheromone is designed for a specific insect. No trap will be effective for all insects.

- Some pheromones may be lower risk than conventional pesticides.

- Lures need to be replaced often to better trap new insects.

- Pheromone traps attract pests. Do not place them in high traffic areas near people or gardens, especially for wasps, hornets, or crop-damaging insects.

- Traps can be affected by weather events, check them after storms to see if they need repair or replacement.

- Keep traps out of reach of children and pets.

- Wash your hands after using pheromones, as pests may follow you if you smell like one of their own.

- Don't use outdoor products indoors. Always follow the label about how and where to use the product.

Pesticides as Last Resort

Chemical control typically refers to the use of synthetic chemical pesticides (Pimentel, 2009). However, to be technically accurate, chemical control should include synthetic chemicals as well as chemicals of microbial or botanical origin. Although botanical extracts such as azadirachtin and pyrethrins, and microbe-derived toxic metabolites such as avermectin and spinosad are regarded as biologicals (Dodia, *et al.*, 2010), they are still chemical molecules, similar to synthetic chemicals, and possess many of the human and environmental safety risks as chemical pesticides. Chemical pesticides are categorized into different groups based on their mode of action (IRAC, 2018) and rotating chemicals from different mode of action groups is recommended to reduce the risk of resistance development (Sparks and Nauen, 2015). Government regulations restrict the time and amount of certain chemical pesticides and help mitigate the associated risks.

There are several alternatives to chemical pesticides that have to be evaluated before deciding to use chemical pesticides. Alternatives include biological pest control agents (BPCA) that include microbial and botanical pesticides, as well as semiochemicals; all of which can be valuable components of IPM. The risk associated with these biological pest control agents are favorable in comparison with the conventional synthetic chemicals and in some cases they might even be acceptable in organic production systems. Application of these pesticides might be limited by their access and price, both of which can be promoted by more favorable policies. In order to facilitate alternative methods of pest control, a good understanding of each pest species and there natural predator should be available for farmers. This can be promoted by leaflets/

flyers that include pictures of each species, together with possible treatment measures. The economic justification should also not only include the current cropping system, but also alternative cropping systems that might be more resistant to pesticides (e.g. crop rotations and intercropping) and that have been proven to be functioning in similar agro-ecological zones. A value should also be attached to health risks and labor requirement.

Conclusion

All the pest management options need careful consideration and application to avoid potential risks. For example, several pests developed resistance to transgenic crops with *Bacillus thuringiensis* toxic proteins (Tabashnik, *et al.,* 2013) and planting non-transgenic plants along with resistant plants has been recommend, among other strategies, to reduce the resistance development (Huang *et al.,* 2011). Pesticide resistance in arthropod pests is a longtime problem in pest management and one of the key factors for developing IPM strategies. As required by the crop and pest situation, one or more of the control options can be used throughout the production period for effective pest management. When used effectively, nonchemical control options delay, reduce, or eliminate the use of chemical pesticides.

Although pest management decisions are supposed to be based on economic injury levels and thresholds, in many situations they are either not available, difficult to determine, not applicable to all geographic regions or seasons, or existing ones need revalidation. Some of the established thresholds are also questionable. Because crop production is highly precise due to modern technologies on one hand, as well as highly variable depending on a myriad of biotic and abiotic factors and the proprietary practices of different farming operations, information management and decision-making parts play a critical role in IPM. One cannot offer a one-size-fits-all solution and the pest control efficacy depends on several factors in addition to the option used.

References

Dara, S.K. (2018). Safe, profitable, and practical label for sustainable production and food security. *Progressive Crop Consultant,* 3: 20–23.

Dodia, D.A., Patel, I.S. and Patel, G.M. (2010). Botanical pesticides for pest management. Scientific Publishers (India), Jodhpur, India.

Gotyal, B.S., Satpathy, S., Ramesh Babu, V. and Selvaraj, K. (2019). Insect Pests of jute: Identification and Management, *Technical Bulletin No. 02/ 2019.* ICAR- Central Research Institute for Jute and Allied Fibres, Barrack pore. Kolkata. p. 30.

Gotyal, B.S., Selvaraj, K., Meena, P.N. and Satpathy, S. (2015). Host Plant Resistance in cultivated jute and its wild relatives towards jute hairy caterpillar, *Spilosomaobliqua* (Lepidoptera: Arctiidae*), Florida Entomologist,* 98(2): 721-727.

Hajek, A.E., and Eilenberg, J. (2018). Natural enemies: an introduction to biological control. Cambridge University Press, Cambridge, United Kingdom.

Huang, F., Andow, D.A. and Buschman, L.L. (2011). Success of the high-dose/refuge resistance management strategy after 15 years of *Bt.* crop use in *North American Entomology Experiment Applicata*, 140: 1–16.

IRAC (Insecticide Resistance Action Committee) (2018). IRAC mode of action classification scheme. https://www.irac-online.org/documents/moa-classification/?ext = pdf.

Klassen, W. and Curtis, C.F. (2005). History of the sterile insect technique, pp. 3–36. *In* V.A. Dyck, J. Hendrichs and A. Robinson (eds.), sterile insect technique. Springer, Dordrecht, The Netherlands.

Lacey, L.A. (2017). Microbial control of insect and mite pests: from theory to practice. Academic Press, London, United Kingdom.

Morrison, W.R., Lee, D.H. Short, B.D., Khrimian, A. and Leskey, T.C. (2016). Establishing the behavioral basis for an attract-and-kill strategy to manage the invasive *Halyomorphahalys* in apple orchards. *Journal of Pest Science,* 89: 81–96.

Nelson, R., Wiesner-Hanks, T., Wisser, R. and Balint-Kurti, P. (2018). Navigating complexity to breed disease-resistant crops. *Nature Reviews Genetics,* 19: 21–33.

Nielsen, A.L., Dively, G., Pote, J.M., Zinati, G. and Mathews, C. (2016). Identifying a potential trap crop for a novel insect pest, *Halyomorphahalys* (Hemiptera: Pentatomidae), in organic farms. *Environmental Entomology,* 45: 472–478.

Painter, R.H. (1951). Insect Resistance in Crop Plants, The University Press of Kansas, Lawrence.

Pimental, D. (2009). Pesticides and pest control pp. 83–87. *In* R. Peshin and A.K. Dhawan (eds.), integrated pest management: innovation-development process. Springer, Dordrecht, The Netherlands.

Satpathy, S., Gotyal, B.S. and Selvaraj, K. (2014). Record of *Aenasius bambawalei* Hayat on Phenacoccus solenopsis, Tinsley in jute ecosystem., *Jaf News,* 12 (1): 20-21.

Selvaraj, K., Gotyal. B.S., Satpathy, S. and Ramasubramanian, T. (2015). First record of *Protapantales obliquae* (Wilkinson) (Braconidae: Hymenoptera) on *Spilosoma obliqua* on jute crop. *Journal of Biological Control*, 29(3): 169-170.

Selvaraj, K., Satpathy, S., Gotyal, B.S. and Ramesh Babu, V. (2013). Record of larval parasitoid parasitoid of Bihar hairy caterpillar, *Spilosoma oblique* Walker (Lepidoptera: Arctiidae) in jute ecosystem in India. *Journal of Biological Control*, 21(1): 56-57.

Smith, C.M. (1989). Plant resistance to insects: A fundamental approach. John Wiley & Sons, Inc. New York.

Sparks, T.C. and Nauen, R. (2015). IRAC: mode of action classification and insecticide resistance management. *Pesticide Biochemistry and Physiology*, 121: 122–128. Stenberg, J.A. 2017. A conceptual framework for integrated pest management. *Trends in Plant Science,* 22: 749–769.

Tabashnik, B.E., Brévault, T. and Carrière, Y. (2013). Insect resistance to Bt crops: lessons from the first billion acres. *Nature Biotechnology*, 31: 510–521.

Wiseman, B.R. (1994). Plant resistance to insects in integrated pest management. *Plant Disease,* 78: 927-932

Zalom, F.G., Bolda, M.P., Dara, S.K. and Joseph, S. (2018). UC IPM pest management guidelines: strawberry (insects and mites). University of California Statewide IPM Program, Oakland, CA, Publication Number 3468.

15

Scope and Potential of Precision Farming in Conservation Agriculture for Improving Input Use Efficiency

K.M. Hati, J. Somasundaram, R.S. Chaudhary, R.K. Singh
A.K. Biswas, N.K. Sinha and M. Mohanty

ICAR-Indian Institute of Soil Science, Berasia Road, Nabibagh
Bhopal-462 038, Madhya Pradesh, India

Introduction

Attaining food and nutritional security for a growing population and reducing poverty while sustaining agricultural production under the current scenario of depleting natural resources, increasing cost of inputs and impending negative impacts of climatic change are the major challenges the country is currently facing (FAO, 2009). In addition to these challenges, decline in soil organic matter, consequent deterioration of soil health, large scale erosion of fertile top-soil are the other bottle-necks for increasing agricultural productivity from a finite land resource base. These constraints are evolving mainly due to: (1) intensive tillage induced soil organic matter decline, soil structural degradation, water and wind erosion, reduced water infiltration, surface sealing and crusting, soil compaction, (2) insufficient return of organic material to soil owing to prevalent crop residue removal and burning (Prasad *et al.*, 1999) and (3) mono-cropping. Conservation agricultural (CA) is a panacea to address these challenges the farming community of this country is presently facing. The CA system incorporates a holistic approach to agriculture that results in improved soil health and water quality, lower water and nutrient consumption, reduced soil loss, strategic land use and healthy biodiversity. It involves minimum soil disturbance, permanent soil cover through crop residues or cover crops, and crop rotations for achieving higher productivity (Hobbs *et al.*, 2008). CA is a sustainable production approach that not only optimizes crop yields, but also reaps economic and environmental benefits as well. The successful adoption of

CA technologies could result in energy savings, higher organic matter content and improved biotic activity in soil, increased crop-water availability and thus resilience to drought, improved recharge of aquifers, less erosion, and reduced impacts from the weather associated risks arising due to climate change. In India, efforts to develop, refine and disseminate CA-based technologies have been underway for nearly two decades and made significant progress since then even though there are several constraints another that affect adoption of CA (Erenstein and Laxmi, 2008). Precision farming, another emerging system of crop production designed to optimize agricultural production through the application of crop information, soil variability, advanced technology and management practices system, could complement the CA system (Basso, 2003). Precision farming comprises a set of technologies that combines sensors, information systems, enhanced machinery, and informed management to optimize production by accounting for variability and uncertainties within agricultural systems. Adapting production inputs site-specifically within a field allows better use of resources to maintain the quality of the environment while improving the sustainability of the food supply (Gebbers and Adamchuk, 2010). There is a tremendous scope for application of precision farming technologies in CA as both the systems strive to lessen environmental pollution, improve water quality and reduce nutrient runoff. Application of the knowledge of precision farming in CA practices have great potential in further improving the input use efficiency, profitability and sustainability of the agricultural production system and ecosystem services (Lal, 2014). The objective of this article is to assess the complementarity of precision farming/site specific input management techniques and CA for sustaining productivity and improving input use efficiency and profitability.

Conservation Agriculture: A Sustainable Approach for Crop Production

Conventional tillage-based soil management for intensive crop production followed over large swath of the country commonly leads to soil degradation and eventual loss of crop productivity. Moreover, farmers following this conventional system have to bear the high costs for fuel, labour, agrochemicals, and other production inputs required by intensive cropping (Hobbs *et al.*, 2008). Besides this, intensive tillage causes a greater loss of soil organic carbon (SOC) and increases greenhouse gas (GHG) emission, mainly CO_2, that not only impacts soil productive capacity but also impacts atmospheric quality that is responsible for climate change. CA provides a viable system for sustainable crop production and agricultural development. CA practices encompasses a series of land management practices that include crop residues retention, cover crops, appropriate cropping system rotation, integrated pest management to minimize land degradation. The Food and Agricultural Organization of the United Nations (FAO) defines CA as an approach to manage agro-ecosystems for improved and sustained productivity, increased profits and food security while preserving and enhancing the resource base and the environment. CA is based on enhancing natural biological processes above and below the ground.

Principles of CA are not confined to a single set of practices or prescription recommended to all types of crops, soil and environment. CA encompasses a gamut of agricultural practices which will be specific to crops, areas and socio-economic condition of the people keeping in view the sustainability of the system from economic as well as from environmental perspective (Palm *et al.*, 2014). CA promotes a series of principles to achieve conservation objectives, rather than a particular technology. This is in recognition of the fact that global agriculture, practiced in many different ecosystems with varied technologies, has to be carefully tailored and fine-tuned for its successful adoption under CA. In the recent past CA has been gaining acceptance in many parts of the world as an alternative to both conventional agriculture and to organic agriculture (Friedrich *et al.*, 2012). The concept of CA or more precisely of conservation tillage was emanated in the great plains of US during the late thirties of the last century to address the problem of rapid progress of dust storms belts in the mid-western plains of US Corn Belt. The major reason then identified for the dust storms was excessive inversion tillage operations for making clean and pulverized seed beds for ease of sowing of maize seeds by mechanical seed planters. Although the practice of CA on a large scale have emerged out from US, Brazil and Argentina over the last three to four decades, similar developments are occurring in many other areas of the world, notably North America in zero tillage, and Africa and Asia with technologies such as agroforestry, crop rotation, bed planting and reduced tillage (Friedrich *et al.*, 2012). In India, conservation tillage system has been practiced in an area of about two million hectares in the Indo-Gangetic plains under rice-wheat system (Erenstein and Laxmi, 2008). Major practices adopted in those areas are zero-tillage, bed planting and laser levelling. Besides this, in rainfed regions different forms of conservation agricultural practices and their benefits are being tested and propagated for their adoption (Hati *et al.*, 2015; Somasundaram *et al.*, 2019). CA is based on the principles of rebuilding the soil, optimizing crop production inputs, including labour, and optimizing profit. CA stimulates soils' natural ability to recycle its resources and maintaining its health through the activities of its flora and fauna living within the ecosystem.

Precision Farming: A New Generation High-tech Agriculture for Efficient Fertilizer and Agro-chemical Management

Precision farming/agriculture, a popular new concept in crop production, can be defined as a comprehensive system designed to optimize agricultural production with high input use efficiency through the application of crop information, soil variability, state of the art technology and management practices. As indicated, understanding soil variability by using state-of-the-art techniques is a key component of precision agriculture. Understanding this variability at field and series level is utmost important as large differences exist in the topography, slope, and management practices followed by individual farmers in this country. Precision farming includes precise micro-management of every step of the farming process. It allows the farmer to produce more efficiently, thereby realizing gains through economical and efficient use of resources (Basso, 2003). The concept of precision farming emerged in the 1980s with the advent of affordable geographic positioning systems (GPS), and has further developed with access to an array of affordable soil and crop sensors, improved computer power and software, and equipment with precision application control, e.g. variable rate fertiliser applicators and irrigation systems. An important effect of precision farming is the high environmental benefit from using chemical treatments such as fertilizers, manure and plant protection chemicals only where and when they are necessary (Belmans *et al.*, 2018). Precision farming reduces the loss of the valuable input from the crop rhizospere and thereby minimizes their flow to the ground and surface water resources. From the view point of an average farmer, the main question is not whether precision fertilization is useful or not but whether it is worthwhile (Blackmore, 1994). Precision fertilization can be efficient and profitable where intra field variability can be assessed reliably and economically and ameliorative measures could be taken. It will not be profitable where the diagnostic assessment remains expensive and unreliable and also where high level of uniformity is neither required nor brings about significant yield increase. Adoption of precision farming tools in CA will promote precision placement of crop inputs to reduce input costs, optimize efficiency of operations, and prevent environmental damage (Zhang *et al.*, 2010). This principle is based on treating the problems at the field location where they occur, rather than blanket treatment of the field, as with conventional systems. The benefits are, increased economic and field operation efficiencies, improved environmental protection, and reduced (optimized) input costs. Here, precision could be exercised at many levels, like, for placement of seed, fertilizer and spray; permanent wheel placement to stop random compaction; individual weed killing with spot-spraying rather than field spraying, etc. GPS are sometimes used to enhance precision, but farmer sensibility in problem diagnosis and precise placement of treatments is the principal basis. In small scale farming systems and horticultural systems, it also includes differential plantings on hills and ridges to optimize soil moisture and sunshine conditions. Combination of conservation agriculture with precision farming also reduces the fuel consumption or overall energy use owing to considerable reduction of overlaps of machine operations and the possibility of variable distribution of inputs (Bertocco *et al.*, 2008; Marquez *et al.*, 2011).

The success in precision agriculture in CA system depends on the accurate assessment of the soil variability, its efficient management and evaluation in space-time continuum for enhanced crop production. It basically depends on measurement and understanding of variability (Blackmore, 1994). Thus information about the variability of different soil attributes within a field forms an essential component for the decision-making process in precision farming. Application of the knowledge of precision farming in CA practices have great potential in further improving the input use efficiency, profitability and sustainability of the production system and ecosystem services. However, the inability to obtain soil characteristics rapidly and inexpensively remains one of the biggest challenges for precision farming. There are two basic methods of implementing precision or site-specific management (SSM) of crop production inputs namely, map-based and sensor-based for the variable-rate application (VRA) (Zhang *et al.*, 2010). The first site-specific management method is based on the use of maps to represent crop yields, soil properties, pest infestations, and based on these input decisions on variable-rate application plans are taken. Different strategies based on soil type, colour and texture, topography (high ground, low ground), and remotely sensed images, are used to implement it. The second SSM method provides the capability to vary the application rate of crop production inputs based on simultaneous estimation of soil conditions on the go using sensors with no mapping involved (Nie *et al.*, 2017). The sensor-based method utilizes sensors to measure the desired properties, usually soil properties or crop characteristics, on-the-go. Measurements made by such a system are then processed and used immediately to control a variable rate applicator. Standard procedures for measuring soil properties are complex and time consuming. These methods are expensive when large numbers of samples are needed. To overcome this problem, various on-the-go soil sensors to measure mechanical, physical and chemical soil properties have been developed and are under the process of testing and fine-tuning (Nawar *et al.*, 2017). These sensors are based on electrical and electromagnetic, optical and radiometric, mechanical, acoustic, pneumatic, and electrochemical measurement concepts. While only electric and electromagnetic sensors are mostly used at this time.

Direct on-the-go measurement of soil chemical characteristics, such as pH or nutrient content by electrochemical methods could be successfully used to directly evaluate soil fertility. This is usually done by either an ion-selective electrode (glass or polymer membrane), or an ion-selective field effect transistor (ISFET). In both cases, measured voltage (potential difference) between sensing and reference parts of the system is related to the concentration of specific ions (H^+, K^+, NO_3^-, etc.). However, ISFET have several advantages over ion selective electrodes, such as its small dimensions, low output impedance, high signal-to-noise ratio, fast response and the ability to integrate several sensors on a single electronic chip. The rapid response of the system allowed samples to be analyzed within few seconds and the low sample volumes required by the multi-sensor ISFET system makes it a potential candidate for use in a real-time soil nutrient sensing system particularly for sensing nitrate and K^+ in the field. But evaluation and calibration of the system is required for improving their practicable application in precision site specific nutrient management system. Besides this, the spectral features of soil materials in the NIR and MIR spectral region are associated with vibration modes of functional groups that are direct absorption bands, overtones or the combination of vibration bands of light atoms that have strong molecular bonds, for example, chemical bonds that contain H attached to atoms such as N, O, or C (Nie *et al.*, 2017; Nawar *et al.*, 2017). It may, therefore, be possible to measure soil constituent such as moisture, organic C, and N using the infra-red spectroscopic technique. Algorithms need to be developed for different soil types for the analysis of soil properties using these nascent techniques (Zhang *et al.*, 2010). Besides this, some of the commercially available sensors use reflected light at two or more wavelengths as proxy variables for vegetative biomass, plant nutrient status, and indicators of crop health and yield. Some of these sensors have been bundled with variable rate application equipment and are commercially available. The GreenSeeker® (NTech Industries), Hydro N-Sensor (Yara International ASA) and Veris EC sensors are some examples of sensors marketed specifically for precise real-time measurements of plant and soil properties, identification and subsequent control of weed population between the crop rows through directed spray of herbicides on identified weed biomass. These techniques have great potential and complementarity to be integrated with the CA practices. As, CA also considerably reduces the loss of nutrients and agro-chemicals from the rhizospere by reducing the runoff and consequent sediment loss from the farmland particularly in rolling topographies and consequently minimizes contamination of surface water bodies. A study is being conducted at ICAR-Indian Institute of Soil Science, Bhopal to evaluate the potential of leaf colour chart (LCC) based precision nitrogen management system in wheat under a conservation agriculture in Vertisols. Encouraging result with increased nitrogen use efficiency was recorded with LCC based nitrogen application in CA system.

Precision Farming Tools Application in Conservation Agriculture for Efficient Water Management

Precision agriculture has mostly emphasized variable-rate nutrients, seeding, and pesticide application, but at several research sites, variable-rate irrigation equipment has been developed to explore the potential for managing irrigation spatially. CA can adapt itself to these precision irrigation water management techniques to further improve its efficiency while reducing the environmental impact of farming on surface and ground water pollution. Soil moisture monitoring at spatial scale in contrast to traditional point based measurement can lead towards proper amount of water utilization thereby improving the use efficiency of this costly natural resource and preventing the large scale depletion of ground water. The term precision irrigation predates site specific agriculture. Its meaning in the irrigation industry connotes a precise amount of water applied at the correct time, but uniformly across the field (Evans *et al.*, 2000). Precision irrigation includes a spatially variable capability to irrigate a field at the right place and right time based on continuous monitoring of water availability in the field on a spatial scale. To achieve such capability, an otherwise conventional irrigation/ fertigation machine would need variable-rate sprinklers, position determination, variable-rate water supply, variable-rate nutrient injection (probably), and variable rate pesticide application (possibly). Potential for conservation accrues from not irrigating non-cropped areas, reducing irrigation amounts to adapt to specific problems, or fully optimizing the economic value of the water applied through irrigation. Bausch and Delgado (2003) reported that we can use remote sensing to manage this variability and increase input use efficiencies. Another way to achieve the desired level of control or automatization of precision irrigation system in CA would be the use of real-time soil water and micrometeorological sensors distributed across a field for continuously re-initializing various decision-making model parameters during irrigation events. Coupling real-time micro-weather stations, plant-based sensors (e.g., reflectance, infrared temperatures or video) and numerous real time soil water sensors scattered around the field at critical locations with a set of good predictive models into a decision support system also minimizes the need for continuous and expensive agronomic oversight. Moisture sensors are one of the important sensors used for precision irrigation water management. Moisture sensors measure the level of water in the soil. Based on the water level feedback system automatic irrigation management system calculates on how much water is needed to irrigate a crop. Too much water in the soil would affect the usage of nitrogen by the roots, availability of oxygen to plant roots. Each type of crops would require various level of water. Different types of sensors such as temperature, humidity and water level are used for agriculture monitoring and crop production. Automatic irrigation system reduces the water consumption to a greater extent. The system is very useful in areas where water scarcity is a major problem. The crop productivity increases and the wastage of water is very much reduced using this irrigation system. Similarly, conservation agriculture reduces water loss through evaporation and increases soil profile water availability through higher infiltration and greater moisture retention capacity of the soil. Thus in CA system, sensor based real time spatial and temporal monitoring of water content in the profile and site specific and timely application of water through algorithm based automatic irrigation management system practiced in precision farming could exploit the complementarities of the systems and further improve the use efficiency of water.

Precision Laser Leveling Technology Use in Conservation Agriculture

Laser-assisted land-leveling is a resource-conserving precision farming technology introduced mainly to the Indo-Gangetic Plains. Traditionally, farmers level their fields in the Indo-Gangetic Plains using scrapers and wooden boards. Now, laser-guided tractors, operated by private contractors, offer more precise leveling of fields at prices smallholders can afford. The technology reduces water losses by as much as 40 percent, improves the efficiency of fertilizer, and boosts yields by from 5 to 10 percent (FAO, 2016). Merging this precision land leveling technology with CA considerably enhance water productivity and improve crop yield.

Conclusion

Conservation agriculture technologies are the future of sustainable agriculture. The benefits range from nano-level (improving microbial diversity, soil properties) to micro-level (saving inputs, reducing cost of production, increasing farm income), and macro-level by reducing poverty, improving food security, alleviating global warming. Despite several advantages there is relatively little and very slow adoption of this sustainable agricultural system on smallholder farms under Indian conditions, as it involves a major change in multiple components of conventional farming system and as such is more difficult for smallholders given their socio-economic constraints. However, moving from conventional to CA based technologies involve paradigm shift in key elements including approaches to develop component technologies of cultivar choices, nutrient, water, weed and pest management while optimizing all component technologies through precision farming. Precision agriculture includes precise micro-management of every step of the farming process. It allows the farmer to produce more efficiently, thereby realizing gains through economical and efficient use of resources such as water, nutrient and agro-chemicals. An important effect of precision farming is the high environmental benefit from using fertilizers, manure and plant protection chemicals only where and when they are required. However, adoption of precision farming under CA practices may not result solely from the biophysical feasibility and benefits of the technology alone but needs support and incentives from institutional arrangements (mechanization, markets, credit, policies etc.) that enable the adoption by farmers with different socio-economic strata.

References

Basso B. (2003) Perspectives of precision agriculture. In L. Garcia Torres *et al.*, (eds.). Conservation Agriculture, pp. 281-288. Kluwer Academic Publishers.

Belmans, E., Campling, P., Dupon, E., Joris, I., Kerselaers, E., Lammens, S., Messely, L., Pauwelyn, E., Seuntjens, P. and Wauters, E., (2018). The Multiactor Approach Enabling Engagement of Actors in Sustainable Use of Chemicals in Agriculture. *Advances in Chemical Pollution, Environmental Management and Protection*, 2: 23-62.

Bertocco, M., Basso, B., Sortori, L., Martin, E.C. (2008). Evaluating the energy efficiency of site specific tillage in maize in NE Italy. *Bioresource Technolgy*, 99: 6957-6965.

Blackmore, S. (1994). Precision farming: an introduction. *Outlook on Agriculture*, 23(4): 275-280.

Evans, R.G., Buchleiter, G.W., Sadler, E.J., King, B.A. and Harting, G.B. (2000). Controls for precision irrigation with self-propelled systems. In: Evans, R.G., B.L. Benham and T.P. Trooien. Proceedings of the 4th Decennial National Irrigation Symposium. American Society of Agricultural Engineers. St. Joseph, MI. November 14-16. pp. 322-332.

Erenstein O, and Laxmi V (2008). Zero tillage impacts in India's rice-wheat systems: a review. *Soil & Tillage Research*, 100: 1–14.

FAO (2009) 'Global agriculture towards 2050.' (FAO: Rome, Italy)

FAO (2016) Save and Grow in practice: maize, rice, wheat. A guide to sustainable cereal production. ISBN 978-92-5-108519-6, http://www .fao.org/3/a-i4009e.pdf.

Friedrich T, Derpsch R, Kassam A. (2012). Overview of the global spread of conservation agriculture, http://factsreports.revues.org/ 1941 Published 12 September 2012.

Gebbers, R., Adamchuk, V.I. (2010) Precision Agriculture and Food Security. *Science*, 327(5967): 828-831.

Hati, K.M., Chaudhary, R.S., Mandal, K.G., Bandyopadhyay, K.K., Singh, R.K., Sinha, N.K., Mohanty, M., Somasundaram, J., Saha, R. (2015). Effects of tillage, residue and fertilizer nitrogen on crop yields, and soil physical properties under soybean-wheat rotation in Vertisols of Central India. *Agricultural Research*, 4 (1): 48-56.

Hobbs, P.R., Sayre, K. and Gupta, R. (2008). The role of conservation agriculture in sustainable agriculture. Philosophical Transactions of the Royal Society of London. Series B, Biological Science, 363: 543–555. doi:10.1098/rstb.2007.2169.

Lal R. 2014. Societal value of soil carbon. Journal of Soil & Water Conservation 69: 188a–192a.

Marquez, F., Sanchez, E.G., Aguera, J., Blanco, G and Gilribes, J.A. (2011) Conservation agriculture and precision agriculture as a method to reduce energy consumption in agricultural systems. *Journal of Soil & Water Conservation*, 69(6): 188a–192a.

Nawar, S., Corstanje, R., Halcro, G., Mulla, D. and Mouazen, A.M., (2017). Delineation of soil management zones for variable-rate fertilization: A review. *Advances in Agronomy*, 143: 175-245

Nie, P., Dong, T., He, Y. and Qu, F., (2017). Detection of soil nitrogen using near infrared sensors based on soil pretreatment and algorithms. *Sensors*, 17(5): 1102.

Palm, C., Blanco-Canqui, H., DeClerck, F., Gatere, L. (2014). Conservation agriculture and ecosystem services: an overview. *Agriculture Ecosystem & Environment*, 187: 87–105.

Prasad, R., Gangaiah, B. and Aipe, K.C., (1999). Effect of crop residue management in a rice–wheat cropping system on growth and yield of crops and on soil fertility. *Experimental Agriculture*, 35(4): 427-435.

Somasundaram, J., Salikram, M. , Sinha, N.K., Mohanty, M., Chaudhary, R.S., Dalal, R.C, Mitra, N.G., Blaise, D., Coumar, M.V., Hati K.M., Thakur, J.K., Neenu, S, Biswas, A.K., Patra, P.K. and Chaudhari, S.K. (2019). Conservation agriculture effects on soil properties and crop productivity in a semiarid region of India. *Soil Research*, 57(2): 187-199.

Zhang, X., Shi, L., Jia, X., Seielstad, G. and Helgason, C., (2010). Zone mapping application for precision-farming: a decision support tool for variable rate application. *Precision Agriculture*, 11(2): 103-114.

16

Role of GIS, Remote Sensing and Agro Advisory in Conservation Agriculture

D. Barman, R. Saha, Tania Bhowmick, Abhishek Bagui Girindrani Dutta and Shikhasri Das

ICAR-Central Research Institute for Jute & Allied Fibres, Barrackpore West Bengal -700 121, India

Introduction

The resource conservation technologies (RCTs) primarily focus on resource-savings management systems for agricultural production through minimal tillage operation, maintaining soil health for available plant nutrients and conserving soil moisture through crop residues and/or cover crops, and following cropping sequence in spatial and temporal scale. The integration of resource-conserving technologies working in synergy is commonly referred to as conservation agriculture (CA). The CA is "a concept for resource-saving agricultural crop production that strives to achieve acceptable profits together with high and sustained production levels while concurrently conserving the environment" (FAO 2007). The CA envisages four principles: (i) minimizing mechanical soil disturbance and seeding directly into untilled soil, (ii) maximizing crop residue retention and/or growing cover crops, (iii) optimizing diversification of cropping sequences and rotations, and (iv) minimizing mechanical soil compaction through controlled traffic. The CA technologies involving no- or minimum-tillage with direct seeding and bed planting, residue management and crop diversification have potential for improving productivity and soil quality, mainly by soil organic matter build-up. Conservation agriculture systems appear to be appealing options to achieve sustainable and intensive crop production under different agroecological environments because they use available resources efficiently and maintain soil fertility. However, there is a need for wider scale testing of these new technologies under diverse production systems, as the CA technologies are site specific and therefore appraisal of CA is important to have significant adoption (Ramesh *et al.*, 2016)

The conservation agriculture (CA)-based RCTs, practiced over an estimated 100 M ha area worldwide and across a variety of climatic, soil and geographic zones (Derpsch and Friedrich, 2009), have proved to be energy and input efficient, besides addressing the emerging environment and soil health problems (Saharawat *et al.*, 2010). The RCTs bring many possible benefits including reduced water and energy use (fossil fuels and electricity), reduced greenhouse gas (GHG) emissions, soil erosion and degradation of the natural resource base, increased yields and farm incomes, and reduced labor shortages (Pandey *et al.*, 2012).

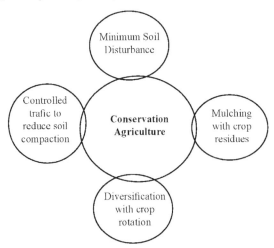

Fig. 1: Diagrammatic representation of the components of conservation agriculture

The future growth in agriculture must come from new technologies which are not only cost-effective but also in conformity with natural climatic regime of the country; technologies relevant to rain-fed areas specifically; continued genetic improvements for better seeds and yields; data improvements for better research, better results, and sustainable planning; bridging the gap between knowledge and practice; and judicious land use resource surveys, efficient management practices and sustainable use of natural resources. Sustainable agricultural production depends on the judicious use of natural resources (soil, water, livestock, plant genetic, fisheries, forest, climate, rainfall, and topography) in an acceptable technology management under the prevailing socio-economic infrastructure. Technology plays an important role in the rapid economic growth and social transformation in developing countries.

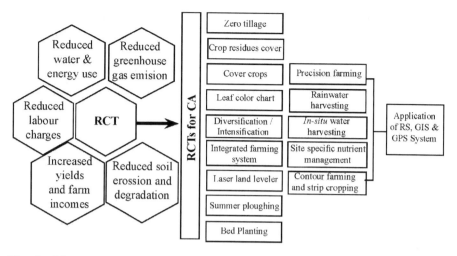

Fig. 2: Diagrammatic representation of resource conservation technology (RCTs) for conservation agriculture and application of remote sensing (RS), GIS and GPS.

Agricultural Mapping

Agricultural mapping is day by day becoming crucial for monitoring and management of soil and irrigation of farmlands. It is facilitating agricultural development and rural development. Accurate mapping of geographic and geologic features of farmlands is enabling scientists and farmers to create more effective and efficient farming techniques. As farmers are able to take more corrective actions in the form of better utilization of fertilizers, treating pest and weed infestations, protecting the natural resources etc., we are bestowed with more and higher quality food production.

Technological innovations and geospatial technology help in creating a dynamic and competitive agriculture which is protective of the environment and capable of providing excellent nutrition to the people. While natural inputs in farming cannot be controlled, they can be better understood and managed with GIS applications. GIS can substantially help in effective crop yield estimates, soil amendment analyses and erosion identification and remediation. More accurate and reliable crop estimates help reduce uncertainty.

A central issue in agricultural development is the necessity to increase productivity, employment, and income of poor segments of the agricultural population, and by applying GIS in agriculture, this situation can be addressed. GIS tools and online web resources are helping farmers to conduct crop forecasting and manage their agriculture production by utilizing multispectral imagery collected by satellites. The ability of GIS to analyze and visualize agricultural environments and workflows has proven to be very beneficial to those involved in the farming industry. GIS has the capability to analyze soil data and determine which crops should be planted where and how to maintain soil nutrition so that the plants are best benefitted. GIS in agriculture helps farmers to achieve increased production and reduced costs by enabling better management of land resources. The risk of marginalization and vulnerability of small and marginal farmers, who constitute about 85% of farmers globally, also gets reduced. Agricultural Geographic Information Systems using Geomatics Technology enable the farmers to map and project current and future fluctuations in precipitation, temperature, crop output etc.

Precision Farming

Precision agriculture refers to the application of precise and correct amounts of inputs like water, fertilizers, pesticides, etc. at the correct time to the crop for increasing its productivity and maximizing its yields. The benefits are of two-fold, i.e. i) the reduction of cost of crop production; ii) the reduction of risk of environmental pollution from overdose of agrochemicals (Earl et al., 1996). Thus, it helps to improve input use efficiencies, economy, and sustainable use of natural resources, because it minimizes wastage of inputs. In other words, it may also be referred to 'Site-Specific Farming'. The practice of precision farming is viewed as comprising of four stages, information acquisition related to variability in environmental and biophysical parameters, their interpretation for input application, evaluation and control. To support precision farming, the important information technology tools are Global Positioning System (GPS), Geographical Information System (GIS), remote sensing, Simulation Modeling for Decision Support System (DSS), yield monitor and variable rate technology. Global Positioning System (GPS) provides accurate site information and is highly useful in locating the spatial variability.

The heterogeneity in agricultural production system is the prevalent constraint for management to get maximum productivity and profit. To reduce the vulnerability of production system to heterogeneity, to conserve the land degradation due to injudicious use of inputs and to recover the soil fertility and productivity; spatial mapping of soil nutrients or generation of homogeneous management units is the pre-requisite for resource conservation and sustainability. Soil is inherently heterogeneous and therefore its information is one of the important inputs in determining crop suitability to different soils (Tamgadge et al., 1999). Moreover, the variability in soil properties like soil fertility causes uneven crop growth, confounds treatment effects in field experiments, and decreases the effectiveness of uniformly applied fertilizer on a field scale (Mulla et al., 1992). Since soil nutrient characteristics vary not only between regions and between farms but also from field to field (Ladha et al., 2000), and within a field, there is a need to take into account such variability while applying fertilizers to a particular crop. An outcome of this concept is precision farming, the site-specific management system to increase farm profits by using remote sensing, global positioning system (GPS) and geographic information system (GIS) (Palmer, 1996). Spatial mapping of soil nutrients is one of the pre-requisites for adoption of precision farming; one can improve the productivity or reduce the cost of production and diminish the chance of environmental degradation caused by excess use of inputs (Pierce and Nowak, 1999).

The concept of Homogeneous Soil Fertility Mapping

The conventional blanket and injudicious use of fertilizer not only reduces nutrient use efficiency but also causes nutrient imbalance in soil resulting in decreased crop yield (Ladha *et al.*, 2005), *e.g.*, declining or stagnating of wheat yield since mid-1980s (Duxbury *et al.*, 2000), and declining trend of rice yields in many parts of south Asia (Pathak *et al.*, 2003a). Estimates of fertilizer requirements through field experiments at several researcher-managed sites, primarily used for rice by Tandon and Tandon (1995) and for wheat by Pathak *et al.*, (2003b) cannot be extrapolated to farmers' field in a wide range of area, because of large variability of soil, climatic and management conditions prevailing in the farms. This problem can be overcome by introducing simulation modeling approach (Das *et al.*, 2009). But this is cumbersome and time-consuming process. Therefore, Barman *et al.*, (2013) recently suggested that the simulation modeling on the basis of homogeneous fertility area is more suitable for estimation of fertilizer requirement for various agri-production systems, which consider variability in soil and plant parameters and would be applicable to larger area with minimum time as well.

Process of Spatial Mapping of Soil Nutrients

Spatial mapping is broadly a two steps process, viz. (i) the collection of spatial soil samples by using GPS, and (ii) the creating continuous surface by interpolation technique in GIS. The collection of spatial soil samples is always a challenging task in soil science research. Soil data sources are commonly observations at sparsely distributed representative point samples from a uniform field in perception by considering its landform. Spatial soil sampling by using GPS gives the spatial soil data, which are usually interpolated in a GIS environment to get regular grids and can be displayed as colour or grey scale maps or by contour lines. Cluster analysis and spatial statistical tools can usefully be employed in classification of field plots into different categories based on data on several characteristics, eg., fertility contour mapping using the uniformity trial data by cluster analysis (Nigam *et al.*, 2004) and spatial mapping of large field variability in hilly and salt affected soil regions by spatial statistics (Bhatia and Prasad, 2003).

Spatial Soil Sampling using GPS

Global positioning system (GPS) is a satellite based navigational aid. Essentially it is a radio-positioning navigation and time transfer system. It provides accurate information on position, velocity and time of an object or a platform at any moment, anywhere on the globe. The system's service is available worldwide with all-weather capability. However, this positioning service can be obtained by any user, only if the user has a suitable GPS receiver. GPS receiver is generally used for collection of soil samples in different types of spatial sampling for the interpolation. Interpolation involves making estimates of the value of an attribute variable at points for which we have no information using the data for a sample of points where we do have information. In general, the more sample points we have the better. Likewise, it is best to have a good spread of sample points across the study area.

Regular sampling guarantees a good spread of points, but it can result in biases if the attribute to be mapped has regular fluctuations (e.g. soil moisture where there are regular spaced drains). It is therefore preferable to have some form of random sampling, where the co-ordinates of the points are selected using random numbers. Pure random sampling tends to produce a pattern with little clusters in some areas and a sparse coverage in other areas, so a stratified random sample in which points are randomly allocated within a regular spaced lattice provides a good compromise. The other three types of sampling are: cluster (or nested) sampling provides detailed information on selected areas and is sometimes used to examine spatial variation at different scales (e.g. you could compare within the amount of variation within a cluster with the variation between clusters); transect sampling is used to survey profiles; and contour sampling is sometimes used to sample printed maps to make a digital elevation model (DEM).

Methods of Interpolation

Interpolation refers to the process of estimating the unknown data values for specific locations using the known data values for other points. Methods of interpolation can be divided into two groups, called global and local interpolators. Global interpolators use all available data to provide predictions for the whole area of interest, while local interpolators operate within a small zone around the point being interpolated to ensure that estimates are made only with data from locations in the immediate neighbourhood, and fitting is as good as possible (Burrough and McDonnell, 1998). Global interpolators are often used to remove the effects of major trends before using local interpolators to analyse the residuals. The global methods include, (i) classification methods, (ii) regression model on surrogate attributes, and (iii) trend surfaces on geometric coordinates; and the local methods include, (i) Thiesen polygon and pycnophylactic methods, (ii) linear and inverse distance weighting, (iii) thin plate splines. Kriging is a particular type of local interpolation using more advanced geostatistical techniques. Detail discussions of these methods is out of the scope of this article.

Spatial Mapping of Soil Nutrients Using GIS

Geographic information system (GIS) is a useful tool for storing the land resource information as a set of thematic maps (Bisht and Kothyari, 2001) which provides a congenial environment for integrating the information, in order to facilitate decision making process a dynamic one (Saxena *et al.*, 2000). This can be used for deriving soil suitability zones by using 'Multi Criteria Decision Making' tool (Sys *et al.*, 1993). Spatial analysis in GIS provides a broad range of powerful spatial modeling and terrain analysis. GIS users can create, query, map, and analyze cell-based raster data; perform integrated raster/vector analysis; derive new information from existing data; query information across multiple data layers; and fully integrate cell-based raster data with traditional vector data sources. GIS integrates real-world variables such as elevation into the geospatial environment to help solve complex problems.

Spatial mapping of soil nutrients by using GIS and GPS was done by several researchers. This was done for the states of Haryana by Kalra *et al.* (2001), for Karnal district of Haryana by Barman *et al.* (2013), for Dhenkanal district of Odisha by Mishra *et al.*, (2014). But the correlation between different information layers is required for site-specific management in agricultural production system. Barman *et al.*, (2011), therefore, suggested homogeneous land unit-based management system by correlating different agri-informative layers. Generation of homogeneous soil fertility units, indeed, is needed for better fertilizer management to optimize crop yield besides other management practices (Barman *et al.*, 2011).

Practical Example of Spatial Mapping of Soil Nutrients Using GIS

Soil fertility map of Karnal was generated in GIS environment on the basis of plant available nitrogen, phosphorus and potassium in the soils of Karnal. Organic carbon content in soils was used as an indicator of plant available nitrogen. By using the inverse distance squared weighted (IDSW) interpolation technique, the representative point data of organic carbon content, phosphorus and potassium of soils were worked out and used to make thematic map. Availability of each nutrient was categorized in two classes. Organic carbon content of the soils of Karnal district was classified into two grades, viz. very low (0.2 to 0.3%) and low (0.31 to 0.5%) (Fig. 3a). Area containing very low soil organic carbon accounted for 61.45% and low organic carbon accounted for 38.55% of the district. For phosphorus, it was very low (12 to 16 kg P_2O_5

ha^{-1}) and low (17 to 20 kg P$_2$O$_5$ ha^{-1}) (Fig. 3b), but in case of potassium, it was medium (235 to 300 kg K$_2$O ha^{-1}) and high (301 to 340 kg K$_2$O ha^{-1}) (Fig. 3c). In the Karnal district, nearly 70% area contains very low phosphorus and 78% area contains high potassium. Overlaying of the organic carbon, phosphorus and potassium thematic maps in the GIS environment resulted one soil fertility map which has eight fertility units (Fig. 3d). The map clearly shows that Assandh and Nissing blocks have very low organic carbon and phosphorus and medium potassium content in the soils due to intensive cultivation. The soil fertility map of Karnal district generated in the GIS environment showed homogeneous zones of eight different classes consisting of OC content, and available P and K. The OC content, and availability P and K in soils were categorized in eight different classes viz. (1) very low - very low - high; (2) very low - very low - medium; (3) very low - low - medium; (4) very low - low - high; (5) low - very low - medium; (6) low - very low - high; (7) low - low - medium and (8) low - low - high, respectively. Each homogenous zone or class contained defined values of individual soil fertility parameters.

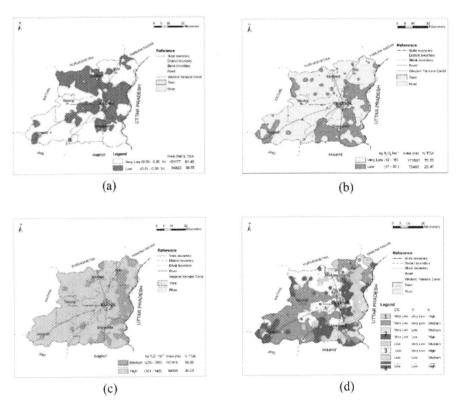

Fig 3. GIS based maps of **(a)** Organic carbon content, **(b)** Available phosphorus content, **(c)** Available potassium content, **(d)** Soil fertility of the Karnal district.

Smart Farming

Sensors in fields, on tractors and on satellites high above farms are constantly collecting data. Advanced technologies are able to turn this data into information that farmers and land managers can use to make more informed and timely decisions. This, in turn, boosts productivity and reduces environmental impacts. Farming is getting smarter with the availability of advanced technologies like precision equipment, the Internet of Things (IoT), sensors and actuators, geo-positioning systems, Big Data, Unmanned Aerial Vehicles, robotics, etc.

A concept in agriculture that is gaining wide popularity due to the plethora of benefits it offers is that of precision agriculture. It enables farmers to collect timely geospatial information on soil-plant requirements and prescribe and apply site-specific treatments to increase agricultural production and protect the environment. Precision agriculture is tied up with high technology tools that are more accurate, cost-effective and user-friendly. Global Market Insights has launched a market report recently, according to which by 2024, the world's precision farming market size will reach $10 billion by 2024. As the agricultural sector opens its arms to embrace new technologies, in times to come, we hope to see every platter to be filled with more nutritious food.

GIS has the capabilities to collect, manage, analyze, report, and share vast amounts of agricultural data to aid in discovering and establishing sustainable agriculture practices. People working in agribusiness use GIS software for precision farming, land management, business operations, and much more. GIS provides the means to spatially view variables that affect crop yields, erosion and drought risk, and business opportunities. This assists them in making well-informed decisions that help increase production and reduce costs using responsible, sustainable practices. The server-based aspect of GIS allows sharing of important data across the globe and saves valuable time and resources. GIS-based mobile technology provides the means to access and collect agriculturally relevant data in the field for pest management, soil treatment, and weed abatement. The different development agencies use GIS for the following applications:

- Predict drought conditions.
- Monitor water resources.
- Visualize remote-sensing data.
- Model data from many sources.
- Evaluate economic and environmental impact.
- Share data and maps between agencies.

242   Conservation Agriculture and Climate Change

- Comply with planning and reporting regulations.
- Educate and advise communities via online services.

Modern precision farming techniques incorporate geospatial technologies to help farmers increase economic yield. Using data from satellites and aircraft, precision farmers can pinpoint problems with drainage, insects, and weeds. They learn where fertilizers are needed and where they are not. Technology has made fertilizer application a precise science. With these methods, farmers have greater control of crop yields before they plant a seed.

In past generations, farmers tended to spread fertilizer evenly over the entire field. Now spreading methods can be more exact by type, quantity, and location of application. By using GIS and GPS to direct application of fertilizers, farmers reduce the amount of potentially harmful runoff of fertilizers into streams and waterways. At the same time, they are reducing their total expenditures on fertilizers.

Time series data sets—Daily agrometeorological data derived from station and satellite data includes precipitation, minimum and maximum temperatures, snow depth, solar and long wave radiation, and potential and actual evapotranspiration. Daily and decadal VINs were derived from local area coverage (approximately 1.1 km pixels) and global area coverage (approximately 8 km pixels from the National Oceanic and Atmospheric Administration's advanced very high-resolution radiometer [AVHRR] satellite series).

In the past, a field day required a large paper map for navigation, clipboards, and water-resistant paper forms to record observations. Staff carried a separate GPS unit to collect spatial data and had to transfer the data to a paper form. Each day spent in the field required three additional days to analyze data and generate reports. Field observations from one week generated a box full of forms, which in turn required nearly a month to analyze and summarize.

Digital technologies have notably reduced the time taken to collect data, complete analysis, and write reports. India is on route to create an Agricultural Market Information System (AMIS). Incorporating Geospatial technologies into this system would improve basic socio-economic & crop statistics for the overall management of crops and demand & supply. It would also enable an equitable distribution of crop insurance due to more accurate crop forecasts and more precise assessment of crop damage due to disease, natural disasters such as drought & flood etc.

Precision farming (PF) is an integrated, information and agricultural management system that is designed to improve the whole farm production efficiency with the low-cost effect while avoiding the unwanted effects of chemical loading to the environment. PF envisages the application of technologies and principles to manage spatial and temporal variability associated with all aspects of agricultural production (Pierce and Nowak, 1999). Basically, it is "doing the right thing, at the right time, in the right place, in the right way." It applies to virtually every aspect of agriculture, from planting to harvest. The focus under PF is to gather information regarding the soil and crop condition and capture the sequence. The database for precision farming generally includes(Venkataratnam, 2001):

- Crop information such as growth stage, health, nutrient requirement
- Soil physical and chemical properties, depth, texture, nutrient status, salinity and toxicity, soil temperature, productivity potential
- Microclimatic data (seasonal and daily) such as canopy temperature, wind direction and speed, humidity
- Surface and sub-surface drainage conditions
- Irrigation facilities, water availability and planning of other inputs

From mobile GIS in the field to the scientific analysis of production data at the farm manager's office, GIS is playing an increasing role in agriculture production throughout the world by helping farmers increase production, reduce costs, and manage their land more efficiently. While natural inputs in farming cannot be controlled, they can be better understood and managed with GIS applications such as crop yield estimates, soil amendment analyses, and erosion identification and remediation. The integration of GPS, remote sensing, and GIS is indispensable in devising an effective approach for selectively applying pesticides and fertilizers to improve farming efficiency and reduce environmental hazards (Runyon *et al.*, 1994).

Crop Rotation and Cropping System

Use of crop rotations or intercropping is considered vital in CA systems (Calegari, 2001), as it offers an option for higher diversity in plant production and thus in human and livestock nutrition, and pest/weed management that are no longer realized through soil tillage. Legumes grown in rotation can provide a range of benefits to the agro-ecosystem. Remote sensing plays an important role in identifying crop rotation and cropping systems for better management of land resources.

Spatial Decision Support System (SDSS)

An integrated system with the three technologies leads to an intelligent system which is designed to help policy planners, farmers to solve complex spatial problems and to make decision concerning the pest management, hotspots identification, niche areas identification for disease outbreaks, irrigation, fertilization and other chemical usage. The fertilizer dose, irrigation schedule and time of sowing, the scenario can be simulated and the best variety can be opted for better yields.

GIS plays a very important role in identifying the potential sites for irrigation. Using the GIS and remote sensing technology experts can provide the location where the farmers can make use of natural slopes and use the same for retaining water in the form of water storage tanks or rain harvesting tank. With the help of the of GIS intervention farmers can set up new channels for irrigation. The steep gradient in the terrain need such technology for identifying potential sites for trapping ground water as well as run-off water during rainy season.

Hyperspectral and multispectral images, consisting of reflectance from the visible, near infrared and mid-infrared regions of the electromagnetic spectrum, can be interpreted in terms of physical parameters (such as crop cover, crop health and soil moisture) and are useful for operations such as stress mapping, fertilization and pesticide application and irrigation management.

Nationwide land use, land cover, soil and wasteland mapping have helped in expansion and intensification of agricultural activities and also in identification of land capability classes and crop suitability indices (Venkataratnam, 2001). Using the technology climatic girds can be prepared with interpolation. With the help of Digital elevation model the suitable range for crops can be identified. The climatic grids can help in identifying the crop specific parameters.

In 2004, a farmer survey was conducted by the Agricultural University of Himachal Pradesh in Shimla and Solan districts for monitoring the use of pesticides in apple and vegetable crops. GIS combined with a pesticide transport model was used to delineate the presence of pesticides in the soil, plant and environment. The results indicated that pesticides were retained in various media, highlighting the importance of their rational and specific use. Furthermore, hot spots of pesticide abundance were identified using spatial analysis – information which has implications for the future planning of pesticide use in mountain agriculture (Sood and Bhagat, 2005).

Agroadvisory System

Agroadvisory plays a major role in resource conservation in agriculture by disseminating appropriate technologies in proper time and space. Weather-based agroadvisory and weather forecasting also save natural resources and provide knowledge to take decision in agricultural operations.

Jute, the most important bast fibre crop in India is facing multifarious challenges and swinging between its commercial subsistence and climate friendly importance. Due to climate change this rainfed cash crop faces production constraints in one hand but on the other the same climate change awareness invokes its reestablishment as a golden fibre crop against the plastics in similar uses. To address the climate related production constraint, an integrated web-based agrometeorological database management system-cum-agroadvisory system named as JuteMet was developed. JuteMet has capability to store long-term daily climate data vis-à-vis to disseminate advisories about the modern jute cultivation technologies to the end users to mitigate climate uncertainty. The agroadvisory module embedded in JuteMet is operated in client-server web-based interactive mode for site specific climate related advisories. This increases the virtual proximity between the technocrats and farmers as well as the interaction between them to meet the timely climatic needs in jute production. JuteMet is developed using ASP.NET(C#) and SQL Server 12.0 for database management. The web pages are designed and configured with texts, figures, maps, and images to make it user friendly and easily understandable to its end users including farmers. The interface of the integrated system contains different modules (Fig. 4), viz. Home Page–describing the JuteMet, its functioning and usefulness to the end users including farmers; Target Area–describing the targeted cultivation area of jute; Agromet Observatory–showing the real picture of an observatory; Location – containing the locations of the five agrometeorological observatories at Barrackpore, Bamra, Budbud, Pratapgarh, and Sorbhog; Instruments–depicting the functions of the different surface instruments installed at an observatory along with their pictorial representation that help in better understanding of climatic data; Climatic Normal–representing the annual and seasonal (kharif, rabi, and summer) climatic normal for all the five locations, which helps in climate based planning of the targeted areas; Climate and Crop–dealing with the climatic requirement in jute production for its agronomic management; Agroadvisory-having two options such as Dynamic Advise for sending weather based contingency planning through registered e-mail IDs of the stakeholders, and General Advise containing know-how of agronomic management and related operations; Daily Weather-displaying last 60 days weather data for decision-making in short-term weather condition, Weekly Weather–displaying current weekly weather condition; Photo Album-contains climatic and/or agrometeorological information as GIS-based maps; Registration–open for registration for farmers and officers in this web-application in getting weather-based dynamic advice directly in their registered e-mails; Admin Login–for logging in by the administrator to add, edit, modify data and for generating report.

Fig. 4: Web-page of JuteMet– the agrometeorological database management system-cum-agroadvisory system.

Conclusion

Technological innovations and geospatial technology help in creating a dynamic and competitive agriculture which is protective of the environment and capable of providing excellent nutrition to the people. Besides agronomic RCTs, using of geospatial technologies such as geographic information system (GIS), global positioning system (GPS), remote sensing, etc., and agroadvisory play an important role in conserving resources by various ways like precision agriculture. While natural inputs in farming cannot be controlled, they can be better understood and managed with remote sensing and GIS applications. Remote sensing and GIS can substantially help in effective crop yield estimates, soil amendment analyses and erosion identification and remediation.

References

Barman, D., Sahoo, R.N., Chakraborty, D., Kalra, N. and Kamble, K. (2011). Deriving homogeneous land unit of Karnal district through geographic information system. *Indian Journal of Soil Conservation,* 39(2): 117-123.

Barman, D., Sahoo, R.N., Kalra, N., Kamble, K. and Kundu, D.K. (2013). Homogeneous soil fertility mapping through GIS for site specific nutrient management by QUEFTS model. *Indian Journal of Soil Conservation,* 41(3): 257-261.

Bhatia, V.K., and Prasad, R. (2003). Studies on data processing techniques for statistical analysis of large field variability in hilly and salt affected soil regions, pp. 102. *NATP-CGP Project Report,* IASRI, New Delhi.

Bisth, B.S. and Kothyari, B.P. (2001). Land cover change analysis of Garur Ganga watershed using GIS/Remote sensing technique. *Journal of the Indian Society of Remote Sensing,* 29(3): 137-141.

Burrough, P.A. and McDonnell, R.A. (1998). Principles of geographical information systems. Oxford University Press Inc., New York.

Das, D.K., Maiti, D. and Pathak, H. (2009). Site-specific nutrient management in rice in eastern India using a modeling approach. *Nutrient Cycling in Agroecosystems,* 83: 85-94

Derpsch, R. and Friedrich, T. (2009). Global overview of Conservation Agriculture adoption. Invited Paper, 4th World Congress on Conservation Agriculture: Innovations for Improving Efficiency, Equity and Environment. 4-7 February 2009, New Delhi, ICAR. (www.fao.org/ag/ca).

Duxbury, J.M., Abrol, I.P., Gupta, R.K. and Bronson, K.F. (2000). Analysis of long-term soil fertility experiments with rice-wheat rotations in South Asia. In: Abrol I.P., Bronson K.F., Duxbury J.M. and Gupta R.K. (eds.). Long term soil fertility experiments with Rice-Wheat Rotations in South Asia. Rice-Wheat Consortium Paper Series No.6. *Rice-Wheat Consortium for the Indo-Gangetic Plains,* New Delhi, India, p. 7-22.

Earl, R., Wheeler, P. N., Blackmore, B. S. and Godwin, R. J. (1996). Precision farming- the management of variability. Land-wards, *Journal of Agricultural Engineering,* 51: 18-23.

Kalra, N., Aggarwal, P.K., Pathak, H., Sujith Kumar, Bandyopadhyay, S.K., Dadhwal, V.K., Sehgal, V.K., Harith, R., Krishna, M. and Roetter, R.P. (2001). Evaluation of regional resources and constraints. In: Eds P.K. Aggarwal, R.P. Roetter, N. Kalra, H. Van Keulen, C.T. Hoanah and H.H. Van Larr, Land use analysis and planning for sustainable food security: with an illustration for the state of Haryana, India, p.167, *New Delhi.*

Ladha, J.K., Fishcer, K.S., Hossain, M., Hobbs, P.R. and Hardy, B. (2000). Improving the productivity and sustainability of rice-wheat systems of the Indo-Gangetic Plains: a synthesis of NARS-IRRI partnership research. Discussion Paper no. 40. International Rice Research Institute, Philippines, p.31.

Ladha, J.K., Pathak, H., Krupnik, T.J., Six, J., van Kessel, C. (2005). Efficiency of fertilizer nitrogen in cereal production: retrospect and prospects. *Adv. Agron.,* 87: 85–156.

Mishra, A., Pattnaik, T., Das, D. and Das, M. (2014). Soil fertility maps preparation using GPS and GIS in Dhenkanal District, Odisha, India. *International Journal of Plant & Soil Science,* 3(8): 986-994, 2014; Article no. IJPSS.2014.8.005.

Mulla, D.J., A.U. Bhatti, M.W. Hammond and J.A. Benson. (1992). A comparison of winter wheat yield and quality under uniform versus spatially variable fertilizer management. *Agriculture Ecosystem and Environment,* 38: 301-311.

Nigam, A.K., Prasad, R. and Gupta, V.K. (2004). Design and Analysis of On-Station and On-Farm Agricultural Research Experiments: A Revisit. *Project Report, ICAR AP-Cess Fund,* pp. 127. *Joint Publication of IASDS,* Lucknow and *IASRI,* New Delhi.

Palmer, R. J. (1996). In Proc. Site-Specific Management for Agric. Syst., Minneapolis, MN, ASA-CSSA-SSSA, Madison, WI, 27–30 March, pp. 613–618.

Pandey, D., Agrawal, M. and Bohra, J. S. (2012). Greenhouse gas emissions from rice crop with different tillage permutations in rice–wheat system. *Agriculture, Ecosystems & Environment*, 159: 133-144.

Pathak, H., Aggarwal, P.K., Roetter, R., Kalra, N., Bandyopadhaya, S.K., Prasad, S. and van Keulen, H. (2003b). Modelling the quantitative evaluation of soil nutrient supply, nutrient use efficiency, and fertilizer requirements of wheat in India. *Nutrient Cycling in Agroecosystems*, 65: 105–113.

Pathak, H., Ladha, J.K., Aggarwal, P.K., Peng, S., Das, S., Singh, Y., Singh, B., Kamra, S.K., Mishra, B., Sastri, A.S.R.A.S., Aggarwal, H.P., Das, D.K. and Gupta, R.K. (2003a). Trends of climatic potential and on-farm yields of rice and wheat in the Indo-Gangetic plains. *Field Crops Research*, 80: 223–234.

Pierce, F.J., and Nowak, P. (1999). Aspects of precision agriculture. *Advances in Agronomy*, 67:1-85.

Ramesh, Negi SC, and Rana SS, (2016). Resource Conservation Technologies (RCTs)-Needs and future prospects: A review. *Agricultural Reviews*, 37 (4): 257-267

Runyon T, Hammitt R, Lindquist R (1994). Buried danger: integrating GIS and GPS to identify radiologically contaminated sites. *Geographic Information System*, 8(4): 28-36.

Saharawat, Y. S., Singh, B., Malik, R. K., Ladha, J. K., Gathala, M., Jat, M. L. and Kumar, V. (2010). Evaluation of alternative tillage and crop establishment methods in a rice–wheat rotation in North Western IGP. *Field Crops Research*, 116: 260-267.

Saxena, R.K., Verma, K.S., Chary, G.R., Srivastava, R. and Barthawal.A.K. (2000). IRS-1C application in watershed characterization and management. *International Journal of Remote Sensing*, 21(17): 3197-3208.

Sood, C. and Bhagat, R.M. (2005). Interfacing geographical information system and pesticide fate modelling applications: a review. *Current Science*, 89: 1362–1370.

Sys, I.R.C, Van Rast, E., Debaveye, Ir J. and Boernaert, F. (1993). Land evaluation. Part- III Crop requirements, pp 115. Agricultural Publication No. & General Administration for Development Cooperation Place du Champ de Mars 5 bte 57 1050, Brussels, Belgium.

Tamgadge, D.B., Giakwad, S.T.and Gajbhiye, K.S. (1999). Soils of Madhya Pradesh-II: Land use capability, cropping systems and soil degradation. *Journal of the Indian Society of Soil Science*, 47: 114-118.

Tandon, H.L.S. and Tandon, K.S. (1995). State-wise and crop-wise recommendations. In: Tandon H.L.S. (ed.), Fertilizer and integrated nutrient recommendations for Balance and Efficiency. FDCO, New Delhi, India, pp. 35-103.

Venkataratnam, L. (2001), 'Remote sensing and GIS in agricultural resources management', Proceedings of the First National Conference on Agro-Informatics (NCAI), INSAIT. Dharwad, pp. 20-29.

Viet CP and Phuong NM, (1993). Natural resources evaluation by the use of remote sensing and GIS technology for agricultural development. *Advances in Space Research*, 13(11):117-121.

17

Overview of Crop Growth Models as Support System to Conservation Agriculture

Saon Banerjee and Soumen Mondal

Bidhan Chandra Krishi Viswavidyalaya, Mohanpur
Nadia, West Bengal -741 252, India

Introduction

The 21st Century agriculture needs knowledge integration and maximum input efficiency. Crop simulation models offer possibilities to evaluate and target agricultural information towards sustainability (Elliott *et al.*, 2015; Van Oort and Zwart, 2018). All the modern day's crop growth models not only predict the growth and yield, but also they are strong Decision Support System (DSS). Due to the impact of climate change, land degradation and biodiversity loss, the agricultural production system becomes the most vulnerable one (Araya *et al.*, 2017). Crop intensification is inevitable to meet the ever-increasing food demand for the billions of populations. It leads to deterioration of soil physical health, reduction of the organic carbon percentage and micro-environmental pollution. Conservation agriculture (CA) has a great potentiality to enhance the inherent resilience against all sorts of degradation and possess the potentiality to regenerate the degraded lands (Haggblade and Tembo, 2003). CA is a farming system that maintains a permanent soil cover to assure its protection, avoids soil tillage, and cultivates a diverse range of plant-species to improve soil conditions, reduce land degradation and increase water and nutrient use efficiency (Friedrich *et al.*, 2012; Sapkota *et al.*, 2015; Bell *et al.*, 2019; Steward *et al.*, 2019). For last few decades, climate change becomes a serious threat to agriculture in context to influence crop and livestock production, hydrological balances, input supplies and other components of agricultural system (Kabir, 2015; Kumar and Kumar, 2016). Agriculture is inherently sensitive to climatic conditions and conservation agriculture may be the only answer to combat the negative impact of climate change.

In response to the need for well-informed decision making on CA, there is a requirement for large-scale empirical data, which is usually too costly to be obtained. Different kinds of mathematical and conceptual models have been developed by researchers with the goal of minimizing the knowledge gaps (Ngwira *et al.*, 2014; Corbeels *et al.*, 2016; Banerjee *et al.*, 2016). These models seek to provide insight in the biophysical, technical, economic, social as well as institutional factors that influence the CA adoption process (Martinez *et al.*, 2013). Biophysical models, e.g. crop simulation models, developed mostly by agronomic or natural science, have the ability to use long term weather data and the local soil parameters to assess long term effects of CA practices.

Model as Simplified Representation of a System

The term system is used to describe a simplified representation of reality that contains interrelated elements. Natural system can also be represented by a model. A mathematical model is an assembly of concepts in the form of mathematical equations that can represent the understanding of natural phenomena. Simulation may be defined as the art of building mathematical models and the study of their properties in reference to those of the systems. A group of functions or set of equations, integrated into a simulation platform, makes up a (mathematical) model. Mathematical models vary from simple to complex and can be classified in different ways depending on the field of application and the scientific disciplines involved in the development of the models (Table 1).

Table 1: Terms to classify mathematical models

Model classification	Description
Empirical	Uses experimental (observed) data to estimated statistical relationships
Descriptive	Shows the existence of relations of the elements of a system
Stochastic	Incorporates a random component, produces results in terms of the probability distributions and describes a range of likely results
Deterministic	Single results from a given set of parameter values
Mechanistic	Uses functions that closely describe the physical processes. Generally more detailed and complex than functional models
Process-based	Describes, represents and simulates the functioning of real-world processes
Explanatory	Consists of quantitative descriptions of the mechanisms and processes involved that are responsible for the behaviour of the system, based on theoretical concepts and computational methods
Theory-based	Opposite of empirical studies
Functional	Uses simple functions that are empirical or only loosely based on physical processes.

Model classification	Description
Distributed	Represents spatial variability, often up to a medium-to-large catchment scale, with an aggregation of 'uniform' cells or elements
Static	Does not account for time. Identifies the before and after outcomes but does not trace the path that the model takes to move from one equilibrium position to another
Dynamic	Simulates several periods, result of period t influence results in periods t+1

Crop Growth Simulation Models

Models with vivid crop growth component are generally referred to as crop growth models. The main aims of these models are biomass and yield forecasting. Although, the crop growth models are often developed as complex agroecosystem models. The term crop growth model is therefore often misleading as it needs additional specification on the processes that are considered in a model to be able to characterize it. Crop modelling is defined as the dynamic simulation of crop growth by integration of all the processes involved through mathematical equation with the aid of computers (Bouman et al., 1996). Crop growth modelling began in the 1960s with the aim of increasing insight into crop growth processes by a synthesis of knowledge expressed in different equations (Jones et al., 2003).

Understanding the processes of plant growth and development and using them to create models became therefore a major focus of the later research, particularly with the advances in computer technology since the 1990s of the last century. Early crop growth models were limited either by the absence of their capabilities to include soil management orientations (i.e. CERES group models) or by the deficiency of high sensibility to environmental extremes required to analyse climatic risks (i.e. EPIC). Consequently, while the earlier models focused particularly on crop production, the models of the last two decades have increasingly taken into account environmental and climate change aspects (Banerjee, 2008).

In general, later developed crop growth models are explanatory models and seek to explain the functioning of crops as a whole by simulating or imitating the behaviour of a real crop in terms of growth of its components, such as leaves, roots, stems and grains (Table 2). They do not only predict final biomass or harvestable yield, but also contain information about major processes involved in the growth and development of a plant and they often also provide information on externalities, such as evapotranspiration pattern, soil erosion or N-leaching. The time steps are usually daily or sometimes even on an hourly basis and thus reduce the time interval involved considerably when compared to empirical-statistical models, which usually use seasons or years.

The spatial unit or 'system' described in most crop growth models is a crop plus the root zone soil below that crop. Some models operate at a farm, watershed or higher level, but calculations are usually done for points that are considered representative for individual fields, homogeneous soil units or grid-cells in a geographic information system (GIS). Possible applications include the prediction of yields and externalities, to simulate the effects of weather, soil water and nitrogen dynamics in the soil on growth and yield for the specified cultivar, for plant breeding optimization and crop management in general.

Major processes of crop production included in crop growth models include

- Phenological development
- Canopy development
- Organ formation
- Photosynthesis
- Assimilate allocation
- Carbon, water and nitrogen dynamics (soil, plant)

Table 2: Different production situation and processes involved in crop growth models

Production situation (occurrence)	Conditions, limiting factors	Major elements/Processes
Production level 1 (cool climates, field and laboratory experiments, glasshouses, intensive production)	Ample nutrients and soil water, growth is determined by weather conditions (absorbed solar radiation, temperature).	Major elements: Dry weights of the leaves, stems, reproductive of storage organs and of roots, surfaces of photosynthesizing tissues Major processes: CO_2 assimilation, maintenance and growth, assimilate distribution and leaf area development
Production level 2 (non-irrigated, intensively fertilized areas)	Growth is limited by water shortage (at least part of the time), when sufficient water is available, growth rate increases up to maximum set by the weather conditions	Extra elements: Water balances of the plant and Soil; Crucial processes: transpiration and its coupling to CO_2 assimilation and all other processes related to soil water, such as evaporation, drainage and run-off

Production situation (occurrence)	Conditions, limiting factors	Major elements/Processes
Production level 3 (in regions with little use of fertilizer, but also frequent situation at the end of the growing season in general)	Growth is limited by N shortage (at least part of the time) and by water or weather for the remainder of the growth season	Important elements: Forms of N in soil and plant; Important processes: transformations of nitrogenous compounds, leaching, denitrification, N absorption by roots, response of growth to N availability, redistribution of N among old and growing organs
Production level 4 (Heavily exploited areas, no fertilizer use, often poorest parts of the world)	Growth is limited by P and K shortage (at least part of the time), and by N, water or weather for the remainder of the growth season	Important elements: P or other mineral contents of the soils and plants; Processes of their transformation into organic and inorganic forms, absorption by roots and the response of plant growth to their availability, P/N ratio

Source: de Vries *et al.* (1982)

Use of Crop growth Model to Assess the Effect of Conservation Agriculture

CA is a complex and interactive process of a different variable and principle. By integrating experimental and modeling activities, several multidisciplinary teams of scientists have been able to assemble comprehensive models to assess the effect of conservation agriculture. The effect of conservation agriculture can be measured in various terms. A model approach can be useful to assess the contribution of different principles of CA in soil C sequestration (Valkama *et al.*, 2020), response to climate change (Ngwira *et al.*, 2014), on the yield and soil quality of the cropping system (Martinez *et al.*, 2013).

DSSAT model can simulate the effect of different tillage practices, fertilizer combinations, and cropping sequence analysis. The details of DSSAT has been discussed in subsequent section. DSSAT has in-built modules on tillage (XBuild-Management-tillage) and crop residue management (XBuild-Management-Organic Amendments). Both of them can be applied widely in CA practices. The model has a detailed input variable of residue incorporation e.g. i) residue material (crop), ii) Amount kg/ha, iii) N, P, K percentage of residue, iv) Incorporation percentage v) Incorporation depth (cm), vi) Method of incorporation, vii) tillage implement used, viii) tillage depth and many more. The cropping sequence is another aspect of CA which can be analysed through crop model (Fig. 1). Using the calibrated model, simulations can be carried out with projected weather data of future climate scenarios generated from different circulation models. Seasonal analysis is used to compare different treatment combinations in projected future weather scenarios. A well calibrated model can able to assess the performance of conservation agriculture under climate change scenario. But it still need more modification for better prediction of interactive factors of CA.

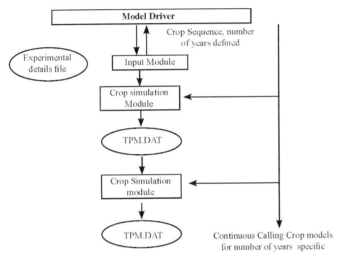

Fig. 1: Flow chart of sequence analysis

ARMOSA is a newly developed model (Peregoet al., 2013) that can be used to simulate different CA components and their interactions and to understand their contributions to SOC increase in comparison to the results under conventional agricultural practices. The tillage module depicts the effects of tillage operations on soil variables, namely BD and organic C pools. Tillage operations are simulated as a function of till depth, timing, degree of soil layers mixing and perturbation, as proposed in the WEPP model (Laflen et al., 1997). The water dynamic is simulated with the Richards' equation resolution used in the SWAP model (van Dam et al., 2008). Carbon and nitrogen related processes are implemented following the approach of the SOILN model (Johnsson et al., 1987). The results of Valkama et al., 2020 clearly demonstrated that, in Southern Kazakhstan and in Southern Finland, cropping systems have a potential to achieve the "4 per 1000" initiative only under CA practices, involving crop rotations (and cover crops).

Sommer et al., (2007) used the Cropstat model to evaluate its capacity to simulate the effect of CA practices ina long-term trial on maize (*Zea mays* L.) and wheat (*Triticum aestivum* L.) crops under primarily rainfed conditions in the Mexican highlandson the CIMMYT research station.Tillage operations provided the closest match with the real operations. Residues incorporated by approximately 90%, (Fischer *et al.*, 2002) for the first years of the trial. Where residues were removed, the 10% of residues remaining was accounted for in the model. However, in this study, Cropstat lacked routines to account for soil crusting, the temporal impact of tillage on soil hydraulic conditions, and the reduction of surface water flow by a residue layer, which are important characteristics for modeling contrasting tillage and residue management regimes.

DSSAT model

The Decision Support System for Agrotechnology Transfer (DSSAT) was originally developed by an international network of scientists, cooperating in the International Benchmark Sites Network for Agrotechnology Transfer (IBSNAT) project to facilitate the application of crop models in a systems approach to agronomic research. The DSSAT is a collection of independent programs, just as independent sub-routines that operate together keeping the simulation control options at its centre (Fig. 2). Databases describe weather, soil, experimental conditions and measurements, and genotype information for applying the models to different situations. Software helps users to prepare these databases and compare simulated results with observations to give them confidence in the models or to determine if modifications are needed to improve accuracy. In addition, programs contained in DSSAT allow users to simulate options for different crop management over a number of years to assess the impacts associated with each option. Thus the impact of different conservation agriculture practices, such as, impact of minimum tillage or mulching can be assessed by this robust model. The changes in soil water, carbon, and nitrogen that take place under a particular cropping system over time can be assessed through this model.

The use of any crop model including DSSAT requires proper calibration and validation for a new set of experiment. The experimental data on crop management practices of CA and their actual results are very important as initial stage of using the model. Moreover, the sequence analysis, in-built in DSSAT, is a helpful tool for assessing various treatment effects in a cropping sequence practiced under CA. The simulation of a cropping sequence requires the continuous calculation of soil-related processes on a daily basis, including the days when no crop is growing. Some variables of preceding crop are needed for the continuous simulation of soil water, carbon and nitrogen processes for the subsequent crop period. The effect of addition of mulching can easily be analyzed through this module. DSSAT enables individual components to be plugged in or unplugged with little impact on the main program or other modules, i.e. for comparison of different models or model components.

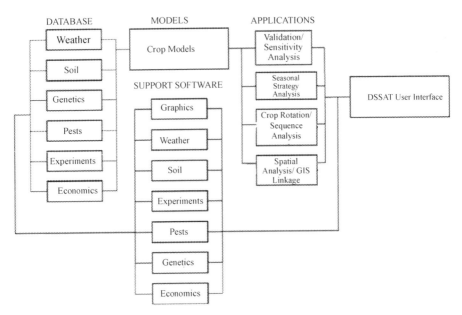

Fig. 2: Diagram showing the components of DSSAT model
Source: Jones *et al.* (2003)

The following four modules (named as 'File') are very important in DSSAT:

Weather File: The model requires daily weather data for the whole crop growing period. Ideally, the weather file should contain data collected before planting to post-maturity period. This would allow simulation to be started before planting, thus providing an estimate of soil conditions during the time of planting. The weather data file contains all the available daily weather data, namely, air temperature (maximum and minimum), rainfall, solar radiations. These are the minimum data required for preparing the weather file. Dew point temperature, wind speed and photosynthetic active radiation are optional input data. This file is independent of crop type. After incorporation of the minimum data set in the Excel file will be then exported into the DSSAT model. Previously it was having to be saved in CSV (Comma Delimited) file type but in DSSAT 4.7 version it can directly read the original format of excel file. For preparing the weather file in DSSAT, latitude, longitude and elevation of that particular station has to be provided and the name of the station which has been assigned for research work must be in four capital alphabets.

Soil file: This file contains detailed information regarding the soil at the experimental site. Soil data required for each soil layer are the layer-thickness (DLAYER, m), saturated water content (SAT, cm3/cm3), drained upper limit of soil water content (DUL, cm^3/cm^3), lower limit of plant extractable water

(LL, cm^3/cm^3), soil bulk density (BD, g/cm^3), root distribution weighing factor (WR, unit-less), and the initial soil water content at start of simulation (cm^3/cm^3). This file is incorporated into the model in ".SOL" format.

Genotype file: The genotype data file contains cultivar-specific genetic parameters needed to predict growth and development. Three morphological and physiological characteristics of a particular genotype are considered. They are: (a) specific characteristics of species, (b) the "ecotype" characteristics within a species and (c) the specific cultivar characteristics within an ecotype grouping.

Output file: The output file contains the overview of input conditions and crop performance, summary of soil characteristics and cultivar coefficients, crop and soil status at the main developmental stages, temporal distribution of crop variables and soil water content. The model predicts the timing of vegetative and reproductive growth stages from emergence to physiological maturity, daily growth of plant components, leaf area index, specific leaf area, root distribution in the soil, % nitrogen in the crop canopy, final yield, yield components and harvest index. In addition, daily soil water balance components, namely soil water evaporation, transpiration, drainage and surface runoff are also estimated and given as output in graphical form.

APSIM Model

The Agricultural Production Systems SIMulator (APSIM) model was produced by the Australian Commonwealth Scientific and Research Organization (CSIRO) and designed around a plug-in, pull-out modular concept. APSIM was developed in early 1990s. The "need for modeling tools that provided accurate predictions of crop production in relation to climate, genotype, soil and management factors, whilst addressing long-term resource management issues in farming systems" was perceived during development of APSIM model (Keating *et al.*, 2003). Its development was influenced by other models, such as CERES, EPIC and CENTURY. APSIM is a modeling environment that uses various component modules to simulate cropping systems dynamically. APSIM can simulate the growth and yield of various crops in response to a variety of management practices, crop mixtures and rotation sequences, including pastures and livestock. Thus the model can be applied to observe the impact of conservation agricultural practices in a successful way. It can do this on the short as well as on the long term, permitting to obtain insight in long-term trends in soil productivity due to organic matter or fertilizer management. APSIM also contains modules that permit the simulation of crop-weed interactions, soil organic matter rundown, nutrient leaching, soil erosion, soil structural decline, acidification, etc.

Evaluation of the long-term effects of rainfall and temperature changes on soil water under CA practices was done by Mwansa *et al.*, 2017. They used APSIM model to simulate soil water under different conservation practices. The R-squared value as observed by them was closer to 100 percent, and this explained that the model was perfectly calibrated. Whilst, the average RMSE was 5.57 and NRMSE was 8.6 percent confirming that the model perfectly predicted the long term effects of rainfall and temperature changes on soil water accumulation for different treatments. The APSIM model showed that rainfall had a positive effect on the soil water accumulation mostly for CA treatments. Increase in annual rainfall had an advantage on CA treatments as the soil water accumulation was equally increased and coupled with the presence of crop residue on the soil surface improved soil water storage. Reducing the annual rainfall in the crop simulation model by 11.3 percent as climate change scenario showed no significant effect on soil water accumulation in the CA treatments. Furthermore, when temperature was raised from 1 to 3°C, there was no significant decrease in simulated soil water accumulation in the CA treatments. The model predicted that there will be only about 0.4 percent increase in maize grain yield on average for CA treatments when 11.3 percent increase in rainfall climate change scenario was applied to the model. Mkonga (2013) also confirmed that the increase in yield on CA treatments does not necessarily depend on the increase in rainfall. Thus like DSSAT, APSIM model can also act as a useful tool to assess the impact of CA practices.

APEX Model

APEX, a biophysical model, is an extension of the Environmental Policy Integrated Climate (EPIC) model (Williams *et al.*, 1998). Agricultural Policy Environmental eXtender (APEX) model is capable of evaluating the effects of various soil and water management practices on the hydrology of the system, crop growth, and other environmental factors. It has the capability of modeling wide ranges of conservation practices (Wang *et al.*, 2008; Tuppad *et al.*, 2010). APEX simulates watershed processes based on weather data, soils characteristics, topography, vegetation, and management practices. Multiple options are available in the APEX model to estimate evapotranspiration, surface runoff, peak runoff rate, and available soil water capacity to derive hydrology of the system. In the recent version released in 2017, a new module "ADDMULCH" was included in the APEX. It can simulate organic mulch cover conservation practices on the soil surface. Detailed description of the APEX model and its major components are shown in Fig. 3. The APEX model has been applied to evaluate the effects of different components of conservation practices by different scientists (Saleh and Gallego, 2007; Yin *et al.*, 2009).

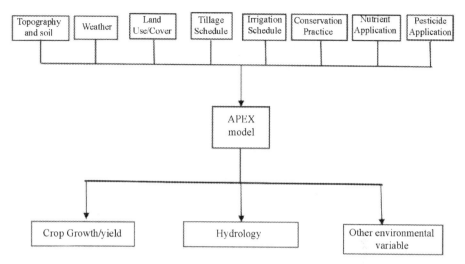

Fig. 3: APEX model major components
Source: Wang *et al.* (2012)

Conclusion

There is a need for well-established decision support system for improvement of CA and such system must be region-specific. The use of crop growth model can indicate the best suited option for soil moisture conservation or soil health restoration. The suitable crop rotation for a region as per the soil and water resources can be obtained by sequential analysis of crop growth model. As per Indian context, many times the farming system may encounter a number of constraints which are typically observed in smallholder agricultural system, which may be overlooked by crop growth model. In small scale farming, the crops are generally grown in a greater variety of configurations with different combination of species, variable plant density, uneven weed competition, etc., which poses a challenge to get reliable results from crop growth model. The calibration and validation are other important tasks behind the successful simulation and application processes. Hence, proper and long term data-base generation should be the priority before the scientific community working on conservation agriculture.

References

Araya, A., Kisekka, I., Lin, X., Prasad, P.V., Gowda, P.H., Rice, C. and Andales, A., (2017). Evaluating the impact of future climate change on irrigated maize production in Kansas. *Climate Risk Management*, 17: 139-154

Banerjee, S. (2008). Possible impact of climate change on rice production in the Gangetic West Bengal, India. Global Issues Paddock Action. Proceedings of the 14th Australian Agronomy Conference (Edited by MJ Unkovich). September 2008, Adelaide South Australia.

Banerjee, S., Mukherjee, A., Das, S. and Saikia, B. (2016). Adaptation strategies to combat climate change effect on rice and mustard in Eastern India. *Mitigation and Adaptation Strategies for Global Change*, 21: 249-261. DOI: 10.1007/s11027-014-9595-y.

Bell, R., Haque, M., Jahiruddin, M., Rahman, M., Begum, M., Miah, M., Islam, M., Hossen, M., Salahin, N., Zahan, T. and Hossain, M., (2019). Conservation Agriculture for Rice-Based Intensive Cropping by Smallholders in the Eastern Gangetic Plain. *Agriculture*, 9 (1): 5.

Bouman, B.A.M., Van Keulen, H., Van Laar, H.H. and Rabbinge, H.H. (1996). The 8 'School of de Wit' crop growth simulation models: pedigree and historical overview. *Agricultural Systems*, 52: 171-198.

Corbeels, M., Chirat, G., Messad, S. and Thierfelder, C., (2016). Performance and sensitivity of the DSSAT crop growth model in simulating maize yield under conservation agriculture. *European Journal of Agronomy*, 76: 41-53.

Elliott, J., Müller, C., Deryng, D., Chryssanthacopoulos, J., Boote, K.J., Büchner, M., Foster, I., Glotter, M., Heinke, J., Iizumi, T. and Izaurralde, R.C., (2015). The global gridded crop model inter-comparison: data and modeling protocols for phase 1 (v1. 0). *Geo-scientific Model Development*, 8 (2): 261-277.

Fischer, R.A., Santiveri, F. and Vidal, I.R., (2002). Crop rotation, tillage and crop residue management for wheat and maize in the sub-humid tropical highlands: I. Wheat and legume performance. *Field Crops Research*, 79(2-3): 107-122.

Friedrich, T., Derpsch, R. and Kassam, A., (2012). Overview of the global spread of conservation agriculture. Field *Actions Science Reports*, Special Issue 6.

Haggblade, S. and Tembo, G. (2003). Development, diffusion and impact of conservation farming in Zambia (No. 1093-2016-87937).

Johnsson, H., Bergstrom, L., Jansson, P.-E., Paustian, K., (1987). Simulated nitrogen dynamics and losses in a layered agricultural soil. *Agricultural Ecosystem Environment*, 18: 333–356.

Jones, J.W., Hoogenboom, G., Porter, C.H., Boote, K.J., Batchelor, W.D., Hunt, L.A., Wilkens, P.W., Singh, U., Gijsman, A.J. and Ritchie, J.T. (2003). The DSSAT cropping system model. *European Journal of Agronomy*, 18: 235-265.

Kabir, H. (2015). Impacts of climate change on rice Yield and variability; an analysis of disaggregate level in the southwestern part of Bangladesh especially Jessore and Sathkhira districts. *Journal of Geography & Natural Disasters*, 5(148): 2167-0587.

Keating, B. Carberry, P.S., Hammer, G., Probert, M.E., Robertson, M.J., Holzworth, Dean, Huth, N.I., Hargreaves, John, Meinke, Holger, Hochman, Z., McLean, G., Verburg, K., Snow, Val, Dimes, John, Silburn, David, Wang, E., Brown, S., Bristow, K., Asseng, S. and Smith, C. (2003). An overview of APSIM, a model designed for farming systems simulation. *European Journal of Agronomy*, 18: 267-288. DOI: 10.1016/S1161-0301(02)00108-9.

Kumar, S. and Kumar, S. (2016). Assessment of impact of climate change on rice and wheat yield in sub humid climate of Bihar. *Journal of Agrometeorology*, 18(2): 249.

Laflen, J.M., Elliot, W.J., Flanagan, D.C., Meyer, C.R., Nearing, M.A., (1997). WEPP-predicting water erosion using a process-based model. *Journal of Soil and Water Conservation*, 52: 96–102.

Martinez, M.S., Lammerding, D.M., Pasamón, J.L.T., Walter, I. and Quemada, M., (2013). Simulating improved combinations tillage-rotation under dryland conditions. *Spanish Journal of Agricultural Research*, 3: 820-832.

Mkonga, Z.J., Tumbo, S. D., Kihupi, N., and Semoka, J. (2013). Extrapolating effects of conservation tillage on dry spell mitigation, yield and productivity of water using simulation modeling. Sokoine University of Agriculture Department of Agricultural Engineering and Land Planning, Morogoro, Tanzania.

Mwansa, F.B., Munyinda, K., Mweetwa, A. and Mupangwa, W. (2017). Assessing the Potential of Conservation Agriculture to Off-set the Effects of Climate Change on Crop Productivity using Crop Simulations Model (APSIM). *International Journal of Scientific Footprints*, 5(1): 9-32.

Ngwira, A.R., Aune, J.B. and Thierfelder, C. (2014). DSSAT modelling of conservation agriculture maize response to climate change in Malawi. *Soil and Tillage Research*, 143: 85-94.

Penning de Vries, F.W.T. and Van Laar, H.H. (editors) (1982). Simulation of plant growth and crop production. Simulation Monographs, Pudoc, Wageningen, The Netherlands.

Perego, A., Giussani, A., Sanna, M., Fumagalli, M., Carozzi, M., Alfieri, L., Brenna, S. and Acutis, M., (2013). The ARMOSA simulation crop model: overall features, calibration and validation results. *Italian Journal of Agrometeorology*, 3: 23-38.

Saleh, A. and Gallego, O. (2007). Application of SWAT and APEX using the SWAPP (SWAT-APEX) program for the upper north Bosque River watershed in Texas. *Trans. ASABE*, 50: 1177–1187.

Sapkota, B. T., Jat, M. L., Aryal, P. J. and Chhetri, K. A. (2015). Climate change adaptation, greenhouse gas mitigation and economic profitability of conservation agriculture: Some examples from cereal system of Indo-Gangetic Plains. *Journal of Intensive Agriculture*, 14(8): 1524-1533.

Sommer, R., Wall, P.C. and Govaerts, B., 2007. Model-based assessment of maize cropping under conventional and conservation agriculture in highland Mexico. *Soil and Tillage Research*, 94(1): 83-100.

Steward, R. P., Thierfelder, C., Dougill, J. A. and Ligowe, I. (2019). Conservation agriculture enhances resistance of maize to climate stress in a Malawian medium-term trial. *Journal Environmental Economics and Management*, 93: 148-169.

Tuppad, P., Santhi, C., Wang, X., Williams, J., Srinivasan, R. and Gowda, P. (2010). Simulation of conservation practices using the APEX model. *Journal Environmental Economics and Management*, 26: 779–794.

Valkama, E., Kunypiyaeva, G., Zhapayev, R., Karabayev, M., Zhusupbekov, E., Perego, A., Schillaci, C., Sacco, D., Moretti, B., Grignani, C. and Acutis, M., (2020). Can conservation agriculture increase soil carbon sequestration? A modelling approach. *Geoderma*, 369: 114298.

Van Dam, J.C., Groenendijk, P., Hendriks, R.F.A., Kroes, J.G., (2008). Advances of modeling water flow in variably saturated soils with SWAP. *Vadose Zone Journal*, 7: 640–653.

Van Oort, P.A. and Zwart, S.J., (2018). Impacts of climate change on rice production in Africa and causes of simulated yield changes. *Global Change Biology*, 24(3): 1029-1045.

Wang, X., Gassman, P., Williams, J., Potter, S. and Kemanian, A. (2008). Modeling the impacts of soil management practices on runoff, sediment yield, maize productivity and soil organic carbon using APEX. *Soil and Tillage Research*, 101: 78–88.

Wang, X., Williams, J., Gassman, P., Baffaut, C., Izaurralde, R., Jeong, J. and Kiniry, J. (2012). EPIC and APEX: Model use, calibration, and validation. *Transactions of the ASABE*, 55: 1447–1462.

Williams, J. R., Arnold, J.G., Srinivasan, R.and Ramanarayanan, T.S. (1998). APEX: A new tool for predicting the effects of climate and CO_2 changes on erosion and water quality. In Modelling Soil Erosion by Water, Springer, New York, USA, 441–449.

Yin, L., Wang, X., Pan, J. and Gassman, P. (2009). Evaluation of APEX for daily runoff and sediment yield from three plots in the Middle Huaihe River Watershed, China. *Transactions of the ASABE*, 52:1833–1845.

Impact of Conservation Agriculture for Natural Resource Management

18

Natural Resource Management Through Conservation Agriculture Under Climate Change Scenario

Debashis Mandal

ICAR- Indian Institute of Soil and Water Conservation, Dehradun Uttarakhand -248 195, India

Introduction

Since the dawn of the civilization when mankind started agriculture, soil erosion has been the single largest environmental problem and has remained so till date (Sullivan, 2004). This is so because removal of the topsoil by any means has, through research and historical evidence, been severally shown to have many deleterious effects on the productive capacity of the soil as well as on ecological wellbeing. Doran and Parkin (1994) captioned the impact of soil erosion in their popular maxim that "the thin layer of soil covering the earth's surface represents the difference between survival and extinction for most terrestrial life". Soil degradation implies decline in its capacity to provide ecosystem services (ESs) of interest to humans and useful to nature's functions. Principal processes of soil degradation are erosion, salinization, nutrient and carbon (C) depletion, drought, decline in soil structure, and tilth. Examples of ESs provided by soil include ecological/supporting (biomass production, nutrient cycling), regulating (water purification and flow, C sequestration, temperature fluctuations), provisional (food, fiber, fuel, and forages), and cultural (aesthetical, spiritual, and cultural). Erosion-induced degradation diminishes soil's capacity to provide ESs, and support ecosystem functions.

Although fertile top soils could be lost when scraped by heavy machineries (Ngwu *et al.*, 2005), the key avenues of topsoil loss include water erosion and wind erosion. Sometimes erosion can be such gradual for so long a time as to elude detection in one's lifetime, thus making its adverse effects hard to detect.

Eswaran *et al.*, (2001) propose an annual loss of 75 billion tons of soil on a global basis which costs the world about US \$400 billion per year. A review of the global agronomic impact of soil erosion identifies two severity groups of continents and reveals that Africa belongs to the more vulnerable group (Biggelaar *et al.*, 2004). Soil erosion by water seems to be the greatest factor limiting soil productivity and impeding agricultural enterprise in the entire humid tropical region. This is evident in many regions of Africa (Dregne, 1990), mainly in the humid and subhumid zones of Sub-Saharan Africa (SSA) where population pressure and deforestation exacerbate the situation and the rains come as torrential downpours, with the annual soil loss put at over 50 t ha [1] (FAO, 1995). In SSA, the problem is not limited to water erosion as wind erosion prevails mainly in the semiarid and arid zones. For instance, soil loss to wind erosion of 58–80 t ha [1] has recently been reported from the West African Sahel (Ikazaki *et al.*, 2011). Both forms of erosion can thus aptly define land degradation in the region. Soil erosion selectively detaches the colloidal fractions of soils and carts them away in runoff (Lal, *et al.*, 2003). These soil colloidal fractions (clay and humus) are needed for soil fertility, aggregation, structural stability, and favourable pore size distribution. The concentration of humus is usually higher in topsoils while that of clay is usually higher in subsoils due to illuviation, and this is mostly true in Ultisols that are widespread in Africa. This implies that humus, which has much greater capacity to hold water and nutrient ions compared to clay, its inorganic counterpart (Esu, 1999) is the more easily eroded.

In spite of the fact that the problem of land degradation is particularly severe in SSA, only little reliable data were available by the end of the 20th century both on its extent (FAO, 1995; Warren *et al.*, 2001) and on the cause-effect relationship between soil erosion and soil productivity (Lal, 1995). A survey highlights the enormous rate of soil erosion and the attendant decline in the productivity of agricultural soils in SSA. It is therefore unsurprising that, in the face of the advances so far made in biotechnology, agricultural productivity in SSA stagnates and remains perennially low as evident in hunger and poverty levels in the entire region (Eejeta, 2010, Abe and Wakatsuki, 2011).

All the adverse impacts on agronomic productivity and environmental quality are respectively due to a decline in land quality and deposition of sediments and have been designated on-site effect and off-site effect, respectively (Lal, *et al.*, 2003) It is widely believed that erosion-induced deposition of sediments occurs in response to topographic gradients and that, since water does not climb hills in agricultural watersheds, the process is hardly reversible. With this in view, we make a case for tackling the agro-ecological problem of soil erosion in the diverse watersheds of SSA offsite rather than onsite. This is a case for the *sawah* ecotechnology, an Asian type system of rice (*Oryza sativa* L.) production that has been adapted in the abundant lowlands in the region. The system can compensate for the loss of upland soil productivity while counteracting the environmental degradation due to soil erosion. It is viewed as the promising option to boosting rice production on a sustainable basis for the realization of the much-awaited Green Revolution in SSA. The impacts of erosion on soil quality, and ecosystem functions and services depend on the rate (Mg/ha/year, mm/year) of soil erosion vis-a-vis the rate of soil renewal (mm/century or millennia). The accelerated soil erosion, when the rate of soil removal exceeds that of its renewal, has adverse on- and off-site effects. The on-site adverse effects of severe erosion are due to loss of the effective rooting depth, reduction in plant-AWC, depletion of SOC and plant nutrients, decline in soil structure, and reduction in soil quality. The off-site effects of erosion are caused by run-on and inundation, sedimentation, non-point source pollution, and emission of greenhouse gases (GHGs) into the atmosphere. The agronomic, economic, and environmental effects of accelerated erosion are colossal at regional and global scales. Protection of soil quality under intensive land use and fast economic development is a major challenge for sustainable resource use in the developing world. The basic assessment of soil health and soil quality is necessary to evaluate the degradation status and changing trends following different land use and smallholder management interventions (Lal and Stewart, 1995). In Asia, adverse effects on soil health and soil quality arise from nutrient imbalance in soil, excessive fertilization, soil pollution and soil loss processes (Hedlund *et al.*, 2003).

Tropical Soils

As many as 1.8 billion people live in areas with some noticeable land and water degradation, which reduces livelihoods and household food security. In the humid tropical ecosystems, the areas with the greatest potential for land and water degradation (those with steep slopes, weathered and acid soils, excess rainfall, and high temperatures) do correspond closely with areas of highest rural poverty and malnutrition (Zhang *et al.*, 1996). Only about 16% of agricultural soils are free of significant constraints, such as poor drainage, poor nutrient status, difficult to work, salinity and alkalinity, or shallowness. Of these favoured soils, 60% are in temperate areas, and only 15% lie within the tropics. Globally, about 54% of the agriculture extent is 'Flat', 20% is on moderate slopes, about 17% on steep slopes and 8% on very steep slopes. All of these sloping lands are at risk of high soil erosion and rainfall runoff without adequate management (Penning de Vries *et al.*, 2002). About 69% of arable land is marginal lands where 65% of the population lives. Most of these are found in settled, densely populated areas.

An estimated 800 million people do not have access to sufficient food to lead healthy, productive lives, mostly in south asia and sub-saharan Africa. Food security is closely associated with poverty. The two-third of the rural population lived on degradation prone marginal agricultural lands and more than one third of all children living in sub-humid and humid tropics are malnourished (Penning de Vries *et al.*, 2002). High level High levels of rural poverty have persisted in many (semi) humid tropical ecosystems, where most of the poor population are rural poor. In the (semi) humid tropical ecosystems, the areas with the greatest potential for land and water degradation (with steep slopes, weathered and acid soils, excess rainfall, and high temperatures) do correspond closely with areas of highest rural poverty and malnutrition. Much of the needed agricultural growth must necessarily come from these lower-quality lands. For Africa, existing data suggest widespread loss of productive potential, due to intensive use of soils types that are highly sensitive to erosion and nutrient depletion, or inherently low in nutrients or organic matter.

Climate Change and Tropical Soils

The Green Revolution and other technologies since mid twentieth century were developed under an assumption of a stable climate. However, current and protected climate change poses severe challenges. Thus, it is pertinent to understand the followings (West *et al.*, 2009):

(i)   climate impacts to soil and ecosystem processes,

(ii) relation between land use/management and soil/ecosystem resilience,

(iii) opportunities for successful implementation of adaptation and mitigation strategies, and

(iv) the process of feasible decision making with consideration of scale and thresholds.

The terrestrial biosphere and soils have been the source of GHGs, (i.e., CO_2, CH_4, and N_2O) for thousands of years as a result of agriculture and the attendant deforestation, biomass burning, soil tillage, cultivation of paddy rice (*Oryza sativa*), and domestication of livestock. (Ruddiman, 2003) argued that cyclic variations in CO_2 and CH_4 caused by Earth's orbital changes caused decreases in atmospheric concentrations of these GHGs throughout the Holocene. However, increase in atmospheric concentration of CO_2 around 8,000 years ago and that of CH_4 around 5,000 years ago corresponded with the onset of early agriculture in Eurasia and of rice cultivation in Asia, respectively. This hypothesis is supported by the argument that despite the low population density, early per capita land use was large because of extensive or extractive farming (Ruddiman and Ellis, 2009). It is also argued that warming caused by these early gaseous emissions, estimated at 320 Pg C during the pre-industrial era compared with 160 Pg C since 1750, reached a global mean value of 0.8 °C (Ruddiman, 2003). Soils of most agro-ecosystems may have lost 30–50 % of the antecedent SOC pool in temperate regions and 50–75 % in the tropics (Lal, 2004). Total emissions from world soils have been estimated at 60–80 Pg C (Lal, 1999). The magnitude of SOC depletion is exacerbated in soils prone to degradation by accelerated erosion, decline in structure, depletion of nutrients, and reduction in plant-AWC. Indeed, there exists a strong link between desertification, accelerated erosion, and climate change. There are several important aspects of the climate change in the context of tropical and semiarid soils (drylands). These are:

(i) decrease in annual rainfall amount,
(ii) change in duration of rainfall events,
(iii) increase in interval between the rainfall events,
(iv) increase in temperature-induced evaporation,
(v) increase in water run-off, and
(vi) decrease in soil water storage.

Increase in frequency of extreme events has been already reported in Africa (Sivakumar *et al.*, 2005). Whereas the equilibrium water infiltration rate may be the dominant process affecting water storage in the Mediterranean type climate, overland flow (surface run-off) is often the control variable in arid regions. Thus, climate-induced effects on water resources and availability to plants are highly complex. However, increase in temperature and decrease in effective rainfall provides a strong positive feedback which accentuates the rate of SOC decomposition and emission of CO_2 into the atmosphere (Fig. 1).

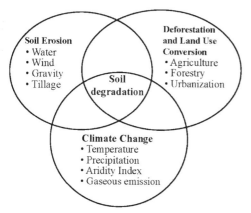

Fig.1: Strong interaction of desertification and drought with accelerated soil erosion, deforestation/land-use conversion, and climate change

Over and above any possible impacts of climate change, soil degradation, and desertification are also caused by long-lasting and perpetual mismanagement by extractive practices. Perpetual mismanagement can replace the climax vegetation in a specific biome because of soil degradation. It has been reported that changes in land-uses, fire regimes, and climate change are replacing the tropical humid forest by savanna (grass) vegetation in the Amazon Basin (Veldman and Putz, 2011). As many as 100 countries are prone to desertification. Increase in the land use under hyperarid (+50.7 Mha or 1.5 %) and arid (+3.1 Mha or 0.1 %) regions between 1931–1960 and 1961–1990 has been linked to climate change. Despite the widespread belief in strong interaction between climate and desertification, it is difficult to state that climate change has caused desertification because of major uncertainties in obtaining credible site-specific data for both independent (climate change) and dependent (desertification, erosion) parameters (Lal, 2001) at the desired temporal and spatial scales. On the contrary, some have hypothesized that desertification (independent variable) may have increased the temperature (dependent variable) of the desertified lands It is also believed that dryland ecosystems are more resilient to climate variability than hitherto presumed (Balling, 1991) probably because of the combination of an opportunistic response of some of its species and prevalence of a wide range of buffering mechanisms. Important among the buffering mechanisms are spatial mosaics of vegetated and bare patches enhancing hydrological links among two. Sediment trapment by vegetated patches enhances redistribution of water and nutrients. Thus, banded vegetation patterns (called "tiger bush" in the Sahel) may be an adaptation to harvest the run-off (Puigdefábregas, 1998). Yet, case studies on dryland degradation in relation to climate change show that some transition-triggered events are caused by combination of anthropogenic and climate factors.

Feddema (1999) hypothesized that, on a continental scale, the impact of global warming on African water resources may be greater than that of soil degradation. A possible increase of 1-3 °C in arid lands corresponding with atmospheric CO_2 abundance of 700 ppmv would increase potential evapotranspiration by 75–225 mm/year (Le Houérou, 1996) and drastically reduce NPP. For example, Oba et al., (2001) reported that both NPP and desertification in Sub-Saharan Africa may be influenced by the global climate variability, with positive feedback on desertification, climate change, and water resources.

Climate change also impacts two among principal processes of desertification—erosion and salinization. Soil erosion hazard depends on climatic erosivity, soil erodibility, and land and crop management practices (Fig. 2). Climate change can impact all of these parameters and greatly accentuate the erosion hazard. Increase in frequency and intensity of extreme events would enhance rainfall intensity and its kinetic energy, wind velocity and its erosivity, and run-off velocity and its shearing and sediment carrying capacity. Furthermore, erosivity of wind-driven rain and that of shallow overland flow impacted by raindrops is more than that of a rain without wind and of a laminar overland flow. Soil erodibility increases with decrease in aggregation and reduction in aggregate strength caused by increase in slaking due to reduction in SOC concentration. Progressive desertification of the dryland tropics may reduce already low amount of SOC stored in these soils. Furthermore, desertification may also alter the emission of GHGs from these ecologically sensitive and fragile ecosystems. Thus, warming-induced decline in SOC pool and aggregation, combined with increase in land conversion to meet the growing human demands, may severely accelerate soil erosion and desertification hazard. Wind erosion hazard, one of the major degradation processes in drylands (Sterk *et al.*, 2001) may also be increased because of the protected climate change in arid regions (Yang *et al.*, 2003). In accord with the argument outlined in Fig. 2, the data in Table 1 shows that the mean rate of continental soil erosion (Mg/ha/year) may increase between 1980 and 2090s in all continents except Europe. The global soil erosion hazard is projected to increase by 14 % with the highest increase in Africa. Yang and co-workers hypothesized that the effects of climate change on accelerated erosion is larger than that of land use conversion, with 9 % increase due to climate change and 5 % to land-use conversion. The regions with increasing trends of population and climate change may experience severe risks of accelerated soil erosion. The global increase in erosion will be 18.2×10^9 Mg/year with $\sim 11.6 \times 10^9$ Mg due to climate change (Table 1).

Table 1: Estimates of soil erosion at continental scale between 1980 and 2090

Continent	Land area (10^6 ha)	Potential soil erosion rate (Mg/ha/Year)		Total increase	
		1980s	2090s	10^6 Mg/Year	%
Africa	2,966	4.4	6.0	+4,746	+36.3
Asia	4,254	12.2	14.4	+9,359	+18.0
South America	1,769	8.5	10.3	+3,184	+3.7
World	13,013	10.2	11.6	+18,218	+13.0

Source: Oldeman and Yang *et al.* (2003)

The serious problem of soil salinization (Rengasamy, 2006) may also be aggravated by the projected aridization caused by an increase in potential evapotranspiration.

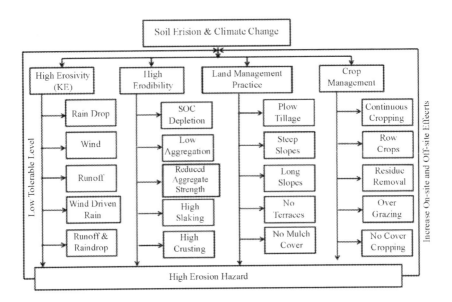

Fig. 2: Increase in risk of soil erosion and desertification due to climate change

Potential of Carbon Sequestration in Degraded Lands

Increasing organic matter inputs via agricultural management is the key to increasing SOM quantity. If the initial level of SOM is below the capacity of a specific soil to store organic matter, then SOM levels increase linearly with increasing input levels although the slope of the line may differ, reflecting the various influences of climate, soil type, and soil management. The major management strategies to increase SOM quantity are increasing primary production (e.g., perennial crops, plant nutrition, and organic amendments) and increasing the proportion of primary production returned to or retained by the soil (e.g., crop residue retention and placement).

Improving agricultural and land-use policies in degraded lands, such as marginal land and eroding landscapes, offers an enormous opportunity for enhancing C sequestration. Proper soil conservation practices that maintain vegetative cover and enhance plant productivity can promote higher SOC input and storage. Because soil C in eroded, marginal lands generally is depleted by a past history of erosion or intensive land use, rotation to minimum tillage or fallow conditions (with a cover of vegetation) is likely to increase the soil's potential to store C. For example, it is estimated that, in some regions, an increase in C storage of 0.2 to 2.2 metric tons per hectare per year may be observed with sustainable soil and water management. Realization of this potential would have significant benefits by reducing atmospheric build-up of CO_2. Moreover, protecting depositional C from oxidation through minimal tillage increases the potential for sequestration. The dependence of NPP and C sequestration on rates of erosion and deposition for sites with and without conservation measures is shown schematically in Fig. 2. If we consider eroding and depositional parts of a watershed separately, under given erosion scenario, as soil erosion increases, NPP decreases; but the C sequestration potential of the soil increases, at least initially, because of the enhanced ability of the degraded upland soils to take up more C compared with undisturbed and undegraded sites. Similarly, at the depositional sites in a given scenario (for example, alluvial plain), actual C sequestration follows a pattern similar to what it was at the eroding site, but with a higher rate of sequestration and a smaller decline after the peak, because the depositional sites continue to receive C-rich eroded soil. The added input of nutrient-rich topsoil at the depositional sites contributes to the maintenance of higher NPP. In the erosion and deposition conditions shown in Fig. 3, proper soil and water conservation measures maintain or increase NPP. If NPP increases at the eroding site, C sequestration in the eroding soils is enhanced and maintained at a higher level, and the depositional sites have the capacity to increase C sequestration until a saturation point is reached. The absence of tillage and other anthropogenic disturbances is critical to achieving the conservation conditions. On a cautionary note, marginal lands are vulnerable by definition, and C sequestration is not permanent. Most of the C stored in floodplains or reservoirs is protected physically by aggregation or burial, and potentially can be mineralized easily at a more rapid rate than it was accumulated if the depositional basins are disturbed by practices such as dredging or dismantling of impoundments (NPP), and carbon sequestration (CS) under given erosion and deposition scenarios and when an appropriate level of soil conservation is applied

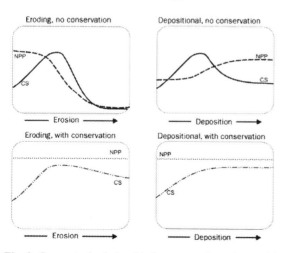

Fig. 3: Conceptual relationship between soil erosion and deposition, net primary productivity

The capacity of a soil to store organic matter is related to the association of SOM with clay and clay plus silt (2–20 μm diam.) particles, soil microaggregates (20–250 μm diam.) and macroaggregates (>250 μm diam.), and the fraction of sand sized macroorganic matter (Tisdall, 1996). Once the clay plus silt is saturated with organic matter, additional SOM would be found in macroaggregates, probably as sand sized macroorganic matter.in well aggeregated soil most of the SOM can be found in macroaggregate complexes. A conceptual model illustrated in Fig. 4 elucidated the SOM content and proportion in soil particles and aggregates at different types of SOM capacity levels related to organic matter inputs and aggregation. clay plus silt serve as a fixed capacity level while the combination of aggregated C and macroorganic C provide an additional variable capacity. The former is soil specific while the latter tends to be contingent on both soil type and management (i.e. C inputs). Coarse textured soils in the sub-tropics possess high C turnover rates because of favourable soil moisture and temperature conditions.

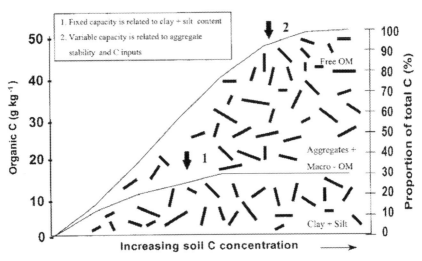

Fig. 4: Conceptual model of soil organic matter (SOM) content and proportion in soil particles and aggregates

Opportunities for C Sequestration in Agro-ecosystems

The agricultural soils in India are low in organic carbon, which may be attributed to excessive tillage, imbalanced fertilizer use, little or no crop residue recycling, and severe soil degradation. Management practices or technologies that increase carbon input to the soil and reduce C loss or both lead to net carbon sequestration in soils. The c input in soil can be increased by a number of ways such as selection of high biomass producing crops, residue recycling or residue retention by lessened tillage intensity, application of organic materials (e.g. animal manure, compost, sludge, green manure etc.), adoption of agroforestry systems, intensification of agricultural through improved nutrient and water management practices, reducing summer or winter fallow, changing from monoculture to rotation cropping, and switching from annual crops to perennial vegetation (Table 2). Soil carbon loss could be decreased by adopting conservation agriculture and minimizing soil disturbance, checking erosion through reduced tillage intensity, and using low quality organic inputs. Agronomic interventions such as changing from monoculture to continuous cropping, changing crop-fallow to continuous cropping, or increasing the number of crops in a rotation can sequester on an average 20 ± 12 g C m^{-2} yr^{-1}. Analysis of results from long-term experiments showed that conservation agricultural technique such as change from conventional tillage to no-till could sequester 57 ± 14 g C m^{-2} yr^{-1} .

Globally, potential for C sequestration in soils over 50 years period has been estimated to be 24-43 Tg C (0.4 to 0.9 Tg C yr^{-1}) through improved management of existing agricultural soils, restoration of degraded lands. The potential of soil carbon sequestration in India is estimated at 39 to 52 Tg (1 Tg= 10^{12} g) which includes restoration of degraded soils (7.2 -9.4 Tg C yr^{-1}), and reduction in erosion-induced emission of C (4.3-7.2 Tg C yr^{-1}).

Table 2: Strategies to sequester soil organic C.

(1) Maximizing C input	(2) Minimizing C loss from soil
• Plant selection	• Reducing soil disturbance
• Species, cultivar, variety	• Less intensive tillage
• Growth habit (perennial/annual)	• Controlling soil erosion
• Diversified crop rotation	• Utilizing available soil water
• Biomass energy crops	• Promotes optimum plant growth
• Greater root biomass	• Reduces soil microbial activity
• Tillage	• Maintaining surface residue cover
• Type	• Increased plant water use and production
• Frequency	• More fungal dominance in soil
• Fertilization	
• Rate, timing,	
• Placement	
• Organic amendments	

C sequestrationThrough Land Reclamation and Management

Some improved practices with favourable impact on soil structure include growing cover crops, sowing crops with conservation tillage, maintaining balance level of soil fertility and converting marginal and degraded lands to restorative land uses. Soil conservation may be achieved through reduction of soil detachment and its transport by agents of erosion. Improving soil's resistance to forces causing detachment and transport involves enhancing soil structure. Any land reclamation practice which improve soil structure and enhance soil quality lead to C sequestration.

Cover crops: Through formation of a quick and protective ground cover, cover crops improve soc content, enhance soil biodiversity, improve soil structure and minimize risks of soil erosion. Experiment conducted at CSWCRTI, Dehradun revealed that cover crop based rotation such as cowpea-wheat, cowpea-lentil and cowpea-mustard added more soil organic carbon than cereal based rotation such as maize-wheat, maize-lentiil and maize-mustard. Annual rate of increase in SOC was highest (734 kg C ha^{-1}) in cowpea-mustard with 120 kg K$_2$O ha^{-1} whereas lowest (97 kg C ha^{-1}) in maize-wheat rotation with 40 kg K$_2$O ha^{-1}

Table 3: Build up of soil organic carbon as affected by cover crop and potassium application

Treatments	Annual rate of SOC (Kg C ha^{-1}) Build-up	Treatments	Annual rate of SOC (Kg C ha^{-1}) Build-up
Maize*-Wheat	217	Cowpea*-Wheat	460
Maize**-Wheat	97	Cowpea**-Wheat	371
Maize*-Lentil	435	Cowpea*-Lentil	583
Maize**-Lentil	295	Cowpea**-Lentil	474
Maize*-Mustard	545	Cowpea*-Mustard	734
Maize**-Mustard	320	Cowpea**-Mustard	524

*=120 kg K$_2$O ha^{-1}, **= 40 kg K$_2$O ha^{-1}
Source: CSWCRTI Annual Report (2004-05)

Conservation tillage: Beneficial impacts of conservation tillage in decreasing runoff and soil erosion are widely recognized. When used in conjunction with crop residue mulch and cover crops, conservation tillage improves soil structure and enhances SOC pool. The data in Table 4 show that use of no-till system increased SOC content and residence time of carbon in soil probably due to encapsulation of SOC within stable aggregates and alteration in soil quality. The benefits of conservation tillage in C sequestration are due to both increases in SOC content, decrease in CO_2 emissions caused by ploughing and to reduction in fuel consumption.

Table 4: Effect of tillage and crop residue on soil organic carbon (SOC) (over 8 years)

Treatments	SOC (%)
T_1 = Conventional tillage	0.65
$T_2 = T_1$ + crop residue	0.68
T_3 = Minimum tillage	0.70
$T_4 = T_3$ + crop residue	0.75
Initial	0.58

Source: CSWCRTI Annual Report (2006-07)

The impact of conservation tillage on soil organic C sequestration may be greater in degraded soils than in fertile soils. The basis for this statement was derived from the observation that the ratio of soil organic C with conservation tillage-to-conventional tillage was logarithmically greater in soils with inherently lower organic C than in soils with inherently higher organic C content (Fig. 5). Therefore, on a relative basis, the improvement in soil organic C was proportionately higher in poorer soils.

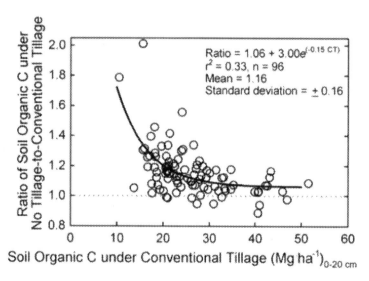

Fig. 5: The ratio of soil organic C under conservation tillage-to-conventional tillage as related to the initial soil organic C content under conventional tillage

Soil fertility management: Nutrient management is essential to increase crop yield. There is a strong relationship between crop yields and the amount of soc in the root zone. Several studies in India have documented a positive relationship between soc concentration in the root zone and yield of a number of crops including wheat, rice, and maize. For example, in alluvial soils of north India, wheat grain yield without fertilizer application increased from 1.4 Mg ha^{-1} at an SOC concentration of 0.2% to 3.5 Mg ha^{-1} in soils with an

SOC concentration of 0.9%. With the application of chemical fertilizers, the effect of soc concentration, on wheat productivity was smaller indicating an interaction between soc and fertilizer use.

Judicious nutrient management is crucial to soil organic C (SOC) sequestration in tropical soils. Adequate supply of nutrients in soil can enhance biomass production and SOC content. Use of organic manure and compost enhances the SOC pool more than application of the same amount of nutrients as inorganic fertilizers. Long-term manure application increases the SOC pool, which not only sequester CO_2 but also enhances productivity of soil. It is, however, argued that SOC sequestration is a major challenge in soils of the tropics and sub-tropics, where climate is harsh and resource-poor farmers cannot afford the input of organic manure and crop residues. The rate of C mineralization is high in the tropics because of high temperature and the humification efficiency is low. Integrated nutrient management involving addition of organic manures/composts along with inorganic fertilizers results in improved soil aggregation and greater carbon sequestration especially in macro aggregates. In corporation of organic manures includes decomposition of organic matter where roots, hyphae and poly-saccharides bind mineral particles into micro aggregates bind to from C rich macro aggregates. This type of C is physically protecte4d with in macro aggregates. The free primary particles are cemented together into micro aggregates by persistent binding agents characterized by humification of organic matter and stimulated accumulation of C aggregates.

Agricultural intensification: Adopting recommended farming practices is an important and effective strategy for soil conservation. Recommended farming practices involve agricultural intensification on prime agricultural land through use of improved varieties adoption of appropriate cropping systems that enhance cropping intensity and elimination of summer fallow. Experiments conducted in CSWCRTI, Dehradun have shown an increase in soc pool at the rate of in 0-15 cm depth. In fact, the SOC pool in 0-15 cm depth increased linearly with increase in cropping intensity. Improvements in crop yield through adoption of recommended technology enhance soc pool and improve soil quality.

Intensive agriculture with improved nutrient and water management results in enhanced C sequestration due to higher crop productivity and greater return of crop residues, root biomass and root exudates to soil results of a 25-year study from Punjab showed that intensive agriculture resulted in improved SOC status by 38%. Enhanced C sequestration was related to increase productivity of rice and wheat, one tonne increase in crop productivity resulted in a C sequestration of 0.85 t ha^{-1}.

Land restoration: Many different strategies have been shown, under controlled conditions, to successfully rehabilitate degraded land, restore land capabilities and enhance the productivity of land. Conversion of marginal land agricultural land to restorative land use such as conversion of degraded farmland into agro-forestry systems, conversion of degraded croplands and pasture to forest and conversion of degraded croplands into grassland reduces soil erosion and increase soc pool. The data in Table 5 from a degraded soil in central India show 1.8 to 3.5 times increase in SOC pool due to adoption of an agripastoral system in a highly degraded soil. The rate of sequestration may differ among soil types, management and ecological factors.

Table 5: Biomass production (Mg ha^{-1}) on degraded lands in Jhansi, India with sylvopastoral system

Treatment	1990	1991	1992	1993	1994	1995	1996
Natural/ Traditional	3.5	3.6	3.5	3.0	3.1	2.1	3.3
Improved	2.0	7.6	7.5	10.4	5.5	7.2	6.8

Crop diversification with legume: Crop diversification with legumes requires minimum fossil energy for production because of less requirement of N fertilizer, where N fertilizer accounts 40 and 70% of the total energy requirement for the non legumes, while it is reported that pea requires 25-50% less energy compared to non legumes due to biological nitrogen fixation (BNF) by legumes. The decrease in the consumption of fossil energy in legume cropping reduces the agricultural contribution to global warming and energy requirement for BNF are met through renewable sources of synthesized carbohydrates by plant using solar energy. The cultivation of a legume saves approximately 0.2 t of fuel per hectare that corresponds to the production of 600 kg of CO_2 per hectare. Hence, legumes are considered as pillar in the development of sustainable agricultural systems.

Long-term experiment conducted since 1976 at ICRISAT showed that improved sorghum/pigeonpea or maize/pigeonpea-based intercropped systems in the watersheds not only produced on an average 5.1 t ha^{-1} grains under rainfed condition over the last 33 years but these systems continued to gain productivity of 82 kg ha^{-1} yr^{-1}. On the other hand, traditional sorghum grown with only application of 5.1 t ha^{-1} FYM once in two years produced average yield of 1.1 t ha^{-1} with average gain of 28 kg grains ha^{-1}yr^{-1}. Improved sorghum/pigeonpea system could support 24 persons as that of 4.6 persons ha^{-1} from the traditional sorghum system. Most importantly improved sorghum/pigeonpea system sequestered 34% more organic C with an overall gain of 7.4 t ha^{-1} C up to 120 cm depth with a gain of 335 kg C ha^{-1}yr^{-1} than the traditional sorghum system. Along with increased C sequestration improved soil biological, physical and chemical properties were also observed. Improved sorghum/

pigeonpea intercropping in vertisols sequestered 34% more organic carbon (C) with an overall gain of 7.4 t ha⁻¹ upto 120 cm depth with a gain of 335 kg C ha⁻¹ (Table 6) than the traditional post rainy continuous sorghum system. Similarly, ICRISAT- Central Research Institute for Dryland Agriculture (CRIDA)-National Bureau of Soil Survey and Land Use Planning (NBSS&LUP) and Indian Institute of Soil Science (IISS) studied 28 benchmark sites in semi arid ecosystem and revealed that legume and horticultural based system has great potential for C sequestration.

Table 6: Organic carbon content (g C/kg soil) of semi- arid tropical Vertisols under improved and conventional systems in a watershed

Soil depth (cm)	Improved system (Sorghum/ pigeonpea)	Conventional system (Sole sorghum)
0-15	4.35 (67%)	2.60
15-30	3.20 (36%)	2.35
30-45	3.35 (39%)	2.40
45-60	3.15 (28%)	2.45
60-75	2.70 (35%)	2.00
75-90	2.50 (21%)	2.05
90-105	2.25 (05%)	2.15
105-120	2.10 (06%)	2.25
Mean	3.11 (34%)	2.32
SE± Location (L): 0.098 Soil depth (SD): 0.16 L X SD: 0.59		

Source: ICRISAT, Patancheru (1998)

Conclusion

There exists a close link between soil erosion/degradation, climate change, and poverty. Soil degradation creates a positive feedback attributed to emission of radiatively active gases depletion of soil organic carbon and nutrient pools, denudation of vegetation cover, and reduction in net primary productivity, increase in frequency and intensity of droughts (especially pedological and ecological droughts), and loss of ecosystem resilience. There is a strong need for prudent management of soil, vegetation, water, and other natural resources. Restoration of degraded soils can increase the ecosystem C pool provided that available water and plant nutrients are adequate. Establishment of tree plantations can reduce stream flow while also decreasing albedo. The strong link between the climate–vegetation–soil–water–continuum and anthropogenic activities necessitate a prudent and a coordinated effort to reverse the downward spiral, restore degraded soils, mitigate climate change, and enhance ecosystem services. There are new tools and techniques available to measure, monitor and verify the status of soils, and natural resources. The community-based natural resources management is a useful strategy for judicious governance. The strategy is to shift from strategic action to strategic governance

Land degradation and the effects of organic matter loss are particularly critical in tropical semiarid regions where the risk of desertification is great, and where precipitation and temperature changes associated with global warming will further undermine ecosystem integrity. If not properly managed all efforts for re-construction of drylands and conservation of organic matter might be doomed in a situation where a rise in average temperature of about 1.4–5.8° C is likely. Because land degradation is accelerating as a serious problem, maintaining or enhancing farmers' soil capital has increasingly become a prime focus for sustainable agriculture and increased food production. To a large extent, this is because farmers collect the crop or plant residue to feed livestock (livestock even graze freely on crop residue) and households use the residue as an energy/fuel source. This removal of crop residue, combined with low levels of fertilizer application, depletes soil fertility and contributes to the deepening poverty in many developing countries. Although resource-saving agricultural crop production is desirable, farmers are not likely to internalize benefits from environmental preservation (e.g., carbon sequestration) unless they are given adequate incentives. Consequently, effective soil conservation strategies have to be based on integrated approaches that combine biophysical techniques with socio-economic and cultural considerations. In addition to erosion control, acceptable soil conservation technologies must have numerous ancillary benefits such as improvement in water quality, enhancement of biodiversity and soil C sequestration mitigating the green house effect. Restoration of eroded soils enhances C sequestration and improves soil quality. There are at least three opportunities for C sequestration through soil conservation;

1. Historic C lost from eroded soils estimated at 24.9 Pg can be resequestered through soil restorative measures including afforestation and establishment of appropriate vegetative cover, land application of biosolids and nutrient management.

2. Effective soil conservation if adopted on a global scale can reduce erosion related emissions of 1.1 Pg C yr[-1].

3. Adaptation of conservation effective farming practices can lead to reduction in soil erosion, improvement in soil quality and enhancement in soc pool. Important agricultural practices with potential for C sequestration include conservation tillage, crop residue mulch, soil fertility management and adoption of recommended cropping systems.

Undoubtedly, there are considerable opportunities for enhancing biotic C sequestration in Indian Agro ecosystems. However, prospects of C sequestration in soil and vegetation should be carefully analyzed before implementation. Information about input and output fluxes of carbon in different agro ecosystem may be generated to develop effective C management strategies. Minimizing soil erosion is vital to protecting natural resources, because accelerated erosion reduces soil quality and depletes the soil resource. Similarly, proper management of already degraded, marginal areas could ensure the environmental benefits of C sequestration resulting from burial and partial replacement of eroded SOC.

References

Abe S.S. and Wakatsuki, T. (2011). "Sawah ecotechnology-a trigger for a rice green revolution in sub-Saharan Africa: basic concept and policy implications," *Outlook in Agriculture*, 40(3): 221–227.

Balling, R.C. Jr (1991). Impact of desertification on regional and global warming. Bulletin of American Meteorological Society, 72: 232–234.

Biggelaar, C. Den, Lal, R., Wiebe, K. and Breneman, V. (2004). "The global impact of soil erosion on productivity. I: absolute and relative erosion-induced yield losses," *Advances in Agronomy*, 81: 1–48.

Doran, J. W. and Parkin, T. B. (1994). "Defining and assessing soil quality," In Defining Soil Quality for a Sustainable Environment, J. W. Doran *et al.*, Ed., vol. 35, Soil Science Society of America Special Publication, Madison, Wis, USA.

Dregne, H.E. (1990). "Erosion and soil productivity in Africa," *Journal of Soil & Water Conservation*, 45(4): 431–436.

Ejeta, G. (2010). African green revolution needn't be a mirage," Science, 327(5967): 831–832.

Esu, I. E. (1999). Fundamental of Pedology, Stirling-Horden Publishers, Ibadan, Nigeria.

Eswaran, H., Lal, R. and Reich, P.F. (2001). "Land degradation: an overview," in Proceedings of the 2nd International Conference on Land Degradation and Desertification, pp. 1–5, Oxford Press, Khon Kaen, Thailand.

FAO, (1995). Land and Environmental Degradation and Desertification in Africa, FAO Corporate Document Repository, T. Ikazaki, K., Shinjo, H., Tanaka, U., Tobita, S., Funakawa, S. and Kosaki, (2011). "Field-scale aeolian sediment transport in the Sahel, West Africa," *Soil Science Society of America Journal*, 75: 1885–1897.

Feddema, J.J. (1999). Future African water resources: interactions between soil degradation and global warming. *Climate Change*, 42: 561–596.

Hedlund, A., Witter, E., An, B.X., (2003). Assessment of N, P and K management by nutrient balances and flows on peri-urban smallholder farms in southern Vietnam. *European Journal of Agronomy*, 20: 71–87.

Lal, R. (1995). "Erosion-crop productivity relationships for soils of Africa," *Soil Science Society of America Journal*, 59(3): 661–667.

Lal, R. (1999). Soil management and restoration for carbon sequestration to mitigate the greenhouse effect. *Progressive Environment Science*, 1: 307–326

Lal, R. (2001). Soil degradation by erosion. *Land Degradation and Development*, 12: 519–539.

Lal, R. (2004). A carbon sequestration in dryland ecosystems of West Asia and North Africa. *Land Degradation and Development*, 13: 45-49.

Lal, R. and Stewart, B.A. (1995). Soil Management: Experimental Basis for Sustainability and Environmental Quality. Advances in Soil Science. CRC Press, Boca Raton, Florida.

Lal, R., Biggelaar, C. Den. and Wiebe, K.D. (2003). "Measuring on-site and off-site effects of erosion on productivity and environmental quality," in Proceedings of the OECD Expert Meeting on Soil Erosion and Soil Biodiversity Indicators, Rome, Italy, March.

Le Houérou, H.N. (1996). Climate change, drought and desertification. *Journal of Arid Environment*, 34: 133–185.

Ngwu, O.E., Mbagwu, J.S.C. and Obi, M.E. (2005). "Effect of desurfacing on soil properties and maize yield—research note," *Nigerian Journal of Soil Science*, 15(2): 148–150.

Oba, G., Post, E., Stenseth, N.C. (2001). Sub-Saharan desertification and productivity are linked to hemispheric climate variability. *Global Change in Biology*, 7:241–246.

Penning de Vries, F.W.T., Molden D., Scherr, S.J. and Valentin, C. (2002). Implications of land and water degradation for food security. Working paper for the comprehensive assessment of water for agriculture. IWMI, Colombo, 55 p.

Puigdefábregas, J. (1998). Ecological impacts of global change on drylands and their implications for desertification. *Land Degradation & Development*, 9: 393–406.

Rengasamy, P. (2006). World salinization with emphasis on Australia. *Journal of Experimental Botany*, 57(5):1017–1023.

Ruddiman, W.F. (2003). The anthropogenic greenhouse era began thousands of years ago. *Climate Change*, 61: 261–293.

Ruddiman, W.F., Ellis, E.C. (2009). Effect of per capita land use changes on Holocene forest clearance. *Quaternary Science Review*. 28: 3011–3015.

Sivakumar M.V.K. (2007). Interactions between climate and desertification. *Agriculture for Meteorology*, 142: 143–155.

Sivakumar, M.V.K., Das H.P., Brunni O. (2005). Impacts of present and future climate variability and change on agriculture and forestry in the arid and semi-arid tropics. *Climate Change*, 70: 31–72.

Sterk, G., Riksen, M., Goossens, D. (2001) Dryland degradation by wind erosion and its control. *Annals of Arid Zone*, 40(3):351–367.

Sullivan, P. (2004). "Sustainable soil management: soil systems guide," Appropriate Technology Transfer for Rural Areas (ATTRA) Fayetteville AR 72702, *National Center for Appropriate Technology* (NCAT).

Tisdall, J.M. (1996). Formation of soil aggregates and accumulation of soil organic matter. In: Carter, M.R. and Stewart, B.A., Eds., Structure and Organic Matter Storage in Agricultural Soils, Boca Raton, FL Lewis.

Veldman J.W. and Putz, F.E (2011). Grass dominated vegetation, not species-diverse natural savanna, replaces degraded tropical forests on the southern edge of the Amazon Basin. *Biological Conservation*. 144(5): 1419–1429.

Warren, A., Batterbury, S. and Osbahr, H. (2001). "Soil erosion in the West African Sahel: a review and an application of a "local political ecology" approach in South West Niger," *Global Environmental Change*, 11(1): 79–95.

West J.M., Julius, S.H, Kareiva, P, Enquist, C., Lawler, J.J., Petersen, B., Johnson, A.E., Shaw M.R. (2009). US natural resources and climate change: concepts and approaches for management adoption. *Environmental Management*, 44(6): 1001–1021.

Yang, D., Kanae, S, Oki T., Koike, T, and Musiake, K. (2003). Global potential soil erosion with reference to land use and climate changes. Hydrological Proceedings 17:2913–2928.

Zhang, W.L., Tian, Z.X, Zhang, N., Li, X.Q. (1996). Nitrate pollution of groundwater in northern China. *Agriculture, Ecosystem & Environment*, 59: 223–231.

19

Conservation Agriculture: An Approach Towards Sustainability of Soil Physical Health

A. Kundu, S. Mukherjee, R. Nandi and P. K. Bandyopadhyay

Bidhan Chandra Krishi Viswavidyalaya, Mohanpur
West Bengal-741 252, India

Introduction

The green revolution is exemplified by too much adoption of high soil disturbance, modern varieties, energy exhaustive technologies, high amount of capital investment and simplification of farming systems (Meeus, 1993), resulting in uniform and featureless landscapes (Nassauer and Westmacott, 1987) in the most intensively farmed arable areas. In post-green revolution epoch, we had experienced a period of remarkable escalation in food productivity, despite rising land values and increasing land scarcity. Farmers from both developing and developed countries get exposed to several production constraints for achieving a sustainable, cost-effective and assured return from their field. Depending upon the circumstances and the scale of farm activities, farmers face varying degrees of climate risk, biotic invasion, and economic uncertainty. Thus, from the context of economical ambiguities, volatile food prices, social pressure to buffer ups and downs of agricultural production and achieving food security for ever growing population and curtailing poverty, farmers must be equipped with novel cultural practices to achieve a sustainable and assured cost-effective grain production. Research experiments throughout the globe found that the principal indicators for unsustainability of agricultural systems are: (1) intensive tillage induced decrement of soil organic matter (OM), structural degradation of soil, accelerated water and wind erosion, reduced water infiltration rates, surface sealing and crusting, soil compaction, (2) insufficient return of organic materials into soil, and (3) monocropping.

Tillage is one of the fundamental agrotechnological operations required for optimum plant growth. But, intensive tillage operations without good soil and crop management practices can increase compaction of surface soil, reduce aggregate stability, shatter surface pore-apertures, downfall in retention and transmission of water and solutes, and enhance the losses due to runoff and erosion. Furthermore, intensive tillage is associated with degradation of natural resources, diminishing wildlife population and biodiversity, exhaustive use of fuel with low efficiency and contributes towards global warming. As a result, rampant change over in tillage practices and soil management is required to achieve desirable productivity while nourishing natural resources. Again, the continuous planting of same crop on the same piece of land has been detrimental to soil fertility, and the sustainability of that system is doubtful (Saha and Mishra, 2009). Depletion of soil organic carbon (SOC), lower soil moisture retention, and lessening of water stable aggregates have been identified as reasons for the unsustainability of monoculture of field crops as well as horticultural crops (Clermont-Dauphin *et al.*, 2004) even with adequate NPK application (Nambiar, 1995; Altieri, 2009). Monoculture boosts harmful soil pathogen population (Olsson and Gerhardson, 1992; Eberlein *et al.*, 2016). Crop rotation can be used to ease such problems, either owing to the nature of the crop itself or in consequence of cultivation methods used.

Hence, Conservation agriculture (CA) has emerged as an alternative to an inefficient tillage-based conventional agriculture. CA is an ecosystem approach to regenerative sustainable agriculture and land management based on three interlinked principles: (1) continuous no or minimum mechanical soil disturbance, (2) permanent maintenance of soil mulch (crop biomass and cover crops) and (3) diversification of cropping system (economically, environmentally and socially adapted rotations including legumes and cover crops), along with other complementary good agricultural production and land management practices (FAO, 2016). CA helps in managing agroecosystems for improved and sustained productivity, increased profits and food security while preserving and enhancing the resource base and the environment. Soil cover can be live cover crops, terminated cover crops, or mulches of crop residues remaining after previous harvests that accumulate on the soil surface. Ultimately it improves soil aggregation, soil biological activity, soil biodiversity, carbon sequestration (Ghosh *et al.*, 2010, Bünemann *et al.*, 2018), water conservation due to reduction in evaporation loss of soil water (Teame *et al.*, 2017; Kumar and Lal, 2012), erosion control (Patil *et al.*, 2013), nutrient addition (Patil *et al.*, 2016), control weeds and ultimately increasing the yield (Das *et al.*, 2018).

Physical condition of a soil has direct as well as indirect effects on crop production and environmental quality. As soil physical properties pithily influence different physico-chemical and biological processes in soil, these should be kept at optimal state (Lal, 1991). Structure of surface soil layer and to a lesser degree subsurface soil, are greatly influenced by management practices and cropping systems. Soil physical quality varies from location to location depending on the tillage system and their intensity, agro-climatic condition and type of soil. A number of soil functions are mediated by specific soil physical attributes (Table 1). Soil texture controls water contents, water intake rates, aeration, root penetration and some chemical properties. Fine textured soils with poor drainage and light textured soils with high percolation need to study with CA to improve physical properties by their affinity to form building blocks for aggregation.

Table 1: Soil functions related to soil physical attributes

Physical attributes	Soil functions
Texture, structure, depth, hydraulic conductivity, infiltration rate, aeration, surface area, bulk density	Water retention, transmission, filtration of pollutants, water cycling and renewability, nutrient adsorption, gaseous exchange etc.

Soil structure is an essential property of the soil system that defines the environment for both plant roots and edaphon. Structure of the surface soil and to a lesser degree, the subsurface soil, are greatly influenced by soil processes and external forces applied upon it (tillage). Consequently, soil structure with environmental behaviour has a direct impact on the processes of water, solute, air and heat transport and all of these processes affect nutrient transport as well as uptake status that trigger root growth. Bulk density, aggregation, aggregate stability and pore size distribution are vital soil physical properties and most likely to alter plant growth indirectly, as their relationship to crop production is articulated through their effect on water, aeration, temperature, and mechanical resistance. Increasing water content cuts aeration which is undesirable but decreases mechanical resistance which is desirable. Physiological metabolic processes are definitely temperature dependent and sharp fluctuation of soil temperature can hamper these processes. The non-limiting water range (NLWR) is also affected by aeration and/or mechanical resistance, particularly in poorly structured soils with high bulk density (Letey, 1958).

Bünemann *et al.*, (2018) compiled a data set for novel physical indicators important for soil quality (Fig. 1) and found that texture, structure, density, penetration resistance, soil water movement and storage are major indicators changing with management practices.

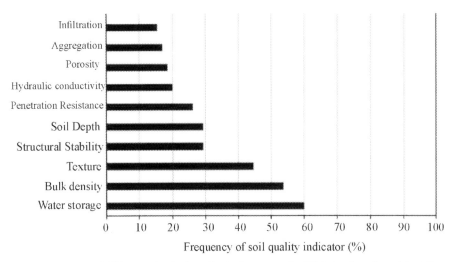

Fig. 1: Frequency of different physical quality indicators (min. 10%) in all reviewed (n=65) soil quality assessment approaches.
Source: Bünemann *et al.* (2018)

Thus, comprising of straw mulching, residue retention, reduced tillage and no tillage with minimal soil disturbances, CA can contribute towards soil and water conservation, increasing water productivity (Wang and Shangguan, 2015), with concomitant reduction in seasonal evapotranspiration (Verhulst *et al.*, 2011; Hou *et al.*, 2012) by increasing water infiltration (Hou *et al.*, 2012) and increasing root zone soil moisture storage (Mohanty *et al.*, 2007; Bandyopadhyay *et al.*, 2016, 2018). Dalal and Bridge (1996) reported that within CA system many constraints such as soil structure degradation, soil compactness, soil crusting etc., related to soil physical health are diminuend. However, there is a long-lasting debate about the effects of CA on soil ecosystem and crop productivity (Pastorelli *et al.*, 2013; Wright *et al.*, 2005).

Impact of Conservation Practices on Soil Physical Health

CA gradually induces desirable changes in many soil physical properties, *viz.* increase in soil water infiltration, improvement in soil structure, porosity, OM content and macro-micro faunal activity on long-term basis, as well as reduction in water runoff, soil loss, and evaporation loss. Tillage and crop residue management can influence soil physical properties as a direct result of alteration of physical matrix of soil or indirectly by altering surface energy partitioning, microbial activity, and soil chemical composition (Verhulst *et al.*, 2010). Some soil properties appear to have little or no response to residue management, including penetration resistance (Unger, 1984), hydraulic conductivity, and bulk density (Skidmore *et al.*, 1986; Gupta *et al.*, 1987).

Verhulst *et al.* (2010) showed in a review on how the tillage [no-till (NT) and conventional tillage (CT)] and residue retention affect the major soil physical properties/ processes where some results were obtained conflicting (Table 2).

Table 2: Effect of tillage and residue management on soil physical properties/ processes

Soil physical properties/ processes	NT compared to CT	Residue retention
Aggregate stability	↑	↑
Bulk density	↑Small no. of studies showing opposite	↓
Total porosity	↓	↑
Macropores	↓↑Average size larger	↑
Mesopores	↑	*
Micropores	↑	*
Hydraulic conductivity	↓Mixed results	↑
Infiltration	↑	↑
Runoff	↓	↓
Evaporation	↓	↓
Plant available water	↑	↑
Erosion	↓	↓

↑,↓ and * denote increasing, decreasing and no information, respectively.
Source: Verhulst *et al.* (2010)

Soil Structure

Soil structure is the definite arrangement pattern of individual soil particles (sand, silt and clay), those are bound together by suitable binding agents to form larger units i.e. soil aggregates, stability of which against different stresses like water, tillage etc. are quite important for assessment of its structural stability. Distribution of various sized soil aggregates gives an idea about physical stabilty of particular soil of a particular site. Reduction of the intensity of tillage and retaining a soil cover are fashioned, which ultimately throw in the major benefits for improving structural stability *vis-a-vis* soil physical quality under CA system (Gregorich and Carter, 1997; Saber and Mrabet, 2002).

Under CT aggregate stability is too low to withstand against impacts of rainwater, irrigation and tillage compared to zero tillage (ZT), no tillage (NT), minimum tillage (MT) with other conservation practices (Zhang *et al.*, 2012). As frequent tilling hamper the aggregate formation process and resulting in higher rate of macroaggregate turnover (Six *et al.*, 2000) in conventional practices. Large macroaggregates may maintain a balance between the destruction and formation of aggregates over the time (Bronick and Lal, 2005; Andruschkewitsch *et al.*, 2013). He *et al.* (2009) demonstrated that after ten years of continous conservation, NT plots contained 13-37% more macroaggregates than CT throughout the soil profile. On the other hand, percentage of micro-aggregates was 25–59% greater in CT (Fig. 2). Tao et al. (2018) noticed that increase in macroaggregates and the carbon (C) concentration within macroaggregates in long-term continuous NT which provide physical protection for soil C sequestration under continuous maize cropping in clayey soil of semiarid regions of northern China.

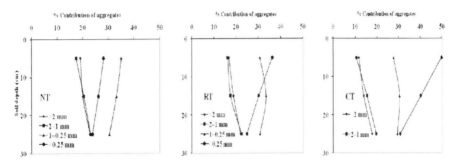

Fig. 2: Contribution of aggregate size under different tillage practices
Source: He et al. (2009)

Crop residue retention at the soil surface not only increases the aggregate formation, but it also decreases the breakdown of aggregates by reducing erosion and protecting the aggregates against raindrop impact. Sometimes it was observed that residue retention may not be possible in all the crops in the rotation due to its other potential uses (for example, as fuel, fodder or thatching material), particularly in poverty-ridden eastern India (Mondal et al., 2019). They observed that CA resulted in 11-12% higher macroaggregation as compared to CT and partial CA (One crop in a double/tripple cropping systems is under ZT) and also enhaced the other structural indices. Fresh crop residues activate the building block for the new aggregate formation as hot spots of microbial activity are developed (Guggenberger et al., 1999; De Gryze et al., 2005). On the other hand, interestingly no significant differences in aggregate structural stability were found between NT+without residue retention and CT (Ferreras et al., 2000). Carbon from crop residue can also promote soil macroaggregation (Bravo-Garza et al., 2010; Zhang et al., 2017) as high linear correlation ($r=0.947$) exists between SOC and aggregate size (Carter, 1992), which proved to be effective in declining the disintegration, slaking of aggregates when exposed to action of water (Blevins et al., 1998).

Crop rotation contributes to the development of soil structure, by returning of crop residues, developing and better distributing biopores and positively affecting dynamics of microbial communities (Ball et al., 2005, Govaerts et al., 2009). Better soil structural quality as result of solely crop rotation and along with conservation tillage have been reported by Askari *et al.*, (2013) where he noticed 14% increase in aggregate size distribution (ASD) with crop rotation than monocropping. But, pronounced enhancement (71%) in ASD was seen with practicing crop rotation in combination with MT and no signigificant effect was seen for crop rotation with CT. Crops can affect soil aggregation by their root systems because plant roots are important binding agents at the scale of macroaggregates. A soil under wheat was found to have more large macroaggregates than a soil under maize (Govaerts *et al.*, 2009). Wheat has a more horizontal growing root system than maize and the plant population of wheat is higher resulting in a denser superficial root network. This denser network could positively influence aggregate formation and stabilization. Mono-cropping with cereals is known to have negative impact on soil structural quality compared to rotations of cereals and legumes (Masri and Ryan, 2006). On the other hand, Bhattacharyya *et al.* (2009) found no effect of crop rotation with soybean-wheat, soybean-lentil and soybean pea on mean weight diameter (MWD) or aggregate ratio (AR) in 0-15 and 15-30 cm depth of soil, however, tillage effect (CT, MT, ZT) was found significant only on the surface layer. Fuentes *et al.* (2012) found that crop rotation with residue retention increased macroaggregated carbon as compared to CT. Inclusion of legumes in crop rotations can influence soil aggregation and their stability because of the higher binding capacity of their fine roots as well as organic substances that released from legumes roots are known to contribute to the formation of new soil aggregate (Jacobs *et al.*, 2009; Verhulst *et al.*, 2011).

Dry and wet aggregate size and its stability is also improved under various types of soil and agro-ecological conditions by performing CA, compared with CT (Sharratt *et al.*, 2006; Li *et al.*, 2007; Lichter *et al.*, 2008; Govaerts *et al.*, 2009), due to the presence of higher volume of OM, root fragments, crop residues and mycorrhizal hyphae, which act as a binding agent for soil particles (Bronick and Lal, 2005). Higher aggregate stability under CA is the result of (i) retention of organic residue on soil surface, which reduces detachment and disintegration of the soil aggregates; (ii) decomposing OM increases the aggregation process; (iii) no soil disturbance increases the fungal hyphae and soil microbes; and (iv) increase in soil density makes aggregates more resistant to changes. Apart from these, it has also been reported that not only soil microbes but macro-fauna populations also decreased in CT, compared to CA, ultimately decreasing the beneficial effect of macrofauna on soil aggregate formation process (Verhulst *et al.*, 2010; 2011).

Bulk Density

Soil bulk density (BD) is an indicator of soil compaction and expressed as the weight of the dry soil in a given volume. It affects the root penetration and porosity that are important for plant growth. Bulk density depends on soil texture, mineralogy, particle size and structure, organic matter, type of crops/varieties and management practices including tillage, intercultural operations and residues retention or incorporation. Conventional tillage decreases soil BD of surface soil (upto 0-10 cm) depending upon the magnitude and intensity of manipulation of soil, as tillage loosen the soil and decreases macroporosity and increases total porosity (Hussain *et al.*, 1998). But, increase in number of traffic during tillage operations for a longer duration of time increase the soil BD and the increase was greater with increase in soil depth (Mamman and Ohu, 1998; Barut and Celik, 2017) due to subsoil compaction and formation of plough pan. Under CA practices BD decreases due to higher SOC content, better soil aggregation, increased root growth and biomass (Li *et al.*, 2007; Parihar *et al.*, 2016). However, it is well established that during the initial years of conservation practices soil BD generelly increases due to less mechanical disturbances than conventional practices (De Vita *et al.*, 2007; Salem *et al.*, 2015). While many others (Hill and Curuse, 1985; Rusu *et al.*, 2009; Haruna *et al.*, 2018; Indoria *et al.*, 2017; Chakraborty *et al.*, 2019) have not found any significant differences in BD between CA and CT. He *et al.* (2009) noticed higher BD during initial years (upto 5 years) under CA, but lower values were recorded in subsequent years, suggesting long term sustainable effect of conservation tillage on soil BD.

Residue retention had a significant impact on soil BD in the in upper soil surface (0-10 cm) while the difference at deeper soil depth (10-20 cm) were not found significant (Blanco-Canqui and Lal, 2008). Mulching with crop residue encourages activity of earthworms and decreases soil BD. Being very much suceptable to cultural/management practces, soil BD decreases for the time after tilling but it again increases due to irrigation and rain (Osunbitan *et al.*, 2005). After seven years of continuous CA in maize rotations, Parihar *et al.* (2016) found 4.3-6.9% decrement of BD in soil profile (0-30 cm) than CT and the lowest and the highest BD were noticed for Maize-Chickpea-*Sesbania* and Maize-Mustard-Mungbean rotations, respectively. Inclusion of pulses in cropping sequences augment more carbon in soil resulting in decrement in BD (Verhulst *et al.*, 2011; Thierfelder *et al.*, 2012; Parihar *et al.*, 2016) as negetive correlartion exists between SOC and BD. In CA, crop rotation also helped in maintaining lower BD compared to conventional farming. Differential chemical composition of crop residues and root biomass brings out differential addition of SOC that leads to difference in soil BD (Congreves *et al.*, 2015). Prominently the main effect of crop rotation on change in BD usually meagre, but interaction effect with tillage and residue retention shows significant effect. Shaver (2010) while experimenting crop residue and soil physical properties obtained that increased quantities of crop residue decrease soil bulk density over time and that 91% of the variability observed in bulk density was explained by the amount of crop residue returned to the system over the 12 year period (Fig. 3).

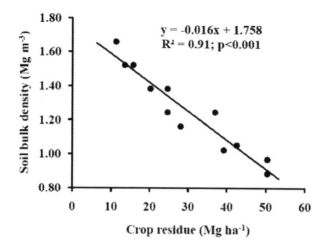

Fig. 3: Soil bulk density as affected by crop residue returned
Source: Shaver (2010)

Soil Porosity

Pores are of different size, shape and continuity and their architecture have significant influence on infiltration, drainage and storage of water, the movement, distribution and exchange of gases, the ease of penetration of soil by growing roots and sheltering soil fauna (Lamandé et al., 2011; Katuwal et al., 2015). Intensive tillage practices increase the proportion of micropores over macropores and biopores (large sized pores >2 mm) as heavy mechanical panipualtion cut short the macropores and turn them into micro range, resulting in higher total soil porosity. Thus, continuous convetional tilling leads to unfavourable BD, penetration resistance, lesser earthworm and root activity (Springett, 1992; Chan, 2001), resulting in poor macropore count and vertical discontinuition in the macropore network through the profile. Macropores may make up only a small portion of total soil porosity but can dominate vertical flow rates during infiltration (Beven and Germann, 1982). Minimum and ZT practices initially lead to a decline in macro-pore volume in soil, which ultimately reduces diffusion of air into soil in comparison to CT. Along with conservation tillage, crop residues and their roots can modify pore properties in numerous ways, by temporal pore clogging (Scanlan and Hinz, 2008), biopore formation upon decaying (Rasse and Smucker, 1998), enhancing wetting–drying cycles, local compaction (Whalley et al., 2004) and hydrophobicity of rooted pore walls (Czarnes et al., 2000). Each 1 t ha⁻¹ of residue addition has been found to enhance effective porosity by 0.3%

(Shaver *et al.*, 2003). Structural porosity is influenced by crop rotation as plant canopy and mulch coverage protect surface aggregates against disruption by heavy rainfall (Blanco-Canqui *et al.*, 2006). Mondal *et al.* (2019) found 6–10% greater total porosity in CT containing 33–159% higher volume of macropores compared to full CA or partial CA in the surface soil layer (0-10 cm), whereas full CA had 12–20% higher mesopore there. Perkons *et al.* (2014) found significantly higher frequency of >5 mm biopores within 0-60cm soil profile under deep rooted perennial chicory over shallow fibrous rooted wheat (annual) crop cultivation.

Cover crops reported to reduce soil bulk density (Blanco-Canqui *et al.*, 2011) and increase macroporosity (Villamil *et al.*, 2006) and microporosity (Auler *et al.*, 2014). Undoubtedly, the maintenance of a continuous residue cover helps to prevent crust formation, contributing to the reduction in surface runoff and in return increases the effective soil porosity. Zero till management induce larger elongated pores (mainly biopore), which are very important for water movement and having better correlation with root density (Pagliai and De Nobili, 1993). Roseberg and McCoy (1992) noted that CT creates greater total porosity reducing macropore continuity. Edwards *et al.* (1988) observed that burrowing earthworms formed large number of continuous pores in the no-till system and these contributed to the high infiltration rates. McGarry *et al.* (2000) studied the effect of zero tillage for 8 years in alluvial vertisols and reported greater volume of largest pore size (1.5-3 mm) in ZT (no soil manipulation) in contrast to CT (34 tillage operations) which has greater volume of pores <1.5 mm in diameter (Fig. 4). According to Katsvairo *et al.* (2002) crop rotation do not significantly influence porosity, however, Mtyobile *et al.* (2019) found significant effect using maize-wheat-maize and maize-wheat-soybean as compared to maize-fallow-maize and maize-fallow-soybean. The interaction of tillage and crop rotation also proved to be established significant to soil porosity.

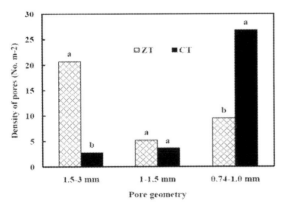

Fig 4: Effect of conservation tillage on abundance of various sized pores

Soil Infiltration and Water Retention

Infiltration, is the downward entry of water from atmosphere into the soil, is governed by inherent soil properties, presence of crop residues or vegetation on soil surface, topography, soil wetness etc. Water retention capacity of soil is defined as the capacity of soils to hold the water against gravitational force. Depending on the level of soil disturbance, different tillage practices could affect the ability of the soil to adsorb and retain water. Moreover, Bradford and Huang (1994) concluded 'surface residue-cover effect' was greater than the 'soil disturbance effect' for maintaining high infiltration rates and reducing soil loss. The residue intercepts rainfall and releases it more slowly afterwards. No-till, zero-till managements, which are leading conservation tillage system for reducing soil erosion may increase (Thierfelder and Wall, 2009; Stone and Schlegel, 2010), reduce (Freebairn *et al.*, 1989; Unger, 1992; Baumhardt *et al.*, 1993) or not affect (Unger, 1992; Pikul and Aase, 1995) water infiltration compared with other tillage systems. Infiltration characteristics of the soil depend on the aggregate size distribution, geometry, continuity, and stability of the pores (Shaver *et al.*, 2002). Decrement of infiltration rate in NT system compared to tilled soil was due to high BD and less amenity to break the crust under NT (Freebairn *et al.*, 1989). Plant available water content was significantly higher with ZT than CT in rice-wheat cropping system (Bhattacharyya *et al.*, 2006; 2008). Again, an increase in 28% soil water as compared to CT and an associated increase of 1.2 t $ha^{-1}y^{-1}$ wheat grain were also noticed by McGarry *et al.*(2000). CA based managements like dry direct-seeded rice (DSR), zero tillage and residue retention positively effect soil infiltration. CT-DSR-CT-maize, ZT-DSR-ZT-maize systems encounter 43 and 47% increase in soil infiltration than transplanted rice-CT-maize system and residue addition shown 24% increase in soil infiltration than without residue treatment (Singh *et al.*, 2016) (Fig. 5). Infiltration rate also increased 65% more with residue retained treatments than bare soil under no-till practice (Baumhardt *et al.*, 2012). Kumar *et al.* (2019) also observed 45 to 65% more infiltration rate in rice-wheat system under CA than CT.

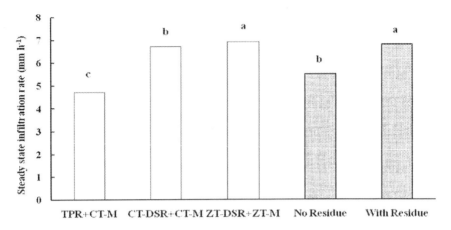

Fig. 5: Steady-state soil water infiltration rate as influenced by tillage and crop establishment techniques and residue management
Source: Singh *et al.* (2016)

Cumulative infiltration increased 63% form total removal of residue to residue incorporation up to 2.5 t ha^{-1} as amount of retention residue significantly ($r=0.88$) triggers the infiltration (Baumhardt and Lascano, 1996). Higher infiltration in ZT with residue retention might be due to direct and indirect influences of residue cover on water infiltration (Verhulst *et al.*, 2010; Jat *et al.*, 2017). Stable aggregates are formed under ZT with residue retention as compared to conventional tillage resulting in lesser breakdown of aggregates and less chances of surface crust formation. McGarry *et al.* (2000) reported higher infiltration rate and cumulative infiltration under ZT with residue retention as compared to conventional tillage. ZT with residue retention might have facilitated formation of continuous soil pores from the soil surface to depth which is attributed to the higher infiltration under CA-based systems (Verhulst *et al.*, 2010; Gathala *et al.*, 2011; Dwivedi *et al.*, 2012). Comparing NT with tilled both with removed surface residue, runoff occurred sooner under NT due to rapid filling as well as clogging of surface pores by impact of raindrops. Thus, runoff is encountered late for tilled conditions because of greater surface roughness and increase of storage volume (Mohamoud *et al.*, 1990). But under CA system, retained straw act as a series of barriers, diminishing the runoff velocity and giving the water more time to infiltrate. Although, NT practice attribute higher bulk density and penetration resistance, infiltration is greater due to higher burrowing earthworm population, roots of previous crops and numerous biopore formation. Rovira *et al.* (1987) observed 83% higher earthworm population under NT compared to CT and 60% higher population under wheat-pasture rotation than wheat-lupin rotation, as dense planting under lupin system would supply lesser food materials to earthworms.

Thierfelder and Wall (2009) postulated that solely maize direct seeding and direct seeding maize with inclusion of legume in the rotation increased water retention and the steady state infiltration from 0.7 to 0.8 mm h^{-1} which were 40-60% more than CT. Higher water retention in surface soil (0-10 cm) under different CA-based practices was also reported by other researchers (Govaerts *et al.*, 2009; Malecka *et al.*, 2012; Wahbi *et al.*, 2014; Jat *et al.*, 2017; Nabayi *et al.*, 2019). Indoria *et al.* (2017) showed that 13–14% higher soil water content is found in no-till compared to other tillage practices. In a 10-year study on a rice–rice cropping system on a Vertisol, Bellakki *et al.*, (1998) showed that application of rice straw incorporation not only meet either 25 or 50% of recommended fertilizer N requirement but also improved 10% more plant water retention capacity. In a no-till relay lentil with rice Bandyopadhyay *et al.* (2016) showed that the drying out pattern during flowering to pod formation stage may be modified by keeping rice stubble height. Zero tillage with residue retention decreases the frequency and intensity of short mid or late season droughts (Bandyopadhyay *et al.*, 2018; Bradford and Peterson, 2000).

Soil Hydraulic Conductivity

Hydraulic conductivity of soil is the measure of soil's ability to transmit water within soil profile. Along with topography, climate and parent materials, type, magnitude and intensity of tillage play a major role in altering saturated hydraulic conductivity (K_{sat}) as BD and porosity are supposed to be altered. K_{sat} is highly dependent upon the size, continuity and arrangement of pores (Unger and Jones, 1998). The 'Till vs No-till' contrast revealed that tillage improved K_{sat} values by about 87% compared with no-till as a results of change in soil BD and coarse mesopores (Haruna *et al.*, 2018). Tillage modifies K_{sat} but the extent depends upon the soil type and pore geometry. Tillage can improve K_{sat} of sandy loam or silty loam soils compared to no-till, because tillage may contribute to a higher proportions of larger pores, especially coarse mesopores (Carter and, Kunelius, 1986; Haruna *et al.*, 2018). In some cases, no-tillage showed decrease in K_{sat} compared to the conventional tillage (Sauer *et al.*, 1990; Karlen *et al.*, 1994). These results are in contrast with Joschko *et al.* (1992) and Bhattachryya *et al.*, (2006) who reported higher K_{sat} values in no-till compared with CT in silty loam and sandy clay loam soils. Bhattachryya *et al.* (2006) suggested reduction of volume fraction of the large pores and increase in volume fraction of the smaller pores with higher pore connectivity under no-tillage, while Joschko *et al.* (1992) suggested that the burrows made by endogeic earthworms contributed to the higher K_{sat} values in no-till compared with tilled treatment. This beneficial effects are suplemented by residue retention and diverse cropping patern. In a review, Bandyopadhyay (2020) observed that by sequestrating 1% SOC, the flow rate will increase to

2 cm min⁻¹ with incorporation/retention of different amounts of residues for continuous 9 years on silt loam soils. Greater amount of water stable aggregates in the reduced tillage system probably contribute to its higher K_{sat} (Singh et al., 1994; Bhattacharyya et al., 2006). Continuous zero till management with residue retention the soil surface, may show higher soil infiltration rates and K_{sat} values due to root channels formed in soil and boosted earthworm activity (Barnes and Ellis, 1979; Rasool et al., 2007), formation of continuous soil bio-pores (Kumar et al., 2012) and presence of continuous macropores under ZT (Loch and Coughlan, 1984). Benjamin (1993) found no effect of crop rotations on K_{sat}, BD, pore size distribution and water retention characteristics. According to Parihar et al. (2016) the K_{sat} increased by 14.3 and 11.2% in permanent bed (PB) plots and 11.1 and 12.0% in ZT plots for 0–15 and 15–30 cm soil layers, respectively, compared to CT plots (Fig. 6a). The increase in K_{sat} under conservation agriculture practices (PB and ZT) was mainly attributed to decrease in bulk density and increase in effective pore volume (Unger and Jones, 1998) because of better soil aggregation in these practices. They also observed that diversified crop rotations significantly affected the K_{sat} at 0–15 and 15–30 cm soil depths (Fig. 6b). The K_{sat} increased by 9.3 and 5.9% in maize-chickpea-*sesbania* (MCS); 8.4 and 5.7% in maize-wheat-mungbean (MWMb); 6.9 and 4.6% in maize-maize-*sesbania* (MMS) in 0–15 and 15–30 cm soil layer, respectively, compared to maize-mustard-mungbean (MMuMb) crop rotation.

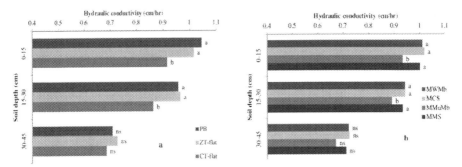

Fig. 6: Effect of long term (after seven years) tillage practices (a) and diversified crop rotations (b) on saturated hydraulic conductivity of soil
Source: Parihar et al. (2016)

Soil Temperature

Crop residue remaining on the soil surface in conservation tillage systems can decrease the rate of soil temperature change because surface residue both increases the refection of incident solar radiation (Li et al., 2013), and acts as an insulating barrier between the soil surface and the warmer (or colder) atmospheric air above. Tillage can affect soil heat capacity and thermal conductivity and therefore thermal difusivity (the ratio of the thermal conductivity to the heat capacity) by changing SOM, BD, inter-aggregate contact and moisture content (Shukla et al., 2003). Field research has shown higher moisture levels, decreased soil temperatures and also more stable soil aggregates, i.e. improved soil structure, under ZT compared to CT (Carter, 1992; Lichter et al., 2008). The soil thermal difusivity was found to be 20–25% higher in the 5–15 cm layer (Johnson and Lowery, 1985) and 37% higher in the 5–25 cm layer (Abu-Hamdeh, 2000) in no-till compared to tilled soil. Terefore, more of the heat absorbed at the surface is transferred into deeper soil in the no-till which leads to lower soil temperature in the near surface soil layers (Sarkar and Singh, 2007). Reduced maize emergence due to lower soil temperature of the seed zone was observed in conservation tillage systems compared to CT systems (Gupta et al., 1988). Moroizumi and Horino (2002) also recorded higher soil temperature up to 15 cm soil depths under tilled than untilled condition.

Mulching had a beneficial effect on soil water and temperature regimes. Residues raise the albedo and restrict air movement, resulting in less energy to warm the soil and there is less evaporation of soil water and greater infiltration. Thus, thermal conductivity of the soil is greater, dissipating heat into a greater depth and moving heat away from the seed zone as the heat capacity of wet soil is greater than that of dry soil, more heat is required to warm the wet, cool soil under the plant residues (Gebhardt et al., 1985). Soil surface evaporation also be reduced, depending upon whether incorporated or surface mulched (Uson and Cook, 1995). Soil temperatures in surface layers can be significantly lower (often between 2 and 8 °C) during daytime (in summer) in zero tilled soils with residue retention compared to conventional tillage (Johnson and Hoyt,1999, Oliveira et al., 2001). Mulches enhance topsoil water retention capacity as compared with no mulch, but no-till maintain higher water contents and as straw application rate increases, water retention increases and temperature decreases, however inorganic mulches increase soil temperature (Cook et al., 2006).

Usually the measured soil temperatures, are correlated strongly with air temperature (Chassot et al., 2001; Sommer et al., 2007; Chen et al., 2011), decreases with increasing depth below the soil surface, and are slightly lower under NT relative to CT and reduced tillage due to accumulated crop residues on the NT surface. Liu et al. (2013) noticed that measured soil temperature averaged across tillages at 0–5 cm was 0.8 °C greater than at 15–30 cm and the measured soil temperature under NT was 0.6–1.9 °C lower than under conventional tillage at each depth; and the measured soil temperature under

RT usually fell in between those under NT and CT. However, the magnitude of variation in soil temperature due to tillage and mulch management was smaller in the lower layer. Retaining rice stubbles on the soil surface decrease daytime soil temperature by 9.4–14.6 °C and modify the crop coefficient (Kc) by increasing transpiration and reducing evaporation, thereby maintaining higher leaf area index (Bandyopadhyay *et al.*, 2016). Residue on the surface can reflect solar radiation and insulate the surface soil from the atmosphere, and thus decrease the warming rate of soil (Rasmussen 1999). The SOC mineralization increased with increase in soil temperature and was higher in ZT than CT (Parihar *et al.*, 2019). In tropical hot soils, mulch cover reduces soil peak temperatures that are too high for optimum growth and development to an appropriate level, favouring biological activity, initial crop growth and root development during the growing season (Acharya *et al.*, 1998, Oliveira *et al.*, 2001).

Soil Compaction and Soil Strength

Soil compaction is the physical compression and/ or consolidation of soil by applied external forces that destroy the structure, choke the porosity, limit water and air infiltration, as a result roots encounter higher resistance for penetration and ultimately yield is hampered. Soils with low OM, clay soils at higher moisture content are very much vulnerable to soil compaction. Aggressive tillage, heavy wheel traction and axle loads are chiefly responsible for soil compaction and the magnitude of compactness depends on frequency as well as appropriate soil moisture conditions (Indoria *et al.*, 2017). Conservation practices can neutralize these negative effects of soil compaction as MT or ZT and residues can modify soil ecosystem alongwith crop rotations. Application of OM would improve BD and porosity of soil and imrove the compaction depending on the quality and qualtity of OM (Ball *et al.*, 2000). Readily oxidizable and less humified OM shows better effect in increrasing aggregate porosity and decreasing tensile strength (Zhang, 1994). Controlled traffic alleviates the compaction problem as under restricted traffic in ZT, water infiltration rate is comparable to that of virgin soil (Li *et al.*, 2001). Several researchers (Chen *et al.*, 2005; Bueno *et al.*, 2006; Rashidi and Keshavarzpour, 2008) obtained higher penetration resistance (PR) in NT systems compared to CT in upper soil layers (0–20 cm). While higher PR was obtained below the tilled layer in CT might be indication of formation of plough hard pan by using the heavy agricultural implements over the years (Yang and Wander, 1999; Doan *et al.*, 2005; Saha *et al.*, 2010). Generally, PR is considered as better indicator of tillage-induced changes in soil behavior and root responses than BD (Hammel, 1989; Singh and Malhi, 2006). Singh and Malhi (2006) further showed after 6 years of study that in NT the PR was 10-79 and 20-95%

higher than CT with and without residue retention. Crop rotation solely does not having significant effect on soil compaction but associated tillage practices have. However, Grant and Lafond (1993) obtained that inclusion of pea crop in the cropping system decreased the PR values, whereas inclusion of flax caused increased soil PR values. Fuentes *et al.* (2009) observed that The ZT+r (residue) treatments, irrespectively of the cropping system (maize/ wheat), had a higher water content and a lower resistance to penetration than the ZT−r treatments. In the ZT+r and CT+r treatments, soil moisture and resistance to penetration were more spatial homogeneous in the plot than in those without residues (ZT−r and CT−r). They also concluded that the variation for resistance to penetration between treatments resulted from only residue management and not tillage, rotation or interaction between these. The variation in moisture content was due to residue management and the interaction between crop residue and the type of tillage.

Conclusion

Optimum soil physical health is very important for efficient utilization of nutrients in any crop production system. Conservation agriculture is beneficial to soil physical improvement as compared to CT. The total impact of CA system on soil physical health varies location-to-location and is dependent on soil inherent properties, period of time under CA system, per cent soil disturbance, nature of the crop, intensity of the crop rotation, type of cover crops, per cent of total surface area covered by crop residues, soil moisture and soil temperature regimes. CA provides some buffer to crop production during drought conditions. The CA practices including residue retention, and certain crop rotations increase SOM in the topsoil that, in turn, influence soil physical properties. The effects of crop rotation have been less investigated that needs to be explored much. Some inconsistent results under CA may be due to soil type, duration of experiments, amount and type of residues retention, and interactions with management. There is a long-lasting debate about the effects of tillage and residue management on soil structure and structure-mediated ecosystems as the residue retention may not be possible in all the crops in the rotation due to its other potential uses. Therefore, a vast research scope is there with retention of suitable amount of residues to improve crop productivity and soil health in a cost effective manner. As soil physical environment imparts direct and indirect effects on soil chemical as well as biological environments significantly, the adoption of systematic CA is a great way out towards sustainability of soil health in the long run.

References

Abu-Hamdeh, N.H. (2000). Effect of tillage treatments on soil thermal conductivity for some Jordanian clay loam and loam soils. *Soil and Tillage Research*, 56: 145-151.

Acharya, C.L., Kapur, O.C. and Dixit, S.P. (1998). Moisture conservation for rainfed wheat production with alternative mulches and conservation tillage in the hills of north-west India. *Soil and Tillage Research*, 46: 153-163.

Altieri, M. A. (2009). The ecological impacts of large-scale agrofuel monoculture production systems in the Americas. *Bulletin of Science, Technology and Society*, 29: 236-244.

Andruschkewitsch, R., Geisseler, D., Koch, H.J. and Ludwig, B. (2013). Effects of tillage on contents of organic carbon, nitrogen, water-stable aggregates and light fraction for four different long-term trials. *Geoderma*, 192: 368-377.

Askari, M.S., Cui, J. and Holden, N.M. (2013). The visual evaluation of soil structure under arable management. *Soil and Tillage Research*, 134: 1-10.

Auler, A.C., Miara, S., Pires, L.F., Fonseca, A.F.D. and Barth, G. (2014). Soil physico-hydrical properties resulting from the management in Integrated Production Systems. *Revista Ciência Agronômica*, 45: 976-989.

Ball, B.C., Bingham, I., Rees, R.M., Watson, C.A. and Litterick, A. (2005). The role of crop rotations in determining soil structure and crop growth conditions. *Canadian Journal of Soil Science*, 85: 557-577.

Ball, B.C., Campbell, D.J. and Hunter, E.A. (2000). Soil compactibility in relation to physical and organic properties at 156 sites in UK. *Soil and Tillage Research*, 57: 83-91.

Bandyopadhyay, P.K. (2020). Functional Behaviour of Soil Physical Parameters for Regulating Organic C Pools. In: *Carbon Management in Tropical and Sub-Tropical Terrestrial Systems* (P. K. Ghosh, S. K. Mahanta, D. Mandal, B. Mandal and S. Ramakrishnan, Eds.), pp. 233-247. Springer, Singapore.

Bandyopadhyay, P.K., Halder, S., Mondal, K., Singh, K.C., Nandi, R. and Ghosh, P.K. (2018). Response of lentil (*Lens culinaries*). to post-rice residual soil moisture under contrasting tillage practices. *Agricultural Research*, 7: 463-479.

Bandyopadhyay, P.K., Singh, K.C., Mondal, K., Nath, R., Ghosh, P.K., Kumar, N., Basu, P.S. and Singh, S.S. (2016). Effects of stubble length of rice in mitigating soil moisture stress and on yield of lentil (*Lens culinaris* Medik). in rice-lentil relay crop. *Agricultural Water Management*, 173: 91-102.

Barnes, B.T. and Ellis, F.B. (1979). Effects of different methods of cultivation and direct drilling, and disposal of straw residues, on populations of earthworms. *Journal of Soil Science*, 30: 669-679.

Barut, Z. and Celik, I. (2017). Tillage effects on some soil physical properties in a semi-arid mediterranean region of Turkey. *Chemical Engineering Transactions*, 58: 217-222.

Baumhardt, R.L. and Lascano, R.J. (1996). Rain infiltration as affected by wheat residue amount and distribution in ridged tillage. *Soil Science Society of America Journal*, 60: 1908-1913.

Baumhardt, R.L., Johnson, G.L. and Schwartz, R.C. (2012). Residue and long-term tillage and crop rotation effects on simulated rain infiltration and sediment transport. *Soil Science Society of America Journal*, 76: 1370-1378.

Baumhardt, R.L., Keeling, J.W. and Wendt, C.W. (1993). Tillage and residue effects on infiltration into soils cropped to cotton. *Agronomy Journal*, 85: 379-383.

Bellakki, M.A., Badanur, V.P. andSetty, R.A. (1998). Effect of long-term integrated nutrient management on some important properties of a Vertisol. *Journal of the Indian Society of Soil Science*, 46: 176-180.

Benjamin, J.G. (1993). Tillage effects on near-surface soil hydraulic properties. *Soil and Tillage Research*, 26: 277-288.

Beven, K. and Germann, P. (1982). Macropores and water flow in soils. *Water Resources Research*, 18: 1311-1325.

Bhattacharyya, R., Kundu, S., Pandey, S.C., Singh, K.P. and Gupta, H.S. (2008). Tillage and irrigation effects on crop yields and soil properties under the rice–wheat system in the Indian Himalayas. *Agricultural Water Management*, 95: 993-1002.

Bhattacharyya, R., Prakash, V., Kundu, S. and Gupta, H.S. (2006). Effect of tillage and crop rotations on pore size distribution and soil hydraulic conductivity in sandy clay loam soil of the Indian Himalayas. *Soil and Tillage Research*, 86: 129-140.

Bhattacharyya, R., Prakash, V., Kundu, S., Srivastva, A.K. and Gupta, H.S. (2009). Soil aggregation and organic matter in a sandy clay loam soil of the Indian Himalayas under different tillage and crop regimes. *Agriculture, Ecosystems and Environment*, 132: 126-134.

Blanco-Canqui, H. and Lal, R. (2008). No-tillage and soil-profile carbon sequestration: an on-farm assessment. *Soil Science Society of America Journal*, 72: 693-701.

Blanco-Canqui, H., Lal, R., Post, W.M., Izaurralde, R.C. and Owens, L.B. (2006). Corn stover impacts on near-surface soil properties of no-till corn in Ohio. *Soil Science Society of America Journal*, 70: 266-278.

Blanco-Canqui, H., Mikha, M.M., Presley, D.R. and Claassen, M.M. (2011). Addition of cover crops enhances no-till potential for improving soil physical properties. *Soil Science Society of America Journal*, 75: 1471-1482.

Blevins, R.L., Lal, R., Doran, J.W., Langdale, G.W. and Frye, W.W. (1998). Conservation tillage for erosion control and soil quality. In: *Advances in Soil and Water Conservation* (F.J. Pierce, Ed.)., pp. 51-68. Routledge.

Bradford, J.M. and Huang, C.H. (1994). Interrill soil erosion as affected by tillage and residue cover. *Soil and Tillage Research*, 31: 353-361.

Bradford, J.M. and Peterson, G.A. (2000). Conservation tillage. In *Handbook of soil science*, (M. E. Sumner, Ed.)., pp G247-G269. Boca Raton, FL, USA: CRC Press.

Bravo-Garza, M.R., Voroney, P., Bryan, R.B., (2010). Particulate organic matter in water stable aggregates formed after the addition of ^{14}C-labeled maize residues and wetting and drying cycles in vertisols. *Soil Biology and Biochemistry*, 42: 953-959.

Bronick, C.J. and Lal, R. (2005). Soil structure and management: a review. *Geoderma*, 124: 3-22.

Bueno, J., Amiama, C., Hernanz, J.L. and Pereira, J.M. (2006). Penetration resistance, soil water content, and workability of grasslands soils under two tillage systems. *Transactions of the American Society of Agricultural and Biological Engineers*, 49: 875-882.

Bünemann, E.K., Bongiorno, G., Bai, Z., Creamer, R.E., Deyn, G.D., Goede, R.D., Fleskens, L., Geissen, V., Kuyper, T.W., Mäder, P., Pulleman, M., Sukkel, W., van Groenigen, J.W. and Brussaard, L. (2018). Soil quality – A critical review. *Soil Biology and Biochemistry*, 120: 105-125.

Carter, M.R. (1992). Influence of reduced tillage systems on organic matter, microbial biomass, macro-aggregate distribution and structural stability of the surface soil in a humid climate. *Soil and Tillage Research*, 23: 361-372.

Carter, M.R. and Kunelius, H.T. (1986). Comparison of tillage and direct drilling for Italian ryegrass on the properties of a fine sandy loam soil. *Canadian Journal of Soil Science*, 66: 197-207.

Chakraborty, D., Mondal, S., Das, A., Paul, A., Thomas, P., Aggarwal, P., Behera, U.K. and Sharma, A.R. (2019). Tillage and crop rotation effects on mechanical properties and structural stability of a sandy loam soil in a semi-arid environment. *Journal of the Indian Society of Soil Science*, 67: 174-182.

Chan, K.Y. (2001). An overview of some tillage impacts on earthworm population abundance and diversity-implications for functioning in soils. *Soil and Tillage Research*, 57: 179-191.

Chassot, A., Stamp, P. and Richner, W. (2001). Root distribution and morphology of maize seedlings as affected by tillage and fertilizer placement. *Plant and Soil*, 231: 123-135.

Chen, Y., Cavers, C., Tessier, S., Monero, F. and Lobb, D. (2005). Short-term tillage effects on soil cone index and plant development in a poorly drained, heavy clay soil. *Soil and Tillage Research*, 82: 161-171.

Chen, Y., Liu, S., Li, H., Li, X. F., Song, C.Y., Cruse, R.M. and Zhang, X.Y. (2011). Effects of conservation tillage on corn and soybean yield in the humid continental climate region of Northeast China. *Soil and Tillage Research*, 115: 56-61.

Clermont Dauphin, C., Cabidoche, Y.M. and Meynard, J.M. (2004). Effects of intensive monocropping of bananas on properties of volcanic soils in the uplands of the French West Indies. *Soil Use and Management*, 20: 105-113.

Congreves, K.A., Hayes, A., Verhallen, E.A. and Van Eerd, L.L. (2015). Long-term impact of tillage and crop rotation on soil health at four temperate agroecosystems. *Soil and Tillage Research*, 152: 17-28.

Cook, H.F., Valdes, G.S. and Lee, H.C. (2006). Mulch effects on rainfall interception, soil physical characteristics and temperature under *Zea mays* L. *Soil and Tillage Research*, 91: 227-235.

Czarnes, S., Hallett, P.D., Bengough, A.G. and Young, I.M. (2000). Root and microbial derived mucilages affect soil structure and water transport. *European Journal of Soil Science*, 51: 435-443.

Dalal, R.C. and Bridge, B.J. (1996). Aggregation and organic matter storage in sub-humid and semi-arid soils. *Structure and organic matter storage in agricultural soils. CRC Press, Boca Raton, FL*, pp. 263-307.

Das, T.K., Saharawat, Y.S., Bhattacharyya, R., Sudhishri, S., Bandyopadhyay, K.K., Sharma, A. R. and Jat, M.L. (2018). Conservation agriculture- effects on crop and water productivity, profitability and soil organic carbon accumulation under a maize-wheat cropping system in the North-western Indo-Gangetic Plains. *Field Crops Research*, 215: 222–231.

De Gryze, S., Six, J., Brits, C. and Merckx, R. (2005). A quantification of short-term macroaggregate dynamics: influences of wheat residue input and texture. *Soil Biology and Biochemistry*, 37: 55-66.

De Vita, P., Di Paolo, E., Fecondo, G., Di Fonzo, N. and Pisante, M. (2007). No-tillage and conventional tillage effects on durum wheat yield, grain quality and soil moisture content in southern Italy. *Soil and Tillage Research*, 92: 69-78.

Doan, V., Chen, Y. and Irvine, B. (2005). Effect of residue type on the performance of no-till seeder openers. *Canadian Biosystems Engineering*, 47: 29-35.

Dwivedi, B.S., Singh, V.K., Shukla, A.K. and Meena, M.C. (2012). Optimizing dry and wet tillage for rice on a Gangetic alluvial soil: Effect on soil characteristics, water use efficiency and productivity of the rice–wheat system. *European Journal of Agronomy*, 43: 155-165.

Eberlein, C., Heuer, H., Vidal, S. and Westphal, A. (2016). Microbial communities in *Globodera pallida* females raised in potato monoculture soil. *Phytopathology*, 106: 581-590.

Edwards, W. M., Shipitalo, M.J. and Norton, L.D. (1988). Contribution of macroporosity to infiltration into a continuous corn no-tilled watershed: Implications for contaminant movement. *Journal of Contaminant Hydrology*, 3: 193-205.

FAO (2016). What is conservation agriculture? http://www.fao.org/ag/ca/

Ferreras, L. A., Costa, J. L., Garcia, F.O. and Pecorari, C. (2000). Effect of no-tillage on some soil physical properties of a structural degraded Petrocalcic Paleudoll of the southern "Pampa" of Argentina. *Soil and Tillage Research*, 54: 31-39.

Freebairn, D.M., Gupta, S.C., Onstad, C.A. and Rawls, W.J. (1989). Antecedent rainfall and tillage effects upon infiltration. *Soil Science Society of America Journal*, 53: 1183-1189.

Fuentes, M., Govaerts, B., De León, F., Hidalgo, C., Dendooven, L., Sayre, K. D. and Etchevers, J. (2009). Fourteen years of applying zero and conventional tillage, crop rotation and residue management systems and its effect on physical and chemical soil quality. *European Journal of Agronomy*, 30: 228-237.

Fuentes, M., Hidalgo, C., Etchevers, J., De León, F., Guerrero, A., Dendooven, L., Verhulst, N. and Govaerts, B. (2012). Conservation agriculture, increased organic carbon in the top-soil macro-aggregates and reduced soil CO_2 emissions. *Plant and Soil, 355*: 183-197.

Gathala, M.K., Ladha, J.K., Saharawat, Y.S., Kumar, V., Kumar, V. and Sharma, P.K. (2011). Effect of tillage and crop establishment methods on physical properties of a medium-textured soil under a seven-year rice–wheat rotation. *Soil Science Society of America Journal*, 75: 1851-1862.

Gebhardt, M.R., Daniel, T.C., Schweizer, E.E. and Allmaras, R.R. (1985). Conservation tillage. *Science, 230*: 625-630.

Ghosh, P.K., Das, A., Saha, R., Kharkrang, E., Tripathi, A., Munda, G. and Ngachan, S. (2010). Conservation agriculture towards achieving food security in North East India. *Current Science*, 99: 915-921.

Govaerts, B., Sayre, K.D., Goudeseune, B., De Corte, P., Lichter, K., Dendooven, L. and Deckers, J. (2009). Conservation agriculture as a sustainable option for the central Mexican highlands. *Soil and Tillage Research*, 103: 222-230.

Grant, C.A. and Lafond, G.P. (1993). The effects of tillage systems and crop sequences on soil bulk density and penetration resistance on a clay soil in southern Saskatchewan. *Canadian Journal of Soil Science*, 73: 223-232.

Gregorich, E.G. and Carter, M.R. (1997). *Soil Quality for Crop Production and Ecosystem Health*. Elsevier, Amsterdam.

Guggenberger, G., Elliott, E.T., Frey, S.D., Six, J. and Paustian, K. (1999). Microbial contributions to the aggregation of a cultivated grassland soil amended with starch. *Soil Biology and Biochemistry*, 31: 407-419.

Gupta, S.C., Schneider, E.C., Larson, W.E. and Hadas, A. (1987). Influence of Corn Residue on Compression and Compaction Behavior of Soils. *Soil Science Society of America Journal*, 51: 207-212.

Gupta, S.C., Swan, J.B. and Schneider, E.C. (1988). Planting depth and tillage interactions on corn emergence. *Soil Science Society of America journal*, 52: 1122-1127.

Hammel, J.E. (1989). Long term tillage and crop rotation effects on bulk density and soil impedance in northern Idaho. *Soil Science Society of America Journal*, 53: 1515-1519.

Haruna, S.I., Anderson, S.H., Nkongolo, N.V. and Zaibon, S. (2018). Soil hydraulic properties: Influence of tillage and cover crops. *Pedosphere* 28, 430-442.

He, J., Kuhn, N.J., Zhang, X.M., Zhang, X.R. and Li, H.W. (2009). Effects of 10 years of conservation tillage on soil properties and productivity in the farming–pastoral ecotone of Inner Mongolia, China. *Soil Use and Management*, 25: 201-209.

Hill, R.L. and Cruse, R.M. (1985). Tillage Effects on Bulk Density and Soil Strength of Two Mollisols 1. *Soil Science Society of America Journal*, 49: 1270-1273.

Hou, X., Li, R., Jia, Z., Han, Q., Wang, W. and Yang, B. (2012). Effects of rotational tillage practices on soil properties, winter wheat yields and water-use efficiency in semi-arid areas of north-west China. *Field Crops Research*, 129: 7-13.

Hussain, I., Olson, K.R. and Siemens, J.C. (1998). Long-term tillage effects on physical properties of eroded soil. *Soil Science, 163*: 970-981.

Indoria, A.K., Rao, C.S., Sharma, K.L. and Reddy, K.S. (2017). Conservation agriculture – a panacea to improve soil physical health. *Current Science*, 112: 52-61.

Jacobs, A., Rauber, R. and Ludwig, B. (2009). Impact of reduced tillage on carbon and nitrogen storage of two Haplic Luvisols after 40 years. *Soil and Tillage Research*, 102: 158-164.

Jat, H.S., Datta, A., Sharma, P.C., Kumar, V., Yadav, A.K., Choudhary, M., Choudhary, V., Gathala, M.K., Sharma, D.K., Jat, M.L. and Yaduvanshi, N.P.S. (2018). Assessing soil properties and nutrient availability under conservation agriculture practices in a reclaimed sodic soil in cereal-based systems of North-West India. *Archives of Agronomy and Soil Science,* 64: 531-545.

Johnson, A.M. and Hoyt, G.D. (1999). Changes to the soil environment under conservation tillage. *Hortechnology,* 9: 380–393.

Johnson, M.D. and Lowery, B. (1985). Effect of Three Conservation Tillage Practices on Soil Temperature and Thermal Properties. *Soil Science Society of America Journal,* 49: 1547-1552.

Joschko, M., Söchtig, W. and Larink, O. (1992). Functional relationship between earthworm burrows and soil water movement in column experiments. *Soil Biology and Biochemistry,* 24: 1545-1547.

Karlen, D.L., Wollenhaupt, N.C., Erbach, D.C., Berry, E.C., Swan, J.B., Each, N.S. and Jordahl, J.L. (1994). Long-term tillage effects on soil quality. *Soil and Tillage Research,* 32: 313–327.

Katsvairo, T., Cox, W.J. and van Es, H. (2002). Tillage and rotation effects on soil physical characteristics. *Agronomy Journal,* 94: 299-304.

Katuwal, S., Norgaard, T., Moldrup, P., Lamandé, M., Wildenschild, D. and de Jonge, L.W. (2015). Linking air and water transport in intact soils to macropore characteristics inferred from X-ray computed tomography. *Geoderma,* 237: 9-20.

Kumar, S., Sharma, P.K., Anderson, S.H. and Saroch, K. (2012). Tillage and rice-wheat cropping sequence influences on some soil physical properties and wheat yield under water deficit conditions. *Open Journal of Soil Science,* 2: 71-81.

Kumar, S.D. and Lal, B.R. (2012). Effect of Mulching on Crop Production under Rainfed Condition: A Review. *International Journal of Research in Chemistry and Environment,* 2: 8-10.

Kumar, V., Gathala, M.K., Saharawat, Y.S., Parihar, C.M., Kumar, R., Kumar, R., Jat, M.L., Jat, A.S., Mahala, D.M., Kumar, L., Nayak, H.S., Parihar, M.D., Rai, V., Jewlia, H. and Kuri, B.R. (2019). Impact of tillage and crop establishment methods on crop yields, profitability and soil physical properties in rice–wheat system of Indo gangetic plains of India. *Soil Use and Management,* 35: 303-313.

Lal, R. (1991). Tillage and agricultural sustainability. *Soil and Tillage Research,* 20: 133-146.

Lamandé, M., Labouriau, R., Holmstrup, M., Torp, S.B., Greve, M.H., Heckrath, G., Iversen, B.V., de Jonge, L.W., Moldrup, P. and Jacobsen, O.H. (2011). Density of macropores as related to soil and earthworm community parameters in cultivated grasslands. *Geoderma,* 162: 319-326.

Letey J. (1958). Relationship between Soil Physical Properties and Crop Production. In: *Advances in Soil Science* (B.A. Stewart, Eds.)., pp 277-294. Springer, NY.

Li, H., Gao, H., Wu, H., Li, W., Wang, X. and He, J. (2007). Effects of 15 years of conservation tillage on soil structure and productivity of wheat cultivation in northern China. *Soil Research,* 45: 344-350.

Li, R., Hou, X., Jia, Z., Han, Q., Ren, X. and Yang, B. (2013). Effects on soil temperature, moisture, and corn yield of cultivation with ridge and furrow mulching in the rainfed area of the Loess Plateau, China. *Agricultural Water Management,* 116: 101–109.

Li, Y., Tullberg, J.N. and Freebairn, D.M. (2001). Traffic and residue cover effects on infiltration. *Soil Research,* 39: 239-247.

Lichter, K., Govaerts, B., Six, J., Sayre, K.D., Deckers, J. and Dendooven, L. (2008). Aggregation and C and N contents of soil organic matter fractions in a permanent raised-bed planting system in the Highlands of Central Mexico. *Plant and Soil,* 305: 237-252.

Liu, S., Yang, J.Y., Zhang, X.Y., Drury, C.F., Reynolds, W.D. and Hoogenboom, G. (2013). Modelling crop yield, soil water content and soil temperature for a soybean–maize rotation under conventional and conservation tillage systems in Northeast China. *Agricultural Water Management*, 123: 32-44.

Loch, R.J. and Coughlan, K.J. (1984). Effects of zero tillage and stubble retention on some properties of a cracking clay. *Soil Research*, 22: 91-98.

Malecka, I., Blecharczyk, A., Sawinska, Z. and Dobrzeniecki, T. (2012). The effect of various long-term tillage systems on soil properties and spring barley yield. *Turkish Journal of Agriculture and Forestry*, 36: 217-226.

Mamman, E. and Ohu, J.O. (1998). The effect of tractor traffic on air permeability and millet production in a sandy loam soil in Nigeria. *Ife Journal of Technology*, 8: 1-7.

Masri, Z. and Ryan, J. (2006). Soil organic matter and related physical properties in a Mediterranean wheat-based rotation trial. *Soil and Tillage Research*, 87: 146-154.

McGarry, D., Bridge, B.J. and Radford, B.J. (2000). Contrasting soil physical properties after zero and traditional tillage of an alluvial soil in the semi-arid subtropics. *Soil and Tillage Research*, 53: 105-115.

Meeus, J.H.A. (1993). The transformation of agricultural landscapes in Western Europe. *Science of the Total Environment*, 129: 171–190.

Mohamoud, Y.M., Ewing, L.K. and Boast, C.W. (1990). Small plot hydrology: I. Rainfall infiltration and depression storage determination. *Transactions of the American Society of Agricultural Engineers*, 33: 1121-1131.

Mohanty, M., Painuli, D. K., Misra, A. K. and Ghosh, P. K. (2007). Soil quality effects of tillage and residue under rice–wheat cropping on a Vertisol in India. *Soil and Tillage Research*, 92: 243-250.

Mondal, S., Poonia, S.P., Mishra, J.S., Bhatt, B.P., Karnena, K.R., Saurabh, K., Kumar, R. and Chakraborty, D. (2019). Short-term (5 years). impact of conservation agriculture on soil physical properties and organic carbon in a rice–wheat rotation in the Indo-Gangetic plains of Bihar. *European Journal of Soil Science*, 1-14. https://doi.org/10.1111/ejss.12879

Moroizumi, T. and Horino, H. (2002). The effects of tillage on soil temperature and soil water. *Soil Science*, 167: 548-559.

Mtyobile, M., Muzangwa, L. and Mnkeni, P.N.S. (2019). Tillage and crop rotation effects on soil carbon and selected soil physical properties in a Haplic Cambisol in Eastern Cape, South Africa. *Soil and Water Research*, 15: 47-54.

Nabayi, A., Girei, A. H. and Abubakar, M. S. (2019). Physical and hydraulic properties of soils under a long-term tillage practices in Hadejia Local Government Area, Jigawa State, Nigeria. *Eurasian Journal of Soil Science*, 8: 267-274.

Nambiar, K.K.M. (1995). Major cropping systems in India. In: *Agricultural sustainability: Economics, environment and statistical consideration* (V. Parnett,R. Range and R. Steiner, Eds.).,pp. 133-169. John Wiley and Sons. Chichester, UK.

Nassauer, J. I. and Westmacott, R. (1987). Progressiveness among farmers as a factor in heterogeneity of farmed landscapes. In: *Landscape heterogeneity and disturbance* (M.G. Turner, Eds.)., pp. 199-210. Springer, NY.

Oliveira, J.C.M., Timm, L.C., Tominaga, T.T., Cassaro, F.A.M., Reichardt, K., Bacchi, O.O.S., DouradoNeto, D. and Camara, G.M.D. (2001). Soil temperature in a sugar-cane crop as a function of the management system. *Plant and Soil*, 230: 61-66.

Olsson, S. and Gerhardson, B. (1992). Effects of long-term barley monoculture on plant-affecting soil microbiota. *Plant and Soil*, 143: 99-108.

Osunbitan, J.A., Oyedele, D.J. and Adekalu, K.O. (2005). Tillage effects on bulk density, hydraulic conductivity and strength of a loamy sand soil in southwestern Nigeria. *Soil and Tillage Research*, 82: 57-64.

Pagliai, M. and De Nobili, M. (1993). Relationships between soil porosity, root development and soil enzyme activity in cultivated soils. In: *International Workshop on Methods of Research on Soil Structure/Soil Biota Interrelationships* (L. Brussaard and M.J. Kooistra, Eds.)., pp. 243-256. Geoderma, Elsevier, Wageningen, The Netherlands.

Parihar, C.M., Yadav, M.R., Jat, S.L., Singh, A.K., Kumar, B., Pradhan, S., Chakraborty, D., Jat, M.L., Jat, R.K., Saharawat, Y.S. and Yadav, O. P. (2016). Long term effect of conservation agriculture in maize rotations on total organic carbon, physical and biological properties of a sandy loam soil in north-western Indo-Gangetic Plains. *Soil and Tillage Research*, 161: 116-128.

Parihar, C.M., Singh, A.K., Jat, S.L., Ghosh, A., Dey, A., Nayak, H.S., Parihar, M.D., Mahala, D.M., Yadav, R.K., Rai, V. and Satayanaryana, T. (2019). Dependence of temperature sensitivity of soil organic carbon decomposition on nutrient management options under conservation agriculture in a sub-tropical Inceptisol. *Soil and Tillage Research*, 190: 50-60.

Pastorelli, R., Vignozzi, N., Landi, S., Piccolo, R., Orsini, R., Seddaiu, G., Roggero, P.P. and Pagliai, M. (2013). Consequences on macroporosity and bacterial diversity of adopting a no-tillage farming system in a clayish soil of Central Italy. *Soil Biology and Biochemistry*, 66: 78-93.

Patil S.S., Kelkar, T.S. and Bhalerao, S.A. (2013). Mulching: A soil and water conservation practice. *Research Journal of Agriculture and Forestry Sciences*, 1: 26-29.

Patil, M.D., Wani, S.P. and Garg, K.K. (2016). Conservation agriculture for improving water productivity in Vertisols of semi-arid tropics. *Current Science*, 110: 1730-1739.

Perkons, U., Kautz, T., Uteau, D., Peth, S., Geier, V., Thomas, K., Holz, K.L., Athmann, M., Pude, U. and Köpke, U. (2014). Root-length densities of various annual crops following crops with contrasting root systems. *Soil and Tillage Research*, 137: 50-57.

Pikul Jr, J.L. and Aase, J.K. (1995). Infiltration and soil properties as affected by annual cropping in the northern Great Plains. *Agronomy Journal*, 87: 656-662.

Rashidi, M. and Keshavarzpour, F. (2008). Effect of different tillage methods on soil physical properties and crop yield of melon (*Cucumis melo*). *ARPN Journal of Agricultural and Biological Science*, 3: 41-46.

Rasmussen, K.J. (1999). Impact of ploughless soil tillage on yield and soil quality: A Scandinavian review. *Soil and Tillage Research*, 53: 3-14.

Rasool, R., Kukal, S.S. and Hira, G.S. (2007). Soil physical fertility and crop performance as affected by long term application of FYM and inorganic fertilizers in rice–wheat system. *Soil and Tillage Research*, 96: 64-72.

Rasse, D.P. and Smucker, A.J. (1998). Root recolonization of previous root channels in corn and alfalfa rotations. *Plant and Soil*, 204: 203-212.

Roseberg, R.J. and McCoy, E.L. (1992). Tillage-and traffic-induced changes in macroporosity and macropore continuity: Air permeability assessment. *Soil Science Society of America Journal*, 56: 1261-1267.

Rovira, A.D., Smettem, K.R.J. and Lee, K.E. (1987). Effect of rotation and conservation tillage of earthworms in a red-brown earth under wheat. *Australian Journal of Agricultural Research*, 38: 829-834.

Rusu, T., Gus, P., Bogdan, I., Moraru, P.I., Pop, A.I., Clapa, D., Marin, D.I., Oroian, I. and Pop, L.I. (2009). Implications of minimum tillage systems on sustainability of agricultural production and soil conservation. *Journal of Food, Agriculture and Environment*, 7: 335-338.

Saber, N. and Mrabet, R. (2002). Impact of no tillage and crop sequence on selected soil quality attributes of a vertic calcixeroll soil in Morocco. *Agronomie*, 22: 451-459.

Saha, R. and Mishra, V.K. (2009). Effect of Organic Residue Management on Soil Hydro-Physical Characteristics and Rice Yield in Eastern Himalayan Region, India, *Journal of Sustainable Agriculture*, 33: 161–176.

Saha, S., Chakraborty, D., Sharma, A.R., Tomoar, R.K., Bhadraray, S., Sen, U., Behera, U.K., Purakayastha, T.J., Garg, R.N. and Kalra, N. (2010). Effect of tillage and residue management on soil physical properties and crop productivity in maize (*Zea mays*). –Indian mustard (*Brassica juncea*). system. *Indian Journal of Agricultural Sciences*, 80: 679-685.

Salem, H.M., Valero, C., Muñoz, M.Á., Rodríguez, M.G. and Silva, L.L. (2015). Short-term effects of four tillage practices on soil physical properties, soil water potential, and maize yield. *Geoderma*, 237: 60-70.

Sarkar, S. and Singh, S.R. (2007). Interactive effect of tillage depth and mulch on soil temperature productivity and water use pattern of rainfed barley (*Hordium Vulgare* L.). Soil and Tillage Research, 92: 79–86.

Sauer, T.J., Clothier, B.E. and Daniel, T.C. (1990). Surface measurements of the hydraulic properties of a tilled and untilled soil. *Soil and Tillage Research*, 15: 359-369.

Scanlan, C.A. and Hinz, C. (2008). A conceptual model to quantify plant root induced changes in soil hydraulic conductivity and water retention. *Geophysical Research Abstracts*, pp. 10 EGU2008–A-05854 (EGU General Assembly 2008).

Sharratt, B., Zhang, M. and Sparrow, S. (2006). Twenty years of tillage research in subarctic Alaska: I. Impact on soil strength, aggregation, roughness, and residue cover. *Soil and Tillage Research*, 91: 75-81.

Shaver, T. (2010). Crop residue and soil physical properties. Proceedings of the 22nd Annual Central Plains Irrigation Conference, Kearney, NE., Available from CPIA, 760 N. Thompson, Colby, Kansas. pp. 22-27.

Shaver, T.M., Peterson, G.A. and Sherrod, L.A. (2003). Cropping intensification in dryland systems improves soil physical properties: regression relations. *Geoderma*, 116: 149-164.

Shaver, T.M., Peterson, G.A., Ahuja, L.R., Westfall, D.G., Sherrod, L.A. and Dunn, G. (2002). Surface soil physical properties after twelve years of dryland no-till management. *Soil Science Society of America Journal*, 66: 1296-1303.

Shukla, M.K., Lal, R. and Ebinger, M. (2003). Tillage effects on physical and hydrological properties of a typic Argiaquoll in central Ohio. *Soil Science*, 168: 802-811.

Singh, B. and Malhi, S.S. (2006). Response of soil physical properties to tillage and residue management on two soils in a cool temperate environment. *Soil and Tillage Research*, 85: 143-153.

Singh, B., Chanasyk, D.S., McGill, W.B. and Nyborg, M.P.K. (1994). Residue and tillage management effects of soil properties of a typic cryoboroll under continuous barley. *Soil and Tillage Research*, 32: 117–133.

Singh, V.K., Dwivedi, B.S., Singh, S.K., Majumdar, K., Jat, M.L., Mishra, R.P. and Rani, M. (2016). Soil physical properties, yield trends and economics after five years of conservation agriculture based rice-maize system in north-western India. *Soil and Tillage Research*, 155: 133-148.

Six, J., Elliott, E.T. and Paustian, K. (2000). Soil macroaggregate turnover and microaggregate formation: a mechanism for C sequestration under no-tillage agriculture. *Soil Biology and Biochemistry*, 32: 2099-2103.

Skidmore, E.L., Layton, J.B., Armbrust, D.V. and Hooker, M.L. (1986). Soil Physical Properties as Influenced by Cropping and Residue Management. *Soil Science Society of America Journal*, 50: 415.

Sommer, R., Wall, P.C. and Govaerts, B. (2007). Model-based assessment of maize cropping under conventional and conservation agriculture in highland Mexico. *Soil and Tillage Research*, 94: 83-100.

Springett, J.A. (1992). Distribution of lumbricid earthworms in New Zealand. *Soil Biology and Biochemistry*, 24: 1377-1381.

Stone, L.R. and Schlegel, A.J. (2010). Tillage and crop rotation phase effects on soil physical properties in the west-central Great Plains. *Agronomy Journal*, 102: 483-491.

Tao, Y., Zhao, C.X., Yan, C.R., Du, Z.L. and He, W.Q. (2018). Inter-annual changes in the aggregate-size distribution and associated carbon of soil and their effects on the straw-derived carbon incorporation under long-term no-tillage. *Journal of Integrative Agriculture*, 17: 2546–2557.

Teame, G., Tsegay, A. and Abrha, B. (2017). Effect of organic mulching on soil moisture, yield, and yield contributing components of sesame (*Sesamum indicum* L.). *International Journal of Agronomy*, 2017: 1-6.

Thierfelder, C. and Wall, P. (2009). Effects of conservation agriculture techniques on infiltration and soil water content in Zambia and Zimbabwe. *Soil and Tillage Research*, 105: 217–227.

Thierfelder, C., Cheesman, S. and Rusinamhodzi, L. (2012). A comparative analysis of conservation agriculture systems: benefits and challenges of rotations and intercropping in Zimbabwe. *Field Crops Research*, 137: 237–250.

Unger, P.W. (1984). Tillage effects on surface soil physical conditions and sorghum emergence. *Soil Science Society of America Journal*, 48: 1423-1432.

Unger, P.W. (1992). Infiltration of simulated rainfall: tillage system and crop residue effects. *Soil Science Society of America Journal*, 56: 283-289.

Unger, P.W. and Jones, O.R. (1998). Long-term tillage and cropping systems affect bulk density and penetration resistance of soil cropped to dryland wheat and grain sorghum. *Soil and Tillage Research*, 45: 39-57.

Uson, A. and Cook, H.F. (1995). Water relations in a soil amended with composted organic waste. In: *Soil Management in Sustainable Agriculture* (H.F. Cook and H.C. Lee, Eds.)., pp. 453–460.Wye College Press, Wye, Ashford, Kent.

Verhulst, N., Govaerts, B., Verachtert, E., Castellanos-Navarrete, A., Mezzalama, M., Wall, P., Deckers, J. and Sayre, K.D. (2010). Conservation agriculture, improving soil quality for sustainable production systems. *Advances in soil science: Food security and soil quality*, pp. 137-208.CRC Press, Boca Raton, FL, USA.

Verhulst, N., Kienle, F., Sayre, K.D., Deckers, J., Raes, D., Limon-Ortega, A., Tijerina- Chavez, L. and Govaerts, B. (2011). Soil quality as affected by tillage-residue management in a wheat-maize irrigated bed planting system. *Plant and Soil*, 340: 453–466.

Villamil, M.B., Bollero, G.A., Darmody, R.G., Simmons, F.W. and Bullock, D.G. (2006). No-till corn/soybean systems including winter cover crops. *Soil Science Society of America Journal*, 70: 1936-1944.

Wahbi, A., Miwak, H. and Singh, R. (2014). Effects of conservation agriculture on soil physical properties and yield of lentil in Northern Syria. In: *EGU General Assembly Conference Abstracts*.EGU General Assembly, Vienna 16, EGU2014-3280-1.

Wang, L.F. and Shangguan, Z.P. (2015). Water–use efficiency of dryland wheat in response to mulching and tillage practices on the Loess Plateau. *Scientific Reports*, 5: 1-12.

Whalley, W.R., Leeds-Harrison, P.B., Leech, P.K., Riseley, B. and Bird, N.R.A. (2004). The hydraulic properties of soil at root-soil interface. *Soil Science*, 169: 90–99.

Wright, A.L., Hons, F.M. and Matocha Jr, J.E. (2005). Tillage impacts on microbial biomass and soil carbon and nitrogen dynamics of corn and cotton rotations. *Applied Soil Ecology*, 29: 85-92.

Yang, X.M. and Wander, M.M. (1999). Tillage effects on soil organic carbon distribution and storage in a silt loam soil in Illinois. *Soil and Tillage Research*, 52: 1-9.

Zhang, H. (1994). Organic matter incorporation affects mechanical properties of soil aggregates. *Soil and Tillage Research*, 31: 263-275.

Zhang, S., Li, Q., Zhang, X., Wei, K., Chen, L. and Liang, W. (2012). Effects of conservation tillage on soil aggregation and aggregate binding agents in black soil of Northeast China. *Soil and Tillage Research*, 124: 196-202.

Zhang, X., Zhu, A., Yang, W., Xin, X., Zhang, J. and Ge, S. (2017). Relationships between soil macroaggregation and humic carbon in a sandy loam soil following conservation tillage. *Journal of Soils and Sediments*, 18: 688–696.

20

Nutrient Dynamics Through Conservation Agriculture Under Climate Change Scenario

A.R. Saha, S.P. Mazumdar and Alka Paswan

ICAR-Central Research Institute for Jute and Allied Fibres Barrackpore West Bengal – 700 121, India

Introduction

Increase in population in our country with decrease in cultivable land is putting the challenges to the scientists and policy planners for increased food supply. There is very little scope for horizontal increase in cultivable land. Hence, the way to meet the need for providing food to ever increasing population is to increase the production of food per unit area of cultivable land.

Farmers, generally apply fertilizer to soil on blanket recommendation. Sometimes, the addition of nutrients is inconsistent. Sometimes, the application of fertilizer is much higher than the recommended dose. Imbalanced application of fertilizer to soil results in low nutrient use efficiency and low crop yield. There is need of site specific nutrient management for increased crop yield and for achieving higher nutrient use efficiency. Proper care should be taken for selection of fertilizer, method of application of fertilizer, rate of fertilizer application and also time of nutrient supply to the soil. Excess application of fertilizer has an adverse effect on soil health.

Traditional tillage may affect physico-chemical as well as biological properties of soil. Change in tillage practice from conventional tillage to conservation agriculture method may have an impact on soil properties and nutrient dynamics. Conservation agriculture affects soil health, proper fertilizer application, both rate and time, along with efficient use of soil moisture improves nutrient use efficiency. For obtaining a good crop yield nutrient should be available for plants in adequate quantity and in right proportion. Nutrients removed by the crop should be replenished by application of inorganic fertilizers alongwith incorporation of organic manures. Soil test is an important tool for determining the quantity of nutrients available in the soil. Based on soil test values and nutrient requirement of the crop rate of fertilizer to be applied in the soil for the plant is determined. Ample quantity of farm yard manure is not available nowadays. Hence, other organic manure such as town waste, green manure etc should be used.

Conventional Tillage

Tillage is used to make soil favourable for crop growth. But doing this, sometimes land becomes degraded, soil becomes compacted with low soil organic carbon and poor drainage facility. Tillage is intensive in nature, hence its continued use increases labour and fuel cost with lower economic return to the farmer. Conventional tillage alongwith deforestation, burning of crop residues and poor management practices result in land degradation and loss of soil fertility. In association with land degradation water and labour crises felt the need of alternative p;ractice for reduction in tillage intensity. maintenance of soil fertility, preservation of soil moisture and retention of crop residue over soil surface.

Conservation Agriculture

Conservation Agriculture is soil management practice that proposes sowing of crops with minimum soil disturbance by direct seeding, maintenance of permanent soil cover with organic mulch and crop rotations which helps in reducing soil degradation and minimizing water pollutions (Sharma and Behera, 2009). Resource conservation technologies (RCTs) include reduced tillage (RT), No-till (NT), integrated nutrient management (INM) and retention of residue. Conservation agriculture not only helps in increasing crop production but also has several benefits in terms of labour saving, land regeneration and protection of environment. Degradation of soil quality in association with scarcity of resources is concern for sustenance of rice-wheat system. Conventional tillage practice should be replaced with conservation management practices for sustainable rice-wheat system and for improvement of soil quality. Declining fertility status of cultivated lands with concomitant decrease in soil organic carbon (SOC) is a major concern. Intensive tillage practices may increase oxidation of SOC resulting in lower SOC in long run. The loss of soil organic carbon may be reduced by adoption of conservation agriculture (CA) which may reduce the oxidation of SOC. Besides this SOC may be accumulated in CA based cropping practice as residue mulching act as a cover against soil erosion. In a field experiment Jat *et al.* (2018) reported that partial CA based rice-wheat-mungbean significantly lowered soil bulk density.

on the other hand, CA based rice-wheat-mungbean system and CA based maize-wheat-mungbean system recorded reduced soil penetration resistance and improved soil infiltration as compared to conventional rice-wheat cropping system. Soil organic carbon, available N and DTPA extractable Zn and Mn were found higher in CA based cropping system in comparison to conventional tillage based rice-wheat cropping system. Hence, CA based cropping system improved soil properties and nutrient availability and proved that it has the potential to minimize the external fertilizer input in the soil. By following CA based management practices nutrient resources may be saved, soil quality may be improved and higher crop productivity may be obtained.

Nutrient Dynamics

Nutrient dynamics of soil is affected by intensity of tillage, residue management and crop rotation through mineralization and recycling of soil nutrients (Galantini *et al.*, 2000). Adoption of conservation tillage instead of conventional tillage of higher intensity may lead to higher buildup of nutrients in the surface top soil because of nutrient mineralization potential of soil (Rasmussen, 1999; Duiker and Beegle, 2006). In conservation agriculture crop residue is retained on the soil cover where as in conventional tillage residue of crop is incorporated in the soil. Higher nutrient availability in conservation agriculture may be due less leaching loss of nutrients with reduced decomposition of nutrients in the top soil.

Integrated Nutrient Management

The projected food grain need of our country by 2050 would be 457 mt. To produce this huge quantity of food, grain a large quantity of nutrient is needed. This huge need of nutrients cannot be met by a single source, be it inorganic fertilizer, organic manure or crop residue. Besides this, the food grain 457 mt will remove nearly 58 mt of nutrients from soil. At present the fertilizer consumption of India is 25.33 mt. If current trend of fertilizer consumption is continuing, then by 2050 there will be addition of 48 mt of nutrients in soil leaving a gap of 10 mt of NPK nutrients in soil per year (Singh *et al.*). Recycling of organic wastes may bridge this gap.

But organic manure alone cannot meet the nutrient required for producing crops which is needed for the sky rocketing human population because of low nutrient content of organic manure. Combined use of inorganic, organic and biological sources of plant nutrients may be capable of meeting the nutrient required for producing crops to feed increased human population. Integrated nutrient management considers combining of nutrients from different sources for sustaining crop yield. Care should be taken for reducing nutrient loss and replenishment of nutrients lost and removed by crops. It should also consider estimating nutrients coming from on farm resources e.g. different organic manures and crop residues.

Soil fertility and soil properties are improved by application of organic manures in combination with inorganic fertilizers. Organic manures may be obtained from different sources. These may be different animanure such as poultry manure, cow dung etc., composts, crop residues, green manure, vermicompost, and waste from different agro-industrial sources, urban wastes, sewage sludge, industrial waste etc. Total nutrient potential from different organic sources is estimated as 32.41 mt by 2025. Sewage and industial wastes are generally discharged in water bodies which may create environmental and health hazards. If properly treated, these wastes also may play a good role in plant nutrition. A very little quantity (5%) of crop residues available in our country is recycled. A large quantity of cow dung is burnt as cow dung cakes.

Total nutrient NPK potential from bio gas slurry obtained as left over residue after biogas digestion is about 4 mt per annum. If this potential of bio gas slurry can be harnessed in proper way, then subsidy given on fertilizer and foreign exchange used for import of chemical fertilizers from outside country. Effect of organic manure on crop yield depends on the crop, soil type and release pattern of nutrients from the manures. Besides this, the management practices involved also determines the crop yield. Organic manures release nutrients slowly and hence, have greater residual effect than inorganic fertilizers. Application of inorganic fertilizer in combination with organic manures give higher crop yield as compared to when fertilizers are applied alone. This improvement in crop yield may be attributed to the effect of organic manure on soil properties. Incorporation of organic manure improves soil structure, water infiltration, porocity etc. Besides this, organic manures add to the soil different nutrients including secondary and micronutrients and beneficial microbes. Organic manures increase nutrient use efficiency. Slow nutrient release pattern of organic manures provides constant, steady and stable supply of nutrients to crops and thus help in increasing crop yield. Hence, time of incorporation of manures before sowing of crop is very much important. Soil temperature and soil moisture also affect decomposition of manures. Hence, time of incorporation of nutrients also affects nutrient release pattern and hence, crop yield. Farmers should apply solid organic manures immediately after land preparation to prevent loss of NH_3. Animal waste if applied too early before sowing and during non-growing season may cause loss of nutrients by surface runoff or drainage. On the other hand, if application of organic manure is delayed and incorporated just before seeding, there is risk of nitrogen deficiency for the next crop. This is because of delayed nitrogen release for use of crop.

Long term experiments evidenced that benefit of organic manure on crop yield depends on the amount of nutrient applied. The beneficial effect depends on the frequency of manure application also. It was observed from different long term fertilizer trials that combined application of inorganic fertilizer and manure, compost or other organic amendments increased the crop yield as compared to organic amendments alone. (Kaur *et al.*, 2008) reported higher maize grain yield when recommended NPK fertilizer is applied alongwith FYM @10 t ha^{-1} for 34 years in maize-wheat cropping system.

Integrated nutrient management should be practiced to increase the crop yield and also to minimize the dependence on chemical fertilizers. Management of crop residues is one important way in this direction. In situ recycling of crop residues with use of happy seeder may be an important resource conservation technology. Besides this, recycling of agro-industrial/municipal wastes should also be considered for nutrient supply to crops.

Yadav *et al.* (2017) studied on effects of conservation tillage and nutrient management practices on soil fertility and productivity of rice-rice system in North Eastern region of India. They reported that replacememt of farmers practice with conservation tillage (no till NT) and integrated nutrient management (INM) practice alongwith residue retention enhanced the straw, root and biomass yield of both rice systems over farmers practice. Adoption of conservation tillage and nutrient management practice involving NT and INM alongwith residue retention can enhance the system productivity, and C and N sequestration in paddy soils is thereby contributing to the sustainability of rice-rice system.

Nutrient Acquisition through Pulse

Nutrient management through integrated use of organic manures and chemical fertilizers is very effective in sustaining crop productivity and soil fertility in pulse and pulse based cropping system. Pulses have the ability of biological nitrogen fixation, carbon sequestration and soil amelioration. With low water requirement pulse play a very important role in sustaining crop productivity. Legumes is found to improve soil physical properties. Legumes also play a very important role in increasing microbial activity and restoration of organic matter in soil. Some of the legumes have the ability to solubilize occluded P and highly insoluble calcium bound P by their root exudates. Chickpea and pigeon pea may be named in this aspect (Venkatesh *et al.*, 2014). In India average consumption of plant nutrients through chemical fertilizer source is very low and there is a great potential for exploiting of direct and residual soil fertility of legumes (Ghosh *et al.*, 2006).

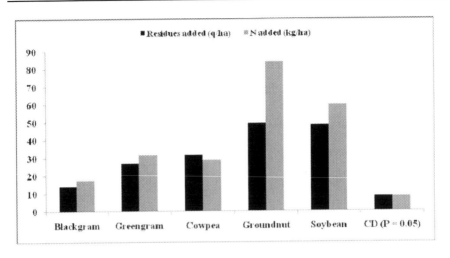

Fig. 1: Residues and nitrogen added by different pulse crops.
Source: Venkatesh *et al.* (2014)

About one third of total crop residues available in India is used for crop production and rest two-third is used as fuel, feed for animals and for other purposes. Crop residues, if incorporated in soil, improves soil physical properties and provides a substrate for deecomposers. Besides this, by incorporating crop residues in soil nutrient status of soil is improved, soil structure is modified through restoration of organic matter, wind and water erosion is reduced and soil productivity is increased. Incorporation of crop residues improves fertilizer use efficiency in addition to supplementing fertilizers. Crop residues, if used judiciously as an important resource, help in conserving soil moisture and also reduce loss of nutrients through leaching, volatilization. Thus it plays a very important role in conserving resources and for increasing crop productivity. Incorporation of rice residue in rice-chickpea cropping sequence yield of chickpea is increased significantly. The influence in increasing seed yield was found better with incorporation of chopped straw with irrigation as compared to removal of rice residue. Crop residues are sources and sink of C and N and influence N cycling. Incorporation of crop residues help to get N available for plants for a longer period because N from crop residues first go through the phase of immobilization and then N gets gradually mineralized. Decomposition of plant residues and their nutrient release pattern help determine the nutrient value of the residues (Adediran *et al.*, 2003).

Nutrient Acquisition through Plant Roots

Plant roots exudate different compounds as rhizodeposition. The root exudates include mucilage, enzymes, organic acids, sugars, amino acids etc. Cereals are reported to transport 20-30% of the total assimilated carbon into the soil. One-half of this amount is found in the roots. One-third of the assimilated carbon is lost as CO_2 by root respiration and microbial utilization of root borne organic substances. The other part of the translocated and underground carbon is incorporated into soil organic matter and soil microorganisms (Ventatesh *et al.,* 2014). Pulse crops have tap root system. This tap root penetrates deep into the soil. Tap root system of pigeon pea is very deep and hence, pigeon pea is termed as plow crop. Legume roots are reported to reach deep upto 2 m in soil profile. Hence, pulse crops are very efficient in nutrient recycling in soil profiles (Ali and Venkatesh, 2009).

Green Manuring

Nutrient recycling as a measure for replenishment of nutrients exhausted due to highly exhaustive cereal based cropping system is very much important for sustaining crop productivity. Recycling of legume residue is an important potential component for integrated nutrient management. High cost of synthetic fertilizer has created awareness and interest for green manuring. Due to lack of resources, farmers are not willing to grow green manure crops. Hence, it is recommended to grow a grain legume crop in the cropping system such as cow pea or green gram. Pods are harvested and biomass of legume crop is turned back into soil. Green manuring legume crops like sesbania or crotalaria were used to be grown by farmers as manuring crops before use of chemical fertilizers in rice based cropping system in rice belt of India. Green manuring increases crop yield and at the same time improves soil fertility also. Saraf and Patil (1995) reported that incorporation and decomposition of green manure has a solubilizing effect of macronutrients (N,P,K) and micronutrients (Zn, Cu, Fe and Mn) in the soil. Recycling of nutrients through green manuring can mitigate deficiency of different nutrients in the soil and in turn increases their availability for the crops. Green manuring improves efficiency of applied plant nutrients by reducing leaching and gaseous losses of nitrogen. Intensive cropping system, availability of chemical fertilizers and irrigation facilities has reduced the area under green manuring. Greengram and cow pea have great potential to add N in the soil when their residues with nodules are incorporated in the soil (Glasener *et al.*, 2002; Shahetal, 2003; Timsina *et al.*, 2006). Hence, green manuring is an important component of integrated nutrient management and acts as a tool in resource conservation technology.

Surface residues of legumes and their roots and nodules are subject to microbial decomposition and a good amount of nitrogen is left for the succeeding crop. Legumes fix atmospheric nitrogen and farmers play an important role in increasing indigenous nitrogen supply in the soil by growing legume crop. Legume crop residues remained after harvesting of grain increase the yield of subsequent crop and reduce the use of synthetic nitrogenous fertilizer. Rekhi and Meelu (1983) while working coarse textured nonacid soils of Punjab observed that 50-60 days old green manure crop when incorporated 1 day before transplanting of rice may substitute a great quantity of nitrogen.

Crop Rotation

Selection of crop rotation in a cropping system for a agro ecological region is an important strategy for conservation of nutrients and improving nutrient use efficiency. Judiciously chosen crop sequences permits in efficient use of nutrient resources and soil resources and helps in proper use of nutrient and water by crops to increase crop yield and sustain soil productivity. In a cropping system, legumes and non-legume crops should be rotated and this is considered as very effective nutrient management practice. Legumes, when included in a cropping system, fix atmospheric nitrogen biologically and reduce nitrogen requirement for the next cereal crop in the rotation. Proper crop rotation has various beneficial effect such as increased water use efficiency, increased nutrient use efficiency, improving soil physical properties and improved soil health. Nutrients are returned to the soil through leaf fall and crop residues from previous crop helps to build organic nutrient pool in the soil and should be considered as an effective resource conserving process. Jute may be mentioned in this context. Jute crop, throughout its growing period, shed 15 tons of green leaves and add enormous amount of nutrients in the soil by which the next crop rice in the jute-rice cropping system is benefitted and nutrient requirement is reduced. In crop diversification crops are included in such a way that different crops may require different nutrient resources. One high N loving crop such as corn may be combined with a low N using crop or a N fixing crop e.g. Soybean or lentil so that nitrogen exhausted from the soil by high N using crop may be replenished by a N fixing crop. Badaruddin and Meyer (1994) reported that wheat yield in wheat-legume sequence was higher as compared to that obtained in wheat-wheat sequence. This might be attributed to the inclusion of N fixing crop in the cropping sequence. Legume, when grown as a green manure crop add considerable amount of nitrogen to the soil which will benefit the succeeding crop. This is because of decomposition of legume residues and contribution of nitrogen to the next crop (Badaruddin and Meyer, 1990; Welty *et al.*, 1988). Hence, legumes when included in the crop rotation plays important role for improving crop productivity and also conservation of nutrient resources. Acquisition of nitrogen through legume incorporation in soil depends on many factors such as water regime, age of legume crop, soil fertility etc. Cassman and Pingali (1995) reported that in continuous rice cropping with two or three rice crop consumption of inorganic fertilizer nitrogen increases but yield rice remains stagnant. low level of soil available nitrogen in continuous rice cropping may be replenished by inclusion of legume in the cropping system and hence, by biological N fixation. Therefore, legume has great importance in nutrient acquisition and in sustain soil fertility. Some workers reported that benefit of yield increase to crops following field pea may be due to mineralization of nitrogen from residues of field pea. In this context it may be inferred that recycling of legume residues play a vital role in resource conservation technology.

Soil Test Crop Response (STCR)

To get expected yield a method of fertilizer recommendation based on soil test values has been developed by AICRP on STCR for different crops for different agro ecological regions. Target yield approach is based on relationship between crop yield, soil test values and nutrient requirement of the crop. Now a days IPNS based target yield approach is considered keeping in mind the contribution of nutrients from organics, soil and fertilizer altogether.

In demonstration of fertilizer application based on soil test values in the farmers' field, target yield is chosen in such a way so that cost of fertilizer remains more or less same as practiced by the resource poor farmers and farmers get expected crop yield. Hence, STCR may play a vital role in resource conservation technology by conserving precious nutrients and at the same time maintaining soil health.

Conclusion

Low use efficiency of nutrients in general and nitrogen in particular continues to be a problem in Indian agriculture. Wide spread yield stagnation and factor productivity decline as well as greater use of N than the recommended levels by the farmers in intensive production systems to maintain the yield levels which previously could be attained with relatively less fertilizers has been documented. Conservation agriculture affects soil health, proper fertilizer application, both rate and time, along with efficient use of soil moisture improves nutrient use efficiency. Soil test is an important tool for determining the quantity of nutrients available in the soil. Based on soil test values and nutrient requirement of the crop rate of fertilizer to be applied in the soil for the plant is determined.

References

Adediran, J.A., De Baets, N., Mnkeni, P.N.S., Kiekens, L., Muyima, N.Y.O. and Thys, A. (2003). Organic waste materials for soil fertility improvement in the border region of the Eastern Cape, South Africa. *Biological Agriculture & Horticulture*, 20: 283–300.
Ali, M. and Venkatesh, M.S. (2009). Improving yield and quality of pulses through secondary and micronutrients. *Indian journal of Fertilisers*, 5(4): 117-123.

Badaruddin, M. and Meyer, D.W. (1994). Grain legume effects on short-term effects of crop rotations on soil nitrogen, grain yield, and nitrogen nutrition of wheat. *Canadian Journal of Soil Science*, 80: 193–202.

Badaruddin, M., and Meyer. D.W. (1990). Green-manure legume effects on soil nitrogen, grain yield, and nitrogen nutrition of wheat. *Crop Science*, 30: 819–825.

Duiker, S.W. and Beegle, D.B. (2006). Soil fertility distributions in long-term no-till, chisel/disk and moldboard plow/disk systems. *Soil & Tillage Research*, 88(1): 30-41.

Galantini, J.A., Landriscini, M.R., Iglesias, J.O., Miglierina, A.M. and Rosell, R.A. (2000). The effects of crop rotation and fertilization on wheat productivity in the Pampean semiarid region of Argentina: 2. Nutrient balance, yield and grain quality. *Soil & Tillage Research*, 53(2): 137-144.

Ghosh, P.K., Bandopadhyay, K.K., Wanjhari, R.H., Manna, M.C., Misra, A.K., Mohanty, M. and Subba Rao, A. (2007). *Journal of Sustainable Agriculture*, 30 (1): 59-86.

Jat, H. S., Datta, Ashim., Sharma, P. C, Kumar, Virender, Yadav, A. K, Choudhary, Madhu Choudhary, Vishu, Gathala, M. K., Sharma D. K., Jat, M. L, Yaduvanshi, N. P. S, Singh, Gurbachan and McDonald, A (2018). Assessing soil properties and nutrient availability under conservation agriculture practices in a reclaimed sodic soil in cereal-based systems of North-West India. *Archives of Agronomy and Soil Science*, 64(4): 531–545.

Rasmussen, K.J. Impact of ploughless soil tillage on yield and soil quality (1999). a Scandinavian review. *Soil & Tillage Research*, 53(1): 3-14.

Rekhi, R.S and Meelu, O.P. (1983). Effect of complementary use of mung straw and inorganic fertilizer N on N availability yield of rice. *Oryza*, 20: 125-129.

Sharma, A.R. and Behera, U.K. (2009). Recycling of legume residue for nitrogen economy andhigher productivity in maize (*Zea mays*)–wheat (*Triticumaestivum*) croppingsystem. *Nutrient Cycling in Agroecosystems*, 83: 197–210.

Venkatesh, M.S., Hazra, K.K. and Katiyar, Rohit (2014). Nutrient acquisition and recycling through pulses. In book: Resource Conservation Technology in Pulses, Edition: First; Chapter, Publisher: Scientific Publisher. Edited by P.K.Ghosh, N. Kumar, M.S.Venkatesh, K.K. Hazra and N. Natarajan, pp. 190-198.

Yadav, M.R., Parihar, C.M., Kumar, Rakesh, Yadav, R.K., Jat, S.L., Singh, A.K., Ram, H., Meena, R.K. Singh, M., Meena, V.K, Yadav, N., Yadav, B., Kumawat, C. and Jat, M.L. (2017). Conservation Agriculture and Soil Quality– An Overview. *International Journal of Current Microbiology and Applied Sciences* ISSN: 2319-7706, 6: 2

21

Conservation Agriculture and Its Impact on Soil Quality in Climate Change Scenario

S.P. Mazumdar, S. Sasmal and Ria Bhattacharya

ICAR-Central Research Institute for Jute and Allied Fibres, Barrackpore West Bengal – 700 121, India

Introduction

There is a widespread degradation of natural resource, e.g., soil erosion, nutrient loss, water-logging, salinity, alkalinity, acidity, decline in soil organic carbon (SOC) content, ground water depletion and micronutrient depletion. The major causes are washing away of topsoil and organic matter due to water erosion, intensive deep tillage and inversion tillage, dismally low levels of fertilizer application, mining and other commercial activities, faulty agricultural practices, no or low use of organics, indiscriminate use of herbicides, pesticides, fungicides etc. Other cause is loss of organic matter with higher temperatures, which increase rate of microbial decomposition of organic matter due to adverse effect of climate change. Climate change due to global warming as a result of increased concentration of GHGs in the atmosphere is a well-established fact. Impacts of climate change are experienced throughout the world. Increases in global mean temperature, rise in sea-level, and occurrence of extreme events are the consequences of climate change. These changes may deplete the SOC pool and soil structural stability, increase soil's susceptibility to water runoff and erosion, disrupt cycles of water, carbon, nitrogen, phosphorus, sulphur and other elements, and cause adverse impacts on biomass productivity, biodiversity and the environment. Climate change is likely to have a variety of impacts on soil quality. In order to restore soil quality, enhance productivity, it is of utmost importance to focus on conservation agriculture practices on long-term basis. Conservation Agriculture (CA) is gaining importance as an alternative to conventional agriculture. Conservation agriculture is defined as a system that combines minimum or no tillage (NT) with permanent soil cover (that leaves at least 30% of the soil covered between harvest and planting) and diversified crop species that include legumes (FAO, 2019). Other practices, such as integrated pest and nutrient management, are also often incorporated into the CA system on a site specific basis to help ensure its success (Lal, 2015; Thierfelder *et al.*, 2018). Conservation agriculture is becoming popular worldwide due to its enhanced C sequestration potential and favourable effects on soil fertility, nutrient dynamics and overall soil quality (Dey *et al.*, 2016, Kumawat *et al.*, 2018, Choudhary, 2018, Parihar *et al.*, 2020). Conservation agriculture improves physical, chemical and biological properties of soils (Hobbs *et al.*, 2008), which in turn, affect the ecosystem services and sustainability of crop production system through counterbalancing the climate variability. It enhances biodiversity and natural biological processes above and below the ground surface, which contribute to increased water and nutrient use efficiency and to improved and sustained crop production. Therefore, it is a complete package of agricultural practices which not only protects the natural resources and environmental quality by slowing down the soil physical, chemical and biological quality degradation but also reduces cost of production and sustains yields (Thierfelder and Wall, 2009; Dikgwatlhe *et al.*, 2014). Resource conservation issues have drawn the attention of scientists to develop innovative tillage and crop establishment techniques for higher crop productivity and agricultural sustainability.

Soil Quality

Research on resource conservation technologies emphasized that conservation agriculture plays an important role in restoring the dynamic attributes of soil quality. Soil quality has been defined by many scientists differently. Some have defined it as "fitness for use", while others as "capacity of the soil to function". However, Soil Science Society of America came up with a wholesome definition for it as "The capacity of a specific kind of soil to function, within natural or managed ecosystem boundaries, to sustain plant and animal productivity, maintain or enhance water and air quality and support human health and habitation". The soil quality can be measured quantitatively using physical, chemical and biological properties of soils as these properties interact in a complex way to give a soil its quality or capacity to function. Thus, soil quality cannot be measured directly, but must be inferred from measuring changes in its attributes or attributes of the ecosystem, referred to as indicators. Again, within each category of attributes, we analyses a number of parameters, such as bulk density, maximum water holding capacity, mean weight diameter under physical; pH, organic carbon, available N, P and K, micronutrients, heavy metals under chemical, and microbial biomass C and

Conservation Agriculture and Its Impact on Soil Quality in Climate Change Scenario 321

N, soil enzymes, mineralizable C and N, soil biodiversity, soil fauna under biological. Some of the important soil quality indicators can be influenced by conservation agriculture practices which in turn can help in moderating the ill effects of climate change

Conservation Agriculture on Soil Physical Quality

Continuous adoption of conventional tillage results in soil compaction, which is undesirable and is associated with increased bulk density (BD) and decreased porosity. Practicing conservation agriculture would result in reduction in compaction over time, thereby decreasing bulk density. The decrease in soil BD under CA is due to higher SOC, better aggregation, increased root growth and biomass (Salem *et al.*, 2015). For example, after long-term adoption tillage practices (24 years) the soil BD was 28.2% higher in conventional tillage (CT) than zero tillage (ZT) (Utomo *et al.*, 2013). Parihar *et al.*, 2016 also reported that after adoption of CA practices for seven years, the soil BD was lowered by 4.3 to 6.9 % in 0-30 cm soil profile than CT. A recent global meta-analysis found that NT with residue retention, on average increased bulk density by 1.4%. Inclusion of legume crops in rotation cycles lowers soil BD (Verhulst *et al.*, 2011 and Thierfelder *et al.*, 2012). Numerous studies have reported the effects of conservation agriculture on BD. In some cases, soils under CA can have higher, similar or lower BDs than under conventional tillage (Indoria *et al.*, 2017; Blanco-Canqui and Ruis, 2018; Somasundaram *et al.*, 2019).

A considerable amount of research has been conducted to study the influence of tillage practices and crop residue management on soil aggregation (Wright and Hons, 2005). Experiments suggest that zero or no-tillage usually favors the formation of soil aggregates and minimizes the potential for rapid oxidation of SOC due to reduced soil disturbance (Six *et al.*, 2000; Kumari *et al.*, 2012). It has been widely reported that conservation agriculture practices improve soil structure, aggregate stability and other physical parameters (Alam *et al.*, 2017; Gathala *et al.*, 2011; Bhattacharya *et al.*, 2013; Parihar *et al.*, 2016). Jat *et al.* (2009) reported that the steady-state infiltration rate, soil aggregation (>0.25 mm) and mean weight diameter (MWD) were higher under permanent beds and double ZT and lower in the CT system. Inclusion of legumes in crop rotation and incorporation of crop residues played a major role in soil aggregation and stability (Govaerts *et al.*, 2009; Srinivas rao *et al.*, 2013). DSR combined with zero tillage in wheat along with residue retention has the highest potential to secure sustainable yield increment (8.3%) and good soil health by improving soil aggregation (53.8%) and SOC sequestration (33.6%) with respect to the conventional tillage with transplanted rice after five years of continuous rice–wheat cropping in sandy loam reclaimed sodic soil of hot semi-arid Indian sub-continent (Gupta Choudhury *et al.*, 2014). Improvement in soil aggregation, aggregate associated C and particulate organic carbon (POC) was observed with full CA-based management practices under cereal based systems of semi-arid Northwest India (Jat *et al.*, 2019). Adoption of CA practice (ZT with 4 t ha^{-1} crop residue retention)in pearl millet – mustard cropping system resulted in higher crop productivity, infiltration rate, mean weight diameter and proportion of macro-aggregates in the soil under rainfed semi-arid regions besides maintaining higher soil water content (Mukesh Choudhary *et al.*, 2019).A recent worldwide meta-analysis observed that, on average, the number of water stable aggregates in NT systems are 31% greater compared to conventionally tilled systems without residue retention (Li *et al.*, 2019).

Many researchers (Kuotsu *et al.*, 2014; Karunakaran and Behera 2015) have also reported higher steady state infiltration rate under ZT with residue retention compared to CT and removal of residues. Zero tillage and permanent beds improved saturated and unsaturated hydraulic conductivity due to greater number of macropores, little soil disturbance, and presence of litter of well decomposed residues (Bhattacharya *et al.*, 2006, Parihar *et al.*, 2016). Double no till and minimum tillage resulted in retention of crop residue on the soil surface was found to maintain higher soil moisture and reduced evaporation as compared to conventional tillage (Ghosh *et al.*, 2010). Others also reported the benefits of CA on greater soil water storage (Verhulst *et al.*, 2011; Page *et al.*, 2019).

Conservation Agriculture on Soil Chemical and Biological Quality

Agricultural activities affect mainly the SOC pool, which constitutes a potential source of greenhouse gasses with estimated current carbon content in the 1 m top layer 2.0 times greater than the atmospheric pool. Organic matter content, which is quite low (<1.0%) in the tropical soils would become still lower and climate change may affect its quality (Lal, 2004). Soil biology and microbial populations are expected to change under changed climatic conditions. Although agriculture is not a large emitter of greenhouse gases (GHGs), it can assist in reducing the overall carbon emissions by sequestering this in large area of land. Options to mitigate CO_2 emissions from agriculture and land use changes include the reduction of emissions from present sources and the creation and strengthening of carbon sinks, by sequestering C in soils. Management to build up SOC requires increasing the C input, decreasing decomposition, or both (Paustian *et al.*, 1997). Soil carbon sequestration is a natural, cost-effective and environmental friendly process. Important basic strategies to enhance carbon sequestration in soil through sustainable management practices include conservation tillage, use of crop residues, manure, compost, biosolids, mulch farming, diverse cropping, integrated nutrient management, improved grazing, forestry and growing energy crops on spare lands. Thus, the management systems that cause minimal soil disturbance by reducing tillage intensity and frequency, addition of higher amount of biomass, conserve soil and water, improve soil biodiversity and aggregation will lead to increased SOC sequestration. Estimates of the total potential of C sequestration in world soils vary widely from a low of 0.4 to 0.6 Gt C yr^{-1} to a high of 0.6 to 1.2 Gt C yr^{-1}. India is having large land area and diverse eco-regions and there is a considerable potential of soil carbon sequestration.

Intensive tillage, especially ploughing accentuates mineralization and CO_2 emissions by mixing crop residues into the soil, bringing it closer to microbes, increasing O_2 concentrations in the soil and disrupting aggregates and exposing physically protected organic matter to microbial and enzyme activity. Therefore, conservation tillage offers a strategy to conserve and sequester carbon in soils by reducing the intensity and frequency of ploughing and leaving crop residues on the soil surface as mulch for increasing SOC content and enhancing carbon sequestration in soil. This helps, improves in soil organic matter quantity and quality and provides food for micro-organisms leading to improved biological health and microbial diversity. Adoption of no-till conserves carbon and improves soil health. Doran (1980) reported that organic C concentration in the surface 15 cm of non-tilled soils is greater than for tilled soil. Adoption of conservation agriculture can result in higher potential rates of carbon sequestration in soils. Conversion to no-till sequesters C in the soil and long term use of NT is an effective technology for sequestering more atmospheric CO_2 in soil. The rates of carbon sequestration through conversion to no-till are 0.12-0.29, 0.09-0.29, 0.12-0.29 and 0.14-0.56 Mg C $ha^{-1} yr^{-1}$ in Asia, Africa, America and USA, respectively. Studies indicated that average rate of carbon sequestration under NT is 57 ± 14 g C $m^{-2} yr^{-1}$ with peak sequestration rates was within 5-10 years after conversion (West and Post, 2002). West and Post (2002) concluded that a move from conventional tillage systems to zero tillage (both with residue retention) can sequester on average 48 ± 13 g C $m^{-2} yr^{-1}$. Alvarez (2005) found that the accumulation of SOC under reduced and zero tillage was an S-shape time-dependent process, which reached a steady state after 25–30 years. Averaging out SOC differences in all of the 161 experiments under reduced and zero tillage, there was an increase of 2.1 t C ha^{-1} over plowing. However, when only those cases that had apparently reached equilibrium were included (all zero tillage vs. conventional tillage comparisons from temperate regions), mean SOC increased by approximately 12 t C ha^{-1}. Pacala & Socolow (2004) estimated that conversion of plough tillage to no-till farming on 1600 M ha of cropland along with adoption of conservation-effective measures could lead to sequestration of 0.5–1 Pg C yr^{-1} by 2050. Surface residue management and reduced tillage provide a favourable for soil environment and native micro flora and fauna. Among the intermediate products of decomposition, many polysaccharides including microbial gum are especially significant in the formation of soil aggregates by their binding action on soil particles. In a 23-year field study, Dolan *et al.* (2006) found that no-tillage managed soil had 30% more soil organic C (SOC) and soil organic N (SON) than the mould board plow and chisel plows soils. Tillage practice can also influence the distribution of SOC in the profile with higher soil organic matter (SOM) content in surface layers with zero tillage than with conventional tillage, but a higher content of SOC in the deeper layers where residue is incorporated through tillage (Jantalia *et al.*, 2007; Thomas *et al.*, 2007). Purakayastha *et al.* (2008) reported that greater SOC stock in 0-20 cm Paulose silt loam soil occurred in Native Prairire (63.7 Mg ha^{-1}) and CT has lowest SOC stock which is 56% less than NP. Varvel and Wilhem (2011) conducted a long term experiment under rainfed conditions with six tillage systems and three cropping systems and reported that soil organic carbon and nitrogen were sequestered deeper in the soil profile. Liu *et al.* (2014) showed that 17 years of ZT had ~8.3% higher cumulative SOC stocks at depths of 0–60 cm compared with CT.

In India under rice-wheat system until 1998, results in northern India show that no-tillage is not a successful technology and is only an option for late sown wheat. But the importance of no-tillage has now been realized when the increase in CO_2 content in earth's atmosphere is voiced loudly. In India the area under zero-tillage under rice-wheat system has been increased tremendously over the years. Using zero tillage (ZT) for wheat had a positive effect on soil quality regardless of the treatments used for rice in rice–wheat cropping on a Vertisol in India (Mohanty *et al.*, 2007). Reports also suggested that that zero-tillage practice may be successful in maintaining sustainability and improving soil biological properties, in hill agro-ecosystems (Mina *et al.*, 2008). Results available in India showed that the reduced tillage practices, in general, could bring about an increase in SOC content in upper layer of the soils (Mandal, 2011). Bhattacharyaa *et al.* (2012) reported that the soil C storage was impacted to a maximum in the resource conservation techniques like application of rice straw+ green manure in rice-fallow cropping system in the tropical flooded soils planted to rice. Bhattacharya *et al.* (2012) reported that adoption of continuous no-tillage is the best management option for improving SOC and aggregation under a rainfed lentil-finger millet cropping system of Indian Himalayas and observed that SOC concentration increased by 33.6%, water stable macro-aggregates by 53.8% and macro-aggregate associated C by 20.8%. Das *et al.* (2014) reported that in North East India, carbon sequestration was higher in zero tillage as compared to conventional tillage. Modak *et al.* (2020) assessed the impacts of zero tillage and residue retention on soil aggregation and carbon stabilization within aggregates in subtropical India. Overall, results indicated that that SOC content increased by ~48, 42 and 36%, respectively, in wheat residue + soyabean residue (WR + SR) SR and WR plots compared with no residue plots and SOC content within macro aggregates were ~30 higher in the zero tillage plots than conventional tillage.

The effectiveness of conservation tillage in SOC sequestration is enhanced by the use of cover crops, such as clover and small grains. Frequent use of sod type legumes and grasses in rotation with food crops is an important strategy to enhance SOC and soil quality. Gains in SOC by growing cover crops in rotation with food crops have been reported throughout the world including Haryana, India (Chander *et al.,* 1997), south-western Nigeria (Juo *et al.,* 1996), Syria (Jenkinson *et al.,* 1999). Research was conducted to evaluate the potential of cover crops to accumulate SOC (Amado *et al.,* 2001) and observed that soil under maize + velvet bean (cover crop) system had 5.42 t ha[-1] more C than the fallow/maize system in the 0 to 20 cm soil layer. Cover crops can be grazed and well maintained pastures with controlled grazing at appropriate stocking rate can enhance SOC. Neill *et al.* (1997) observed that SOC increased in 14 of 18 pastures and the magnitude of increase in some cases was as much as 18 Mg C ha[-1] to a 30 cm depth. Strategies that increase the cropping intensity such as the use of rotations with winter cover crops to increase the amount of biomass C returned to the soil can affect the size, turnover and vertical distribution of both active and passive pools of SOC (Franzluebbers *et al.,* 1995). Sainju *et al.* (2002) reported that concentrations of soil organic C was greater with rye, hairy vetch and crimson clover cover crops than without a cover crop.

For sustainability of intensive cropping system, it is desirable not to grow a particular crop or a group of crops on the same soil for a long period. Varying the types of crop grow can increase the level of soil organic matter. However, effectiveness of crop rotation depends on the type of crops and crop rotation times. Cropping systems provide an opportunity to produce more biomass C than in a monoculture system and to thus increase SOC sequestration. Chander *et al.* (1997) studied soil organic matter under different crop rotations for 6 years and found that the inclusion of green manure crop of *Sesbania aculeata* in the rotation improved soil organic matter status and microbial C increased from 192 mg kg[-1] soil in pearl millet-wheat fallow rotation to 256 mg kg[-1] soil in pearl millet-wheat green manure rotation. Benefits of crop rotations in managing SOC are documented by several long-term experiments in the United States. Legume-based cropping systems could help to increase crop productivity and soil organic matter levels, thereby enhancing soil quality as well as having the additional benefit of sequestering atmospheric C (Gregorich *et al.,* 2001). The quantity of C below the plough layer in legume-based rotation was 40% greater than that in monoculture. The soil organic matter below the plough layer in soil under the legume-based rotation appeared to be in a more biologically resistant form (i.e., higher aromatic C content) compared with that under monoculture. Effects of rotation with legume crops had the highest total C and biomass levels reported. Similar evidence of the effects of crop rotations on soil aggregation and SOC storage comes from studies on soils of USA (Wright and Hons, 2005), Eastern Canada, Europe (Alvaro-Fuentes *et al.*, 2008) and soils of Indian Himalayas (Verma and Sharma, 2007; Bhattacharya *et al.*, 2015).

In India, most of the commonly practiced cropping systems are cereal-cereal (rice-rice), cereal-cereal-cereal (rice-wheat-maize), cereal-cereal-legume (maize-wheat-green gram), finger millet-wheat-gram, or pearl millet-wheat-gram. Singh *et al.* (1996) show that over a period of five years the net change in SOC is negative under cereal-cereal sequences, whereas in other sequences having legume component the changes are positive. Mixed or intercropping systems are also advantageous in terms of OC sequestration. In a typical black soil, continuous cropping and manuring increase organic carbon content by 20-40% over a period of three years or introducing summer green manure crop *dhaincha* after harvest of wheat and before planting of rice. The maintenance of organic matter in rice-wheat cropping system is extremely important. In Indo-Gangetic plains rice-wheat sequence needs to be tested by introducing legume crops in the system in different ways, viz., replacing rice or wheat crop by legume i.e. rice by pigeon pea in summer and wheat by lentil in winter. The beneficial effect of growing summer mung as a catch crop in rice-wheat rotation on enhancing SOC pool and the total productivity of the crops in the system in Mollisols of Pantnagar has been reported by Ghosh and Sharma (1996).

Integrated nutrient management is also essential for C sequestration. On a long-term basis, increased crop yield and organic matter returned to the soil with judicious fertilizer application result in higher SOC content and biological activity than under controlled conditions (absence of fertilizers). Lal *et al.* (1998) summarized the results of a number of studies and concluded that fertility management practices can enhance the SOC content at the rate of 50-150 kg ha^{-1}yr^{-1}. Balanced application of fertilizer and manure substantially improve the SOC under different soils and cropping systems (Nayak *et al.*, 2012; Sandeep *et al.*, 2016; Mazumdar *et al.*, 2018). In semi-arid regions of India, utilizing the wastes through compost, amended with minerals such as rock phosphates, pyrites and N application have been recognized for improving the crop yields and SOC (Manna *et al.*, 1997). Liebig *et al.* (2002) observed that high N rate treatments increased SOC sequestration rates by 1.0–1.4 Mg C ha^{-1} yr^{-1} compared with unfertilized controls. Long-term manurial experiment on vertisol of central India under soybean-wheat cropping system showed an enrichment of organic carbon ranged from 85-739 kg C ha^{-1} at 0-15 cm and 54-149 kg C ha^{-1} at 15-30 cm soil depth (Kundu *et al.*, 2001). They also reported that a minimum of 888 kg C ha^{-1} yr^{1} is required to maintain the SOM content of the experimental soil at equilibrium level.

Choudhury *et al.* (2014) studied the influence of different combinations of tillage and residue management on C stabilization in different sized soil aggregates in a continuous rice-wheat cropping system on a sandy loam reclaimed sodic soil in Northern India. Mohanty *et al.* (2015) observed that the tillage reduction in association with residue retention increased the total SOC by approximately 20% over conventional systems in an alluvial soil of Odisha. Adoption of CA also increases different pools of soil N and subsequently total soil N (TSN) with a concomitant increase in total SOC, mostly in surface soils (Dikgwatlhe *et al.*, 2014; Sainju *et al.*, 2015; Dey *et al.*, 2016).Ghosh *et al.* (2016) reported that conservation measure with Palmarosa as vegetative barriers along with organic amendments (FYM + Vermi compost + Poultry manure) plus weed mulch application three times under minimum tillage is effective in decreasing runoff and soil erosion, increasing system productivity and improving soil quality. In an irrigated rice-wheat system of North Wastern Indo-Gangetic Plain, two-years of CA showed significant effect of ZT, greenmanuring (GM; Green gram residue retention) and brown manuring (BM; Dhaincha residue retention) on very labile SOC in surface layer (Dey *et al.*, 2016). CA based cropping systems improved soil properties and availability of nutrients (N, P, K, Zn, Fe and Mn) in surface soil layer compared to conventional farmer's practice. In addition to this, appreciable amount of N and K fertilizers to the tune of 30% and 50% can be saved under CA-based management system after 4 years of continuous cultivation as revealed through nutrient omission study in a reclaimed sodic soil in cereal-based systems of North-West India (Jat *et al.*, 2018). Adaptation of CA based practices (tillage, residue retention and inclusion of legumes in cropping system) increased the availability of macro and micronutrients (Amaresh Chaudhary *et al.*, 2019; Pheap *et al.*, 2019; Sithole and Magwaza, 2019). Nandan *et al.* (2019) reported that zero–till crop establishment practices (ZTTPR–ZT and ZTDSR–ZT) in rice–based systems had a positive impact on soil organic C–pools, macro–aggregate formation, and carbon stock in aggregates. Many literatures on CA recognized that CA systems is associated with greater acidity and greater decrease in pH was observed in observed in treatments with legume-based crop rotations and mineral N fertilization (Vieira *et al.*, 2009).CA also influenced cation exchange capacity and the change is variable (Sa *et al.*, 2009; Sithole and Magwaza, 2019; Williams *et al.*, 2018).

Long term CA resulted in microbial diversity (both bacterial and fungal) and had positive effects on microbial activities (Wang *et al.*, 2016). Reduced tillage resulted into higher numbers of nematodes and mites compared with conventional tillage and residue retention resulted in higher number of nematodes as compared to residue removal Zhang *et al.* (2012). CA significantly enhanced soil macro fauna such as earthworms, termites and beetles. Significant enhancement in earthworm activity and diversity under CA as compared to conventional systems (Kladivko 2001; Briones and Schmidt, 2017). Increase in the populations of arbuscular mycorrhizal fungi have been observed under NT which may improve P nutrition ((Mbuthia *et al.*, 2015). No-till also recorded higher soil microbial biomass, dehydrogenase and earthworm population which resulted in higher yield of crops under zero tillage. Conservation tillage recorded higher soil organic carbon in Vertisols of Central India, and had a positive effect on microbial activity, also enhancing biological health of the soil (Kumar *et al.*, 2018). Microbial biomass and soil enzymatic activities were enhanced by CA based management practices (Singh *et al.*, 2009). Choudhary *et al.*, 2018 concluded that the SQI was higher by 22% in ZT compared to CT, and 100% in residue recycling compared with residue removal. Conservation tillage treatments increased the stabilization of residue C–input compared to conventional CTTPR–CT. Zero tillage with residue retention under efficient crop sequence increases the soil microbial properties (Organic carbon, MBC, MBN, APA, DHA and β-glucosidase) and thus increases soil fungal diversity and richness (Choudhary *et al.*, 2018).

Conclusion

It is evident that CA not only improves SOC content of the soil but at the same time lead to improvement in physical, chemical and biological properties of soil, ensuring long term sustainability to the agro-ecosystem. In order to restore soil quality, enhance productivity, it is of utmost importance to focus on conservation agriculture practices on long-term basis.

References

Alam, M.K., Salahin, N., Islam, S., Begum, R.A., Hasanuzzaman, M., Islam, M.S., Rahman, M.M. (2017). Patterns of change in soil organic matter, physical properties and cropproductivity under tillage practices and cropping systems. Bangladesh. *Journal of Agricultural Science*, 155:216-238.

Alvaro-Fuentes, J.L., Gracia, A.R., Lo´ Pez, M.V. (2008). Tillage and cropping intensification effects on soil aggregation: temporal dynamics and controlling factors under semiarid conditions. *Geoderma*, 145: 390–396.

Bhattacharya, R., Prakash, V., Kundu, S. and Gupta, H.S. (2006). Effect of tillage and croprotations on pore size distribution and soil hydraulic conductivity in sandy clay loamsoil of the Indian Himalayas. *Soil and Tillage Research*, 86:129–140.

Bhattacharyya, R., Das, T.K., Pramanik, P., Ganeshan, V., Saad, A.A., Sharma, A.R., (2015). Conservation agriculture effects on soil organic carbon accumulation and crop productivity under a rice–wheat cropping system in the western Indo-Gangetic Plains. *European Journal of Agronomy*, 70: 11-21.

Bhattacharyya, R., Pandey, S.C., Bisht, J.K., Bhatt, J.C., Gupta, H.S., Titi, M.D., Mahanta, D., Mina, B.L., Singh, R.D., Chandra, S., Srivastva, A.K., Kundu, S. (2013). Tillage and irrigation effects on soil aggregation and carbon pools in the Indian sub-Himalayas. *Agronomy Journal*, 105:101–112.

Bhattacharyya, R., Tuti, M.D., Bisht, J.K., Bhatt, J.C., Gupta, H.S. (2012). Conservation tillage and fertilization impacts on soil aggregation and carbon pools in the Indian Himalayas under an irrigated rice–wheat rotation. *Soil Science*, 177:218–228.

Blanco-Canqui, H., Schlegel, A.J., and Heer, W.F. (2011). Soil-profile distribution of carbon and associated properties in no-till along a precipitation gradient in the central Great Plains. *Agriculture, Ecosystems & Environment*, 144, 107–116. doi: 10.1016/j.agee.2011.07.004

Briones, M.J.I. and Schmidt, O. (2017). Conventional tillage decreases the abundance and biomass of earthworms and alters their community structure in a global meta-analysis. *Global Change Biology*, 23: 4396–4419. doi: 10.1111/gcb.13744

Chander, K., Goyal, S., Mundra, M. C. and Kapoor, K. K. (1997). Organic matter, microbial biomass and enzyme activity of soils under different crop rotations in the tropics. *Biology and Fertility of Soils*, 24: 306-310.

Chaudhary, A., Meena, M.C., Dwivedi, B.S., Datta, S.P., Parihar, C.M., Dey, A., Sharma, V.K. (2019). Effect of conservation agriculture on soil fertility in maize (Zea mays)-based systems. *Indian Journal of Agricultural Sciences*, 89 (10): 1654-1659.

Choudhary, M., Datta, A., Jat, H.S., Yadav, A.K., Gathala, M.K., Sapkota, T.B. (2018). Changes in soil biology underconservation agriculture-based sustainable intensification of cereal systems in Indo-Gangetic Plains. *Geoderma*, 313:193-204.

Choudhary, M., Rana, K.S., Meena, M.C., Bana, R.S., Jakhar, P., Ghasal, P.C. and Verma, R.K. (2019). Changes in physico-chemical and biological properties of soil under conservation agriculture based pearl millet–mustard cropping system in rainfed semi-arid region. *Archives of Agronomy and Soil Science*, 65:7, 911-927, DOI: 10.1080/03650340.2018.1538556

Choudhary, M., Sharma, P.C., Sahay Jat, H., McDonald, A., Lal Jat, M., Choudhary, S., Garg, N. (2018). Soil biological properties and fungal diversity under conservation agriculture in Indo-Gangetic Plains of India. *Journal of Soil Science and Plant Nutrition*, 18 (4): 1142-1156.

Dey, A., Dwivedi, B.S., Bhattacharyya,R., Datta, S.P., Meena, M.C., Das, T.K.and Singh, V.K. (2016). Conservationagriculture in a rice-wheat croppingsystem on an alluvial soil of north-western Indo-Gangetic plains: Effecton soil carbon and nitrogen pools. *Journal of the Indian Society Soil Science*, 64: 246-254

Dikgwatlhe, S.B., Chen, Z., Lal, R., Zhang, H., Chen, F. (2014). Changes in soil organic carbon and nitrogen as affected by tillage and residue management under wheat–maize cropping system in the North China Plain. *Soil and Tillage Research*, 144: 110–118.

Dolan, M.S., Clapp, C.E., Allmaras, R.R., Baker, J.M., Molina, J.A.E. (2006). Soil organic carbon and nitrogen in a Minnesota soil as related to tillage, residue and nitrogen management. *Soil and Tillage Research*, 89(2): 221-231

Doran, J.W. (1980). Soil microbial and biochemical changes associated with reduced tillage. *Soil Science Society of America Journal*, 47: 102-107.

FAO (2019). Conservation Agriculture. Available online at: http://www.fao.org/conservation-agriculture/overview/what-is-conservation-agriculture/en/ (accessed August 2019).

Franzluebbers, A.J., Hons, F.M. and Zuberer, D.A. (1995). Soil organic carbon, microbial biomass, and mineralizable carbon and nitrogen in sorghum. *Soil Science Society of America Journal*, 59(2): 460-466.

Gathala, M.K., Ladha, J.K., Saharawat, Y.S., Kumar, V., Kumar, V., Sharma, P.K., (2011). Effect of tillage and crop establishment methods on physical properties of a medium textured soil under a seven-year rice–wheat rotation. *Soil Science Society of America Journal*, 75: 1851–1862.

Ghosh, B.N., Meena, V.S., Alam, N.M., Pradeep Dogra, Bhattacharyya, Ranjan, Sharma, N.K., Mishra, P.K. (2016). Impact of conservation practices on soil aggregation and the carbon management index after seven years of maize–wheat cropping system in the Indian Himalayas Agriculture. *Ecosystems and Environment*, 216: 247–257.

Govaerts, B. Conservation agriculture as a sustainable option for the central Mexican high lands (2009). *Soil and Tillage Research*, 103: 222–230.

Gupta Choudhury, S., Srivastava, S., Singh, R., Chaudhari, S.K., Sharma, D.K., Singh, S.K., Sarkar, D. (2014). Tillage and residue management effects on soil aggregation, organic carbon dynamics and yield attribute in rice-wheat cropping system under reclaimed sodic soil. *Soil and Tillage Research*, 136: 76-83.

Hobbs, P.R., Sayre, K., Gupta, R. (2008). The role of conservation agriculture in sustainable agriculture. *Philosophical Transactions of the Royal Society B: Biological Sciences*, 363 (1491): 543-555. doi:10.1098/rstb.2007.2169

Indoria, A.K., Srinivasa Rao, Ch., Sharma, K.L., Reddy, K.S. (2017). Conservation agriculture – a panacea to improve soil physical health. *Current Science*, 112: 52–61.

Jantalia, C.P., Resck, D.V.S., Alves, B.J.R., Zotarelli, L., Urquiaga, S. and Boddey, R.M. (2007). Tillage effect on C stocks of a clayey Oxisol under a soybean-based crop rotation in the Brazilian Cerrado region. *Current Science*, 95: 97–109.

Jat, H.S., Datta, A., Choudhary, M., Yadav, A.K., Choudhary, V., Sharma, P.C., Gathala, M.K., Jat, M.L., McDonald, A. (2019). Effects of tillage, crop establishment and diversification on soil organic carbon, aggregation, aggregate associated carbon and productivity in cereal systems of semi-arid Northwest India. *Soil Tillage Res.*, 190:128-138. doi: 10.1016/j.still.2019.03.005.

Jat, H.S., Datta, A., Sharma, P.C., Kumar, V., Yadav, A.K., Choudhary, M., Choudhary, V., Gathala, M.K., Sharma, D.K., Jat, M.L., Yaduvanshi, N., Singh, G. and McDonald, A. (2018). Assessing soil properties and nutrient availability under conservation agriculture practices in a reclaimed sodic soil in cereal-based systems of North-West India. *Archives of Agronomy and Soil Science*, 64(4): 531–545.

Jat, M.L., Gathala, M.K., Ladha, J.K., Saharawat, Y.S., Jat, A.S., Kumar, V., Sharma, S.K., Gupta, R. (2009). Evaluation of precision land leveling and double zero-till systems in the rice-wheat rotation: Water use, productivity, profitability and soil physical properties. *Soil and Tillage Research*, 105 (1): 112-121

Juo, A.S.R., Franzluebbers, K., Dabiri, A. and Ikhile, B. (1996). Soil properties and crop performance on a Kaolinitic Alfisol after 15 years of fallow and continuous cultivation. *Plant and Soil*, 180: 209-217.

Karunakaran, V., Behera, U.K. (2015). Influence of sequential tillage and residue managementpractices on soil and root parameters in soybean (Glycine max)-wheat (Triticumaestivum) cropping system. *Indian Journal of Agricultural Sciences*, 85:182-188.

Kladivko, E.J. (2001). Tillage systems and soil ecology. *Soil and Tillage Research*, 61(1): 61-76.

Kumar, A., Mishra, V.N., Biswas, A.K., Somasundaram, J. (2018). Soil organic carbon, dehydrogenase activity, and fluorescein diacetate as influenced by contrasting tillage and cropping systems in Vertisols of Central India. *Journal of Environmental Biology*, 39: 1047-1053.

Kumari, M., Chakraborty, D., Gathala, M.K., Pathak, H., Dwivedi, B.S., Tomar, R.K., Garg,R.N., Singh, R., Ladha, J.K. (2011). Soil aggregation and associated organic carbonfractions as affected by tillage in a rice–wheat rotation in North India. *Soil Science Society of America Journal*, 75(2): 560–567.

Kumawat, C., Sharma, V.K., Meena, M.C., Dwivedi, B., Barman, M., Kumar, S., Chobhe, K.A. and Dey, A. (2018). Effect of crop residue retention and phosphorus fertilization on P use efficiency of maize (Zea mays) and biological properties of soil under maize-wheat (Triticumaestivum) cropping system in an Inceptisol. *Indian Journal of Agricultural Sciences*, 88: 1184-1189.

Kuotsu, K., Das, A., Lal, R., Munda, G.C., Ghosh, P.K., Ngachan, S.V. (2014). Land forming and tillageeffects on soil properties and productivity of rainfed groundnut (*Arachishypogaea* L.)–rapeseed (*Brassica campestris* L.) cropping system in northeastern India. *Soil and Tillage Research,* 142:15–24.

Lal, R (2004). Soil carbon sequestration in India Climate Change. 65: 277-296.

Lal, R. (2015). Sequestering carbon and increasing productivity by conservation agriculture. *Journal of Soil and Water Conservation,* 70, 55A–62A. doi: 10.2489/jswc.70.3.55A

Li, Y., Li, Z., Cui, S., Jagadamma, S., and Zhang, Q. P. (2019b). Residue retention and minimum tillage improve physical environment of the soil in croplands: a global meta-analysis. *Soil and Tillage Research,* 194: 104292.

Liebig, M.A., Varvel, G.E., Doran, J.W. and Wienhold, B.J. (2002). Crop sequence and nitrogen fertilization effects on soil properties in the western Corn Belt. *Soil Science Society of America,* https://doi.org/10.2136/sssaj2002.5960

Liu, E., Teclemariam, S.G., Yan, C., Yu, J., Gu, R., Liu, S. and Liu, Q. (2014). Long-term effects of no-tillag management practice on soil organic carbon and its fractions in the northern China. *Geoderma,* 213: 379-384.

Mazumdar, S.P., Kundu, D.K., Saha, A.R., Majumdar, B., Saha, R., Singh, A.K, Barman, D. Ghosh, D., Dey R., Behera, M.S., Mitra, S. & Bhattacharyya, P. (2018). Carbon and nutrient dynamics under long-term nutrient management in tropical rice-wheat-jute system. *Archives of Agronomy and Soil Science,* 64(11): 1595-1607.

Mbuthia, L.W., Acosta-Martínez, V., DeBruyn, J., Schaeffer, S., Tyler, D., Odoi, E., *et al.* (2015). Long term tillage, cover crop, and fertilization effects on microbial community structure, activity: Implications for soil quality. *Soil Biology and Biochemistry,* 89: 24–34. doi: 10.1016/j.soilbio.2015.06.016

Mina, B.L., Saha, S., Kumar, N., Srivastva, A.K.., Gupta, H.S. (2008). Changes in soil nutrient content and enzymatic activity under conventional and zero-tillage practices in an Indian sandy clay loam soil. *Nutrient Cycling in Agroecosystems,* 82:273–281.

Modak, K, Biswas,D.R., Ghosh ,A,Pramanik,P, Das,TK, Das,S, Kumar, S, Krishnan, P., and Bhattacharyya, R. (2020). Zero tillage and residue retention impact on soil aggregation and carbon stabilization within aggregates in subtropical India. *Soil and Tillage Research,* 202:104649

Mohanty, A., Mishra, K.N., Roul, Dash P.K., S.N. and Panigrahi, K.K. (2015). Effects of conservation agriculture production system (CAPS) on soil organic carbon, base exchange characteristics and nutrient distribution in a tropical rain fed agroecosystem. *International Journal of Plant, Animal and Environmental Sciences,* 5:310-314

Mohanty, M., Painuli, D.K., Misra, A.K. and Ghosh, P.K. (2007). Soil quality effects of tillage and residue under rice–wheat cropping on a Vertisol in India. *Soil & Tillage Research,* 92: 243–250

Nandan, R., Singh, V., Singh, S. S., Kumar, V., Hazra, K. K., Nath, C. P., Poonia, S., Malik, R. K., Bhattacharyya, R., & McDonald, A. (2019). Impact of conservation tillage in rice-based cropping systems on soil aggregation, carbon pools and nutrients. *Geoderma,* 340, 104–114. https://doi.org/10.1016/j.geoderma.2019.01.001

Nayak, A.K., Gangwar, B., Shukla, A.K., Mazumdar, S.P., Kumar, A., Raja, R., Kumar, A., Kumar, V., Rai, P.K. & Mohan, U. 2012. Long-term effect of different integrated nutrient management on soil organic carbon and its fractions and sustainability of rice–wheat system in Indo Gangetic Plains of India. *Field Crops Research,* 127, 129– 139.

Pacala, S. and Socolow, R. (2004). Stabilization wedges: solving the climate problem for the next 50 years with current technologies. *Science*, 305: 968–972.

Page, K.L., Dang, Y.P., Dalal, R.C., Reeves, S., Thomas, G., Wang, W., *et al.* (2019). Changes in soil water storage with no-tillage and crop residue retention on a Vertisol: Impact on productivity and profitability over a 50-year period. *Soil and Tillage Research*, 194:104319.

Parihar, C.M., Singh, A.K., Jat, S.L., Dey, A., Nayak, H.S., Mandal, B.N., Saharawat, Y.S., Jat, M.L., Yadav, O.P. (2020). Soil quality and carbon sequestration under conservation agriculture with balanced nutrition in intensive cereal-based system. *Soil and Tillage Research*, 202: 104653

Parihar, C.M., Yadav M R, Jat S L, Singh A K, Kumar B, Pradhan S, Chakraborty D, Jat M L, Jat R K, Saharawat Y S, Yadav O P. (2016). Long term effect of conservation agriculture in maize rotations on total organic carbon, physical andbiological properties of a sandy loam soil in north-westernIndo-Gangetic Plains. *Soil and Tillage Research*, 161: 116–128.

Paustian, K., Andrén, O., Janzen, H., Lal, R., Smith, P., Tian, G., Tiessen, H., van Noordwijk, M. and Woomer, P. (1997). Agricultural soil as a C sink to offset CO_2 emissions. *Soil Use Manage*, 13: 230–244.

Pheap, S., Lefevre, C., Thoumazeau, A., Leng, V., Boulakia, S., Koy, R., *et al.* (2019). Multifunctional assessment of soil health under Conservation Agriculture in Cambodia. *Soil Soil and Tillage Research*, 194: 104349.

Purakayastha, T.J., Huggins, D.R., Smith, J.L. (2008). Carbon sequestration under native Prairie, conservation reserve and no tillage in Paluose region. *Soil Science Society of America Journal*, 72: 534-540.

Sa, J.C.D., Cerri, C.C., Lal, R., Dick, W.A., Piccolo, M.D., and Feigl, B.E. (2009). Soil organic carbon and fertility interactions affected by a tillage chronosequence in a Brazilian Oxisol. *Soil and Tillage Research*, 104: 56–64.

Sainju, U.M., Allen, B.A., CaesarTonThat, T. and Lenssen, A.W. (2015). Dryland soil carbon and nitrogenafter thirty years of tillage andcropping sequence combination. *Agronomy Journal*, 107: 1822-1830.

Sainju, U.M., Singh, B.P. and Whitehead, W.F. (2002). Long-term effects of tillage, cover crops and nitrogen fertilization on organic carbon and nitrogen concentrations in sandy loam soils in Georgia, USA. *Soil and Tillage Research*, 63: 167-179.

Salem, H.M., Valero, C., Muñoz, M., Rodríguez, M.G, Silva, L.L. (2015). "Short-term effects of four tillage practices on soil physical properties, soil water potential, and maize yield", *Geoderma*, 237-238: 60-70.

Sandeep, S., Manjaiah, K.M., Pal. S., Singh. A.K. (2016). Soil carbon fractions under maize-wheat system: effect of tillage and nutrient management. *Environmental Monitoring and Assessment*, 188(1): 14.

Singh, G., Marwaha, T.S. and Kumar, D. (2009). Effect of resource-conserving techniques on soil microbiological parameters under long-term maize (Zea mays)–wheat (Triticum aestivum) crop rotation. *Indian Journal of Agricultural Research*, 79(2): 94–100.

Sithole, N. J., and Magwaza, L. S. (2019). Long-term changes of soil chemical characteristics and maize yield in no-till conservation agriculture in a semi-arid environment of South Africa. *Soil and Tillage Research*, 194: 104317.

Six, J., Elliott, E.T., Paustian, K., (2000). Soil macroaggregate turnover and microaggregateformation: a mechanism for C sequestration under no–tillage agriculture. *Soil Biology and Biochemistry*, 32: 2099–2103.

Somasundaram, J., Salikram, M., Sinha, N.K., Mohanty, M., Chaudhary, R.S., Dalal, R.C., *et al.* (2019). Conservation agriculture effects on soil properties and crop productivity in a semiarid region of India. *Soil Research*, 57: 187–199.

Srinivasarao, C., Venkateswarlu, B., Lal, R., Singh, A.K., Kundu, S. (2013). Sustainable management of soils of dryland ecosystems of India for enhancing agronomic productivity and sequestering carbon. *Advances in Agronomy,* 121:253-325.

Thierfelder C, Wall, P.C. (2009). Effects of conservation agriculture techniques on infiltration and soil water content in Zambia and Zimbabwe. *Agronomy for Sustainable Development,* 105: 217–227.

Thierfelder, C., Baudron, F., Setimela, P., Nyagumbo, I., Mupangwa, W., Mhlanga, B., (2018). Complementary practices supporting conservation agriculture in southern Africa. A review. *Agronomy for Sustainable Development,* 38:16.

Thierfelder, C., Cheesman, S., Rusinamhodzi, L. (2012). A comparative analysis of conservation agriculture systems:Benefits and challenges of rotations and intercropping in Zimbabwe. *Field Crops Research,* 137 (20): 237-350

Thomas, G.A., Dalal, R.C., and Standley, J. (2007). No-till effects on organic matter, pH, cation exchange capacity and nutrient distribution in a Luvisol in the semi-arid subtropics. *Soil and Tillage Research,* 94: 295–304.

Utomo, M., Banuwa, I.S., Buchari, H., Anggraini, Y., Berthiria (2013). Long-term tillage and nitrogen fertilization effects on soil properties and crop yields. *Journal of Tropical Soils,* 18(20): 21–30.

Varvel, G.E., Wilhelm, W.W. (2011). No-tillage increases soil profile carbon and nitrogen under long-term rainfed cropping systems. *Soil and Tillage Research.* 114: 28-36.

Verhulst, N., Nelissen, V., Jespers, N., Haven, H., Sayre, K. D., Raes, D., *et al.* (2011). Soil water content, maize yield and its stability as affected by tillage and crop residue management in rainfed semi-arid highlands. *Plant Soil,* 344: 73–85.

Verma, S. and Sharma, P.K. (2007). Effect of long-term manuring and fertilizers on carbon pools, soil structure, and sustainability under different cropping systems in wet-temperate zone of northwest Himalayas. *Biology and Fertility of Soils,* 44: 235–240.

Vieira, F.C.B., Bayer, C., Zanatta, J. and Ernani, P.R. (2009). Organic matter kept Al toxicity low in a subtropical no-tillage soil under long-term (21-year) legume-based crop systems and N fertilisation. *Australian Journal of Soil Research,* 47: 707–714.

Wang, Z., Chen, Q., Liu, L., Wen, X., and Liao, Y. (2016). Responses of soil fungi to 5-year conservation tillage treatments in the drylands of northern China. *Applied Soil Ecology,* 101: 132–140.

West, T.O. and Post, W.M. (2002). Soil organic carbon sequestration rates by tillage and crop rotation: a global data analysis. *Soil Science Society of America Journal,* 66: 1930–1946.

Williams, A., Jordan, N.R., Smith, R.G., Hunter, M.C., Kammerer, M., Kane, D. A., et al. (2018). A regionally-adapted implementation of conservation agriculture delivers rapid improvements to soil properties associated with crop yield stability. *Scientific Reports,* 8: 8467.

Wright, A.L. and Hons, F.M. (2005). Tillage impacts on soil aggregation and carbon and nitrogen sequestration under wheat cropping sequences. *Soil and Tillage Research,* 84(1): 67–75.

Zhang, X., Li, Q., Zhu, A., Liang, W., Zhang, J. and Steinberger, Y. (2012). Effects of tillage and residue management on soil nematode communities in North China. *Soil and Tillage Research,* 13(1): 75-81.

22

Impact of Conservation Agriculture Practices on Soil Microbial Diversity

B. Majumdar, S. Sarkar, Lipi Chattopadhyay and Shrestha Barai

ICAR-Central Research Institute for Jute and Allied Fibres Barrackpore, West Bengal – 700 121, India

Introduction

The main aim of conservation agriculture is to sustain and improve the crop productivity and to provide protection against biotic and abiotic stress, while at the same time protecting and enhancing the biological activities of the soil. The essential principles of conservation agriculture are no-tillage (and direct seeding) or reduced tillage, the maintenance of a cover of live or dead vegetative mulch on the soil surface and the wise use of crop rotations. The crop sequences are planned in such a way that which will discourage the build-up of pests or diseases and will help to optimize plant nutrient use by synergy between different crops of the rotation. While deciding the crop rotation, the locally important crops should be considered and there should be deep rooted and shallow rooted crops in the sequence so that the utilization of soil nutrients is maximized. At least one leguminous crop should be there in the crop sequence. The use of plant residues influences the soil physical, chemical and biological properties by reducing surface temperature, rate of evaporation and maintaining water content, nutrient load and rate of organic matter decay. The soils under conservation agriculture are happened to be highly active and diverse biologically and can supply plant nutrients for a longer period of time because of their higher nutrient loading capacities.

Soil biodiversity comprises of all organisms whether single-cell organisms or multi-cell animals or plants that live in the soil. Soil microbial diversity includes "genetic diversity, that is the amount and distribution of genetic information within microbial species, diversity of bacterial and fungal species in microbial communities, and ecological diversity, that is variation in community structure, complexity of interactions, number of trophic levels and number of guilds (Nannipieri *et al.*, 2003).

Soil biodiversity is the driver of healthy soil for sustainable crop production in conventional as well as in conservation agriculture system. The importance of soil microbial diversity felt because of ill effect of conventional agriculture system using deep tillage with crop residue removal causing soil erosion, degradation of soil structure, loss of soil organic matter and plant nutrients and non- sustainability of crop yield. The conservation agriculture came into existence to minimize the ill effects of conventional agriculture by maintaining soil microbial diversity which takes care of soil physical properties, soil structure and nutrient supplying capacity on a sustainable basis. While studying the microbial diversity under conservation agriculture, we should also study the abundance of plant pathogens build up in the soil, which should not cross the threshold limit; otherwise the whole benefits of conservation agriculture may be questionable in long run. Crop residue management is a burning issue in the states like Punjab, Haryana, Uttar Pradesh etc.; where burning of crop residues not only polluting the environment, but affecting the beneficial soil microbes to a great extent. So, conservation agriculture can play a great role by using crop residue in a systematic way without polluting the environment, side by side enhancing soil fertility and microbial diversity of the cultivated soil. The measurement of soil microbial biodiversity in a particular cropping system or under different land use is more important for the knowledge of soil health and to detect the loss of soil biodiversity. Use of appropriate methodology and tools are more important in the measurement of soil microbial biodiversity.

There are several ways to measure soil microbial biodiversity like

- Enumeration of microbial population (total bacterial, fungal, actinomycetes etc. and specific microbes like N fixing microbes, P solubilizers etc.)

- Measurement of soil enzymatic activities (dehydrogenase, fluroscein dehydrate hydrolyzing capacity (FDHA), β-glucosidase, urease, phosphatase, caboxymethylcellulase, xylanase, pectinase etc.)

- Measurement of carbon dioxide evolution rate and basic soil respiration rate (BSRR)

- Measurement of microbial biomass C, N, P and S through fumigation technique

- Use of 'Biolog' to observe microbial C substrate utilization pattern

- Assessment of multitude of aerobic and anaerobic microbes in soil system through metagenomics approach

- Taxonomic and functional characterization of ^{13}C-DNA using ^{13}C source

Soil enzymes and microbial biomass have been considered as major indicators of soil quality due to their relationship to soil fertility and high sensitivity to changes originated by the management and environmental factors (Diosma *et al.*, 2006). Basal soil respiration rate is the most important tool for assessment of side effects of chemicals such as heavy metals, pesticides etc. (Alef, 1995). On the other hand, the FDHA is considered as a suitable tool for measuring the early detrimental effect of pesticides on soil microbial biomass, as it is a sensitive and non-specific test able to depict the hydrolytic activities of soil microbes. It also depicts the contribution of several enzymes, mainly involved in the decomposition of organic matter in soil. Dehydrogenase is an oxidoreductase enzyme present in all viable microbial cell, which is considered as a sensitive indicator of soil quality (Nannipieri, 1994) and a valid biomarker to indicate changes in total microbial activity due to changes in soil management practices (Roldan *et al.*, 2004). The β-glucosidase enzyme is an important indicator of the ability of a given soil ecosystem to degrade crop residues as it provides simple sugars for the heterotrophic microbial population (Stott *et al.*, 2010). The Carboxymethylcellulase, hemicellulases/xylanases are required for biodegradation of celulose, hemicellulose and xylan present in plant materials. Acid and alkaline phosphatases play a critical role in P-cycle of soil ecosystem and hence, apart from being good indicators of soil fertility, also play key role in soil system (Makoi and Ndakidemi, 2008).

The metabolic characteristics of soil microbial communities can be measured using ECO Biolog system microplates (Biolog Inc. Hayward, CA), which were used to determine the C source utilization pattern. The 96-well Eco microplate comprised three replicate wells; each replicate comprised 31 C substrates and a control well without C substrates. An average well color development (AWCD) can be calculated to determine the rate of color development on Biolog plates for each plate at each reading. Data at 96 h are used to calculate AWCD. Shannon index can be calculated to evaluate microbial metabolic diversity. This index is calculated as follows: $H = -\Sigma[pi \times ln(pi)]$, where pi is the ratio of activities on each substrate to the sum of activities on all substrates.

Importance of Soil Microbial Biodiversity in Agriculture

Soil is a home for various microorganisms like protozoa, algae, bacteria, fungi, actinomycetes etc. These microorganisms play a significant role by their active role in organic material decomposition, breakdown of toxic compounds, inorganic transformations, nitrogen fixation, as rhizobacteria or plant growth promoting rhizobacteria and as bio-control agents in plant protection. All these microbial activities are useful and helpful for continuation of agricultural activities in the soil. Microbes play an active role in soil fertility as a result of their involvement in the cycle of nutrients like carbon and nitrogen, which are required for plant growth. The major function of microbes in the carbon cycle is as decomposers of complex organic molecules like cellulose, hemicelluloses, chitin, lignin etc. that would otherwise permanently sequester carbon, keeping it from being useful to organisms. The biological nitrogen fixation by symbiotic, non-symbiotic and associative symbiotic soil microbes and the roles of microbes in nitrogen mineralization process including ammonification and nitrification are very important from the soil fertility point of view.

Soil microorganisms are also responsible for the decomposition of the organic matter entering the soil and therefore for the recycling of nutrients in soil. Another important role played by soil microbes is the breakdown of toxic compounds including both metabolic by-products of organisms and agrochemicals used for weed and pest control purposes. Soil microorganisms improve the entry and storage of water, resistance to erosion, plant nutrition, and breakdown of organic matter. Soil microorganisms are not only responsible for enhancement of crop productivity but also play beneficial role in crop protection and minimization of environmental pollution. The help rendered by soil microorganisms can be summarized in following three kinds of benefits.

- **Economic benefits**: Soil microorganisms reduce input costs by enhancing resource use efficiency. The value of 'ecosystem services' (organic waste disposal, soil formation, bioremediation, N_2 fixation and bio control) provided each year by soil biota in agricultural systems worldwide may exceed US$ 1,542 billion (Pimental *et. al.*, 1997).

- **Environmental protection**: Soil receives lots of agro-chemicals like herbicides, pesticides applied during crop production. Besides that, lots of other inorganic and organic wastes discharged from factories also reached to the soil as pollutants. Soil organisms played a great role in filtering and detoxifying these agro-chemicals and pollutants from factory waste and also absorb the excess nutrients that would otherwise become pollutants when they reach groundwater or surface water.

- **Food security**: Soil biological management can improve crop yield and quality, especially through controlling pests and diseases (bio-control agents) and enhancing crop growth (plant growth promoting rhizobacteria).

Effect of Conservation Agriculture on Soil Microbial Diversity

Soil microbial community structure in conservation agriculture

The changes in the composition of microbial community in the soil under conservation agriculture have been reported by several researchers. Wang *et al.* (2014) found higher soil bacterial abundance and diversity under no tillage

(NT) and residue incorporation treatments than under conventional tillage (CT), and suggested NT plus 100% crop residue incorporation was the best agricultural strategy for improving soil microbial communities. In a recent study, the effect of no tillage and incorporation of maize straw as mulching at different rates was studied after 10 years of maize crop on microbial diversity and abundance of two pathogenic fungi (*Fusarium graminearum* and *Fusarium moniliforme*) in northeast China. The results suggested that, no tillage and residue incorporation enhance the fungal biodiversity but had little effect on bacterial diversity and encourages the dominance of two pathogenic fungi *F. graminearum* and *F. moniliforme* which may increase the incidence of root rot of maize in near future (Wang *et al.*, 2020). From this study, it was clear that although NT and residue management is beneficial for several soil properties but proper amount of residue for field application should be decided to avoid abundance of pathogens.

In a very significant study, Wachira *et al.* (2009) reported that a general build-up of soil pathogens (*Pythium, Fusarium* and *Rhizoctonia*) and a decline in beneficial microorganisms, like entomopathogenic nematodes except for nematode-destroying fungi, was observed with increase in land intensification in Kenya. The ratio of plant parasitic nematodes to free living nematodes under vegetable land use (0.21) was lower than the ratio recorded in natural forest, which was 4.01 (Table 1). Comparing land uses under crops, it was observed that land uses with perennial crops like coffee recorded high population and diversity of soil microorganisms compared to land uses under annual crops and the undisturbed land uses had higher soil biodiversity both in abundance and diversity compared to the disturbed soil.

Table 1: Comparison of PPN, FLN, and ratio of FLN: PPN in soils under different uses in Taita Taveta, Kenya

Land Use	PPN (Plant parasitic nematodes)	FLN (Free-living nematodes)	FLN:PPN
Natural forest	189	759	4.01
Planted forest	353	526	1.49
Fallow	796	458	0.57
Coffee	948	419	0.44
Maize	938	120	0.13
Vegetable	915	191	0.21
LSD<P 0.05	122	137	

Source: Wachira *et al.* (2009)

Organically managed soils can harbor microbial communities that are distinct in structure and function compared to conventionally managed soils, fungi and actinobacteria were more dominant utilizers of crop residue derived –C during early and late stages of decomposition in organically managed soils, and microbial activity and abundance was more responsive to residue additions (Arcand *et al.*, 2016). In a 30-year long fertility experiment, continuous addition of maize straw @ 4.5 to 9 t/ha changed the microbial community structure and increased the activities of β-glucosidase, β-xylosidase and N-acetyl-glucosaminidase in soil over no addition of maize straw (Zhao *et al.*, 2016).

The effect of conservation tillage (zero tillage/no tillage) is not visible significantly on soil microbial diversity or their activity as compared with conventional tillage practices at very early stage of cultivation. In a significant study by Diosma *et al.* (2003) showed that the zero tillage practice did not produce any significant changes in the activity of the soil microorganisms (soil microbial biomass, dehydrogenase activity) studied detectable early on compared with the conventional tillage practice. The lack of significant biomass response to tillage practices was reflected by the wheat biomass and grain yield and which was same under conventional and conservation tillage practices.

Microbial community structure differed between cultivated and uncultivated soil, and it has been reported by several researchers (Buckley and Schmidt 2001, Bossio *et al.*, 2005, Chaer *et al.*, 2014) depending on the terminal restriction fragment length polymorphism (T-RFLP) profiles, community level physiological profiles (CLPPs) and phospholipid fatty acid (PLFA)-based assessment of microbial community structure. Conservation tillage found to increase significantly microbial substrate utilization (including amino acids, carboxylic acids, polymers, phenolic compounds, and carbohydrates) pattern of >0.25 and <0.25 mm aggregates assesses by using Biolog. Higher Shannon index was also observed in >0.25 and <0.25 mm aggregates in the upper soil layer under conservation tillage (Guo *et al.*, 2016).

Effect of crop residue addition on soil microbial diversity

Crop residue burning is a vital issue of environmental pollution in recent years particularly in the Indian states like Punjab, Haryana, Uttar Pradesh and to a limited extent in the East Bardhhaman and Birbhum districts of West Bengal. Crop residue management plays a great role in maintenance of soil microbial biodiversity. Burning not only encourages air pollution but also accelerates the loss of soil biodiversity to a large extent.

In a field study, Gaind and Nain (2011) measured the effect of bio-augmented paddy straw compost on wheat yield and various soil properties including microbial diversity in soil at IARI, New Delhi and concluded that inoculation of paddy straw @ 3 t/ha with cellulolytic and lignolytic fungi (*Aspergillus awamori* (F 18), *Trichoderma reesei* (MTCC 164) and *Phanerochaete chrysosporium* (MTCC 787) enhanced the activities of native and added microorganisms as indicated by increase C, N and humus content of soil under wheat crop. They also reported that the in-situ incorporation of paddy straw amended with $N_{60} P_{60}$ was at par with fungal inoculated paddy straw treatments in increase of dehydrogenase, alkaline phosphatase, cellulase, cellobiase and urease activity in soil. This residue management technology may prove effective in protecting the soil from further impoverishment and can provide the farmer with an option of low cost, eco friendly technology for effective utilization of their farm waste.

In a recent study at IARI, New Delhi (Table 2), addition of paddy straw in combination with microbial priming with two promising fungus isolates in a wheat growing soil helps in increase in alkaline phosphatise activity and carbon dioxide evolution rate, i.e., increase in microbial activity and thereby decomposition of paddy straw compared with paddy straw removal and only paddy straw addition. Microbial priming could accelerate the decomposition process of straw which is visible from the higher alkaline phosphatise activity and CO_2 evolution rate at different time intervals compared with only straw retained and straw removal treatments, which can be a breakthrough for management of straw in near future instead of burning of straw. The addition of only straw takes a longer time for acceleration of decomposition process, hence straw addition with proper microbial priming can help in the faster decomposition of the added biomass (Kumar *et al.*, 2019).

Table 2: Effect of addition and removal of paddy straw in wheat growing soil

Treatments	Alkaline phosphatase activity (µg PNP/g/h)				Carbon dioxide evolution (mg CO_2 evolved/100 g soil/day)			
	Days							
	7	15	30	60	7	15	30	60
Straw removed	164.12	169.94	145.76	69.44	4.03	6.23	6.60	7.70
Straw retained	171.13	221.36	156.33	74.01	4.40	6.31	6.97	9.17
Straw retained + microbial priming[*]	179.04	242.98	225.37	73.84	5.20	11.73	20.90	25.50
CD (P=0.05)	12.5	21.96	50.93	NS	NS	1.19	1.62	1.86

[*]*Coprinopsis cinerla LA2 + Cyanthus stercoreus ITCC 3745*

Source: Kumar *et al.* (2019)

Higher soil fertility in terms of soil microbial activity (soil respiration, soil microbial biomass carbon and soil dehydrogenase activity) in conservation tillage practice (zero tillage) was reported compared to the conventional tillage practices receiving the same irrigation and nutrients addition (only organic or inorganic + organic sources) in a two years field experiment in wheat crop.

This study revealed that microbial activity could be regulated by no tillage, water and nitrogen management in the soil in a sustainable manner to save energy, water, and cost of cultivation and to maintain high productivity of crops in an ecologically balanced manner (Sharma *et al.*, 2011).

Effect of agro-chemicals on soil microbial biodiversity

The factors behind the loss of soil microbial biodiversity are: population increase, national food insufficiency, internal food production imbalances, progressive urbanization, and a growing shortage of land suitable for conversion to agriculture. Other practices that lead to loss of biodiversity are continuous cultivation of land without even a brief period of rest, monoculture, removal of crop residues by burning or transfer for use as fodder, soil erosion, soil compaction due to degradation of the soil structure, and repeated application of pesticides.

In a three years field experiment under jute crop, seed treatments with carbendazim and mancozeb, the dehydrogenase and fluorescein diacetate hydrolysing activities in jute soil were adversely affected even at a lower dose and both the enzymatic activities (Table 3) could not cope up with control and two hand weeding even at harvest after 120 days of pesticide application (Majumdar *et al.*, 2010). The application of herbicides also affected the soil enzymatic activities drastically which replenished almost to the initial status at the time of harvest of jute crop. The microbial biomass carbon and basic soil respiration rate content in herbicides and fungicides treated plots decreased initially and started recovering after 15 days of their application and recovered almost to the extent of respective initial level at harvest of jute crop.

Table 3: Loss of enzymatic activities as a result of herbicide and fungicide application in a jute soil.

Treatment	Dehydrogenase (mg tpf/kg soil)				FDHA (mg fluoroscein/kg soil)			
	*1	2	3	4	1	2	3	4
Control	6.50	6.60	6.60	6.65	105.2	106.2	110.5	115.6
Two hand weeding	6.55	6.60	6.65	6.75	104.0	106.3	115.0	119.5
Carbendazim as seed treatment + 2 hand weeding	0.90	3.00	5.10	5.60	65.0	85.5	95.7	102.8
Mancozeb as seed treatment + 2 hand weeding	0.95	2.90	5.15	5.55	63.5	83.5	92.7	103.1
Trifluralin @1.5 kg a.i./ha	1.00	2.65	5.00	5.45	54.4	72.0	90.8	102.0
Fluchloralin @1.5 kg a.i./ha	0.85	2.50	4.85	5.30	52.0	68.5	85.9	100.7
CD (P=0.05)	0.86	0.88	0.73	0.57	7.40	5.90	6.80	7.20

1, 2, 3 & 4 respectively after 7, 15, and 30 and at harvest (120 days)

Source: Majumdar *et al.* (2010)

Application of different herbicides molecules recoded a reduction in the total microbial population within 7 to 30 days of their application in the soil, and also affected soil enzymatic activity, cellular membrane composition, protein biosynthesis, and the amount of plant growth regulators and the symbiotic N fixation by *Rhizobium* (Milosevic and Govedarica, 2002). Herbicides application also found to adversely affect the microbial biodiversity indirectly by altering the physiology or biosynthetic mechanisms (Kremer and Means, 2009).

The population of N fixing bacteria and the biological N fixation by them are affected severely by most copper (Cu) based fungicides and fungicidal residues of apron, arrest, captan, tendto (Van-Zwieten *et al.*, 2003; Kyei-Boahen *et al.*, 2001). Fungicides mancozeb and chlorothalonil were found to decrease the process of nitrification and denitrification at an incubation period of ≥ 48 h (Kinney *et al.*, 2005) and the soil enzymatic activity of dehydrogenase, urease, and phosphatase were inhibited by fungicides benomyl, mancozeb, and tridemorph (Shukla, 2000). Insecticides were found to have variable adverse impact on soil microbes related to N transformation in soils depending on the type of insecticides used (Das and Mukherjee, 2000). The important biochemical process of nitrification and denitrification are reduced in soils contaminated with insecticides like monocrotophos, lindane, dichlorvos, endosulfan, malathion, and chlorpyrifos with lower to very higher doses of their application (Madhaiyan *et al.*, 2006).

To overcome the harmful effects of herbicides, insecticides and fungicides on soil microbial biodiversity in conservation agriculture, different management practices like use of biopesticides viz. *Bacillus thuringiensis* (Bt), *Baculoviruses, Trichoderma, Azadirachta indica*, and use of plant growth-promoting rhizobacteria (PGPR) like *Pseudomonas fluorescence* and use of transgenic herbicides resistant crops are suggested (Meena *et al.*, 2020).

Effect of INM and IPM on soil microbial biodiversity

Integrated nutrient management (INM) and integrated pest management (IPM) are a part of conservation agriculture. Continuous application of only inorganic fertilizer or pesticides excluding bio-control measures in conservation agriculture may spoil the beneficial effects of conservation agriculture in long run. In a field study under jute crop, after three years of experimentation, the population of beneficial microbes and enzymatic activities, viz. dehydrogenase, urease, fluorescein diacetate hydrolyzing activity, SMBC and BSRR (Table 4) in jute rhizosphere were higher in 100% NPK + 10t FYM/ha treatment compared with 100% and 150% NPK. So, integration of recommended dose of fertilizer with 10 t FYM/ha proved to be the best possible option not only for jute fibre production but also for the maintenance of soil microbial health and fertility status and thus sustainability (Majumdar *et al.*, 2014).

Table 4: Effect of various fertilizer treatments on enzymatic activities, soil microbial biomass carbon (SMBC) and basic soil respiration rate (BSRR) in jute rhizosphere

Treatments	Dehydrogenase (mg tpf/kg soil)	FDHA (mg fluoroscein/kg soil)	Urease (mg urea hydrolyzed/g oven dry soil/h)	SMBC (mg C/g oven dry soil)	BSRR (mg CO_2-C/g oven dry soil/h at)
Control	5.2	180.5	120.0	229.5	0.80
50% NPK	5.5	184.7	132.5	239.3	0.90
100% NPK	6.3	210.0	133.4	350.5	1.30
100% NPK + 10 t FYM/ha	7.2	236.1	169.4	368.5	1.50
150% NPK	6.2	185.7	152.0	360.5	1.00
CD (P=0.05)	1.4	7.4	8.0	12.2	0.20

Source: Majumdar *et al.* (2014)

Table 5: Effect of integrated nutrient management on microbial properties of mango orchard soil (0-15 cm)

Treatments	Dehydrogenase (µg TPF/g soil)	FDHA (mg fluorescein/g soil)	Microbial Biomass C (mg/kg soil)	Microbial biomass N (mg/kg soil)	Microbial biomass P (mg/kg soil)
T1- FYM @ 40 kg/tree + Azotobacter + Azospirillum+ PSB (10^8 cfu/g) + Mycorrhiza	2.39	516.33	148.67	133.23	0.78
T2-Biodynamic compost (30 kg/tree) + bio-enhancers cow pat pit 100g	4.57	944.33	750.27	181.43	4.58
T3-Neem cake + FYM (20 kg/tree) + Azotobacter + Azospirillum+ PSB (10^8 cfu/g)	3.25	595.00	237.40	154.77	1.47
T4-Vermi compost (30 kg/tree+ Azotobacter + Azospirillum+ PSB (10^8 cfu/g)	3.16	539.67	287.13	140.77	1.97
T5-FYM @ 40 kg/tree + bio-enhancer (Amritpani 5% soil application)	2.81	525.0	311.63	133.60	2.80
T6-FYM @ 40 kg/tree + green manuring (sunnhemp) + Azotobacter + Azospirillum+ PSB (10^8 cfu/g)	4.08	832.0	438.27	188.87	1.12
T7-1000 g N, P, K/tree	2.73	403.50	274.33	112.57	0.36
CD(P=0.05)	1.31	236.74	186.01	41.16	1.09

Source: Ram *et al.*, (2019)

In a two years field experiment conducted on 35 years old mango orchard (mango cv. Mallika), the effect of various organic amendments and inorganic application of NPK on microbial properties of soil was studied after two years of their application. All the organic amendments treatments improved the dehydogenase, FDHA, microbial biomass C, N and P in soil (Table 5) compared with inorganic supply of NPK and the biodynamic compost with bio-enhancers cow pat pit performed better among all the treatments under study (Ram *et al.*,2019).

Manna *et al.* (2005) showed that applying only nitrogen or nitrogen + phosphorous led to a decline in particulate organic matter (>53 mm fraction) and soil respiration, microbial biomass C and N, which were however improved significantly on addition of NPK or NPK+ organics.

In a two decades old long-term field experiment, organic and conventional farming system showed the richness of distinct microbiomes. Systems receiving organic manures were characterized by specific microbial groups, primarily involved in biodegradation of manure and composts, whereas, conventional farming not receiving manures showed a versatile microbial community structure characterized by oligotrophic microorganisms which are more adaptable to nutrients limited environments (Hartmann *et al.*, 2015).

Balanced fertilization with FYM and NPK in recommended dose sustain yield in long term of rice-wheat-jute cropping system in tropical lower Indo-Gangetic plains. Incorporation of jute in cropping sequence entrusted sustainability both in respect of C build up and nutrient dynamics as it provides considerable amount of biomass in very less time and relatively drier period of the year. Balanced fertilization maintains soil quality as indicated by higher CPI and LI and enhanced SMBC, WBC content in soil. Application of FYM and inorganic fertilizer in balanced proportion helps to enhance the C sequestration and promote food security on long term basis in this region (Mazumdar *et al.*, 2018).

Soil microbes are capable of reclamation of degraded soil. There are instances that degraded soils were reclaimed by different means like afforestation by acid, alkaline or salt tolerant plant spp. There are reports to support that, in spite of slower growth of various plant spp., afforestation can reclaim the barren sodic soil by the process of bioamelioration (Singh *et al.*, 2016). The root exudates secreted by these plant species are the home for activities of various microorganisms. Further, the biomass production by various tree spp. Also help to increase the microbial activities in the soil. These microbes with their activities improve the soil physical, chemical and biological properties in a continuous manner and the soils become normal because of normalization of soil pH, organic carbon status, water holding capacity, porosity, CEC etc.

In a long term study, different tree species were planted along with natural fallow in a highly degraded sodic soil (pH 10.5, EC 1.2 dS/m and ESP 89.0) with an aim for reclamation of the degraded soil. After ten years of their plantation, it was quite visible that the tree species like *Casurina equisetifolia, Acacia nilotica, Prosopis julifloria* could improve the soil biological properties (MBC, MBN and MBP) compared with natural fallow may be due to addition of organic matter and increased biological activities in the soil (Fig.1). There was also improvement in pH, EC and ESP under various tree spp. compared to initial status /natural fallow (Singh *et al.*, 2019).

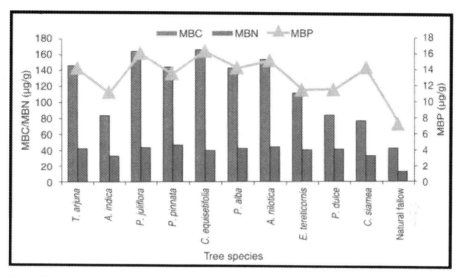

Fig. 1: Biological properties of rehabilitated and natural fallow sodic land at 0-15 cm soil depth
Source: Singh *et al.* (2019)

In India, intensive agriculture has given more emphasis on yield maximization and total production, including application of chemical fertilizers, pesticides, and extensive use of machinery, without any consideration for soil biodiversity. These practices have been identified as unsustainable and incompatible with natural balance, and their consequence is loss of soil biodiversity. Microbial communities in conventional soils were obviously sufficient and active enough to respond to organic inputs. Consequently, the rates of decomposition of organic matter were not limited in conventional soils, suggesting that microbial inoculation may not be necessary during the transition from conventional to organic farming systems.

In conclusion, minimum tillage, proper residue management and limited use of chemicals for pest control or judicious use of inorganic fertilizer with organic inputs, bio-control measures like use of biopesticides, PGPR etc. will not only main the sustainable crop productivity will also take care of soil biodiversity and conservation of soil for future generation.

References

Alef, K. (1995). Soil Respiration. *In* "Methods in Applied Soil Microbiology and Biochemistry" (K. Alef and P Nannipieri Eds.) pp 214-219. Academic Press, London.

Arcand, M.M., Helgason, B.L., Lemke, R.L. (2016). Microbial crop residue decomposition dynamics in organic and conventionally managed soils. *Applied Soil Ecology,* 107: 347–359.

Bossio, D.A., Girvan, M.S., Verchot, L., Bullimore, J., Borelli, T., Albrecht, A., Scow, K.M., Ball, A.S., Pretty, J.N. and Osborn, A.M. (2005). Soil microbial community response to land use change in an agricultural landscape of Western Kenya. *Microbial Ecology,* 49:50–62.

Buckley, D.H., and Schmidt, T.M. (2001). The structure of microbial communities in soil and the lasting impact of cultivation. *Microbial Ecology,* 42:11–21.

Chaer, G.M., Fernandes, M.F., Myrold, D.D. and Bottomley, P. J. (2014). Shifts in Microbial Community Composition and Physiological Profiles across a Gradient of Induced Soil Degradation. *Soil Science Society of America Journal, 73*:1327-1334.

Das, A.C. and Mukherjee, D. (2000). Influence of Insecticides on Microbial Transformation of Nitrogen and Phosphorus in Typic Orchragualf Soil. *Journal of Agricultural and Food Chemistry,* 48: 3728–3732.

Diosma, G., Aulicino, M., Chidichimo, H. and Balatti, P.A. (2006). Effect of tillage and N fertilization on microbial physiological profile of soils cultivated with wheat. *Soil and Tillage Research,* 91:236-243.

Diosma, G., Golik, S.I., Chidichimo, H.O. and Balatti, P.A. (2003). Nitrification potential, dehydrogenase activity and microbial biomass in an argiudol soil cultivated with wheat under two tillering methods. *Spanish Journal of Agricultural Research,* 1: 111–119.

Gaind, S. and Nain, L. (2011). Soil health in response to bio-augmented paddy straw compost. *World Journal of Agricultural Sciences,* 7(4): 480-488.

Guo, L.J., Lin, S., Liu, T. Q., Cao, C.G. and Li, C.F. (2016). Effects of conservation tillage on top soil microbial metabolic characteristics and organic carbon within aggregates under a rice (*Oryza sativa* L.) - wheat (*Triticum aestivum* L.) cropping system in central China. *PLOS ONE,* 11(1): e0146145.

Hartmann, M., Frey, B., Mayer, J., Mader, P. and Widmer, F. (2015). Distinct soil microbial diversity under long-term organic and conventional farming. *The ISME Journal,* 9: 1177-1194.

Kinney, C.A., Mandernack, K.W. and Mosier, A.R. (2005). Laboratory investigations into the effects of the pesticides mancozeb, chlorothalonil, and prosulfuron on nitrous oxide and nitric oxide production in fertilized soil. *Soil Biology and Biochemistry,* 37: 837–850.

Kremer, R.J. and Means, N.E. (2009). Glyphosate and glyphosate-resistant crop interactions with rhizosphere microorganisms. *European Journal of Agronomy,* 31: 153–161.

Kumar, A., Singh, S., Shivay, Y.S., Das, S., Pal, M. and Nain, L. (2019). Microbial priming for *in situ* management of paddy straw and its effect on soil microbiological properties under rice-wheat cropping system. *Indian Journal of Agricultural Sciences,* 89(9): 1503-1508.

Kyei-Boahen, S.; Slinkard, A.E. and Walley, F.L. (2001). Rhizobial survival and nodulation of chickpea as influenced by fungicide seed treatment. *Canadian Journal of Microbiology,* 47: 585–589.

Madhaiyan, M., Poonguzhali, S., Hari, K., Saravanan, V.S. and Sa, T. (2006). Influence of pesticides on the growth rate and plant-growth promoting traits of Gluconacetobacter diazotrophicus. *Pesticide Biochemistry and Physiology,* 84: 143–154.

Majumdar, B., Saha, A.R., Ghorai, A.K., Sarkar, S.K., Chowdhury, H., Kundu, D.K. and Mahapatra, B. S. (2014). Effect of fertilizer treatments on jute (*Corchorus olitorius*), microbial dynamics in its rhizosphere and residual fertility status of soil. *Indian Journal of Agricultural Sciences,* 84 (4): 503-508.

Majumdar, B., Saha, A.R., Sarkar, S., Maji, B. and Mahapatra, B.S. (2010). Effect of herbicides and fungicides application on fibre yield and nutrient uptake by jute (*Corchorus olitorius*), residual nutrient status and soil quality. *Indian Journal of Agricultural Sciences*, 80 (1): 878-83.

Makoi, J. and Ndakidemi, P. (2008). Selected soil enzymes: examples of their potential roles in the ecosystem. *African Journal of Biotechnology*, 7(3): 181-191.

Manna, M.C., Swarup, A., Wanjari, R.H., Ravankar, H.N., Mishra, B., Saha, M.N., Singh, Y. V., Sahi, D. K. and Sarap, P. A. (2005). Long-term effect of fertilizer and manure application on soil organic carbon storage, soil quality and yield sustainability under sub-humid and semi-arid tropical India. *Field Crops Research*, 93: 264-280.

Mazumdar, S.P., Kundu, D.K., Saha, A.R., Majumdar, B., Saha, R., Singh, A.K., Barman, D., Ghosh, D., Dey, R., Behera, M.S., Mitra, S. & Bhattacharyya, P. (2018). Carbon and nutrient dynamics under long-term nutrient management in tropical rice-wheat-jute system. *Archives of Agronomy and Soil Science*, 64 (11): 1595–1607.

Meena, R.S., Kumar, S., Datta, R., Lal, R., Vijayakumar, V., Brtnicky, M., Sharma, M.P., Yadav, G. S., Jhariya, M.K., Jangir, C.K., Pathan, S. I., Dokulilova, T., Pecna, V. and Marfo, T. D. (2020). Impact of agrochemicals on soil microbiota ans management: A Review. *Land* 9, 34, doi:10.3390/land9020034.

Milosevic, N. and Govedarica, M. (2002). Effect of herbicides on microbiological properties of soil. *Matica Srpska Proceedings for Natural Sciences*, 102: 5–21.

Nanipierri, P. (1994). The potential use of soil enzymes as indicator of productivity, sustainability and pollution. *In*: Soil Biota, Management in Sustainable Farming System. Pankhurst, C.E., B.M. Doulse, V.V. Gupta, S.R. Gupta and P.R. Grace (Eds.), CSIRO, Melbourne, pp: 218-244.

Nannipieri, P., Ascher, J., Ceccherini, M.T., Landi, L., Pietramellara, G. and Renella, G. (2003). Microbial diversity and soil functions. *European Journal of Soil Science*, 54(4): 655-670.

Piemental, D., Wilson, C., McCullum-Gomez, C., Huang, R., Dwen, P., Flack, J., Tran, Q., Saltman, T. and Cliff, B. (1997). Economic and environmental benefits of biodiversity. *Bioscience*, 47(11): 747-757.

Ram, R. A., Singha, A. and Sigh, V.K. (2019). Improvement in yield and fruit quality of mango (*Mangifera indica*) with organic amendments. *Indian Journal of Agricultural Sciences*, 89(9): 1429-1433.

Roldan, S., Herrera, D., Santa-Cruz, I., O'Connor, A., Gonzalez, I. and Snaz, M. (2004). Comparative effects of different chlorhexidine mouth-rinse formulations on volatile sulphur compounds and salivary bacterial counts. *Journal of Clinical Periodontology*, 31 (12): 1128-1134.

Sharma, P., Singh, G. and Singh, R.P. (2011). Conservation tillage, optimal water and organic nutrient supply enhance soil microbial activities during wheat (*Triticum aestivum* L.) cultivation. *Brazilian Journal of Microbiology*, 42: 531-542.

Shukla, A.K. (2000). Impact of fungicides on soil microbial population and enzyme activities. *Acta Botanica Indica*, 28: 85–88.

Singh, Y.P., Mishra, V.K., Sharma, D.K., Singh, G., Arora, S., Dixit, H. and Cerda, A. (2016). Harnessing productivity potential and rehabilitation of degraded sodic lands through *Jatropha* based intercropping systems. *Agriculture, Ecosystems and Environment*, 233: 121-129.

Singh, Y.P., Singh, G., Mishra, V.K., Arora, S., Singh, B. and Gupta, R.K. (2019). Restoration of ecosystem services through afforestation on degraded sodic lands in Indo-Gangetic plains. *Indian Journal of Agricultural Sciences*, 89 (9): 1492-1497.

Stott, P.A., Gillet, N.P., Hegerl, G.C., Karoly, D.J., Stone, D.A., Zhang, X. and Zwiers, F. (2010). Detection and attribution of climate change: a regional perspective. *Wires Climate Change*, (2): 192-211.

Van-Zwieten, L., Ayres, M.R. and Morris, S.G. (2003). Influence of arsenic co-contamination on DDT breakdown and microbial activity. *Environment Pollution,* 124: 331–339.

Wachira P. M., Kimenju, J. W., Okoth, S. A. and Mibey, R. K. (2009). Stimulation of nematode-destroying fungi by organic amendments applied in management of plant parasitic nematode. *Asian Journal of Plant Sciences,* 8:153–159.

Wang, Huanhuan., Li, Xiang., Li, Xu., Wang, Jian., Li, Xinyu., Guo, Qiucui., Yu, Zhixiong., Yang, Tingting. and Zhang, Huiwen (2020). Long-term no-tillage and different residue amounts alter soil microbial community composition and increase the risk of maize root rot in northeast China. *Soil and Tiillage Research,* 196: 104452. DOI: 10.1016/j. still.2019.104452.

Wang, J.J., Zhang, H.W., Li, X.Y., Su, Z.C., Xu, M.K. (2014). Effects of tillage and residue incorporation on composition and abundance of microbial communities of a fluvoaquic soil. *European Journal of Soil Biology,* 65:70–78.

Zhao, S., Li, K., Zhou, W., Qiu, S., Huang, S., He, P. (2016). Changes in soil microbial community, enzyme activities and organic matter fractions under long-term straw return in north-central China. *Agriculture, Ecosystem and Environment,* 216: 82–88. DOI: 10.1016/j.agee.2015.09.028.

23

Impacts of Conservation Agriculture Practices on Soil Water Dynamics

S. Mitra, R. Saha and N.M. Alam

ICAR- Central Research Institute for Jute & Allied Fibres, Barrackpore West Bengal -700 121, India

Introduction

Conservation tillage is a model of sustainable agriculture as it leads to profitable food production while protecting and even restoring natural resources. Conservation agriculture benefits farmers because it reduces production costs and increases yields, but it also has positive impacts on the whole society: enhancement of food security thanks to a better soil fertility, improvement of water quality, reduction of erosion and mitigation of climate change by increasing carbon sequestration. Conservation agriculture (CA) systems are also less sensitive to extreme climatic events and therefore contribute to the adaption of climate change and the resilience of agricultural systems. Hence, CA becomes a fundamental element of sustainable production intensification, combining high production with the provision of environmental security.

CA is based on healthy functioning of the whole agro-ecosystem with a maximum attention and focus on the soil. The soil is the entry point and it has to be considered not only as simple physical support to roots and plants, but as a living entity with its physical, chemical and biological characteristics. The focus of CA embraces not only the nutrient contents of the soil but also its biological and structural status, which are determinants of sustained productivity.

The paradigm of CA is that an undisturbed soil has the opportunity to develop and produce healthier plants. Soil life can develop in a stable habitat in quantity and quality better than on tilled soils, the structural integrity of soil is maintained, so continuous vertical macro-pores are not destroyed and remain as drainage channels for rainwater into the soils. Seeding under conditions of minimum soil disturbance is achieved by direct seeding through the mulch cover without tillage.

The water balance is an accounting of the inputs and outputs of water. The water balance of a place, whether it be an agricultural field, watershed, or continent, can be determined by calculating the input, output, and storage changes of water at the Earth's surface. The major input of water is from precipitation and output is evapotranspiration.

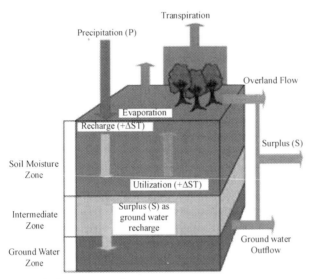

The water balance equation:

$$ET = P + I + C_p - D_p - R_p - \Delta ST$$

Where P is the precipitation, I is the depth of irrigation water applied, C_p is the contribution through capillary rise from ground water-table, D_p is deep percolation loss, R_p is surface water runoff and ΔST is change in soil water profile. In the present study, no irrigation is applied. The C_p was assumed to be negligible as the water-table was below 4 m throughout the crop season. The soil surface was covered with crop biomass and grasses to reduce surface run-off or deep drainage. This was confirmed by the periodic monitoring of the soil-water profile. So, all these parameters (I, C_p, D_p and R_p) was ignored during calculation.

The change in soil water profile (ΔS) was measured by:

$$\Delta S = S_i - S_f$$

Cover Crop and its Effect

A permanent organic soil cover (from a crop, a cover crop or a vegetative mulch) ensures the protection of the soil surface from wind, rain, sun and from drying out and provides a regular supply of organic matter, which is key feature for soil fertility. It increases the water storage by minimizing the surface sealing and enhancing infiltration, as well as by directly reducing the evaporation. In addition, the mulches suppress the germination of weeds, provides habitat for beneficial fauna and feed for soil organisms. The effectiveness of soil cover/ residue depends upon the proportion of soil covered and hence the quantity present, but, as with fallow performance generally, the response depends on the complex interactions between weather, surface conditions and soil water storage capacity. The effects are discussed below:

Organic matter enrichment

Organic management systems attempt to increase soil organic matter through additions of plant biomass generated by cover cropping practices, additions of manure, compost, and other organic amendments, and conservation of crop residue. Soil organic matter enhances the formation of aggregates, which stabilizes soil and reduces runoff and erosion (Sainju *et al.*, 1997). Increased aggregation and porosity can promote root growth by decreasing soil bulk density (weight per unit volume) and reducing resistance to root penetration. Soil organic matter improves soil tilth, reduces crusting, and increases the rate of water infiltration. Addition of organic matter to soil can also increase the populations of soil microbes, micro- and macroarthropods, and earthworms, all of which contribute to efficient nutrient cycling and improvements in soil structure. Roberson *et al.* (1991) found that cover cropping increased the heavy fraction of soil carbohydrates and increased aggregate stability. The heavy fraction carbohydrates are enriched in extracellular polysaccharides produced by soil microorganisms. These extracellular polysaccharides can be described as "glue" that binds individual soil particles together in aggregates.

Soil moisture conservation

Cover crop residues help conserve soil moisture during the summer growing season by reducing water evaporation from the soil surface before full crop canopy has been established and by increasing water infiltration (Smith *et al.*, 1987). With corn (*Zea mays*), a crop highly sensitive to moisture stress at critical stages of development, a greater reservoir of available water can substantially increase yields. It is needed to monitor early spring conditions to maximize biomass production without severely depleting soil moisture before planting time. While some N production by legumes will be sacrificed, in dry years the best strategy for managing cover crops is to kill the cover crops approximately two weeks before planting (depending on weather forecasts). In most seasons, sufficient rainfall for adequate crop emergence will occur during the two-week preplant period or within the week immediately following planting. In wet years, the cover crop can be killed immediately before soil preparation (if any) and planting.

Weed management

Cover crops can reduce weeds in subsequent cash crops. While the cover crop is growing, it can suppress the germination and growth of some early spring weeds through competition and shading. Cover crop residues remaining on the soil surface can physically modify conditions for seed germination by altering the seed environment (through changes in light availability, soil temperature, and soil moisture) and through other types of interference, primarily allelopathy. After cover crop desiccation, it is important to prevent soil disturbance to maintain maximum soil cover from cover crop residues (Burgos and Talbert, 1996).

Nitrogen fixation

Leguminous cover crops can "fix" significant amounts of N for use by subsequent crops. Through a symbiotic association with legumes, Rhizobia bacteria convert atmospheric N_2 into an organic form that the legume uses for growth. Non-leguminous cover crops, typically grasses or small grains, do not fix N_2. Nonetheless, they can be effective in recovering mineral N from soil after crops are harvested. Plant available N that might otherwise be lost to leaching or runoff during the fall and winter months is retained as "biomass N". Cover crop biomass and N accumulation depend on the length of the growing season, local climate, and soil conditions. Leaf fall during legume development can contain up to 35 lbs N per acre (Bergersen *et al.*, 1989). Although roots and leaf fall can contribute N to the potentially available soil N pool, in most circumstances this contribution varies and is difficult to calculate. Therefore, the N credit provided by cover crops is normally estimated by the live crop biomass present and the N concentration of this biomass when the cover crop is killed. The portion of total cover crop N that is contained in the aboveground biomass is approximately 75% for cereal rye, 90% for hairy vetch, and 80% for crimson clover (Shipley *et al.*, 1992).

Selection of Cover Crop

Generally, cover crop selection is based on individual enterprise or production goals. For example, if the purpose of a cover is to provide readily available, biologically fixed N for subsequent crops, then the grower should choose a legume such as hairy vetch or cowpea. If the cover crop will be managed as surface mulch for weed suppression or incorporated to improve soil quality, then the grower should choose a grass cover crop such as cereal rye or sorghum-sudangrass. Both of these grass cover crops can produce large amounts of biomass with high C:N ratios at maturity, and both are reported to suppress (some) weeds.

Mixtures of cover crop species can be planted to optimize the benefits associated with cover crop use. Grass species establish ground cover more quickly than legume monocultures, and roots are more physiologically active in autumn. Mixtures that include grasses can therefore more effectively prevent soil erosion and reduce soil N concentration. Aboveground cover crop biomass N content can be increased with a mixture that is spatially more efficient in utilization of nutrients and water. For example, a deep-rooted cover crop that is grown with a shallow rooted cover crop can utilize water and resources throughout more of the soil profile. Planting mixtures of cover crops can take advantage of the allelopathic potential of the cover crops to suppress weeds. Allelopathic suppression of weeds has been shown to be a species specific phenomenon; therefore, a broader spectrum of weed control may be possible by growing a mixture of cover crop species, each contributing allelopathic activity towards specific weed species (Creamer and Bennett, 1997). Mixtures can also be planted to influence insect populations. Cover crop species, regardless of biomass or biomass N production potential, could be included a mixture if they were known to attract important beneficial insects into the cropping system.

Cover Crop Residue Management

In organic systems, cover crops may be destroyed by tillage, mowing, undercutting, or rolling. In a wet growing season, incorporating legumes may produce the highest yields. However, under relatively dry growing conditions, legume residue left on the surface helps conserve soil moisture and yields are higher with conservation tillage. In no-till organic production systems, cover crops are normally killed mechanically and left on the surface as a mulch. Of the three methods for mechanically killing cover crops, undercutting, mowing, and rolling, the first is the most effective. Mowing with a flail mower leaves the finely chopped residue evenly distributed over the bed, and, unless packed well, the residue tends to decompose quickly. Rolling the cover crop damages, the plants by lodging them severely and by successive crimping of cover crop stems. Rolling keeps the above ground part of the plant attached to the root system. Rolled plants decompose more slowly than those killed by mowing and, consequently, control weeds for a longer period of time (Lu *et al.*, 2000).

Mulching Effect

The practice of mulching has been widely used as a management tool in many parts of the world. Mulch can be defined as any material on the soil surface through which continuous liquid water films from the soil are not present. Organic mulch adds nutrients to soil when decomposed by microbes and help in carbon sequestration. The traditional mulch consists of a well-aerated, and therefore poorly conducting, surface cover, such as straw, leaf litter or gravel. Crop residues affect soil water content by having a direct effect on the evaporation from the soil surface and on the amount of water that infiltrates into the soil. Decreases in infiltration due to sealing of the pores caused by raindrop impact are reduced by residue cover. Crop residues significantly influence evaporation by affecting: 1) net radiation due to changes in surface albedo, 2) aerodynamic vapour conductance due to changes in wind speed, 3) resistance to vapour diffusion, which is dependent on mulch thickness, tortuosity and volumetric air fraction.

Mulch materials dampen the influence of environmental factors on soil by increasing soil temperature controlling diurnal/seasonal fluctuations in soil temperature. The surface mulch favourably influences the soil moisture regime by controlling evaporation from the soil surface, improves infiltration, soil water retention, decreases bulk density and facilitates condensation of soil water at night due to temperature reversals. Modification of the soil microclimate by mulching favours seedling emergence and root proliferations and suppress weed population.

Mulching of cropped soil caused a substantial reduction in soil evaporation component of water loss primarily due to cover effect (Prihar et al., 1996). Straw mulching (6 t/ha) reduced soil evaporation under medium and coarse textured soil by 18.5 and 13.1 cm in maize, 23.8 and 16.6 cm in cotton and 23.6 and 17.6 cm in sugarcane crops, respectively. Bond and Willis (1969) showed that evaporation rates from soil columns treated with various rates of rye straw, measured during the initial constant drying stage, decreased with increasing residue rates. Reduction in soil water evaporation compared to a bare soil was around 20% and 90% for the 560 kg/ha and 17920 kg/ha straw rates, respectively. The effect lasted only a few days for the lower straw rate and 60 days for the higher rate. Mulches modify the following evaporation parameters: 1) the drying coefficient (k), an index of relative soil drying rate; 2) the soil boundary coefficient (kb), an index of relative evaporation; and 3) the diffusivity (Dw), the effective rate of transport of water to the surface. More evaporation was measured from a bare surface as compared to a wheat stubble covered surface (3400 kg/ha), until the bare surface was dry. Then the stubble covered surface had larger evaporation due to its greater available water. The monitored daily net radiation was always greater for the bare soil surface. The bare soil had higher temperatures than the stubble covered surface except for overcast conditions with wet surfaces when they were about equal. In a comparison of various surface residue-tillage treatments, maximum temperature differences were 12 and 19°C, respectively for fall and spring, between no residue and surface residue treatments for the same tillage condition (Gupta *et al.*, 1983).

Field experiments (Chakraborty *et al.*, 2008) were conducted in a sandy loam soil to evaluate the soil and plant water status in wheat under synthetic (transparent and black polyethylene) and organic (rice husk) mulches with limited irrigation and compared with adequate irrigation with no mulch. Though all the mulch treatments improved the soil moisture status, rice husk (RH) was found to be superior in maintaining optimum soil moisture condition for crop use. The residual soil moisture was also minimum, indicating effective utilization of moisture by the crop under RH. The plant water status, as evaluated by relative water content (RWC) and leaf water potential (LWP) were favourable under RH. RWC in leaves was maintained higher than 90% (except on 57 DAS) till 98 DAS (Fig. 1), decreased sharply and recorded 75–80% thereafter. Rice husk maintained a better leaf water status for most of the time with maximum seasonal average RWC (89.5%) closely followed by TP mulch (88.2%). Similar to soil water potential, average LWP over the season was also higher under RH treatment (-1.72 MPa). Black and transparent polyethylene mulch recorded the minimum average LWP (-1.82 and -1.81 MPa, respectively) (Fig. 2).

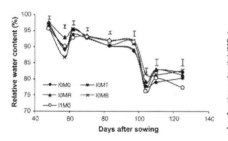 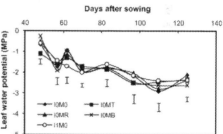

Fig. 1: Relative water content in leaves of wheat under different mulches

Fig. 2: Leaf water potential in wheat under of wheat under different mulches

Tomar *et al.* (1992) experimented on the effect of various perennial mulches comprised of leaves of *Leucaena leucocephala, Eucalyptus* hybrid, *Shorea robusta, Broussonatia paprifera* or *Puerarua hirsuta* (chopped) @ 4 t/ha, applied just after sowing of wheat and studied the residual effect of applied mulch on moisture conservation. At various crop growth stages the highest amount of water was found in the plots mulched with *S. robusta* followed by *E.* hybrid. The distribution of profile moisture revealed that at the time of sowing of wheat the soil water content did not differ with depth (0–90 cm). With the advancement of time, the magnitude of moisture distribution changed.

No-tillage system without crop residue left on the soil surface has no particular advantage because of water loss from the surface. It is evident from soil moisture data (Fig. 3) of a study under hilly eco-system conducted by Ghosh et al. (2010). In this study, three treatments (only panicle harvested, 50% crop residue cut and complete removal of residue) were maintained with the hypothesis that residue kept in the field could maintain soil moisture required for pea. No-tilled peas were sown by hand dibbler in all the plots. In rice fallow, better crop performance was found under 75% rice residue retention (only panicle harvested), followed by 50% rice residue retention.

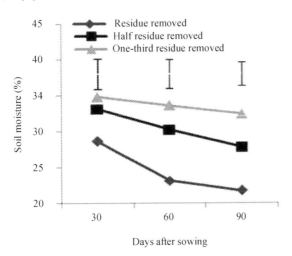

Fig 3: Soil moisture at different growth stages of pea under varying degrees of residue retention

Table 1: Effect of locally available organic residue management on water losses and profile water storage in acid soils.

Treatments	Water loss at different stages (mm/hr)					Profile water storage (m^3/m^3)
	Transplanting	Active tillering	Maximum tillering	Flowering	Maturity	
Control	2.62	2.82	2.76	2.78	2.90	0.31
Jungle grass*	1.79	1.91	1.97	2.03	2.12	0.35
Ambrosia sp.	1.93	1.93	2.00	2.12	2.19	0.36

*Jungle grass consisted of goat weed (Ageratum conyzoides), quick weed (Galinsoga parviflora), eupatorium (Eupatorium odoratum), beggartricks (Bidens pilosa), thatch grass (Imperata cylindrica), etc.

The incorporation of organic residues resulted in significant increase in crop yield coupled with improvement in soil physical properties. Study on locally available organic residues on soil health under hilly agro-climatic condition showed that incorporation of jungle grass or *Ambrosia sp.,* continuously for 5 years in puddled rice soil improved soil organic carbon (SOC) by 21.1%; the stability of micro-aggregates, moisture retention capacity, and infiltration rate of the soil by 82.5, 10, and 31.3%, respectively (Saha and Mishra, 2009). Addition of organic residues resulted in decrease of water loss as compared with control, indicating that the sealing of pores and broken capillaries under puddled conditions had inhibited the percolation rate resulting in reduced water loss. It is evident through higher profile water storage (0.35 to 0.39 m^3/m^3) than the control (0.31 m^3/m^3) up to 30 cm depth (Table 1).

Conclusion

Many of the advantages of the conservation tillage systems currently in use derive from the presence of cover crop and crop residue on the soil surface. Crop cover/residues reduce soil erosion losses, therefore increase soil water infiltration, and decrease soil water loss by evaporation. In addition to the soil water conserving properties of a cover crop, effects such as a decrease in soil temperature variations can be an advantage for the growth of the subsequent crop, and should be considered in the management practices of the cover crop.

References

Bergersen, F.J., Brockwell, J., Gault, R.R., Morthorpe, L., Peoples, M.B. and Turner, G.L. (1989). Effects of available soil nitrogen and rates of inoculation on nitrogen fixation by irrigated soybeans and evaluation of delta 15N methods for measurement. *Australian Journal of Agricultural Research*, 40:763-780.

Bond, J.J., Willis, W.O. (1969). Soil water evaporation: surface residue rate and placement effect. *Soil Science Society of America Journal*, 33: 445-448.

Burgos, N.R., and Talbert, R.E. (1996). Weed control by spring cover crops and imazethapyr in no-till southern pea (*Vigna unguiculata*). *Weed Technology*, 10:893-899.

Chakraborty, D., Nagarajan, S., Aggarwal, P., Gupta, V.K., Tomar, R.K., Garg, R.N., Sahoo, R.N., Sarkar, A., Chopra, U.K., Sundara Sarma, K.S. and Kalra, N. (2008). Effect of mulching on soil and plant water status, and the growth and yield of wheat (*Triticum aestivum L.)* in a semi-arid environment) *Agricultural Water Management*, 95 (12): 1323-1334.

Creamer, N.G., and Bennett, M.A. (1997). Evaluation of cover crop mixtures for use in vegetable production systems. *HortScience*, 32: 866-870.

Ghosh, P.K., Das, A., Saha, R., Kharkrang, E., Tripathi, A.K., Munda, G.C. and Ngachan S.V. (2010) Conservation Agriculture towards Achieving Food Security in North East India. *Current Science*, 99(7): 915-921.

Gupta, S.C., Larson, W.E., Linden, D.R. (1983). Tillage and surface residue effects on soil upper boundary temperatures. *Soil Science Society of America Journal,* 47: 1212-1218.

Lu, Y.C., Watkins, K.B., Teasdale, J.R., and Abdul-Baki, A.A. (2000). Cover crops in sustainable food production. *Food Reviews International*, 16: 121-157.

Prihar, S.S., Jalota, S.K. and Steiner, J.L. (1996). Residue management for reducing evaporation in relation to soil type and evaporativity. *Soil Use Management*, 12: 150-157.

Roberson, E.B., Sarig, S. and Firestone, M.K. (1991). Cover crop management of polysaccharidemediated aggregation in an orchard soil. *Soil Science Society of America Journal*, 55: 734-739.

Sainju, U.M., and Singh, B.P. (1997). Winter cover crops for sustainable agricultural systems: Influence on soil properties, water quality, and crop yields. *HortScience*, 32: 21-28.

Saha, R. and Mishra, V.K. (2009). Effect of organic residue management on soil hydro-physical characteristics and rice yield in eastern Himalayan region, India. *Journal of Sustainable Agriculture*, 33(2): 161-176.

Shipley, P.R., Meisinger, J.J. and Decker, A.M. (1992). Conserving residual corn fertilizer nitrogen with winter cover crops. *Agronomy Journal*, 84: 869-876.

Smith, M.S., Frye, W.W. and Varco, J.J. (198). Legume winter cover crops. *Advances in Soil Science*, 7: 95-139.

Tomar, V.S., Narain, P. And Dadhwal, K.S. (1992). Effect of perennial mulches on moisture conservation and soil-building properties through agroforestry. *Agroforestry Systems*, 19(3): 241-252.

24

Conservation Agriculture for Enhancing Soil Health and Crop Production

J. Somasundaram, A.O. Shirale, N.K. Sinha, B.P. Meena
K.M. Hati, M. Mohanty, A.K.Naorem[1], A.K. Biswas and A.K.Patra

ICAR –Indian Institute of Soil Science, Nabibagh, Bhopal
Madhya Pradesh-462 038, India
[1]ICAR– Central Arid Zone Research Institute, Regional Research
Station-Kukma, Bhuj, Gujarat-370 105

Introduction

India agriculture is now at turning point as we have travelled a long way after the advent of green revolution. Over the past four to five decades our strategies, policies and actions were guided by goals of 'self-sufficiency' in foodgrains production via green revolution. Indian agriculture has been successful in achieving increased food grains production albeit at a low level of satisfaction. While the mission of increasing foodgrains production stands somehow achieved, these gains were accompanied by widespread problems of resource degradation and high factor productivity, which now pose a serious challenge to the continued ability to meet the demand of an increasing population and lifting our people above the poverty line (Tan *et al.*, 2010). Indian agriculture has reached a point where it must seek new directions – those by way of strategies, policies and actions which must be adopted to move forward addressing sustainable intensifications. The past strategies to increase food grains production, however, have resulted in massive exploitation of natural resources, contributing to unsustainable growth; there is need to change this approach in the future. This will call for strategies, which are different than the ones we adopted in the 'green revolution' era. High levels of fertilizer use and decreasing resource use-efficiency are increasingly contributing to groundwater pollution and increased emissions of green-house gases (GHGs) (Gupta, 2004). High level uses of pesticides in many areas have become a major health hazard. Thus, with continuously deteriorating resources, widespread problem of soil and water contamination and eroding ecological foundation, sustainability of agriculture is becoming highly questionable (Conway and Barbier, 1990). There is need to seriously debate on the strategies required to ensure that agriculture continues to play a critical role in the overall development.

Soil is a non-renewable natural resource on which agrarian activities such as agriculture, livestock and forestry are carried out. It is interconnected with other natural resources, which are also essential for human life such as the air, water, fauna and flora. Soil acts as the most important intermediate and regulating factor in most agricultural processes.Hence, it can be said that if the soil is well managed, the effects of agriculture on the environment will be acceptable and conversely, if it is badly managed. Traditional or conventional agriculture bases most of its operations or practices on soil tillage; i.e., inversion tillage such as mould board (MB) ploughing or disk harrow, or vertical tillage such as chisel, "spiked" harrow and other tools. Soil tillage drastically alters its original structure, breaking up its natural aggregates and burying the residues of the previous crop. So the bare soil becomes unprotected and exposed to the action of the wind and rain. Under these circumstances water and soil erosion and sediment runoff are likely to occur. Furthermore, with tillage, soil organic matter and biodiversity content are reduced and unnecessary emissions of CO_2 into the atmosphere take place.

Fortunately, the past few decades has seen the development of conservation agriculture (CA) which attempts to alter the soil profile as little as possible, leaving it permanently protected from the action of the wind and rain with the plant residues from the previous crop (stubble) and/ or with "cover crops", whose mission is precisely the protection of the soil in the periods or phases occurring between the crops planted for economic purposes. Direct sowing, reduced or minimal (minimum) tillage and cover crops in tree plantations are diverse modalities of conservation agriculture (Garcia-Torres *et al.*, 2003).

Conservation Agriculture – The Way Forward

Conservation agriculture has emerged as a new paradigm to achieve goals of sustainable agricultural production (Sangar and Abrol, 2005; Somasundaram *et al.*, 2020). It is a major step towards transition to sustainable agriculture. The term CA refers to the system of raising crops without tilling the soil while retaining crop residues on the soil surface (Fig 1). The key elements which characterize CA include (Fig 2):

Fig. 1: Crop residue retention under CA

- Minimum soil disturbance by adopting no-tillage with minimum traffic for agricultural operations,
- Retention and management of crop residues on the soil surface
- Adopt spatial and temporal crop sequencing/crop rotations to derive maximum benefits from inputs and minimize adverse environmental impacts.

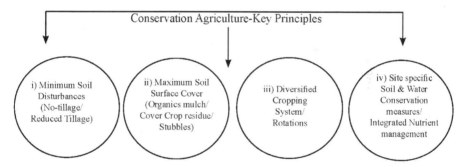

Fig. 2: Key principles of Conservation Agriculture

Combining the above elements with improved land-shaping (e.g. through laser aided levelling, planting crops on beds, etc.) further enhances the opportunities for improved resource management. In conventional systems, while soil tillage is a necessary requirement to produce a crop, tillage does not form a part of this strategy in CA. Intensive tillage in conventional system causes gradual decline in soil organic matter content through accelerated oxidation, resulting in reduced capacity of the soil to regulate water and nutrient supplies to plants. Burning of crop residues, a common practice in many areas (e.g. rice–wheat cropping system) further causes pollution, GHG emission and loss of valuable plant nutrients. When crop residues are retained on the soil surface in combination with no tillage, it initiates processes that lead to improved soil quality and overall resource enhancement (Abrol and Sangar, 2006; Somasundaram *et al.*, 2018).

Benefits of CA are manifold which includes direct benefits to farmers through reduced cost of cultivation through savings in labour, time and farm power, and improved use efficiency resulting in reduced use of inputs. More importantly, CA practices to reduce resource degradation due to minimal soil disturbances and crop residue retention. Gradual decomposition of surface residues improves soil organic matter status (Somasundaram *et al.*, 2017, 2019), biological activity and diversity and contributes to overall improvement in soil quality. CA is a way to reverse the processes of degradation as compared to conventional agricultural practices involving intensive cultivation, burning and/or removal of crop residues, etc. CA leads to sustainable improvements in efficient use of water and nutrients by improving nutrient balance and availability, infiltration and retention by the soil, reducing water loss due to evaporation and improving the quality and availability of ground and surface water (Singh and Sidhu, 2014).

Conservation Agriculture Success World Over

Conservation agriculture has emerged as an effective strategy to achieve goals of sustainable agriculture worldwide. It has the potential to address increasing concerns of serious and widespread problems of natural resource degradation and environmental pollution, while enhancing system productivity. According to current estimates, CA systems are being adopted in some 180 million ha, largely adopted in rainfed areas and the area under CA is expanding rapidly (Harrington and Olaf, 2005; Kassam *et al.*, 2018). USA has pioneered research and development efforts and currently CA is being practiced in more than 18 million ha of land. Other countries where CA practices are being widely adopted include Australia, Argentina, Brazil and Canada.

In many countries of Latin America, CA systems are finding rapid acceptance by farmers. Many countries have now policy decision to promote CA. In Europe, France and Spain, CA was being adopted in about 1 m ha area under annual crops (Sangar, Abrol and Gupta Raj, 2005). In Europe, the European Conservation Agriculture Federation, a regional lobby group uniting national associations in UK, France, Germany, Italy, Portugal and Spain, has been founded. CA is also being adopted to varying extents in countries of Southeast Asia, *viz.* Japan, Malaysia, Indonesia, the Philippines, Thailand, etc. A unique feature which has triggered widespread adoption of CA systems in many countries is the community-led initiative strongly supported by R&D

organizations rather than as aresult of the usual research-extension system efforts (Sangar, Abrol and Gupta Raj, 2005).

Conservation Agriculture and Soil Health

There are several reports on the influence of conservation agricultural management practices comprising of tillage practices, residue recycling, application of organic manures, green manuring and integrated use of organic and inorganic sources of nutrients, soil water conservation measures, integrated pest management, organic farming, etc., on soil quality (Prasad *et al.*, 1999; Bijay Singh *et al.*, 2008; Anikwe and Ubochi 2007; Wang *et al.*, 2015; Balota *et al.*, 2004). Most of the soil quality assessment studies, do not deal with comprehensive data set comprising of all the physical, chemical and biological indicators (Sharma *et al.*, 2008). However, the results pertaining to the effect of CA practices on soil quality are given below: The basic objective of integrated nutrient use is to reduce the inorganic fertilizer requirement, to restore organic matter in soil, to enhance nutrient use efficiency and to maintain soil quality in terms of physical, chemical and biological properties.

Prasad *et al.* (1999) showed that incorporation of crop residue also improved soil fertility status as judgedby organic carbon and available P and K contents. Residues retention improves soil physical (e.g., structure, infiltration rate, plant available water capacity), chemical (e.g., nutrient cycling, cation exchange capacity, soil reaction), and biological (e.g., SOC sequestration, microbial biomass C, activity and species diversity of soil biota) quality (Beri *et al.*, 1992, 1995, Bijay-Singh *et al.*, 2008, Power *et al.*, 1986, Singh *et al.*, 2005).

Studies reveal that adoption of CA practices leads to significant improvement in soil physical environment and thereby soil quality over time (Verhulst *et al.*, 2010). However, effects of CA on soil physical properties can vary from location to location depend on the tillage system and their intensity, agro-climatic condition and type of soil. For example, ZT systems which maintain residue retention over soil surface resulted in significant change in soil physical environment, especially in upper few centimetre of the soil (Anikwe and Ubochi 2007). For example intensive tillage in case of CT disrupts soil aggregates, fragments root and mycorrhizal hyphae, which are act as major binding agents for micro-aggregates leading to lower soil aggregation and structural stability in CT over CA (Wang *et al.*, 2015). Moreover, residue removal besides intensive disturbance in CT may also leads to interruption of different aggregate formation process over time (Six *et al.*, 2001). Many researchers have found that ZT significantly improved saturated and unsaturated hydraulic conductivity owing to either continuity of pores or flow of water through very few large pores (Bhattacharya *et al.*, 2006). Greater number of macro-pores, little soil disturbance to soil and presence of litter of well decomposed residues formed by accumulated organic matter is main cause of better hydraulic conductivity under CA practice over CT (McGarry *et al.*, 2000; Osunbitan *et al.*, 2005). Likewise, Kahlon *et al.*, 2013 and Chen *et al.*, 2014 found that reduced tillage significantly increased the initial soil infiltration capacity over CT which resulted in higher saturated and unsaturated hydraulic conductivity of the soil as compared to CA. Among the CA practices, over last three decades, ZT have been intensively tested, applied and demonstrated in order to maintain or improve the stock of organic carbon in soil and reduce CO_2 emissions (Dimassi *et al.*, 2014). A comparative analysis of SOC under different medium and long term studies revealed that ZT accounted higher SOC in the tune of 3.86- 31.0% over conventional tillage (Balota *et al.*, 2004; Govaerts *et al.*, 2009). They also suggested that to achieve the beneficial effect of ZT in terms of higher SOC, its long term implementation is essential. Likewise, Machado *et al.* (2001) found that adoption of ZT resulted into significant increase in total carbon (30%), active carbon pool (10%), and passive carbon pool (18%) compared to CT (Aziz *et al.*, 2015). The intensive tillage in case of CT increased organic matter decomposition and enhanced its oxidation (Balasdent *et al.*, 2001; Balota *et al.*, 2003; Thomas *et al.*, 2007) that leads to lower SOC under CT compared to CA (Fig. 3) (Somasundaram *et al.*, 2018). Tillage, residue management and crop rotation can strongly affect the nutrient dynamics of any soil through thier effect on mineralization and recycling of soil nutrients (Galantini *et al.*, 2000). It is believed that long-term adoption of CA practices can lead to higher buildup of nutrients in top soil due to larger nutrient mineralization potential of soil as compared to CT (Rasmussen 1999; Duiker and Beegle 2006). The benificial effect of CA in terms of higher nutrient availability partly may be due to crop residues retention over soil surface in comparison with incorporation of crop residues with CT (Ismail *et al.*, 1994) and partly due to arresting their leaching losses by reducing decomposition of surface placed residues (Balota *et al.*, 2004, Kushwaha *et al.*, 2000). However, the response of soil chemical fertility to different tillage practices is reported to be vary with location to location, soil type, cropping systems, climate, fertilizer and other agronomic management practices (Rahman *et al.*, 2008). Govaerts *et al.*, 2006 found that after long term adoption of CA (26 cropping seasons) in a input responsive well irrigated production system , the N mineralization was higher in PB with residue retention than in CT with residues incorporation. ZT with residue retention over long run can enhances the supplying power of soil that leads to higher soil available N as compared to CT. Numerous studies also showed P stratification in soil under different tillage systems where ZT system associated with higher concentration of P due to preferential movement of P in the soil (Abdi *et al.*, 2014). Piegholdt *et al.* (2013) also reported 15% higher total P content in the top soil (0-5 cm) of ZT plots as compared to CT due to larger P addition from decomposition of residues retained on the soil surface. Likewise, higher P levels in ZT than in CT were reported by other researchrs (Du Preez *et al.*, 2001;Duiker and Beegle 2006). Micronutrient (Zn, Fe, Cu and Mn) tend to be present in higher levels under ZT with residue retentions compared to CT, especially near the soil surface (Franzluebbers and Hons 1996). In contrast, Govaerts *et al.* (2007) reported that tillage practice had no significant effect on the concentration of extractable Fe, Mn and Cu, but that the concentration of extractable Zn was significantly higher in the 0-5 cm layer of PB planting compared to CT with full residue retention. Similar results were reported by Du Preez *et al.,* 2001 and Franzluebbers and Hons 1996. Residue retention significantly decreased concentrations of extractable Mn in the 0-5 cm layer in PB compared to CT (Govaerts *et al.*, 2007). However, according to Peng *et al.*, 2008, the Mn concentration was higher under CA as compared to CT due to increased SOC content. Sharma *et al.* (2008) observed significant improvement in soil physical, chemical and biological properties after long term CT and gliricidia lopping in sorghum and green gram. Long term adoption of CA supposed to be enhancing not only the biological activity but also their biomass and diversity compared to CT (Lupwayi *et al.*, 2001). Dong *et al.*, 2009 also reported that the mean annual MBC was highest in the ZT with residue, while lowest in CT without residue. Similarly, Silva *et al.*, 2010 consistently found higher values of MBC and microbial biomass nitrogen up to more than 100 % under NT in comparison to CT. Likewise, Singh *et al.* (2009) reported that the dehydrogenase enzyme activity of soil under permanent bed planting method registered significantly higher (62%) than CT. Based on the network tillage experiment being carried out since 1999 at various centers of All India Coordinated Research Project on Dryland Agriculture (AICRPDA), it was observed that in arid (< 500 mm rainfall) region, low tillage was almost comparable to conventional tillage and the weed management was not so difficult, whereas, in semi-arid (500 –1000 mm) region, conventional tillage was found superior (Kumar *et al.*, 2011).

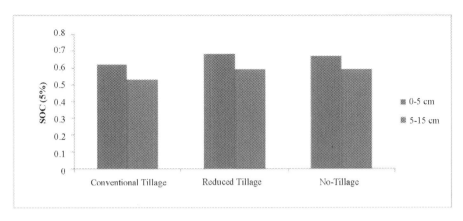

Fig. 3: Soil organic carbon under different tillage system
Source: Somasundaram *et al.* (2018)

Importance of Cover Crops in Conservation Agriculture

Keeping the soil covered is a fundamental principle of CA. Crop residues are left on the soil surface, but cover crops may be needed if the gap is too long between harvesting one crop and establishing the next. Cover crops improve the stability of the CA system, not only on the improvement of soil properties but also for their capacity to promote an increased biodiversity in the agro-ecosystem (Singh *et al.*, 2009).While commercial crops have a market value, cover crops are mainly grown for their effect on soil fertility or as livestock fodder. In regions where smaller amounts of biomass are produced, such as semi-arid regions or areas of eroded and degraded soils, cover crops are beneficial as they:

- Protect the soil during fallow periods.
- Mobilize and recycle nutrients.
- Improve the soil structure and break compacted layers and hard pans.
- Permit a rotation in a monoculture.
- Can be used to control weeds and pests

Climate Change and Conservation Agriculture

Lal (2005) suggested that by adopting improved management practices on agricultural land (use of NTand crop residues), food security would not only be enhanced but also offset fossil fuel emissions at the rate of 0.5 Pg C yr. Climate change is likely to strongly affect rice–wheat, rice–rice and maize-based croppingsystems that, today, account for more than 80% of thetotal cereals grown on more than 100 M ha of agricultural lands in South Asia.

Global warming maybe beneficial in some regions, but harmful in thoseregions where optimal temperatures already exist; anexample would be the rice–wheat mega-environments in the Indo-gangetic plains that account for 15% of global wheatproduction. Agronomic and crop management practiceshave to aim at reducing CO_2 and other greenhouse gas emissions by reducing tillage and residueburning and improving nitrogen use efficiency. Methane (CH_4) emissions that have a warming potential 21 times thatof CO_2 are common and significant in puddledanaerobic paddy fields and also when residues areburnt. A reviewof the other benefits of direct seeding and NT in RWareas of South Asia can be found in. Nitrous oxide (N_2O) has 310 times the warming potential ofcarbon dioxide, and its emissions are affected by poornitrogen management. Sensor-based technologies formeasuring normalized differential vegetative index andmoisture index have been used in Mexico and SouthAsia to help improve the efficiency of applied nitrogenand reduce nitrous oxide emissions Grace *et al.* (2003).

Conclusion

Crop production in the coming decade will have to produce more food from less land by making more efficient use of natural resources and with minimal impact on the environment. Only by doing this food production keep pace with demand and the productivity of land could be preserved for future generations. Use of sustainable management practices like conservation tillage (NT and ZT) and crop residue management practices described in this paper can help resolve this problem. Crop and soil management systems that help improve soil health parameters such as physical, biological and chemical as well as reduce farmer costs are essential. Development of appropriate equipments to allow these systems to be successfully adopted by farmers is a prerequisitefor success. Overcoming traditional mindsets about tillage by promoting farmer experimentationwith this technology in a participatory way will help accelerate adoption of CA practices.

References

Abdi D, Cade-Menun BJ, Ziadi N, Parent LE (2014). Long-term impact of tillage practices and phosphorus fertilization on soil phosphorus forms as determined by P nuclear magnetic resonance spectroscopy. *Journal of Environmental Quality,* 43(4):1431-1441.

Abrol IP, Sunita Sangar (2006). Sustaining Indian agriculture – conservation agriculture the way forward.*Current Science,* (91):1020-1025.

Anikwe MAN, Ubochi JN (2007). Short-term changes in soil properties under tillage systems and their effect on sweet potato (*Ipomea batatas* L.) growth and yield in an Ultisol in south-eastern Nigeria. *Australian Journal of Soil Research,* 45:351–358.

Balota EL, Colozzi Filho A, Andrade DS, Dick RP (2004). Long-term tillage and crop rotation effects on microbial biomass and C and N mineralization in a Brazilian Oxisol. *Soil and Tillage Research,* 77(2):137-145.

Beri V, Sidhu BS, Bahl GS, Bhat AK (1995). Nitrogen and phosphorus transformations as affected by crop residue management practices and their influence on crop yield. *Soil Use and Management,* 11:51-54.

Beri V, Sidhu BS, Bhat AK, Singh BP (1992). Nutrient balance and soil properties as affected by management of crop residues. In: Proceedings of the international symposium on nutrient management for sustained productivity. Vol II (Ed: Bajwa M S) pp. 133-135. Department of Soils, Punjab Agricultural University, Ludhiana, Punjab, India.

Bhattacharya R, Prakash V, Kundu S, Gupta HS (2006). Effect of tillage and croprotations on pore size distribution and soil hydraulic conductivity in sandy clay loam soil of the Indian Himalayas. *Soil and Tillage Research,* 86:129– 140.

Bijay-Singh, Shan YH, Johnson-beeebout SE, Yadvinder-Singh and Buresh RJ (2008). Crop residue management for lowland rice-based cropping systems in Asia. *Advances in Agronomy,* 98:118-199.

Conway Gordon R, Barbier EB (1990). After the Green Revolution: Sustainable Agriculture for Development, EarthscanPublications Ltd, London, p. 205.

Das SK, Rao ACS, Sharma KL (1991). Legume based crop rotation on a drylandAlfisol. *Indian Journal of Dryland Agricultural Research and Development,* 6 (1&2):46-59.

Dong W, Hu C, Chen S, Zhang Y (2009). Tillage and residue management effects on soil carbon and CO2 emission in wheat–corn double cropping system. *Nutrient Cycling in Agroecosystems,* 83: 27–37.

Du Preez CC, Steyn JT, Kotze E (2001). Long-term effects of wheat residue management on some fertility indicators of a semi-arid Plinthosol. *Soil and Tillage Research,* 63(1):25-33.

Duiker SW, Beegle DB (2006). Soil fertility distributions in long-term no-till, chisel/disk and moldboard plow/disk systems. *Soil and Tillage Research,* 88(1):30-41.

Eagle AJ, Bird JA, Hill JE, Horwath WR, van Kessel C (2001). Nitrogen dynamics and fertilizer use efficiency in rice following straw incorporation and winter flooding. *Agronomy Journal,* 93:1346-1354.

Franzluebbers AJ, Hons FM (1996). Soil-profile distribution of primary and secondary plantavailable nutrients under conventional and no tillage. *Soil and Tillage Research,* 39(3):229-239.

Galantini JA, Landriscini MR, Iglesias JO, Miglierina AM, Rosell RA (2000). The effects of crop rotation and fertilization on wheat productivity in the Pampean semiarid region of Argentina: 2. Nutrient balance, yield and grain quality. *Soil and Tillage Research,* 53(2):137- 144.

Garcia-Torres L, Benites J, Martinez-Vilela A, Holgado-Cabrera A (2003). Conservation agriculture: environment, farmers experiences, innovations, socio-economy, policy. Kluwer Academia Publishers; Dordrecht, The Netherlands; Boston, Germany; London, UK.

Govaerts B, Mezzalama M, Sayre KD, Crossa J, Nicol JM, Deckers J (2006). Long-term consequences of tillage, residue management, and crop rotation on maize/wheat root rot and nematode populations in subtropical highlands. *Applied Soil Ecology,* 32(3):305- 315.

Govaerts B, Sayre KD, Goudeseune B, De Corte P, Lichter K, Dendooven L, Deckers J (2009). Conservation agriculture as a sustainable option for the central Mexican highlands. *Soil and Tillage Research,* 103:222-230.

Govaerts B, Sayre KD, Lichter K, Dendooven L, Deckers J (2007). Influence of permanent bed planting and residue management on physical and chemical soil quality in rain fed maize–wheat system. *Plant and Soil,* 291: 39–54.

Grace PR, Harrington L, Jain MC, Robertson GP (2003). Long-term sustainability of the tropical and subtropical rice–wheat system: an environmental perspective. In Improving the productivity and sustainability of rice– wheat systems: issues and impact (eds J. K.

Ladha, J. Hill, R. K. Gupta, J. Duxbury & R. J. Buresh). ASA special publications 65, ch. 7, pp. 27–43. Madison, WI: ASA.

Gupta PK (2004). Residue burning in rice–wheat cropping system: Causes and implications. *Current Science,* (87): 1713–1717.

Harrington L, Olaf E (2005). Conservation agriculture and resource conserving technologies – A global perspective. In *ConservationAgriculture – Status and Prospects* (edsAbrol, I. P., Gupta, R. K. and Mallik, R. K.), Centre for Advancement of Sustainable Agriculture, New Delhi, pp. 1–12.

Ismail I, Blevins RL, Frye WW (1994). Long-term no-tillage effects on soil properties and continuous corn yields. *Soil Science Society of America Journal,* 58(1):193-198.

Kahlon MS, Lal R, Varughese M (2013). Twenty two years of tillage and mulching impacts on soil physical characteristics and carbon sequestration in Central Ohio. *Soil and Tillage Research,* 126:151–158.

Kassam A, Friedrich T, Derpsch R (2018). Global spread of Conservation Agriculture, *International Journal of Environmental Studies.* doi.org/10.1080/00207233.2018.149492 7.

Kumar S, Sharma KL, Kareemulla K, Ravindra Chary G, Ramarao CA, SrinivasaRao Ch, Venkateswarlu B (2011). Techno-economic feasibility of conservation agriculture in rainfed regions of India. *Current* Science, 101(9):1171-1181.

Kushwaha CP, Tripathi SK, Singh KP (2000). Variations in soil microbial biomass and N availability due to residue and tillage management in a dry land rice agro-ecosystem. *Soil and Tillage Research,* 56(3): 153-166.

Lal R (2005). Enhancing crop yields in the developing countries through restoration of the soil organic carbon pool in agricultural lands. *Land Degradation & Development,* 17:197–209.

Lupwayi NZ, Arshad MA, Rice WA, Clayton GW (2001). Bacterial diversity in waterstable aggregates of soils under conventional and zero tillage management. *Applied Soil Ecology,* 16(3):251-261

McGarry D, Bridge BJ, Radford BJ (2000). Contrasting soil physical properties after zero and traditional tillage of an alluvial soil in the semi-arid subtropics. *Soil and Tillage Research,* 53:105-115.

Osunbitan JA, Oyedele DJ, Adekalu KO (2005). Tillage effects on bulk density, hydraulic conductivity and strength of a loamy sand soil in south-western Nigeria. *Soil and Tillage Research,* 82:57–64.

Power JF, Doran JW, Wilhelm WW (1986). Crop residue effects on soil environment and dryland maize and soybean production. *Soil and Tillage Research,* 8:101-111

Prasad R, Gangaiah B, Aipe KC (1999). Effect of crop residue management in a rice–wheat cropping system on growth and yield of crops and on soil fertility. *Experimental Agriculture,* (35):427-435.

Rahman MA, Chikushi J, Saifizzaman M, Lauren JG (2005). Rice straw mulching and nitrogen response of no-till wheat following rice in Bangladesh. *Field Crops Research,* 91:71-81

Rahman MH, Okubo A, Sugiyama S, Mayland HF (2008). Physical, chemical and microbiological properties of an Andisol as related to land use and tillage practice. *Soil and Tillage Research,* 101(1):10-19.

Rasmussen KJ (1999). Impact of ploughless soil tillage on yield and soil quality: a Scandinavian review. *Soil and Tillage Research,* 53(1):3-14.

Sangar S, Abrol IP (2005). Conservation agriculture for transition to sustainable agriculture. *Current Science,* (88) 686–687.

Silva AD, Babujia LC, Franchini JC, Souza RA, Hungria (2010). Microbial biomass under various soil and crop management systems in short and long-term experiments in Brazil. *Field Crops Research,* 119: 20–26.

Singh G, Jalota SK, Sidhu BS (2005). Soil physical and hydraulic properties in a rice-wheat cropping system in India: effects of rice-straw management. *Soil Use and Management*, 21:17-21.

Singh, G., Marwaha, T.S. and Kumar, D. (2009). Effect of resource-conserving techniques on soil microbiological parameters under long-term maize (*Zea mays*) – wheat (*Triticum aestivum*) crop rotation. *Indian Journal of Agricultural Research*, 79(2):94–100.

Six J, Guggenberger G, Paustian K, Haumaier L, Elliott ET, Zech W (2001). Sources and composition of soil organic matter fractions between and within soil aggregates. *European Journal of Soil Science*, 52:607-618.

Somasundaram J, Chaudhary RS, Awanish Kumar D, Biswas AK, Sinha NK, Mohanty M, Hati KM, Jha P, Sankar M, Patra AK, Dalal RC (2018). Effect of contrasting tillage and cropping systems on soil aggregation, carbon pools and aggregate-associated carbon in rainfed Vertisols. *European Journal of Soil Science*, 69(5): 879-891.

Somasundaram J, Reeves S, Wang W, Heenan M, Dalal RC (2017). Impact of 47 years of no-tillage and stubble retention on soil aggregation and carbon distribution in a vertisol. *Land Degradation & Development*, 25:1589–1602.

Somasundaram J, Salikram M, Sinha NK, Mohanty M, Chaudhary RS, Dalal RC, Mitra NG, Blaise D, Coumar MV, Hati KM, Thakur JK, Neenu S, Biswas AK, Patra AK, Chaudhari SK (2019). Conservation agriculture effects on soil properties and crop productivity in a semiarid region of India. *Soil Research*, 57: 187-199.

Somasundaram, J., Sinha, N.K., Dalal, R.C., Lal, R., Mohanty, M., Naorem, A.K., Hati, K.M., Chaudhary, R.S., Biswas, A.K., Patra, A.K., Chaudhari, S.K. (2020). No-till farming and conservation agriculture in South Asia – Issues, challenges, prospects and benefits. *Critical Reviews in Plant Sciences*, 39(3): 236-279. DOI: 10.1080/07352689.2020.1782069..

Thuy NH, Yuhua S, Bijay-Singh, Wang K, Cai Z, Yadvinder- Singh, Buresh R J (2008). Nitrogen supply in rice-based cropping systems as affected by crop residue management. *Soil Science Society of America Journal*, 72:514-523

Verhulst N, Govaerts B, Verachtert E, Castellanos-Navarrete A, Mezzalama M, Wall P, Deckers J, Sayre KD (2010). Conservation agriculture, improving soil quality for sustainable production systems. *Advances in Soil Science: Food Security and Soil Quality*, 137-208.

Wall P (2011) What drives the conservation agriculture revolution? *FertAgric*May Issue 12. *Int FertInd Assoc* (IFA), Paris, France.www.fert.org

Wang X, Zhikuan J, Lianyou L (2015). Effect of straw incorporation on the temporal variations of water characteristics, water – use efficiency and maize biomass production in semi-arid China. *Soil and Tillage Research*, 153:36–41.

25

Energy Budgeting and Farm Mechanization in Conservation Agriculture

R.K. Naik and Alka Paswan

ICAR-Central Research Institute for Jute and Allied Fibres, Barrackpore West Bengal -700 121, India

Introduction

The basic needs of life are food, clothing and shelter. In the 21st century a fourth need added to mankind is energy. According to the laws of physics, energy is defined as capacity to do work. Energy, economics and the environment are mutually dependent. Also, there is very close relationship between agriculture and energy. The level and pattern of energy use in agricultural production system depends on variety of agronomic and socio-economic factors. Energy use in agriculture encouraged maximization of yields from limited available arable land, minimization of labour intensive operations and improving standard of living. Efficient and effective energy use in agriculture is one of the conditions for sustainable agricultural production, since it provides financial savings.

Classification of Energy Use in Agriculture

a) Basis of source

On the basis of source, the energy can be classified as direct and indirect energy.

Direct source of energy

The direct sources of energy are those which release the energy directly-like human labour, bullocks, stationary and mobile mechanical or electrical power units, such as diesel engines, electric motor, power tiller and tractors. The direct energy may be further classified as renewable and non-renewable sources of energy depending upon their replenishment.

Renewable direct sources of energy: The energy sources, which are direct in nature but renewable. Ex. Human beings, animals, solar and wind energy, fuel wood and agricultural waste etc.

Non-renewable direct sources of energy: The energy sources, which are direct in nature but not renewable. Ex. Coal, fossil fuels etc.

Indirect sources of energy

The indirect sources of energy are those which do not release energy directly but release it by conversion process. Some energy is invested in producing indirect sources of energy. Seeds, manures, chemicals, fertilizers and machinery are indirect sources of energy. The indirect energy may be further classified as renewable and non-renewable sources of energy depending upon their replenishment.

Renewable indirect sources of energy: Seed and manure termed as renewable indirect source of energy as they can be replenished in due course of time.

Non-renewable indirect sources of energy: The sources which are non-replenished are come under non-renewable indirect sources of energy. Ex. Chemicals, fertilizers and machinery etc.

b) Basis of comparative economic value

On the basis of comparative economic value, the energy may be classified as commercial and non-commercial energy.

Non-commercial energy

The energy sources which are available cheaply are called non-commercial sources of energy. Ex. Human labour, animal, fuel wood, agricultural waste, farmyard manure etc.

Commercial energy

The energy sources which are capital intensive are called commercial sources of energy. Ex. Petroleum products, electricity, chemicals, fertilizers, machinery, seeds etc.

The productivity and profitability of crop production depend upon energy consumption. In fact, energy use in agricultural production has become more intensive due to the use of energy inputs in the form of fossil fuel, chemical fertilizers, pesticides and electricity (Devsenapathy *et al.*, 2009). The amount of energy used depends on the mechanization level, quantity of active agricultural work and cultivable land. It was realized that crop yields and food supplies are directly linked to energy use (Stout, 1990). Energy input-output relationships in cropping systems vary with the crops grown in succession, type of soils, nature of tillage operations for seed bed preparation, nature and amount of organic manure and chemical fertilizers, plant protection measures, harvesting and threshing operations, yield levels and biomass production. Efficient use of energy in one of the principal requirements of sustainable agriculture for minimizing environmental problems, preventing destruction of naturals resources and promoting economical production system as well as sustainability. Thus energy auditing is one of the most common approaches to examining energy efficiency and environmental impact of the production system.

Conventional agriculture has largely been characterized by tillage, which include soil loosening and leveling for seed bed preparation, mixing fertilizer into soil, weed control and crop residue management for higher crop yields. It is an energy-intensive farm operation due to use of fossil fuel and contributes to about 30% of the total energy use in crop production Whereas, conservation agriculture (CA) principles are minimal soil disturbance (Zero tillage/ no tillage), retention of crop residue mulch and a rational use crop rotation along with profitability at farm level.

Energy Budgeting

Assessment of energy requirement and yield in relation to cropping systems is an efficient approach to gauge, quantify and determine the relationships between action and reaction, inputs and outputs and to augment energy use efficiency as well as crop productivity. Energy budgeting is necessary for efficient management of scare resources in improved agricultural production system. Also, it is not only helpful in the development of energy efficient technologies but also useful in life cycle assessment in the perspective of environment impact and ecological implications.

Energy index has been in use in modern agriculture worldwide for the purpose of evaluation and assessment of inputs and outputs.

Energy Calculation

In order to calculate the energy inputs and energy output, the entire data from land preparation to harvesting and post-harvest operation to be converted into output and input energy levels using equivalent energy coefficients given in Table 1.

376 Conservation Agriculture and Climate Change

Table 1: Equivalents for various sources of energy

Particulars	Units	Equivalent Energy (MJ)
INPUTS		
Farm operation		
Human labour (Adult man)	Man-hour	1.96
Human labour (Adult woman)	Man-hour	1.57
Operator	Man-hour	1.05
Machinery	Hour	62.70
Electricity	kWh	11.93
Electric motor	Hour	64.80
Diesel fuel	Litre	56.31
Petrol	Litre	48.20
Oil	Litre	50.23
Water	m^3	1.02
Transportation	Km ton	4.50
Agro chemicals		
Nitrogen	kg	60.60
Phosphorus	kg	11.10
Potassium	kg	6.70
FYM	kg (Dry mass)	0.30
Superior chemicals	kg	120.00
Inferior chemicals	kg	10.00
Seed		
Rice, wheat, lentil, pea, French bean	kg	15.20
Mustard	kg	25.50
Jute	kg	10.50
Potato/ramie/sisal	kg	5.10
Outputs		
Cereals	kg	14.70
Pulses	kg	14.70
Oilseeds	kg	25.00
Sugarcane	kg	5.30
Vegetables	kg	3.60
Fibre crops	kg	11.8
Jute stick	kg	18.40

*Source:*Mittal and Dhawan, (1988); Devasenapathy *et al.* (2009)

Input and Output Energy

The following equationsare used for computing input energy of various operation and inputs involved:

*Input Energy (MJ ha⁻¹) = Input energy of BE + ChE + Field operation
FOE* $\qquad$ (1)

Where, BE is biological energy, ChE is chemical energy, FOE is field operation energy

BE = Labour x Hours of work ha⁻¹ x Energy equivalent (MJ ha⁻¹) (1a)

ChE = Fertilizer energy (FE) + Toxin energy (TE) (1b)

FE = WF X EM XE

Where WF is recommended dose of fertilizer, EM is pure fertilizer percentage and E is energy required to produce fertilizer.

TE = wt x et x E, where wt is toxin weight (kg), et is pure toxin (%) and E is energy required for pure production.

FOE (MJ L⁻¹) = (FC x E) + (MO x E) (1c)

Where FC is fuel consumption (Lh⁻¹); MO is machinery operation h⁻¹; E is energy equivalent

Output Energy (MJ kg⁻¹) = Product yield + Ee (2)

Energy Efficiency

The various energy efficiency viz. net energy return, energy ratio, energy profitability, energy productivity, specific energy and energy intensiveness are to be calculated using the following equations.

$$\text{Net energy (MJ ha}^{-1}) = \text{Energy output (MJ ha}^{-1}) - \text{Energy input (MJ ha}^{-1}) \quad (3)$$

$$\text{Energy ratio} = \frac{\text{Energy output (Mj/ha)}}{\text{Energy intput (Mj/ha)}} \quad (4)$$

$$\text{Energy profitability (MJ ha}^{-1}) = \frac{\text{Net energy return (Mj/ha)}}{\text{Energy intput (Mj/ha)}} \quad (5)$$

$$\text{Energy productivity (kg MJ}^{-1}\text{ha}^{-1}) = \frac{\text{Total output (kg/ha)}}{\text{Energy input (Mj/ha)}} \quad (6)$$

$$\text{Specific energy (MJ kg}^{-1}\text{ha}^{-1}) = \frac{\text{Energy input (Mj/ha)}}{\text{Total output (kg/ha)}} \quad (7)$$

$$\text{Energy intensiveness (MJ Rs}^{-1}) = \frac{\text{Energy input (Mj/ha)}}{\text{Cost of cultivation (Rs/ha)}} \quad (8)$$

Energy Use Pattern

The pattern of energy input in agriculture is mainly divided into three heads based on operations, source and nature of energy.

378 Conservation Agriculture and Climate Change

a) Operation wise energy use pattern: Field preparation, fertilizer application, sowing, weeding, irrigation, plant protection, harvest and processing.

b) Source wise energy use pattern: Seed, fertilizers, diesel, chemicals, machinery and human.

c) Nature wise energy use pattern: Direct non-renewable, indirect non-renewable, direct renewable and indirect renewable.

Economic Analysis

In order to analyze the economic condition of agriculture system, the amount of payments as labour wages or purchasing fertilizer, chemical and seed have to be taken for total cost of production. The respective crop yield to be multiplied by its price as fixed by Government of India for different crops and considered as gross income. The benefit-cost ratio to be calculated as:

B:C ratio = Gross value of production (Rs ha^{-1})/ Total cost of production (Rs ha^{-1})

Farm Mechanization in Relation to Conservation Agriculture

Conservation Agriculture refers to a range of soil management practices that minimize effects on composition, structure and natural biodiversity and reduce erosion and degradation. Such practices include direct sowing / no-tillage, reduced tillage / minimum tillage, surface incorporation of crop residues and establishment of cover crops in both annual and perennial crops. As such the soil is protected from rainfall erosion and water runoff; the soil aggregates, organic matter and fertility level naturally increase and soil compaction is reduced. Furthermore, less contamination of surface water occurs, water retention and storage is enhanced, which allows recharging of aquifers.

Issues in Conservation Agriculture Mechanization

- Reduced tillage and direct seeding with appropriate machines that reduce costs of labour, Fuel and system of machinery.

- Increased opportunities for crop diversification which allow response to market opportunities as well as to the reduced supply of water. Example: Vegetables could be grown more profitably than rice and prevent soil structural degradation and hardpan formation from intensive rice cultivation.

- Mechanical weeding and inter-culture operations in dryland crops and other non-rice crops that reduce labour costs.

- Mechanical placement of fertilizers below the soil surface, leading to improved fertilizer use efficiency.

- Reduced seed requirement of a range of crops compared with conventional practices.

- Greater yields of all crops that can be obtained in response to better soil structure and surface drainage, more timely sowing because of direct seeding and other management operations.

Conservation Tillage

The term conservation tillage refers to a number of strategies and techniques for establishing crops in a previous crop's residues, which are purposely left on the soil surface. The principal benefits of conservation tillage are improved rain water conservation and the reduction of soil erosion. Additional benefits include reduced fuel consumption, reduced soil compaction, planting and harvesting flexibility, reduced labour requirements and improved soil health.

Developing a crop residue management system to conserve resources to fit in one's farm operations requires careful consideration of various factors such as:

- Capability of the planting machine to handle crop residue.

- Spreading of straw and chaff in the field.

- Total crop residue and its condition, particularly after heavy crop or green and damp straw or lodged stalk.

- Tall standing stubble reduces erosion but may cause plugging during planting.

- A balance is needed between standing stubble and chopped residue on surface.

Table 2: Considerations in various types of crop residue management

Crop	Points to remember
Pea/ Gram	➢ Most difficult residue to handle ➢ All vines must be picked up and then completely chopped and spread. ➢ The tillers / cultivators usually just bunch the pea stubble into piles
Cereals	➢ Cereals straw relatively easy to chop and spread ➢ Special attention is needed when lodged or damp. ➢ Routine harrowing after combine harvesting improve the spread of straw. ➢ Stubble taller than 40cm generally requires a subsequent field operation to chop and spread residue.

Three of the most common conservation tillage systems, Ridge tillage, No-till and Mulch till, are discussed below.

Ridge Tillage

Ridge tillage is a form of conservation tillage that uses specialized planters and cultivators to maintain permanent ridges on which row crops are grown. After harvest, the crop residue is left until planting time. The planter places the seed in the top of the ridge after pushing residue out of the way and slicing off the surface of the ridge-top. Ridges are re-formed during the last cultivation of the crop. Often, a band of herbicide is applied to the ridge-top during planting. With banded herbicide applications, two cultivations are generally used: one to loosen the soil and another to create the ridge. Because ridge tillage relies on cultivation to control weeds and reform ridges, this system allows farmers to further reduce their dependence on herbicides as compared to either conventional till or strict no-till systems.

Maintenance of the ridges is critical to a successful ridge tillage system. The equipment must accurately reshape the ridge, clean away crop residue, plant in the ridge center, and leave a viable seedbed (Fig. 1 & 2). The ridge-tillage cultivator not only removes weeds but also builds up the ridge. To harvest grain in ridged fields, one may need to put tall, narrow, dual wheels on the combine / harvesting machine. This modification permits the combine to straddle several rows, leaving the ridges undisturbed. Maintenance of the ridge becomes a consideration for each process.

Fig. 1 & 2: Ridge tillage and planting system

No Tillage

No-tillage systems, as the name implies, do not use tillage for establishing a seedbed. Crops are simply planted into the previous year's crop residue. No-till planters are equipped with coulters that slice the soil, allowing a double disc opener to place the seed at a proper depth. The slot is closed with a spring press wheel. Herbicides are typically used as the sole means for weed control in no-till systems. No-till methods have been criticized for a heavy reliance on chemical herbicides for weed control. Additionally, no-till farming requires careful management and expensive machinery for specially for seeding and harvesting applications (Fig. 3, 4 & 5.).

Also, increased insect and rodent pest problems have been reported. On the positive side, no-till methods offer excellent soil erosion control and require fewer trips across the field. No-till methods take several years to prove themselves. It's best to allow at least five years and remember that proper tracking of the system will be necessary from time to time. Soil quality improvements will come slowly, as earthworms and other soil organisms increase in number.

Fig. 3: T.O. zero till drill

Fig. 4: Power tiller-mounted Happy Seeder   **Fig. 5:** Tractor-mounted Happy Seeder

Mulch Tillage

The soil is disturbed prior to planting. Tillage tools such as chisels, field cultivators, disks, sweeps, or blades are used. The cropping practices survey assumes any system with 30 percent or more residue after planting that is not a no-till or ridge-till system is a mulch-till system.

Crop Stubble, Cover Crop Management under No Tillage

The objective of cover crop, residue management is to prepare the land for planting the subsequent commercial crop and to manage the weeds so that they cannot interfere with the crop growth (Fig.6). In conservation agriculture systems, the management should facilitate penetration of direct seeding equipment into the soil and into a favorable environment for seed germination, without obstructing the implement performance.

It is desirable that the residues from a cover crop should protect the soil for quite some time against the impacts of rainfall and that liberates allelopathic chemicals to suppress the germination of weeds. The release of these chemicals should be slow and gradual until the commercial crop is able to compete with the weeds. One of the factors influencing the release of allelopathic chemicals is the decomposition of organic matter.

Mechanical management of residues and cover crops can be done by using machetes, knives or sickles, knife rollers, slahers, mowers etc. or any similar implement. A common practice is to slash the weeds and residues of previous crops with a suitable device / machine before sowing. The residues are left on the surface and the subsequent crop is sown into it.

Fig. 6: Cover crops under No-till system

Manually carried motorized device and tractor operated crop stubble slasher

The motorized device and T.O. crop stubble slasher is recommended where well designed mechanized equipment is not available. In the first instance the crop stalk / straw is cut to knee height using motorized device after the cob / ear head harvest. In the second instance the cut stubble is slahed using tractor operated machine prior to planting of the rabi crop, usually resulting in the uniform spread of the previous crop residue. It is usually a combination of operations to manage the residue to avoid intensive tillage operations to burry the crop residue.

This way, there is no need to apply a high dose of herbicide to desiccate the vegetative cover, and will substantially reduce the crop production cost. If the crop stalk is cut, the stubble might re-sprout. Mechanical planting is also easier with specially developed planters, if the residues are slashed and spread but still in contact with the soil (Fig. 8).

Fig. 7: Stubble slasher

Fig. 8: T.O. stubble slasher

Knife rollers or chopping rollers

The knife roller is used to bend over and crush the weed or cover crop vegetation prior to planting the commercial crop, usually resulting in the death of the cover crop. It is usually known as a tool for animal traction or for tractors (Fig. 9). This operation is best carried out after flowering but before maturity of the seeds of the cover crop. This way, there is no need to apply a herbicide to desiccate the vegetative cover, and will substantially reduce the cost of production. In this case it is important that the knife roller only breaks and crushes but does not cut the cover crop plants so that they dry out and die. If the plants are cut, the stubble might re-sprout. Mechanical planting is also easier if the residues are not cut but still in contact with the soil.

Fig. 9: Cover crop chopping roller

Conclusion

Mechanization of agriculture is an essential input in modern agriculture. It enhances productivity, besides reducing human drudgery and cost of cultivation. Mechanization also helps in improving utilization efficiency of other inputs, safety and comfort of the agricultural worker, improvements in the quality and value addition of the produce. Efficient machinery helps in increasing production and productivity, besides enabling the farmers to raise a second crop or multi crop making the Indian agriculture attractive and a way of life by becoming commercial instead of subsistence. Energy budgeting is necessary for efficient management of scare resources in improved agricultural production system.

References

Borkar, U.N., Shambhu, V.B. and Naik, R.K. (2008). Jute and allied fibre crops: Energy analysis and mechanization *In*: Jute and Allied fibre Updates-Production and Technology, (Eds.; Karmakar, P.G., Hazra, S.K., Ramasubramanian, T., Mandal, R.K., Sinha, M.K. and Sen, H.S.), ICAR-CRIJAF, Barrackpore, Kolkata. pp 251-263.

Chaudhary, V.P., Gangwar, B., Pandey, D.K. and Gangwar, K.S. (2009). Energy auditing of diversified rice-wheat cropping systems in Indo-Gangetic plains. *Energy*, 34: 1091-1096.

Devasenapathy, P., Senthilkumar, G., and Shanmugam, P.M. (2009). Energy management in crop production. *Indian Journal of Agronomy*, 54(1): 80-90.

Mittal, J.P. and Dhawan, K.C. (1988). Research manual on energy requirements in agricultural sector. New Delhi: ICAR, 20-23.

Sims, B. and Kienzle, J. (2015). Mechanization of conservation agriculture for smallholders: Issues and options for sustainable intensification. *Environments*, 2: 139-166.

Stout, B.A. (1990). Handbook of energy for world agriculture. London: *Elsevier Applied Science*.

Climate Change Mitigation Strategies
& Socio Economic Impact of Conservation Agriculture

26

Mining Genetic Resources for Plant Traits Suited to Changing Climatic Conditions

Pratik Satya, Suman Roy, Laxmi Sharma, Soham Ray Amit Bera and Srinjoy Ghosh

ICAR-Central Research Institute for Jute and Allied Fibres

Barrackpore Kolkata – 700 121, West Bengal

Introduction

The inevitability of detrimental effects of climate change is looming large not only on the sustenance of agriculture, but also on the survival and progress of mankind. Estimates by the Organization for Economic Cooperation and Development (OECD) reveal that the atmospheric concentration of GHG would exceed 680 parts per million (ppm) CO_2-equivalents by 2050, which is about 50% higher than baseline year 2010 (OECD, 2012). Additionally, global temperature is predicted to increase by 2°C within this period, resulting in meansea level rise and erraticweather conditions. Consequently, the catastrophic effects on food consumption pattern by the end of 2050 will create malnutrition in at least 290 million people. Agricultural crops provide 63% of the human protein, and additionally supply almost 100% protein to the livestock, which are the second major sources of protein in human diet. Altogether, the detrimental effect of climate change on agriculture would make about 300 million additional people malnourished. Add to that a 2.7 billion increase in human population by 2050, coupled with shortage of water availability that would affect 3.9 billion people by 2050, and one can easily perceive the grim scenario for our kind in not so distant future.

It has been debated for a long time how increased CO_2 concentration would affect crop growth. Several studies have shown that the biomass of the crop would increase under a high-CO_2 concentration scenario, particularly for cereal crops. Other studies indicated that although net biomass would increase, nutrient content of the biomass would decrease. This is evident, since increased CO_2 concentration would channel biomass attributed to more carbon fixation, but would reduce all other major nutrients like nitrogen, zinc, iron, calcium, magnesium and most micronutrients. A report published in The Lancet Planetary Health (2019) (https://www.thelancet.com/journals/lanplh/home) shows that the global nutrient availability would be decreased by 19·5% for protein, 14·4% for iron, and 14·6% for zinc. This has led to reorientation of earlier agriculture scenario which projected a boost in agriculture productivity by 2050 under climate change. Therefore, the global investigations have aimed for probing options not only to stabilize, but to increase crop yield under these changing conditions. Overall, a consensus has been achieved for developing strategies to change crop architecture for combating the changing climate including generic as well as crop specific approaches. In this chapter, we will discuss some of the traits and look for the approaches used for mining genetic resources for identification of such traits.

Climate Resilient Traits

Development of climate resilient (CR) variety with specific traits requires not only in-depth understanding of genetics and physiology of CR traits but also genetic resources harboring those traits. Since the traditional yield-targeting breeding programmes provide little attention to generation of breeding lines that carry the CR traits, genetic stocks present in various germplasm repositories are the primary sources for these traits. Further, identification of CR traits is more challenging because the traits have to be identified and validated through long-term selection experiments under simulated conditions.

Higher Photosynthesis

Modified crop canopy

Breeding crops for achieving higher yield potential majorly targets maximization of canopy photosynthesis. Shoot architecture like plant height, branching angle, leaf area, leaf angle and phyllotaxy play important role in increasing the photosynthesis. Therefore, optimization of plant as well as canopy architecture is the key to accelerate photosynthesis in plants. The increase in crop yield till date has been achieved mainly by exploiting leaf area and active photosynthetic duration but not the rate of photosynthesis (Richards, 2000). Selection has also been emphasized on early canopy coverage to maximize interception of solar radiation (Joggi *et al.*, 1983). It can be thus be inferred that breeding along with modeling for better canopy architecture is important to gain the achievable yield targets.

Conversion of C_3 to C_4 plants

Photosynthetic pathways have been known to be correlated with carbon concentrating mechanisms and thus fixing more carbon in C_4 types. Majority of crop plants like rice, wheat and potato have C_3 photosynthetic pathway, while some like maize and sugarcane follow the C_4 route. The C_4 photosynthetic pathway increases the photosynthetic efficiency along with water and nutrient use. Additionally, enzymes like phosphoenol pyruvate carboxylase (PEPCase) along with ribulose-1,5-bisphosphate carboxylase/oxygenase (RuBisCO) can act more efficiently in C_4 plants under high light, increased temperature and dry conditions (Sage, 2002). In this context, integration of C_4 photosynthesis into C_3 crops is expected to increase the yield. But the complexity of introducing Kranz anatomy in C_3 crops as well as the metabolic engineering is a daunting task (Covshoff and Hibberd, 2012). However, worldwide initiative has already been taken to meet this grand challenge in rice and a fruitful outcome is expected in the near future.

Stay green traits

Senescence or ageing is a developmental process in crop plants characterized by loss of chlorophyll, nutrient remobilization and decrease in photosynthetic rate. The signal for initiation of senescence may be genetic or inducible through biotic or abiotic stress. Besides, this process is also subjected to hormonal control. Alteration in such signals ensuring delay in senescence will thus allow the plants to maintain its chlorophyll and photosynthesis and therefore create the 'stay green trait'. This trait is one of the important factors for improving productivity under normal as well as under stress condition. Although it has been known to be a complicated trait, breeding efforts have led to success in sorghum (Subudhi *et al.*, 2000), maize (Belicuas *et al.*, 2014) and wheat (Kumar *et al.*, 2010). Further understanding of the underlying physiological and molecular mechanisms of stay green trait is important for yield improvement especially under abiotic stress conditions.

Better Plant Growth and Development

Increased grain filling

Grain yield and quality is determined by the filling rate and duration of grain filling. Grain filling is known to be largely governed by quantitative trait loci. Physiological mechanisms underlying poor grain filling may be due to low source ability, limited sink capacity, hormonal imbalance, limitation in transportation of assimilates or reduced activity of enzymes involved in carbon metabolism. Several candidate genes, apart from genes related to sucrose-starch conversion, have been reported to improve the grain filling under normal as well as stress condition. For example, over expression of *GIF1* (*GRAIN INCOMPLETE FILLING 1*) in rice has been known to improve grain filling (Wang *et al.*, 2008). In addition, delay in complete senescence of whole plant also hampers proper grain filling. Environmental factors have been known to govern the grain filling efficiency. Soil moisture content and nutrient status are important factors that determine grain filling. Therefore, improved source-ink relationship, transportation of assimilates and root activity and signals may be emphasized while improving the grain filling in crop plants.

Tolerance to Heat Stress

Global temperature rise by 2-4°C has been predicted by the end of 21st century (Pachauri and Reisinger, 2007). This increase in temperature will incur major impact on plant growth and development throughout their life cycle. High temperature stress influence the plant development process right from seed germination to grain filling stages. High temperature stress negatively affects various developmental and physiological processes leading to induction of membrane injury, reduction in photosynthetic rate, hormonal signaling, accumulation of osmo-protectants and impaired stability of different proteins. All these may lead to alteration in vegetative and reproductive processes eventually resulting in yield loss.

Terminal heat tolerance

High temperature stress in plants refers to the increase in temperature during the growth stages that influence the growth and development process. However, some crops face this consequence at later stages of growth, particularly at reproductive stage that severely affect grain yield (Asthir, 2015). Such heat stress is known as 'terminal heat stress'. Terminal heat stress is a serious concern, as the reproductive phase of crop plants are exposed to high temperature. Thus, to minimize yield loss, the plants undergo various morphological, physiological and biochemical alternations. One of the plant's response strategies is to complete life cycle early (early maturity) as an adaptive mechanism. Canopy temperature depression is yet another strategy to reduce yield loss under higher temperature and has been found in crop plants like rice (Takai *et al.*, 2010), wheat (Ayeneh *et al.*, 2002) and lentil (Biju *et al.*, 2018).

Tolerance to UV Radiation

Secondary metabolite production

Depletion of ozone layer in recent years has increased the UV radiation stress on crop plants. UV radiation, especially UV-B is detrimental to life on earth precisely affecting the photosynthetic organisms. Productions of secondary metabolites like anthocyanin, flavonoids, phenols, ferulic acid, tannins etc are important plant response against UV-B stress (Liu *et al.*, 1995). Production of some chemicals like phytoalexin imparts tolerance to UV irradiation in rice (Park *et al.*, 2013). In addition, signaling cascades of UV-B have been well documented in maize which leads to the modification of different biosynthetic pathways (Casati *et al.*, 2011).

Elevated ROS detoxification

UV radiation increases reactive oxygen species (ROS) content and alters the redox state of crop plants, thereby creating oxidative stress. Non-enzymatic (Ascorbate; Asa, Glutathione; GSH, phenols, flavonoids, carotenoids like xanthophylls, beta-carotene, lycopene etc and tocopherols) and enzymatic antioxidants (superoxide dismutase; SOD, catalase; CAT, guaiacol peroxidase; GPX, enzymes of ascorbate-glutahione; AsA-GSH cycle such as ascorbate peroxidase; APX, mono-dehydroascorbate reductase; MDHAR, dehydroascorbate reductase; DHAR, and glutathione reductase; GR) play crucial role in arresting the build-up of ROS during oxidative stresses. In *Arabidopsis*, an AsA-deficient mutant *vtc1* showed more sensitivity to UV-B treatment than its wild-type plants (Gao and Zhang, 2008). Transgenic tomato with increased expression of cytosolic ascorbateperxoidase (cAPX) enhanced tolerance to UV-B radiation (Wang *et al.*, 2006).

Increased Water use Efficiency

Root architecture modification

Enhancing water use efficiency through modification in root traits has been a long desired target for researchers. Root system architecture (RSA) refers to spatial arrangement of root morphological characters like primary root length, lateral root numbers and length, root angle etc. and has been already modified throughout the course of breeding process without being significantly identified. RSA is majorly regulated by water and nutrient availability. QTLs have been extensively identified for RSA under water deficit conditions in various crops. Enormous genetic variability for RSA has been identified among species to optimize water use efficiency (Sanguineti *et al.*, 2007) under water limited conditions. Specific adaptations to the hydrology of the target environment should be taken into consideration for breeding with root traits (Tron*et al.*, 2015). Overexpression of *ANGUSTIFOLIA3* (*AN3*) in *Arabidopsis* improved water use efficiency by development of longer primary roots and more lateral roots (Meng and Yao, 2015). Deep root structures have been identified as drought avoidance mechanisms in rice (Wang *et al.*, 2019). In addition to this, proper genetic improvement for RSA inevitably requires proper phenotyping under target environment. Hence targeting the root agronomic traits through precise phenotyping in diverse genetic background (germplasm, breeding materials, mapping populations etc.) will enable to develop elite cultivars with improved water use efficiency.

Increased Nutrient Use Efficiency

Introduction of biological nitrogen fixation capacity

Biological nitrogen fixation (BNF) is well characterized in symbiotic root nodules of leguminous plants. Fixation of nitrogen biologically reduces the high energy and cost involved in synthetic nitrogen production (Glendining *et al.*, 2009). In addition to leguminous plants, several nitrogen fixing bacteria also known as diazotrophs have been identified to associate with non-leguminous plants symbiotically or as free living organisms for fixing nitrogen in rhizosphere (Santi *et al.*, 2013; Rilling *et al.*, 2018). They make association with the host plant by forming nodules on roots and N_2 fixation is assisted by the nitrogenase enzyme complex. Introduction of biological N-fixing capacity in non-symbiotic plants is an ambitious approach to mitigate external nitrogenous fertilizer application. In recent years, overexpression of nitrogenase genes and root nodule symbiosis development has been the two important approaches to introduce BNF capacity in cereals (Buren and Rubio, 2018).

Resistance to Diseases

Elevated horizontal resistance

Unpredictable change in climate also affects life cycle of the pathogens. Under such conditions, pathogens mutate more rapidly to develop new strains, the result of which are sometimes devastating to crop production. For example, the appearance of UG-99 strain of stem rust pathogen in wheat in the last decades temporarily paralyzed wheat cultivation in many countries. Under such condition, development of resistant cultivar is a major challenge, which requires quick assessment of pathogen diversity, screening germplasm, identification of effective chemicals/biological controlling agents and methods to restrict spread of the disease. While gene-specific vertical resistance is always desirable, horizontal broad-base, contributed by higher phenolics, flavonoids, ROS, phytoalexins or other mechanical traits that delay disease progress are expected to be more successful in combating such diseases under changing climate.

Secondary metabolite production and ROS detoxification

Plants defense mechanism against disease include the production of reactive oxygen species (ROS), changes in the cell wall constituents, production of secondary metabolite, production of defense peptides and proteins. Plants produce around 100,000 natural products or secondary metabolites (Dixon, 2001). Phenolic compounds like phenols, flavonoids, alkaloids, phenolic acids play important role against fungal diseases (Ashry *et al.*, 2011). Phytoalexins and phytoanticipants also functions actively against fungus. Breeding for specific higher secondary metabolite production in plant is becoming a primary area of crop improvement in changing climatic conditions.

Tolerance to Heavy Metal Toxicity

Various anthropogenic activities have increased the load of heavy metal in our environment thereby contaminating the soil and water bodies. Heavy metals interrupt the normal physiological development of plants by replacing the existing metals in pigments and enzymes, often reacts with thiol group of proteins and alter their configurations, induces oxidative stress and alters the redox state of plants. Conventional approaches for ameliorating soil and water from these contaminations are expensive, intrusive in nature, often not feasible and destroy the soil structure and its carbon exchange capacity. Therefore, cleaning such contaminated environment using plants known as phytoremediation is an important and potential tool for removing heavy metals without disturbing the soil structure. Phytoremediation includes various approaches like accumulation of heavy metals in a particular tissue (phytoextraction), plants and microbes mediated degradation of organic pollutants (phytodegradation), roots mediated accumulation of heavy metals (rhizofiltration), volatilization of pollutants (phytovolatilization) and reducing the bioavailability of heavy metals to plants (phytostabilization) (Tangahu *et al.*, 2011). The plants involved in phytoremediation act both as accumulators and excluders. Various mechanisms for heavy metal tolerance in plants include binding of metals to apoplast, compartmentalization, reduction of metal transport, chelation and active efflux.

Resistance to Pests

Climate change not only affects the crop but also the insects that survive on the crop. As a result, the insect-pest as well as the beneficial insects suffer a change in their life-cycle, rate of birth and mortality, larval growth, pupation behavior and mating behavior. In many cases, these changes result in appearance of new pests or conversion of a minor pest into major one. A study by Lehmann *et al.* (2020) shows that 41% of the insects are expected to cause higher damage in a warmer environment, while 55% gave a mixed response to climate change, which makes breeding for tolerance to insect-pest damage more complex. Several combination of traits, such as increased volatile organic compounds, avoidance, Insect non-preference, mechanical resistance and toxin production should be considered based on the response of the insect to the environment and crop. Volatile organic compounds (VOCs) are naturally occurring metabolites involved in defense against herbivores, pathogens as well as pest. Use of VOCs is very important in terms of sustainable agriculture. Natural volatile organic compounds do not affect the natural enemies of pest in the field (Lamy *et al.*, 2017). But these VOCs has a high rate of evaporation when applied in open field, which is a major challenge for popularization of this strategy.

Resistance to Hypoxia

Survival under limited oxygen

Excessive rainfall or excess irrigation with poor drainage facility imparts hypoxia or low oxygen availability to the crop plants. Hypoxia causes a considerable reduction in crop yield (Linkemer *et al.*, 1998). One of the prominent modifications in root structure under low oxygen environment is the formation of aerenchyma which improves the transport of gas inside submerged plant tissues. Aerenchyma, develops in rice, maize and wheat on imposed oxygen deficiency stress (Yamauchi *et al.*, 2018). Hypoxia has been well understood through hormonal regulations. It has been found that gaseous hormone ethylene plays major role in imparting these modifications like cell elongation, adventitious root formation and carbohydrate metabolism (Fukao and Serres, 2008). Gibberellic acid (GA) and abscisic acid (ABA) in coordination with ethylene plays crucial physiological role under low oxygen deficiency stress (Steffens *et al.*, 2006).

Tolerance to Salinity

Modified root architecture

Soil salinity is yet another important abiotic stress expanded over more than one billion hectares around the world (FAO, 2015). Munns *et al.*, (2020) showed that it is crucial for plants to maintain a low net rate of Na^+ and Cl^- uptake while taking up water from saline soils. Hence root architecture plays an important role here. Studies show that relative growth rate of primary roots in *Arabidopsis*is more drastically affected than the lateral roots under salt stress (Julkowska *et al.*, 2014) which indicates modification of root system architecture under salt stress. The elongation of primary root has been one of the major effects of salinity (Munns and Tester, 2008). Further studies showed that in wheat and barley, it is the cell division that is inhibited in primary root under salinity stress (Rahnama *et al.*, 2011; Shelden *et al.*, 2013) which leads to reduction in root elongation.

Salt-exclusion capacity

Another mechanism to overcome salt stress in plants is exclusion mechanism. Plants are categorized as glycophytes, xerophytes and halophytes with varying degree of salt-tolerant capacities. Halophytes are salt tolerant whereas glycophytes are unable to complete their life cycle when grown above 200 mM of NaCl (Flowers *et al.*, 2015). The halophytes are capable to withstand salinity due to one of the inherent ability of excluding salts from the roots. Salt exclusion mechanism prevents the entry of Na^+ ions to the vascular system. This is achieved either by selective ion transporters or by ion exporters (Rus *et al.*, 2001; Chen *et al.*, 2020).The well-known SOS1 Na^+/H^+ antiporter system has been exclusively studied in various crops and is known to exclude the uptake of Na^+ ions at the root surface thereby reducing the Na^+ uptake (Wu, 2018). Again, selective uptake of K^+ over Na^+ reduces sodium toxicity and potassium deficiency simultaneously (Rus *et al.*, 2001). Other transporters like HKT1 transporters, ions channels and vacuole localized Na^+/H^+ exchangers also play important role in Na exclusion mechanism.

Tolerance to Drought

Modified root architecture

Roots are the initial site of perception for water deficit stress. The response is then mediated via the stress hormone abscisic acid (ABA) to the closure of stomata (Zhu, 2002). Along with this the root system also modified in terms of length, distribution of root hairs, radial expansion, root angle etc. Deeper root system has been reported as an adaptive trait under water deficit condition in crops like wheat (Wasson *et al.*, 2012); chickpea (Kashiwagi *et al.*, 2006)and maize (Krishnamurthy *et al.*, 1999). Increase in root surface area by more production of root hairs is sometimes beneficial under dry soils (Mackay and Barber, 1985). Overexpression of *OsERF1* in rice roots lead to the development of larger aerenchyma and radial root growth. This modification in root architecture gives tolerance to water deficit stress in rice (Lee *et al.*, 2017). More expression of *DEEPER ROOTING 1* (*DRO1*) in rice imparts drought tolerance by increase in root growth angle and helping the root to grow deeper into the soil (Uga *et al.*, 2013).

Early maturity for drought escape

Plants adopt a mechanism to complete their life cycle before the onset of drought. This is known as drought escape. Drought escape mechanism involves high metabolic rate, rapid cell expansion and division, open stomata, high rate of gas exchange, higher photosynthesis with low water use efficiency. Early flowering is the most common mechanism of drought escape in cereals like wheat (Shavrukov *et al.*, 2017), rice (Xu *et al.*, 2005), and maize (Dahlan *et al.*, 1996). Legume crops shrink their life cycle to avoid the exposure of dry spells (Siddique *et al.*, 1993). Kooyers (2015) reviewed critically the evolution of drought escape mechanisms in natural herbaceous populations.

Approaches for Mining Genetic Resources for CR Traits

Genetic resources of crops are the primary sources for the CR traits. These resources can be subdivided into primary, secondary and tertiary gene pools. Additionally, a quaternary gene pool encompasses all the other sources excluding these gene pools. The major approaches for identification of climate resilient traits are-

Phenotypic screening of genetic resource

Phenotypic screening of germplasm repository is the first approach for identification of genotypes carrying the CR traits. Although most of the International Germplasm repositories publish reports on morphological descriptor traits, flowering and few other specific traits like disease or pest resistance, most of these did not concentrate on screening for CR traits like root morphology, photosynthesis potential, canopy structure or tolerance to abiotic stresses. However, some repositories have generated extensive data which can be very helpful for identification of CR traits. For example, the U.S. National Plant Germplasm System maintains and shares information on several such traits. In addition, the crop specific international institutes also provide morphological data of germplasm, including information on resistance to pest and diseases. If no such information is available or retrievable, one has to obviously screen genetic resources available for specific traits or create new variations. It is also desired that the National and Regional Germplasm repositories of a country should develop catalogs for CR traits for assisting breeding programmes.

Allele mining

Allele mining is the process of identification of allelic variants of a gene from a genetically related population or a collection of genetic resources. With availability of gene sequence data, one can design specific primers for cloning and sequencing of target genes from one or more genotypes. The new sequence may be same (monomorphic, same allele) or different (polymorphic, a new allele) from the original (wild-type) gene sequence. A population or collection of genetic resources thus may have several alleles that may differ from each other by single nucleotide difference (single nucleotide polymorphism or SNP) or may have differences in blocks of sequences (haplotype polymorphism). However, allele mining for a single gene in a large population is cost-intensive and would yield limited information. Therefore, strategies have been designed to identify multiple SNPs of different genes at a single experiment. Two such approaches are eco-TILLING and genome re-sequencing. TILLING (Targeting Induced Local Lesions in Genomes) is a mutation-based gene discovery approach, where variations in a genotype are created by mutation followed by gene-specific PCR-based identification of allelic variants that were created by the mutation (Koornneeff *et al.*, 1982). Once the mutant line has been identified that carries the desired mutation, it is critically phenotyped to determine gene effects. Eco-TILLING is a variation of TILLING where the initial population is a collection of germplasm or genetic population created by extensive hybridization (Till *et al.*, 2006). Another approach is low-depth resequencing of whole genome of all the genotypes present in the germplasm/population using a reference genome of the same species.

Mining genome sequence database

Decades of genomic researches have produced extensive digital sequence data from genome and transcriptome of various genetic stocks of crop genetic resources, which are curated in large, mostly open-access databases around the world. Standard genome information can be obtained from several sites including NCBI's RefSeq and Phytozome and species specific databases. A good number of the genes listed in these resources have been functionally annotated, linking the phenotypic variations with particular mutations in gene sequences. Moreover, experimental data published in research journals are continuously integrated in these databases to produce high quality annotations providing various information on gene effect such as gene structure, allelic variations, metabolic pathways, cellular locations and site of expressions.

Climate Resilient Crops: A Few Examples

Stay-green trait in sorghum

Stay-green is referred to a genetically controlled trait of delaying foliar senescence longer than normal. In Sorghum (*Sorghum bicolor* L. Moench), stay-green is considered as a desirable trait in consequence of drought stress tolerance and is characterized by delaying of premature senescence or death of plant due to drought at post-flowering stage (Xu *et al.*, 2000). Stay-green and drought response traits depict close association in sorghum and even genetically the quantitative trait loci (QTLs) of these two often tend to coincide (Thomas and Ougham, 2014). Besides it has also been observed at numerous occasions that selection for stay-green trait improves drought tolerance (Jordan *et al.*, 2012). Sorghum is very susceptible to drought at post-flowering stage as drought at this stage cause premature leaf senescence and hampers grain development as a consequence. Introduction of stay-green trait in such situation maintains leaf chlorophyll content which boosts photosynthesis and hence aid in normal grain fill under drought conditions (McBee *et al.*, 1983). Vis-à-vis it also maintains high carbohydrate level in stem, reduces lodging and imparts tolerance to charcoal stem rot (Borrell *et al.*, 2000; Burgess *et al.*, 2002). Stay-green trait in cultivated sorghum is supposed to have introgressed form East African perennial land races (Thomas *et al.*, 2000). Line B35, a derivative of Ethiopian durra and Nigerian landraces, is the source of stay-green trait used in most of the sorghum breeding programmes (Mahalakshmi and Bidinger, 2002). Genetic mapping studies identified that B35 contains four major stay-green QTLs (*Stg1, Stg2, Stg3*,and *Stg4*) which contribute 54% of the phenotypic variance among which *Stg2* being the important (Subudhi *et al.*, 2000). A study by Borrell *et al.* (2014) suggested, these QTLs modulates basic plant architecture (*viz.* root growth, leaf anatomy and canopy structure) in such a way that pre-flowering water demand is reduced and hence the conserved moisture can be utilized by the plant latter during grain development stage.

Terminal heat tolerance in wheat

High temperature stress during grain filling stage is termed as terminal heat stress. Being a temperate crop it is a serious threat in case of wheat cultivation. Under the present scenario of global warming, incorporating terminal heat tolerance seems to be one of the most important objectives for wheat breeding. Heat stress during pre-flowering or anthesis stage may reduce pollen number (Xuan *et al.*, 2013). Besides, terminal heat stress also affects grain quality (Spiertz *et al.*, 2006). Terminal heat tolerance is a quantitative trait (Bohnert *et al.*, 2006) and chromosomes 3A and 3B seems to be the hotspot for heat tolerance QTLs in wheat (Sun and Quick, 1991; Xu *et al.*, 1996). A novel QTL (*qDHY.3BL,* containing 22 genes) imparting heat and drought tolerance has been fine mapped and cloned form chromosome 3B (Thomelin *et al.*, 2016). Heat shock proteins (HSPs), which work as molecular chaperon during protein folding, play critical role in achieving thermotolerance (Majoul *et al.*, 2004; Peng *et al.*, 2006; Kamal *et al.*, 2010). Besides global gene expression profiling revealed that the heat responsive genes work in diverse biological pathways (Qin *et al.*, 2008). An array of wheat genes related to heat stress sensing and response has been characterized by overexpression screen (summarized in Ni *et al.*, 2018)which also includes genes for phytohormone- and microRNA-regulation besides genes encoding HSPs. Recently, epigenetic regulation of heat stress responses is coming up as an interesting area of research (Liu *et al.*, 2015) and genes like *TaGCN5*, which encodes a histone acetyl trasferase facilitating H3K9 and H3K14 acetylation, is showing promise results (Ni *et al.*, 2018). With current improvement in wheat transformation technology, transgenesis is also coming up as a viable option in this regard. For example *ZmEF-Tu1* and *ZmPEPC* overexpression has shown to provide increased tolerance to terminal heat stress (Fu *et al.*, 2008; Qi *et al.*, 2017).

Nitrogen fixation in cereals

Suboptimal nitrogen availability is the most prominent limiting factor for plant growth and development. The issue has long been addressed in conventional agriculture through applications of synthetic nitrogen-rich fertilizers which stabilize yield but also escalate input cost and contribute to environmental pollution, as well (Beatty and Good, 2011). Hence, alternate source of nitrogen is the need of the hour, especially in case of cereals because of their high nitrogen demand (Rosenblueth *et al.*, 2018). Biological nitrogen fixation (BNF) is the only solution to this problem; but it is mainly associated in legume crops. However, some soil-, seed- or water-borne diazotrophs (nitrogen-fixing bacteria) have been found to form associations with cereals, either as endophyte or via plant-root but without nodulation (Stoltzfus *et al.*, 1997, Rosenblueth *et al.*, 2018). Studies have been conducted for enhancing colonization of these diazotrophs in cereals (Elbeltagy *et al.*, 2001; Rosenblueth *et al.*, 2004; Kutter *et al.*, 2006). A few biofertilizers are commercially available which utilizes some these diazotrophic bacterial formulations. Apart from this, there has been a long-standing interest in introducing nitrogenase-encoding bacterial genes into plants so that they can fix their own nitrogen. Achieving this goal will require, either engineering the complete biosynthetic pathway of the nitrogenase enzyme into cereals (Beatty and Good, 2011) or engineering cereals to develop legume-like root-nodule (Rogers and Oldroyd, 2014). Some success has been achieved in expressing components of nitrogen fixing machinery (NifH and nifM) in chloroplasts of model plant (tobacco) but in cereals it remains elusive (Ivleva *et al.*, 2016, Bloch *et al.*, 2020). Engineering several *nif* genes simultaneously in a single background and maintaining their balanced expression in spatial/temporal manner in such a way so that oxygen toxicity of Nif proteins can be avoided represents major challenge in engineering nitrogen fixing cereals (Rosenblueth *et al.*, 2018). Recently, a "*fuse-and-cleave* virus-derived polyprotein strategy" devised by Yang *et al.* (2018) has shown some promise in this regard.

Converting C_3 plants to C_4 plants

Ribulose-1, 5-bisphosphate carboxylase/oxygenase (RuBisCO; EC 4.1.1.39), the most abundant protein in biosphere, is a promiscuous enzyme (Bar-On and Milo, 2019). It performs critically important and the first common step of carbon assimilation from inorganic- to organic-form during photosynthesis through its carboxylase activity. But under hot and arid climates, when CO_2

concentration drops, it favors oxygenase-activity and over carboxylase-activity and initiate energetically wasteful process of photorespiration (Parry *et al.*, 1999). Condensation of Ribulose-1,5-bisphosphate (RuBP) and CO_2 to 3-Phosphoglycerate (3-PG) by carboxylase-activity of RuBisCO is known as C_3 cycle as the first intermediate, 3-PG is a three-carbon compound. C_4 carbon fixation, on the other hand, is an extension C_3 carbon fixation which has specifically evolved in some plants of hot and arid zone to reduce wasteful oxygenase activity of RuBisCO which ultimately manifested as photorespiration (Gowik and Westhoff, 2011). There are morphophysiological variations in types of C_4 photosynthesis but basically all the variants tend to increase CO_2 concentration around RuBisCO (Gowik and Westhoff, 2011; Schuler *et al.*, 2016). Classical C_4 plants display Kranz anatomy which spatially separates RuBisCo for oxygen exposure and thus increase effective CO_2 concentration around the enzyme (Edwards and Voznesenskaya, 2010). CAM plants, on the other hand, temporally separate the process by collecting CO_2 at night and concentrate the collect CO_2 around RuBisCO during day time for carrying out efficient photosynthesis (Males and Griffiths, 2017). Based on current knowledge an elaborate multistep engineering strategy is required to install C_4 photosynthesis system in a C_3 plant. Fortunately, however, all enzymes required for C_4 photosynthesis are present in C_3 plant and there are some plants which naturally perform C_3-C_4 intermediate photosynthesis (Schuler *et al.*, 2016). Initial efforts of engineering C_4 photosynthesis in C_3 background were targeted to construct single-cell CO_2 concentrating machinery by overexpressing multiple C_4 enzymes but distinct subcellular organizations of the photosynthetic components in the single-celled was found to be problematic and cumbersome (Miyao *et al.*, 2011). The slow progress in this field of genetic engineering is partially due to morpho-physiological complication and partially due tomissing engineering tools and knowledge gaps (Schuler *et al.*, 2016). The field, even after 25 years of concentrated research efforts, still remains extremely challenging. However, it is still worth trying since any success in this field might warrant second green revolution.

Future Outlook

Climate change is predicted to elevate atmospheric concentration of GHG 680 parts per million (ppm) CO_2-equivalents along with increased temperatures across the world in the range of 1.6°C to as much as 6°C by 2050. Although rainfall is predicted to increase globally, some areas will receive less annual rainfall, while others may receive much more. These predicted changes in climate are expected to have fairly widespread impacts on agriculture, with poor countries highlighted as being particularly vulnerable, having already weak economies and limited institutional capacities to adapt.Attainment of food security in the climate change scenario requires climate smart varieties, which are capable of producing more even with harsh climatic conditions. A consensus has been reached for developing strategies to change crop architecture for combating the changing climate through climate resilient traits. There is a need to reorient the breeding programs to develop multipurpose crop varieties having high input use efficiency, improved nutritional qualities, and varieties that fit to ecosystem based cultivation system for sustainable development.

References

Ashry, N. A., and Mohamed, H. I. (2011). Impact of secondary metabolites and related enzymes in flax resistance and or susceptibility to powdery mildew. *World Journal of Agricultural Sciences*, 7(1): 78-85.

Asthir, B. (2015). Protective mechanisms of heat tolerance in crop plants. *Journal of Plant Interactions*, 10(1): 202-210.

Ayeneh, A., Van Ginkel, M., Reynolds, M. P., and Ammar, K. (2002). Comparison of leaf, spike, peduncle and canopy temperature depression in wheat under heat stress. *Field Crops Research*, 79(2-3): 173-184.

Bar-On, Y. M., and Milo, R. (2019). The global mass and average rate of rubisco. *Proceedings of the National Academy of Sciences*, 116(10), 4738-4743.

Beatty, P. H., and Good, A. G. (2011). Future prospects for cereals that fix nitrogen. S*cience*, 333(6041): 416-417.

Belícuas, P. R., Aguiar, A. M., Bento, D. A. V., Câmara, T. M. M., and de Souza Junior, C. L. (2014). Inheritance of the stay-green trait in tropical maize. *Euphytica*, 198(2): 163-173.

Biju, S., Fuentes, S., and Gupta, D. (2018). The use of infrared thermal imaging as a non-destructive screening tool for identifying drought-tolerant lentil genotypes. *Plant Physiology and Biochemistry*, 127: 11-24.

Bloch, S. E., Ryu, M. H., Ozaydin, B., and Broglie, R. (2020). Harnessing atmospheric nitrogen for cereal crop production. *Current Opinion in Biotechnology*, 62: 181-188.

Bohnert, H. J., Gong, Q., Li, P., and Ma, S. (2006). Unraveling abiotic stress tolerance mechanisms–getting genomics going. *Current Opinion in Plant Biology*, 9(2): 180-188.

Borrell, A. K., Hammer, G. L., and Henzell, R. G. (2000). Does maintaining green leaf area in sorghum improve yield under drought? II. Dry matter production and yield. *Crop Science*, 40(4): 1037-1048.

Borrell, A. K., Mullet, J. E., George-Jaeggli, B., van Oosterom, E. J., Hammer, G. L., Klein, P. E., and Jordan, D. R. (2014). Drought adaptation of stay-green sorghum is associated with canopy development, leaf anatomy, root growth, and water uptake. *Journal of Experimental Botany*, 65(21): 6251-6263.

Burén, S., and Rubio, L. M. (2018). State of the art in eukaryotic nitrogenase engineering. *FEMS Microbiology Letters*, 365(2): fnx274.

Burgess, M. G., Rush, C. M., Piccinni, G., and Schuster, G. (2002). Relationship between charcoal rot, the stay-green trait, and irrigation in grain sorghum. *Phytopathology*, 92, S10.

Casati, P., Morrow, D. J., Fernandes, J., and Walbot, V. (2011). UV-B signaling in maize: transcriptomic and metabolomic studies at different irradiation times. *Plant Signaling and Behavior*, 6(12): 1926-1931.

Chen, T. W., Gomez Pineda, I. M., Brand, A. M., and Stützel, H. (2020). Determining ion toxicity in cucumber under salinity stress. *Agronomy*, 10(5): 677.

Covshoff, S., and Hibberd, J. M. (2012). Integrating C4 photosynthesis into C3 crops to increase yield potential. *Current Opinion in Biotechnology*, 23(2): 209-214.

Dahlan, M., Mejaya, M. J., andSlamet, S. (1996). Maize losses due to drought in Indonesia and sources of drought tolerance and escape. *Drought-and Low N-Tolerant Maize*, 103.

Dixon, R. A. (2001). Natural products and plant disease resistance. *Nature*, 411(6839): 843-847.

Edwards, G. E., andVoznesenskaya, E. V. (2010). C 4 photosynthesis: Kranz forms and single-cell C 4 in terrestrial plants. In *C4 photosynthesis and related CO2 concentrating mechanisms* (pp. 29-61). Springer, Dordrecht.

Elbeltagy, A., Nishioka, K., Sato, T., Suzuki, H., Ye, B., Hamada, T., and Minamisawa, K. (2001). Endophytic colonization and in planta nitrogen fixation by a Herbaspirillum sp. isolated from wild rice species. *Applied and Environmental Microbiology*, 67(11): 5285-5293.

FAO, I. (2015). Status of the world's soil resources (SWSR)–main report. *Food and agriculture organization of the United Nations and intergovernmental technical panel on soils, Rome, Italy, 650.*

Flowers, T. J., Munns, R., andColmer, T. D. (2015). Sodium chloride toxicity and the cellular basis of salt tolerance in halophytes. *Annals of Botany*, 115(3): 419-431.

Fu, J., Momčilović, I., Clemente, T. E., Nersesian, N., Trick, H. N., and Ristic, Z. (2008). Heterologous expression of a plastid EF-Tu reduces protein thermal aggregation and enhances CO 2 fixation in wheat (*Triticumaestivum*) following heat stress. *Plant Molecular Biology*, 68(3): 277-288.

Fukao, T., and Bailey-Serres, J. (2008). Ethylene—a key regulator of submergence responses in rice. *Plant Science*, 175 (1-2): 43-51.

Gao, Q., and Zhang, L. (2008). Ultraviolet-B-induced oxidative stress and antioxidant defense system responses in ascorbate-deficient vtc1 mutants of Arabidopsis thaliana. *Journal of Plant Physiology*, 165(2): 138-148.

Glendining, M. J., Dailey, A. G., Williams, A. G., Van Evert, F. K., Goulding, K. W. T., and Whitmore, A. P. (2009). Is it possible to increase the sustainability of arable and ruminant agriculture by reducing inputs? *Agricultural Systems*, 99(2-3): 117-125.

Gowik, U., andWesthoff, P. (2011). The path from C3 to C4 photosynthesis. *Plant Physiology*, 155(1), 56-63.

Ivleva, N. B., Groat, J., Staub, J. M., and Stephens, M. (2016). Expression of active subunit of nitrogenase via integration into plant organelle genome. *PLoS One*, 11(8): e0160951.

Joggi, D., Hofer, U., andNösberger, J. (1983). Leaf area index, canopy structure and photosynthesis of red clover (Trifoliumpratense L.). *Plant, Cell and Environment*, 6(8): 611-616.

Jordan, D. R., Hunt, C. H., Cruickshank, A. W., Borrell, A. K., andHenzell, R. G. (2012). The relationship between the stay green trait and grain yield in elite sorghum hybrids grown in a range of environments. *Crop Science*, 52(3): 1153-1161.

Julkowska, M. M., Hoefsloot, H. C., Mol, S., Feron, R., de Boer, G. J., Haring, M. A., andTesterink, C. (2014). Capturing Arabidopsis root architecture dynamics with ROOT-FIT reveals diversity in responses to salinity. *Plant Physiology*, 166(3): 1387-1402.

Kamal, A. H. M., Kim, K. H., Shin, K. H., Choi, J. S., Baik, B. K., Tsujimoto, H., and Woo, S. H. (2010). Abiotic stress responsive proteins of wheat grain determined using proteomics technique. *Australian Journal of Crop Science*, 4(3): 196.

Kashiwagi, J., Krishnamurthy, L., Crouch, J. H., andSerraj, R. (2006). Variability of root length density and its contributions to seed yield in chickpea (*Cicer arietinum* L.) under terminal drought stress. *Field Crops Research*, 95(2-3): 171-181.

Koornneeff, M., Dellaert, L. W. M., and Van der Veen, J. H. (1982). EMS-and relation-induced mutation frequencies at individual loci in *Arabidopsis thaliana* (L.) Heynh. Mutation Research/Fundamental and Molecular Mechanisms of Mutagenesis, 93(1): 109-123.

Kooyers, N. J. (2015). The evolution of drought escape and avoidance in natural herbaceous populations. *Plant Science*, 234: 155-162.

Krishnamurthy, L., Johansen, C., andSethi, S. C. (1999). Investigation of factors determining genotypic differences in seed yield of non irrigated and irrigated chickpeas using a physiological model of yield determination. *Journal of Agronomy and Crop Science*, 183(1): 9-17.

Kumar, U., Joshi, A. K., Kumari, M., Paliwal, R., Kumar, S., andRöder, M. S. (2010). Identification of QTLs for stay green trait in wheat (*Triticumaestivum* L.) in the 'Chirya 3'×'Sonalika'population. *Euphytica*, 174(3): 437-445.

Kutter, S., Hartmann, A., andSchmid, M. (2006). Colonization of barley (Hordeumvulgare) with Salmonella enterica and Listeria spp. *FEMS Microbiology Ecology*, 56(2): 262-271.

Lamy, F. C., Poinsot, D., Cortesero, A. M., andDugravot, S. (2017). Artificially applied plant volatile organic compounds modify the behavior of a pest with no adverse effect on its natural enemies in the field. *Journal of Pest Science*, 90(2): 611-621.

Lee, D. K., Yoon, S., Kim, Y. S., and Kim, J. K. (2017). Rice OsERF71-mediated root modification affects shoot drought tolerance. *Plant Signaling and Behavior*, 12(1): e1268311.

Lehmann, P., Ammunét, T., Barton, M., Battisti, A., Eigenbrode, S. D., Jepsen, J. U. and Økland, B. (2020). Complex responses of global insect pests to climate warming. *Frontiers in Ecology and the Environment*, 18(3): 141-150.

Linkemer, G., Board, J. E., and Musgrave, M. E. (1998). Waterlogging effects on growth and yield components in late-planted soybean. *Crop Science*, 38(6): 1576-1584.

Liu, J., Feng, L., Li, J., and He, Z. (2015). Genetic and epigenetic control of plant heat responses. *Frontiers in Plant Science*, 6: 267.

Liu, L., Gitz Iii, D. C., and McClure, J. W. (1995). Effects of UV-B on flavonoids, ferulic acid, growth and photosynthesis in barley primary leaves. *Physiologia Plantarum*, 93(4): 725-733.

Mackay, A. D., and Barber, S. A. (1985). Effect of soil moisture and phosphate level on root hair growth of corn roots. *Plant and Soil*, 86(3): 321-331.

Mahalakshmi, V., andBidinger, F. R. (2002). Evaluation of stay-green sorghum germplasm lines at ICRISAT. *Crop Science*, 42(3): 965-974.

Majoul, T., Bancel, E., Triboï, E., Ben Hamida, J., andBranlard, G. (2004). Proteomic analysis of the effect of heat stress on hexaploid wheat grain: characterization of heat-responsive proteins from non-prolamins fraction. *Proteomics*, 4(2): 505-513.

Males, J., and Griffiths, H. (2017). Stomatal biology of CAM plants. *Plant Physiology*, 174(2): 550-560.

McBee, G. G., Waskom III, R. M., Miller, F. R., and Creelman, R. A. (1983). Effect of Senescene and Nonsenescence on Carbohydrates in Sorghum During Late Kernel Maturity States 1. *Crop Science*, 23(2): 372-376.

Meng, L. S., and Yao, S. Q. (2015). Transcription co-activator Arabidopsis ANGUSTIFOLIA 3 (AN 3) regulates water-use efficiency and drought tolerance by modulating stomatal density and improving root architecture by the transrepression of YODA (YDA). *Plant Biotechnology Journal*, 13(7): 893-902.

Miyao, M., Masumoto, C., Miyazawa, S. I., andFukayama, H. (2011). Lessons from engineering a single-cell C4 photosynthetic pathway into rice. *Journal of Experimental Botany*, 62(9): 3021-3029.

Munns, R., and Tester, M. (2008). Mechanisms of salinity tolerance. *Annual Review of Plant Biology*, 59: 651-681.

Munns, R., Passioura, J. B., Colmer, T. D., andByrt, C. S. (2020). Osmotic adjustment and energy limitations to plant growth in saline soil. *New Phytologist*, 225(3): 1091-1096.

Ni, Z., Li, H., Zhao, Y., Peng, H., Hu, Z., Xin, M., and Sun, Q. (2018). Genetic improvement of heat tolerance in wheat: recent progress in understanding the underlying molecular mechanisms. *The Crop Journal*, 6(1): 32-41.

Organisation for Economic Cooperation and Development (OECD). (2012). The OECD environmental outlook to 2050: Key findings on climate change. https://www.oecd.org/environment/out look to 2050.

Pachauri, R. K. and Reisinger, A. (2008). Climate change Synthesis report. Contribution of Working Groups I, II and III to the fourth assessment report. *Cambridge University Press, Cambridge*.

Park, H. L., Lee, S. W., Jung, K. H., Hahn, T. R., and Cho, M. H. (2013). Transcriptomic analysis of UV-treated rice leaves reveals UV-induced phytoalexin biosynthetic pathways and their regulatory networks in rice. *Phytochemistry*, 96: 57-71.

Parry, M. A., Keys, A. J., Bainbridge, G., Colliver, S. P., Andralojc, P. J., Paul, M. J. and Madgwick, P. J. (1999). Rubisco: attempts to reform a promiscuous enzyme. In *Regulation of Primary Metabolic Pathways in Plants* (pp. 1-16). Springer, Dordrecht.

Peng, D. H., Peng, H. R., Ni, Z. F., Nie, X. L., Yao, Y. Y., Qin, D. D., and Sun, Q. X. (2006). Heat stress-responsive transcriptome analysis of wheat by using GeneChip Barley1 Genome Array. *Progress in Natural Science*, 16: 1379-1387.

Qi, X., Xu, W., Zhang, J., Guo, R., Zhao, M., Hu, L., and Li, Y. (2017). Physiological characteristics and metabolomics of transgenic wheat containing the maize C 4 phosphoenolpyruvate carboxylase (PEPC) gene under high temperature stress. *Protoplasma*, 254(2), 1017-1030.

Qin, D., Wu, H., Peng, H., Yao, Y., Ni, Z., Li, Z., and Sun, Q. (2008). Heat stress-responsive transcriptome analysis in heat susceptible and tolerant wheat (*Triticumaestivum* L.) by using Wheat Genome Array. *BMC Genomics*, 9(1): 432.

Rahnama, A., Munns, R., Poustini, K., and Watt, M. (2011). A screening method to identify genetic variation in root growth response to a salinity gradient. *Journal of Experimental Botany*, 62(1): 69-77.

Richards, R. A. (2000). Selectable traits to increase crop photosynthesis and yield of grain crops. *Journal of Experimental Botany*, 51(suppl_1), 447-458.

Rilling, J. I., Acuña, J. J., Sadowsky, M. J., andJorquera, M. A. (2018). Putative nitrogen-fixing bacteria associated with the rhizosphere and root endosphere of wheat plants grown in an andisol from southern Chile. *Frontiers in Microbiology*, 9: 2710.

Rogers, C., andOldroyd, G. E. (2014). Synthetic biology approaches to engineering the nitrogen symbiosis in cereals. *Journal of Experimental Botany*, 65(8): 1939-1946.

Rosenblueth, M., andMartínez-Romero, E. (2004). Rhizobium etli maize populations and their competitiveness for root colonization. *Archives of Microbiology*, 181(5): 337-344.

Rosenblueth, M., Ormeño-Orrillo, E., López-López, A., Rogel, M. A., Reyes-Hernández, B. J., Martínez-Romero, J. C. and Martínez-Romero, E. (2018). Nitrogen fixation in cereals. *Frontiers in Microbiology*, 9: 1794.

Rus, A., Yokoi, S., Sharkhuu, A., Reddy, M., Lee, B. H., Matsumoto, T. K., and Hasegawa, P. M. (2001). AtHKT1 is a salt tolerance determinant that controls Na+ entry into plant roots. *Proceedings of the National Academy of Sciences*, 98(24): 14150-14155.

Sage, R. F. (2002). Variation in the k cat of Rubisco in C3 and C4 plants and some implications for photosynthetic performance at high and low temperature. *Journal of Experimental Botany*, 53(369): 609-620.

Sanguineti, M. C., Li, S., Maccaferri, M., Corneti, S., Rotondo, F., Chiari, T., andTuberosa, R. (2007). Genetic dissection of seminal root architecture in elite durum wheat germplasm. *Annals of Applied Biology*, 151(3): 291-305.

Santi, C., Bogusz, D., andFranche, C. (2013). Biological nitrogen fixation in non-legume plants. *Annals of Botany*, 111(5): 743-767.

Schuler, M. L., Mantegazza, O., and Weber, A. P. (2016). Engineering C4 photosynthesis into C3 chassis in the synthetic biology age. *The Plant Journal*, 87(1): 51-65.

Shavrukov, Y., Kurishbayev, A., Jatayev, S., Shvidchenko, V., Zotova, L., Koekemoer, F., and Langridge, P. (2017). Early flowering as a drought escape mechanism in plants: How can it aid wheat production?. *Frontiers in Plant Science*, 8: 1950.

Shelden, M. C., Roessner, U., Sharp, R. E., Tester, M., andBacic, A. (2013). Genetic variation in the root growth response of barley genotypes to salinity stress. *Functional Plant Biology*, 40(5): 516-530.

Siddique, K. H. M., Walton, G. H. and Seymour, M. (1993). A comparison of seed yields of winter grain legumes in Western Australia. *Australian Journal of Experimental Agriculture*, 33(7): 915-922.

Spiertz, J. H. J., Hamer, R. J., Xu, H., Primo-Martin, C., Don, C., and Van Der Putten, P. E. L. (2006). Heat stress in wheat (Triticumaestivum L.): Effects on grain growth and quality traits. *European Journal of Agronomy*, 25(2): 89-95.

Steffens, B., Wang, J., andSauter, M. (2006). Interactions between ethylene, gibberellin and abscisic acid regulate emergence and growth rate of adventitious roots in deepwater rice. *Planta*, 223(3): 604-612.

Stoltzfus, J. R., So, R. M. P. P., Malarvithi, P. P., Ladha, J. K., and De Bruijn, F. J. (1997). Isolation of endophytic bacteria from rice and assessment of their potential for supplying rice with biologically fixed nitrogen. *Plant and Soil*, 194(1-2): 25-36.

Subudhi, P. K., Rosenow, D. T., and Nguyen, H. T. (2000). Quantitative trait loci for the stay green trait in sorghum (Sorghum bicolor L. Moench): consistency across genetic backgrounds and environments. *Theoretical and Applied Genetics*, 101(5-6): 733-741.

Sun, Q. X., and Quick, J. S. (1991). Chromosomal locations of genes for heat tolerance in tetraploid wheat. *Cereal Research Communications*, 431-437.

Takai, T., Yano, M., and Yamamoto, T. (2010). Canopy temperature on clear and cloudy days can be used to estimate varietal differences in stomatal conductance in rice. *Field Crops Research*, 115(2): 165-170.

Tangahu, B. V., Sheikh Abdullah, S. R., Basri, H., Idris, M., Anuar, N., andMukhlisin, M. (2011). A review on heavy metals (As, Pb, and Hg) uptake by plants through phytoremediation. *International Journal of Chemical Engineering*. doi:10.1155/2011/939161

Thomas, H., and Howarth, C. J. (2000). Five ways to stay green. *Journal of Experimental Botany*, 51(suppl_1): 329-337.

Thomas, H., andOugham, H. (2014). The stay-green trait. *Journal of Experimental Botany*, 65(14): 3889-3900.

Thomelin, P. M., Bonneau, J., Taylor, J. D., Choulet, F., Sourdille, P., Langridge, P. and Fleury, D. (2016, January). Positional cloning of a QTL, qDHY. 3BL, on chromosome 3BL for drought and heat tolerance in bread wheat. In Plant and Animal Genomie Conference XXIV (p. np).

Till, B., Zerr, T., Comai, L. and Henikoff S. (2006). A protocol for TILLING and Ecotilling in plants and animals. *Nature Protocols*, 1:2465–2477. https://doi.org/10.1038/nprot. 2006.329.

Tron, S., Bodner, G., Laio, F., Ridolfi, L., andLeitner, D. (2015). Can diversity in root architecture explain plant water use efficiency? A modeling study. *Ecological Modelling*, 312: 200-210

Uga, Y., Sugimoto, K., Ogawa, S., Rane, J., Ishitani, M., Hara, N., and Inoue, H. (2013). Control of root system architecture by DEEPER ROOTING 1 increases rice yield under drought conditions. *Nature Genetics*, 45(9): 1097-1102.

Wang, E., Wang, J., Zhu, X., Hao, W., Wang, L., Li, Q., ... and Ma, H. (2008). Control of rice grain-filling and yield by a gene with a potential signature of domestication. *Nature Genetics*, 40(11): 1370-1374.

Wang, X., Samo, N., Li, L., Wang, M., Qadir, M., Jiang, K., and Hu, Y. (2019). Root Distribution and Its Impacts on the Drought Tolerance Capacity of Hybrid Rice in the Sichuan Basin Area of China. *Agronomy*, 9(2): 79.

Wang, Y., Wisniewski, M., Meilan, R., Cui, M. and Fuchigami, L. (2006). Transgenic tomato (Lycopersicon esculentum) overexpressing cAPX exhibits enhanced tolerance to UV-B and heat stress. *Journal of Applied Horticulture*, 8(2): 87-90.

Wasson, A. P., Richards, R. A., Chatrath, R., Misra, S. C., Prasad, S. S., Rebetzke, G. J., and Watt, M. (2012). Traits and selection strategies toimprove root systems and water uptake in water-limited wheat crops. *Journal of Experimental Botany*, 63(9): 3485-3498.

Wu, H. (2018). Plant salt tolerance and Na+ sensing and transport. *The Crop Journal*, 6(3): 215-225.

Xu, J. L., Lafitte, H. R., Gao, Y. M., Fu, B. Y., Torres, R. and Li, Z. K. (2005). QTLs for drought escape and tolerance identified in a set of randomintrogression lines of rice. *Theoretical and Applied Genetics*, 111(8): 1642-1650.

Xu, R., Sun, Q., and Zhang, S. (1996). Chromosomal location of genes for heat tolerance as measured by membrane thermostability of common wheat cv. Hope. *Hereditas*, 18(4): 1-3.

Xu, W., Rosenow, D. T., and Nguyen, H. T. (2000). Stay green trait in grain sorghum: relationship between visual rating and leaf chlorophyll concentration. *Plant Breeding*, 119(4): 365-367.

Xuan, Y., Xu, T., Baode, C., Zhan, T. I. A. N. and Honglin, Z. (2013). Impacts of heat stress on wheat yield due to climatic warming in China. *Progress in Geography*, 32(12): 1771-1779

Yamauchi, T., Colmer, T. D., Pedersen, O. and Nakazono, M. (2018). Regulation of root traits for internal aeration and tolerance to soil waterlogging-flooding stress. *Plant Physiology*, 176(2): 1118-1130.

Yang, J., Xie, X., Xiang, N., Tian, Z. X., Dixon, R., and Wang, Y. P. (2018). Polyprotein strategy for stoichiometric assembly of nitrogen fixation components for synthetic biology. *Proceedings of the National Academy of Sciences*, 115(36): E8509-E8517.

Zhu, J. K. (2002). Salt and drought stress signal transduction in plants. *Annual Review of Plant Biology*, 53(1): 247-273.

27

Scope of Agroforestry Systems for Climate Change Adaptation

R. Saha, D. Barman, Suman Roy and Pradipta Samanta

ICAR-Central Research Institute for Jute and Allied Fibres, Barrackpore West Bengal– 700 121, India

Introduction

Conservation of biodiversity and mitigation of global warming are two major environmental challenges today. In the context of climate change and the global carbon cycle, the relationship between plant biodiversity and soil organic carbon (SOC) sequestration has become a subject of considerable scientific interest. The Earth's terrestrial vegetation plays a pivotal role in the global carbon cycle. Not only are tremendous amounts of carbon stored in the terrestrial vegetation, but large amounts are also actively exchanged between vegetation and the atmosphere. Anthropogenic perturbations exacerbate the emission of CO_2 from soil caused by decomposition of soil organic matter (SOM) or soil respiration (Schlesinger, 2000). The emissions are accentuated by agricultural activities including tropical deforestation and biomass burning, plowing, drainage of wetlands and low-input farming or shifting cultivation. In addition to its impact on decomposition of SOM, macroclimate has a large impact on a fraction of the SOC pool which is active. Conversion of natural to agricultural ecosystems increases maximum soil temperature and decreases soil moisture storage in the root zone, especially in drained agricultural soils (Lal, 1996). Thus, land use history has a strong impact on the SOC pool (Pulleman *et al.*, 2000). The dynamic relationship between plant biodiversity and SOC depicts that any land use practices that increase vegetative cover, or reduce its removal, could have an influence on the global carbon budget by increasing the terrestrial carbon sink.

Promoting agroforestry is one option many perceive as a major opportunity to deal with problems related to land-use and CO_2-induced global warming. Agroforestry is defined as any land-use system that involves the deliberate retention, introduction or mixture of trees or other woody perennials with agricultural crops, pastures and/or livestock to exploit the ecological and economic interactions of the different components (Nair, 1993 and Young, 1997). Historical evidence showed that agroforestry has been widely practised through the ages as a means of achieving agricultural sustainability and slowing the negative effects of agriculture such as soil degradation and desertification.

Importance of Agroforestry in Resource Conservation

The agroforestry system (AFS) has today become an established approach to integrated land management, not only for renewable resource production, but also for ecological considerations. It represents the integration of agriculture and forestry to increase the productivity and sustainability of the farming system. Agroforestry (also known as multistrata tree gardens or analogue forests) and homegardens are other variants of these complex systems, but involve higher plant diversity. Trees play an important role in soil C sequestration (Takimoto *et al.*, 2009): with an increase in the number of trees (high tree density) in a system, the overall biomass production per unit area of land will be higher, which in turn may promote more C storage in soils. In fact, recent research has reported higher soil C stock (amount of carbon stored in soil) under deeper soil profiles in agroforestry systems compared to treeless agricultural or pasture systems under similar ecological settings (Haile *et al.*, 2008; Nair *et al.*, 2009). Multipurpose trees (MPTs) form an integral component of different agroforestry interventions and models. MPTs, besides furnishing multiple outputs like fuel, fodder, timber, and other minor products, also help in the improvement of soil and other ecological conditions. The trees play various functions, including shading crops to reduce evapotranspiration, erosion control and nutrient cycling (Young, 1997). Some of the potential AFS are agri-horti-silviculture, multistoreyed AFS, home garden, agri-silviculture, horti-pastoral, Agri-horti-silvi-pastoral etc.

C Sequestration Potential in Agroforestry Systems

Agroecosystems play a central role in the global C cycle and contain approximately 12% of the world terrestrial C. The terrestrial (plant and soil) C is estimated at 2000 ± 500 Pg, which represents 25% of global C stocks (DOE, 1999). The sink option for CO_2 mitigation is based on the assumption that this figure can be significantly increased if various biomes are judiciously managed and/or manipulated (Table 1). It is clear that forests have tremendous potential for C sequestration (1-3 Pg yr^{-1}) so as to reduce GHG concentrations in the atmosphere. In this connection, agroforestry systems will have a great impact on the flux and long-term storage of C in the terrestrial biosphere (Dixon, 1995) as the area of the world under agroforestry will increase substantially in the near future undoubtedly. The amount of C sequestered largely depends on the agroforestry system put in place, the structure and function of which are, to a great extent, determined by environmental and socio-economic factors. Other factors influencing carbon storage in agroforestry systems include tree species and system management.

Table 1: Categorisation of biomes and their C sequestration (CS) potential

Biomes	Primary method to increase CS	Potential CS (Pg C yr[1])
Agricultural lands	Management	0.85–0.90
Biomass croplands	Manipulation	0.50–0.80
Grasslands	Management	0.50
Rangelands	Management	1.20
Forests	Management	1–3
Deserts and degraded lands	Manipulation	0.80–1.30

Source: DOE, (1999)

A well-managed agroforestry system has greater amounts of C sequestration potential in and out of the soil. About 20–25% of the total living biomass of trees prevails in the roots and there is constant addition of organic matter to the soil through decaying dead roots, which leads to improvements in the C status of the soil (Balkrishnan and Toky, 1993). Better soil aggregation under natural forest, multi-storied AFSs and silvi-hortipastoral systems maintaining intensive vegetative cover throughout the year could be ascribed to the effect of the higher percentage of organic carbon (Table 2).

Table 2: Effect of agroforestry systems on soil organic carbon and fertility in humid tropics

Agroforestry systems	Organic C (g kg[-1])	Available N (ppm)	Available P (ppm)	Exchangeable Ca	Exchangeable Mg
				[cmol (p$^+$)/ kg]	
Silvi-horti-pastoral	22.5	199.40	0.94	0.31	0.48
Multistoried AFS	24.6	216.90	3.36	0.65	0.71
Natural forest	27.0	167.20	0.63	0.26	0.16

Source: Saha *et al.* (2010)

Role of Agroforestry in Soil Fertility Improvement

Compared to natural, a managed agricultural ecosystem has greater amounts of nutrient flowing in and out, less capacity for nutrient storage and less nutrient recycling. The trees are able to maintain or improve soil health, which is evident from the high fertility status and closed nutrient cycling under natural forest. The processes by which trees improve the soil fertility are:

- Photosynthetic fixation of carbon and its transfer to the soil via litter and root decay

- Nitrogen fixation by all leguminous trees and also by few non-leguminous species (e.g., Alder and Casuarinas)

- Improved nutrient retrieval by tree roots including mycorrhiza from lower horizon

- Erosion control by combination of cover and barrier effect

- Efficient uptake of nutrients that would otherwise lost by leaching,

- Soils under trees have favourable structure and water holding capacity through organic matter maintenance and root action

- Exudation of growth promoting substance

- Pruning of trees synchronize the time of release of nutrients from litter with crop demand

Nitrogen Fixing Tree Species Suitable for RCTs

The N-fixing trees and shrubs play a vital role in maintaining soil fertility by reducing land degradation. These species are *Acacia, Albizia, Alnus, Calliandra, Casuarina, Dalbergia, Gliricidia, Leicaena, prosopis, Robinia, Sesbania, Crotolaria, Desmodium, Flemingia, Indigofera, Tephrosia etc.* These trees fix the atmospheric nitrogen as a result of symbiotic association with Rhizobial bacteria in case of Leguminoseae and in some members of Ulmaceae and actinomycetes, genus Frankia in case of other trees. The rhizobia or Frankia provide amino acids for combined organism. Three genera of rhizobia namely *Rhizobium, Bradyrhizobium* and *Azorhizobium* are involved in N-fixation in nitrogen fixing trees (NFTs). The microbial symbiont infects the roots of nitrogen fixing trees and stimulates the tree to form nodule. Within the nodule the symbiont multiplies and fixes atmospheric nitrogen. The tree contributes to the process of symbiosis by providing suitable environment for N-fixation and energy in the form of carbohydrate from photosynthesis. This fixed nitrogen is in a form that is useable to the individual plant itself or in the soil (through leaching from roots, leaf fall, lopping or green manuring) and available to domestic animal (through fodder). Besides these, there are a number of economically important non-legumes that fix nitrogen e.g. *Alnus, Casurina* etc. On the basis of diagnostic survey and appraisal of the existing agroforestry systems/ practices, Dhyani and Tripathi (2000) have recommended eight nitrogen fixing tree species suitable for agroforestry systems in Meghalaya.

Nutrient Cycling in Agroforestry Production Systems

Leaves of trees generally contain higher nutrient than other parts of the plant. In this respect, nitrogen-fixing trees are far superior to non-nitrogen fixing trees with major role in improving nutrient cycling. Mechanisms involved are (i) uptake from lower soil horizon, (ii) reduction of leaching to lower horizons by deep root systems, (iii) balanced nutrient supply and (iv) improvement in the ratio of available nutrients and the fixed nutrients.

Leaf fall exerts an important influence on physical, chemical and biological characteristics of soils and thus balances the nutrients present in soil. Tree leaves, twigs, flowers etc., are periodically or continuously dropped on the ground to form litter, which decomposes in time to release the nutrient contained therein to the soil from where they are re-circulated. George and Verghes (1990) also observed that in a 10-year-old *Eucalyptus globulus* plantation, total litter production was 8492 kg ha^{-1} of which, leaf litter contributed 40 per cent, twigs 38 per cent and bark 22 per cent. The highest concentration of N, P and K was in leaf with a total return of 58 kg N, 48 kg K and 4.6 kg P ha^{-1}, indicating that major nutrient return was through leaf litter and lowest through bark.

In an undisturbed tree system, if nutrients supply is not met by the biological cycling, the import of nutrients from atmosphere is taken care of by soil microorganisms. The weathering of parent rock also supplies most of the required nutrients. The nutrient cycling in most of the agroforestry systems involves conservation of resources and their partitioning between soil and plants. Although trees draw nutrients, a large portion of absorbed nutrients is returned to soil through crown wash, stem flow and through litter fall. The potential nutrient returns in form of a litter are in the order of 80-120 kg N, 8-12 kg phosphorus, 40-120 kg potassium and 20-60 kg calcium ha^{-1} (Sreemannarayana, 2003). The accumulated litter after decomposition, releases nutrients for reuse by tree stand. Hence additions of external nutrients are not required to sustain tree growth and soil productivity. Thereby closed production system of organic agriculture is ensured in agroforestry system.

Role of Tree Roots in Soil Health Improvement

About 25 per cent of the total living biomass of the trees is in roots and there is a constant addition of organic matter to the soil through dead and decaying roots, although it has been recognized that major addition of organic matter to soils is through litter, which constitute nutrient absorbed by roots from deeper soil layer. Tree roots harbor many useful organisms and encourage beneficial microbial activity in the vicinity of roots or rhizosphere. The nitrogen fixing capacity of nodulative organism like rhizobium, brady rhizobium and frankia in association with leguminous non- leguminous trees is well known and varies widely. The deep-rooted woody perennials contribute to soil organic matter through shedding of their leaves and root and thereby improve soil structures, fertility and water holding capacity. They absorb nutrients from greater depths and deposit them on the surface in organic matter so that they are available to shallow rooted crops in greater quantities.

Alley Cropping

Alley cropping is an agroforestry practice in which fast-growing trees and shrubs are established in hedgerows on arable cropland and annual food crops cultivated in the alleys between the hedgerows. The hedgerow are pruned periodically during cropping cycles to prevent shading of the companion crop, with the pruning applied to the soil as green manure and/or mulch. Hedgerows are allowed to grow freely to cover the land between cropping cycles. Majority of the hedgerow species are nitrogen fixing, leguminous species, and several non N-fixing species have also shown promise. Thus in an alley cropping systems, the cropping and fellow phases can take place concurrently on the same land, allowing the farmer to crop the land for an extended period when socioeconomic conditions do not allow adequately long fellow period for sustaining soil productivity.

Multipurpose Trees

The multipurpose tree species (MPTs) form an integral component of different agroforestry interventions in crop sustainability. The MPTs, besides furnishing multiple outputs such as fuel, fodder, timber and other miscellaneous products, help in the improvement of soil health and other ecological conditions. Multipurpose tree species like *P. kesiya, A. nepalensis, P. roxburghii, M. oblonga* and *G. arboria* with greater surface cover, constant leaf litter fall and extensive root systems increased soil organic carbon by 96.2% (Table 3), helped with better aggregate stability by 24.0%, improved available soil moisture by 33.2%, and in turn reduced soil erosion by 39.5% (Saha *et al.*, 2007). Screening of MPTs is an important prerequisite for determining the suitability of agroforestry models for various agroecological regions. Jha *et al.* (2010) suggested that inclusion of species like *P. Juliflora,* L. *Leucocephala, A. nilotica* and *A. indica* could be a better choice for restoring and rehabilitation of degraded ravine lands in the riparian zone. They reported that the SOC contents in forest systems with these MPTs were twofold higher in comparison to the reference site. Mishra *et al.* (2004) also reported increase of SOC under 6-year-old plantations of *P. juliflora, D. sissoo* and *E. tereticornis.* The poplar based agroforestry system improves aggregation of soil through huge amounts of organic matter in the form of leaf biomass. The extent of improvement may be affected by the age of the poplar trees and the soil type. Gupta *et al.* (2009) reported that the poplar trees could sequester higher soil organic carbon in 0–30 cm profile during the first year of their plantation (6.07 Mg ha^{-1} yr^{-1}) than the subsequent years (1.95–2.63 Mg ha^{-1} yr^{-1}). Similarly, a comparative study on the effect of various MPTs on soil organic carbon pool (Table 4) showed a concomitant rise in SOC in soils under MPTs and a subsequent decline in soils of open space over 4–16 years. Maximum rise in SOC was noticed in soils of *A. indica* (28.6 Mg hm^{-2}) followed by *A. Aurculiformisi* (21.9 Mg hm^{-2}), *G. arborea* (21.8 Mg hm^{-2}), *M. Champaca* (16.7 Mg hm^{-2}), etc. The minimum rise in SOC was noted in soils under *T. grandis*. So an increase of SOC was noted from 3.8 Mg hm^{-2} in soils of open space to 19.5 Mg hm^{-2} in that under MPTs after 16 years. The comparatively high humin carbon present in soils under *A. auriculiformis*, *L. leucocephala* and *G. Arborea* (Fig. 1) indicated the enhanced storage of organic carbon pool in agroforestry systems (Datta and Singh, 2007). Swamy *et al.* (2003) estimated that a six year old *G. arborea* based agri-sivicultural systems in India sequestered 31.4 Mg hm^{-2} carbon.

Table 3: Growth, litter production, fine root biomass of promising MPTs in humid tropics and their contribution on SOC content.

MPTs	Annual litter production (g m^{-2})	Time required for decomposition (days)	Total fine root biomass (g m^{-2})	Organic C (g kg^{-1})
P. kesiya	621.5	718	496.75	35.4
A. nepalensis	473.75	350	435.50	32.2
P. roxburghii	341.75	385	415.50	23.1
M. oblonga	512.25	390	462.00	33.6
G. arboria	431.75	360	419.00	28.6

Source: Saha *et al.* (2010)

Table 4: Changes in SOC (Mg hm^{-2}) over the years under various MPTs in humid tropics

MPTs	Years			
	4	8	12	16
A. auriculiformis	11.1	11.9	17.9	21.9
M. alba	9.9	9.9	9.9	15.9
L. leucocephala	11.5	11.5	12.8	16.7
D. sissoo	13.1	12.5	13.1	13.9
G. maculate	13.1	13.1	13.9	14.9
A. indica	10.9	10.9	14.7	28.6
M. champaca	13.9	13.7	13.9	16.9
E. hybrid	9.9	9.9	14.9	16.1
T. grandis	11.5	11.3	11.5	12.9
G. arborea	12.2	12.2	12.8	21.8
S. saman	10.6	11.3	11.3	13.9
A. procera	13.5	13.1	13.5	14.7
Open space (Control)	11.9	11.9	11.1	9.1

Source: Datta and Singh (2007)

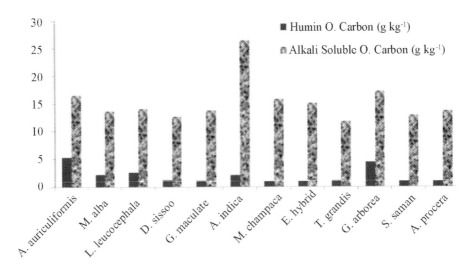

Fig. 1: Humus in soils under MPTs in humid tropics
Source: Datta and Singh (2007)

Hedgerow Intercropping for Green Manuring

Green manuring and intercropping of legumes is important aspect of farming system, not only in regard to weed control, but also in reducing the leaching of nutrient and soil erosion. A green cover throughout the year is one of the main goals of farming systems. Depending on the green manure mixture or the legume used for under sowing, there may be an increase in soil organic matter and soil N as well as in other nutrients. Weed management is one of the main concerns in organic agriculture. The elements to consider in preventing weed problems are crop rotation, green manuring, manure management and tillage. Mulching on a large scale by using manure spreaders may also be useful in weed control.

Hedgerow intercropping (HI) refers to the agroforestry systems where crops are grown between rows of regularly coppiced woody species. Initially, it has been developed to restore the fertility of degraded soils in the humid and sub-humid tropics, HI has later been adopted in other regions not only to ameliorate soils, but also to provide other products (e.g. fodder) and services (e.g. erosion control). Scientific literature showed strong variations in C storage potential in HI (1–37 Mg ha^{-1}) depending on climate, soil type and system management (Kang *et al.*, 1999). However, C storage is only temporary in HI systems since the biomass is continuously harvested for prunings (or fodder) and firewood. In many areas of the tropics, regular addition of prunings and root turnover over the years have contributed to the build-up of SOM and nutrient stocks in the soil. In a 12-year HI trial on a Nigerian Alfisol, *G. sepium* and *L.*

leucocephala increased surface soil organic carbon (SOC) by 15% (2.38 Mg C ha⁻¹) compared to sole crops (Kang *et al.*, 1999). The total productivity of pruned biomass, was recorded highest in case of *C. tetragona* (195.5 q ha⁻¹) and lowest (47 q ha⁻¹) in *F. macrophylla* (Table 5).

Table 5: Production of total pruned biomass (q ha⁻¹) and nutrient content in different hedgerow species

Hedge species	Total biomass*	Nutrient concentration (%) in leaf biomass		
		N	P	K
C. cajan	89.43	3.29	0.67	1.43
C. tetragona	195.49	3.47	0.48	1.63
D. rensonii	70.65	3.63	0.48	1.56
F. macrophylla	47.04	3.23	0.45	1.26
I. tinctoria	120.15	3.86	0.81	1.63
T. candida	108.57	3.57	0.32	1.67

*Fresh weight basis

The pruning of *C. tetragona* added an amount of 80, 11 and 38 kg N, P and K, respectively per hectare per year. On average, pruning of N fixing hedgerow species added 20-80; 3-4 and 8-38 kg ha⁻¹ yr⁻¹ of N, P and K, respectively. Addition of leaf biomass from hedgerow species resulted a significant improvement in the fertility status of the soil and lowered the soil acidity remarkable from its initial level. A significant improvement of soil fertility was observed through addition of pruning of *C. tetragona* and *F. macrophylla* compared to other species (Laxminarayana *et al.*, 2006).

Pasture Management

Restoring degraded grazing lands and improving forage species is important to sequestering SOC and SIC. Furthermore, converting marginal croplands to pastures (by CRP and other set-aside provisions) can also sequester C. Similar to cropland, management options for improving pastures include judicious use of fertilizers, controlled grazing, sowing legumes and grasses or other species adapted to the environment, improvement of soil fauna and irrigation (Follett *et al.*, 2001). Conant *et al.* (2001) reported rates of SOC sequestration through pasture improvement ranging from 0.11 to 3.04 Mg C ha⁻¹ yr⁻¹ with a mean of 0.54 Mg C ha⁻¹ yr⁻¹.

Traditional and Improved Fallow

Traditional fallow refers to the abandoned agricultural phase in which secondary vegetation develops with normal density and it is mainly composed of native trees. This phase lasts a variable period of 10 to more than 50 years, depending on land availability. The definition of improved fallows (IF) is restricted to the agroforestry systems in which one (pure) or a few (mixed) tree species are planted as a substitute to natural fallow, to achieve the benefits of the latter in a shorter time (Young, 1997). Rao *et al.* (1998) distinguished two categories of IF: (1) the short-duration fallows with fast growing leguminous trees or shrubs seeking to replenish soil fertility; and (2) the medium-to-long-duration fallows with diverse species and aimed at rehabilitating degraded and abandoned lands as well as exploiting tree products such as poles and firewood. Lasco and Suson (1999) calculated an average C storage of 16 Mg ha^{-1} over the 6-year period under *L. leucocephala* during the fallow phase of the Naalad system. Several studies have shown increased SOM after a few seasons of tree planting on degraded soils. Considering the changes in bulk density induced by the improved fallow practice and the sampling depth, SOC accretions were estimated between 0.73 and 12.46 Mg ha^{-1}.

Conclusion

It is evident that most of all agroforestry systems have the potential to sequester C. With adequate management of trees in cultivated lands and pastures, a significant fraction of the atmospheric C could be captured and stored in plant biomass and in soils. However, increasing C stocks in a given period of time is just one step; the fate of those stocks is what ultimately determines sequestration. In agroforestry systems C sequestration is a dynamic process and can be divided into phases. At establishment, many systems are likely to be sources of GHGs (loss of C and N from vegetation and soil). Then follow a quick accumulation phase and at maturation period when tons of C are stored in the boles, stems, roots of trees and in the soil. At the end of the rotation period, when the trees are harvested and the land returned to cropping (sequential systems), part of the C will be released back to the atmosphere (Dixon, 1995). Therefore, effective sequestration can only be considered if there is a positive net C balance from an initial stock after a few decades. Realistically, C storage in plant biomass is only feasible in the perennial agroforestry systems (perennial-crop combinations, agroforests, windbreaks), which allow full tree growth and where the woody component represents an important part of the total biomass. The analysis of C stocks from various parts of the world showed that significant quantities of C (1.1–2.2 Pg) could be removed from the atmosphere over the next 50 years if agroforestry systems are implemented on a global scale.

References

Balkrishnan and Toky, O.P. (1993). Significance of nitrogen fixing woody legume trees in forestry. *Indian Forester*, 119: 126-134.

Conant, R.T., Paustian, K. and Elliott, E.T. (2001). Grassland management and conversion into grassland: effects on soil carbon. *Applied Ecology*, 11: 343–355.

Datta, M. and Singh, N.P. (2007). Growth characteristics of multipurpose tree species, crop productivity and soil properties in agroforestry systems under subtropical humid climate in India. *Journal of Forestry Research*, 18: 261-270.

Dhyani, S.K and Tripathi, R.S. (2000). Biomass production of fine and coarse root's of trees under agrisilvicultural practices in north east India. *Agroforestry systems,* 50: 107-121.

Dixon, R.K. (1995). Agroforestry systems: sources or sinks of greenhouse gases? *Agroforestry Systems,* 31: 99-116.

DOE (1999). *Carbon Sequestration: State of the Science*. US Department of Energy (DOE), Washington, DC.

Follett, R.F., Kimble, J.M. and Lal, R. (2001). The Potential of U.S. Grazing Lands to Sequester Carbon and Mitigate the Greenhouse Effect. CRC/Lewis, Boca Raton, FL. p. 442.

George, M. and Verghese, G. (1990). Nutrient cycling in *Eucalyptus globulus* plantation.II. Litter production and nutrients return. *Indian Forester*, 116(12): 962-968.

Gupta, Naveen, Kukal, S.S., Bawa, S.S. and Dhaliwal, G.S. (2009). Soil organic carbon and aggregation under poplar based agroforestry system in relation to tree age and soil type. *Agroforestry Systems,* 80: 437-445.

Haile, S.G., Nair, P.K.R. and Nair, V.D. (2008). Carbon storage of different soil-size fractions in Florida silvopastoral systems. *Journal of Environmental Quality,* 37:1789-1797.

Jha, Pramod, Mohapatra, K. P. and Dubey, S.K. (2010). Impact of land use on physico-chemical and hydrological properties of ustifluvent soils in riparian zone of river Yamuna, India. *Agroforestry Systems.* 80, 437-445.

Kang, B.T., Caveness, F.E., Tian, G. and Kolawole, G.O. (1999). Long-term alley cropping with four species on an Alfisol in southwest Nigeria—effect on crop performance, soil chemical properties and nematode population. *Nutrient Cycling in Agroecosystem*s, 54: 145-155.

Lal, R. (1996). Deforestation and land use effects on soil degradation and rehabilitation in western Nigeria. II: soil chemical properties. *Land Degradation and Development,* 7: 87–98.

Lasco, R.D. and Suson, P.D. (1999). A *Leucaena leucocephala*-based indigenous fallow system in central Philippines: the Naalad system. *International Tree Crops Journal*, 10: 161–174.

Laxminarayana, K., Bhatt, B.P. and Rai, Tulsi. (2006). Soil Fertility Build up through Hedgerow Intercropping in Integrated Farming System: A Case Study. In: *Agroforestry in North East India: Opportunities and Challenges* (eds. B.P. Bhatt and K.M. Bujarbaruah pp. 479-490, ICAR Research Complex for NEH Region, Umiam, Meghalaya

Mishra, A., Sharma, S.D., Pandey, R. and Mishra, L. (2004). Amelioration of a highly alkaline soil by trees in northern India. *Soil Use and Management*, 20: 325–332

Nair, P.K.R. (1993). An Introduction to Agroforestry. Kluwer Academic Publishers, Dordrecht, The Netherlands, p. 499.

Nair, P.K.R., Kumar, B.M. and Nair, V.D. (2009). Agroforestry as a strategy for carbon sequestration. *Journal of Plant Nutrition and Soil Science*, 172: 10–23.

Pulleman, M.M., Bouma, J., van Essen, E.A. and Meijles, E.W. (2000). Soil organic matter content as a function of different land use history. *Soil Science Society of America Journal,* 64: 689–693.

Rao, M.R., Nair, P.K.K. and Ong, C.K. (1998). Biophysical interactions in tropical agroforestry systems. *Agroforestry Systems*, 38: 3-50.

Saha, R., Ghosh, P.K., Mishra, V.K., Majumdar, B. and Tomar, J.M.S. (2010). Can agroforestry be a resource conservation tool to maintain soil health in the fragile ecosystem of north-east India? *Outlook in Agriculture*, 39, DOI: 10.5367/oa.2010.0004.

Saha, R., Tomar, J. M. S., and Ghosh, P. K. (2007). Evaluation and selection of multipurpose tree for improving soil hydrophysical behaviour under hilly eco-system of northeast India, *Agroforestry System*, 69: 239–247.

Schlesinger, W.H. (2000). Carbon sequestration in soil: some cautions amidst optimism. *Agricultural Ecosystems and Environment,* 82: 121–127.

Sreemannarayana, B. (2003). Role of agroforestry in soil fertility. In: Agroforestry: Potential and prospects (Eds. P.S.Pathak and Ram Newaj). Agrobios (India); pp. 65-78.

Swamy, S.L., Puri, S. and Singh, A.K. (2003) Growth, biomass, carbon storage and nutrient distribution in *Gmelina arborea* stands on red lateritic soils in Central India. *Bioresource Technology,* 90: 109–126.

Takimoto, A., Nair, V.D. and Nair, P.K.R. (2009). Contribution of trees to soil carbon sequestration under agroforestry systems in the West African Sahel. *Agroforestry Systems,* 76: 11-25.

Young, A. (1997). *Agroforestry for Soil Management,* 2nd ed. CAB International, Wallingford, UK, p. 320.

28

Conservation Agriculture Approaches for Reducing Carbon Footprints

A.K. Singh

ICAR-Central Research Institute for Jute and Allied Fibres, Barrackpore West Bengal -700121, India

Introduction

Intensive soil tillage, burning of crop residues and over use of fertilizer and irrigation water under current agricultural practices has accelerated the pace of degradation in Indian agriculture. Intensive soil tillage increases soil erosion and nutrient runoff into nearby waterways. Crop residue burning resulting in loss of plant nutrients and release of greenhouse gases (GHGs) in the atmosphere. Imbalanced use of chemical fertilizer leads to soil compaction, slows down the fertilizer utilization rate and contaminates the local environment. Increasing demand of irrigation water causes water shortage and harms the environment in several ways including increased salinity, nutrient pollution, and the degradation of flood plains and wetlands. In the face of these management and environmental challenges, there is a need to formulate such agricultural practices which improve the productivity of natural resources as well as of external inputs and help to prevent soil degradation. Conservation agriculture (CA) practices such as reduced tillage, residue retention and proper crop rotations offer such solutions. CA also helps in making agricultural systems more resilient to climate change and safeguard ecosystem services.

Five key farming strategies that are proven to be effective in increasing crop production while lowering carbon footprint are as follows:

(i) use of reduced tillage in combination with crop residue retention and decomposition to increase soil organic carbon;

(ii) reduction in use of inorganic fertilizer and improvement of nitrogen (N) fertilizer use efficiency including N_2- fixing pulses in rotations to lower the carbon footprints of field crops as N fertilizer contribute about 35 to 50% of the total emissions;

(iii)   use of diversified cropping systems and adopting intensified rotation with reduced summer fallow for lowering the carbon footprint by as much as 150 %;

(iv)   integration of key cropping practices which can increase crop yield (15 to 60 %), reduce emissions (25 to 50 %), and lower the carbon footprint of cereal crops (25 to 35 %), and

(v)   enhancing soil carbon sequestration as the emissions from crop inputs can be partly offset by carbon conversion from atmospheric CO_2 into plant biomass and ultimately sequestered into the soil.

With the adoption of these improved conservation agriculture technology, one can optimize the system performance while reducing the carbon footprint of crop cultivation.

Carbon Footprints from Farm Inputs and Farm Operations

Carbon footprint is a measure of the total emission of greenhouse gases in carbon equivalents (CO_2- Carbon dioxide, CH_4- methane and N_2O- nitrous oxide) from a product across its life cycle from the production of raw material used in its manufacture, to disposal of the finished product (Carbon Trust, 2007). The magnitude of decline or enhancement of carbon due to continuous cultivation depends on the balance between the loss of carbon by oxidative forces and the quantity and quality of crop. The loss of carbon is maximum in tropical and subtropical regions because of high atmospheric temperature (Jenny and Raychaudhuri, 1960). Scientific evidence suggests that 50 to 66 per cent of the cumulative historic carbon loss from soil can be recovered if managed intelligently (Lal, 2004; Velayutham *et al.*, 2000).

GHG emissions from field crop production are mostly derived from (1) crop residue decomposition; (2) inorganic fertilizers applied to the crop; (3) various farming operations such as tillage operations, fossil fuels for running of tractors, electricity for running of water pumps, spraying pesticides, planting and harvesting the crop; (4) soil carbon gains or losses from various cropping systems; and (5) emissions of N_2O from summer fallow areas. Below are more detailed descriptions for the major emission contributing factors (Fig. 1).

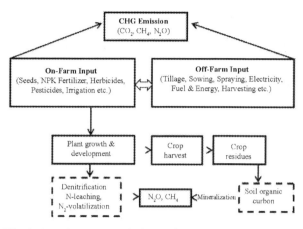

Fig. 1: Greenhouse gas emission during crop production

GHG emission represent per unit of the land used in crop production, per unit weight of the produced yield and per unit of the energy input or output (Soltani *et al.*, 2013). The amount of CO_2 produced is calculated by multiplying the input application rate per hectare (e.g. labour, diesel fuels, chemical fertilizers, herbicides and pesticides) by its corresponding coefficient as given in Table 1.

Table 1: Greenhouse gas (GHG) emission coefficient of inputs

Input	Unit	GHG Coefficient (kg CO_2 eq ha^{-1})	Data Source/ Reference
Mouldboard ploughing	MJ	15.2	Singh *et al.* (1999)
Field Cultivation	MJ	4.0	Pathak and Wassmann (2007)
Seed sowing	MJ	3.20	Pathak and Wassmann (2007)
Machinery	MJ	0.071	Dyer and Desjardins (2006)
Diesel fuel	L	2.76	Dyer and Desjardins (2003)
Manure	kg	0.0032	Pathak and Wassmann (2007)
Nitrogen (N) fertilizer	kg	1.30	Lal (2004a); Pathak and Wassmann, (2007)
Phosphorus (P_2O_5) fertilizer	kg	0.20	Lal (2004a); Pathak and Wassmann, (2007)
Potassium (K_2O) fertilizer	kg	0.20	Lal (2004a); Pathak and Wassmann, (2007)
Herbicide	kg	6.30	Lal (2004a); Pathak and Wassmann, (2007)
Insecticide	kg	5.10	Lal (2004a); Pathak and Wassmann, (2007)
Fungicide	kg	3.90	Lal (2004a); Pathak and Wassmann, (2007)
Water for irrigation	mm	0.05	Pathak and Wassmann, (2007)
Harvesting	MJ	10	Pathak and Wassmann, (2007)
Retting	tonne	434	Banik *et al.* (1993)

Source: Singh *et al.* (2018)

Synthesis of the research data as published by Lal (2004a) shows that estimate of GHG emissions (kg CO_2-eq/ha) for different fertilizer nutrients are 0.9–1.8 for N, 0.1–0.3 for P_2O_5, 0.1–0.2 for K_2O, 0.03–0.23 for lime, 6.3 for herbicides, 5.1 for insecticides and 3.9 for fungicides. Similarly, estimates of C emissions are 1–1.4 for spraying chemicals, 2–4 for seed sowing and 6–12 for harvesting. Irrigation, lifting water from deep wells and using sprinkling systems, emits 129 kg CO_2-eq for applying 25 cm of water and 258 kg CO_2-eq for 50 cm of water. Emission for different tillage methods are 35.3 kg CO_2-eq/ha for conventional tillage, 7.9 kg CO_2-eq/ha for minimum tillage, and 5.8 kg CO_2-eq/ha for no-tillage method. A careful assessment is needed to reduce their use, and to enhance use efficiency of these practices.

Farm Practices Reducing Carbon Footprint

Managing tillage practices

A soil disturbance due to tillage is the dominant factor reducing soil carbon stabilization within micro-aggregates in the soil. Carbon emissions depend on numerous factors like soil properties, tractor size, implement used, depth of tillage and quantity of fuel used. The fuel requirement increases with increase in depth of ploughing, tractor speed and type of equipment used for ploughing. The direct fuel consumption is more for heavy than light textured soils (Collins *et al.*, 1976). As per study (Lal, 2004a), the average C emission is about 15.2 kg CO_2-eq/ha for mouldboard ploughing, 11.3 kg CO_2-eq/ha for sub-soiling, 8.3 kg CO_2-eq/ha for heavy disking, 7.9 kg CO_2-eq/ha for chiselling, 5.8 kg CO_2-eq/ha for standard disking, 4.0 kg CO_2-eq/ha for cultivator and 2.0 kg CO_2-eq/ha for rotary hoeing. Therefore, conversion of conventional tillage (based on mouldboard ploughing) to reduced tillage (disking or chisel till) can lead to drastic reductions in C emissions. Carbon emission in complete tillage (involving ploughing, two disking, field cultivator and rotary hoeing) is 35.3 kg CO_2-eq/ha that can be reduced to 20.1 kg CO_2-eq/ha by elimination of mouldboard ploughing, and merely 7.9 kg CO_2-eq/ha if follow seeding after chiselling. Reduced tillage in combination with additional carbon input from cover crops can significantly improve the soil organic carbon content (Garcia-Franco *et al.*, 2015; Pinheiro *et al.*, 2015). Tillage may influence soil carbon and microbial biomass, but may not necessarily increase soil available nutrients or crop yields (Campbell *et al.*, 2011).

Crop residues retention and decomposition

Crop residues are normally left on the soil surface or are incorporated into the soil after a field crop is harvested. The crop residues serve as an important C and N source in the soil, contributing directly and indirectly CO_2 and N_2O emissions (Forster *et al.*, 2007). The amount of CO_2 and N_2O emission from the decomposition of the straw, stubble, leaves and roots depends on the net productivity of the crop. In a study (Singh *et al.*, 2019), the C-input addition through mixing of root-stubble and leaves, under jute-rice-wheat crop cycle helped in doubling the C-input (5.65 t ha^{-1}). Jute plant contributed maximum C inputs (3.80 t ha^{-1}) through retention of root-stubbles and its shredded leaves (Table 2). These large amounts of required C-input to maintain soil organic carbon resulted from net primary production from original biomass production. In the production of economic crops, the carbon footprint can be reduced by effective management of straw, stubble, leaves and roots.

Table 2: Carbon derived from crop residues.

Crop	C-Input addition (t ha^{-1})		
	Root + stubbles	Other residue (leaves)	All residues
Jute	0.97	2.85	3.80
Rice	0.94	-	0.94
Wheat	0.89	-	0.89
Total	2.80	2.85	5.65

Source: Singh *et al.* (2019)

Field experiments on a rice–wheat cropping system in India showed that incorporation of crop residues as compared to burning or removal increased organic carbon contents (Table 3). Soil texture and soil pH play an important role in decomposition of crop residue. Decomposition is more rapid in soils with less clay content. As clay content increases, soil surface area also increases which results in increased organic C stabilization potential (Ladd *et al.*, 1996). In general, decomposition of crop residues proceeds more rapidly in neutral pH soil than in acidic soils. Application of lime in acid soils enhances the decomposition of plant residues (Condron *et al.*, 1993). Alkaline and saline soil also decreased the decomposition of plant residues (Nelson *et al.*, 1996).

Table 3: Effect of crop residue management on organic carbon in soil

Cropping system	Type of crop residues	Soil	Organic C (%)	Reference
Rice-wheat	Rice and wheat straw	Clay loam	0.86	Dhiman *et al.* (2000)
		Silty clay loam	1.24	Verma and Bhagat (1992)
		Sandy loam	0.50	Singh *et al.* (2005)
Rice-mustard	Rice and mustard straw	Sandy clay loam	0.61	Kumar *et al.* (2000)
Rice-rice	Rice straw	Clayey	1.90	IRRI (1986)

Fertilizing crops based on soil tests and including N$_2$- fixing pulses

More fertilizer doesn't always mean a higher crop production, if there isn't the right mix of nutrients, water and soil health conditions. Soil Health Card (SHC) assessment provide soil nutrient status of each farm and advise corrective dosage of major fertilizers, micronutrient and soil amendments to maintain soil health and obtain a better yield (Singh *et al.*, 2020). Imbalanced use of N, P and K leads to the loss of fertility of the soil over a period of time, which affects efficiency of fertilizer use and crop productivity (Rahman *et al.*, 2012). The problem is more severe in intensively cropped regions where farmers use excessive nitrogen fertilizer to attain yield levels harvested earlier with less fertilizer (Dwivedi, 2017). Excessive use of fertilizer can also harm the environment and human health through emissions of GHG (CH_4, N_2O, CO_2) and eutrophication caused by the deposits of nitrate and phosphorus in water resources (Bumb and Baanante, 1996). Hence, overall impact and mitigation potential of fertilizer management with regard to GHG emission is important (Singh *et al.*, 2018).

Nitrogen fertilizer is the main contributor to greenhouse gas (GHG) emissions especially in cereal production. The average 100% carbon footprint using N fertilizer is up to 3.0 kg CO_2-eq per kg of N (FE, 2014). Increased use of N fertilizer application to the crops increases the total GHG emission. The carbon footprint is a linear function of the rate of N fertilizer applied to the crops, and slope of the linear regression varied with crop species and amount of N fertilizer applied. The decreased use of N fertilization lowers the carbon footprint accordingly. Optimizing N-fertilizer efficiency is key to reduce the GHG emission significantly, depending on soil and weather conditions. A good soil structure increases N-use efficiency and reduces N_2O losses.

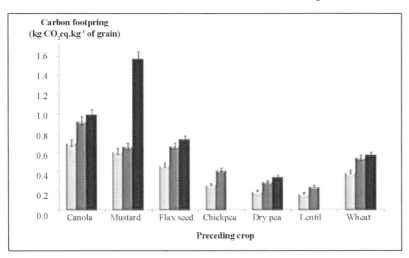

Fig. 2: The carbon footprint in wheat after cultivation of different preceding pulse and non-pulse crop in the crop rotation

Source: Gan *et al.* (2011)

Inclusion of N_2-fixing pulse crops in a crop rotation can significantly decrease GHG emissions and the carbon footprint of the crop grown in the subsequent year (Fig. 2). Symbiotic N_2 fixation by the leguminous crop from the atmosphere significantly decreases the use of synthetic N fertilizer, thus lowering the carbon footprint (Gan *et al.*, 2011).

Enhancing water productivity

In the agricultural production systems, maximum energy is utilized in tillage and irrigation management. Efficient water management is one of the potential options for saving and utilizing energy for productive purposes (Choudhary *et al.*, 2017). Increasing use of groundwater for irrigation is linked to high energy demand, depleting resources and resulting in a high carbon footprint. Overall water application efficiency for different crops in India is in the range of 40–60% (Grant Thornton, 2016). More than 50% of irrigation water remains unused by the crop and returns to the hydrologic system, either as deep percolation or non-productive evaporation. Hence, reducing water delivery to farms and increasing water use efficiency are important to reduce energy consumption and improving water productivity. Using improved on-farm practices help in reducing the need of irrigation water which in turn results in decreased energy consumption and carbon footprint. Adopting the improved schedule at 65 to 75% efficiency, close to irrigation efficiency in the farm will reduce the irrigation water demand and thereby carbon footprint up to 30% (Karimi *et al.*, 2012). Conservation agriculture along with some water saving technological interventions like alternate raised and sunken bed technique in low lands, system of rice intensification (SRI) technique of rice cultivation and mulching etc. can be adopted to increase crop yield, to reduce evapo-transpiration and water footprints (Kar *et al.*, 2014). Crop residue retention help in reducing the water requirement by conserving soil moisture through a reduction in evaporation loss and improvement in the water-holding capacity (Gathala *et al.*, 2013). Higher water productivity (33-56%) under conservation agriculture (zero tillage + crop residues) was observed in cereal based cropping system (Jat *et al.*, 2020) in the North West Indo-Gangetic Plain of India due to better grain yields with less water usage. Planting of crop on ridges, straw mulching and micro irrigation (sprinklers and drips) can also help in reducing water use, and thereby carbon footprint. Rotational irrigation is often recommended to irrigate a large area with a limited water supply which also ensures more effective use of rainfall.

Diversification of crop rotations

Crop diversification has been considered a key agriculture practice for improving agro-ecosystem productivity and lowering the carbon footprint (Gan *et al.*, 2015). Crop diversification help in controlling weeds (Harker

424   Conservation Agriculture and Climate Change

et al., 2009), suppressing plant diseases (Kutcher *et al.*, 2013), and thereby increase production sustainability (Mhango *et al.*, 2013). Researchers found that the total emissions per unit of land varied significantly among the various cropping systems. Average GHG emission and the carbon footprint of biomass based cropping system were found maximum in cereal based cropping system (Table 4). Incorporation of pulse crops in the crop rotation helped in reducing the total GHG and carbon footprint.

Table 4: GHG emission and carbon foot print in biomass based crop rotation

Cropping system	Total emission (kg CO_2 eq kg^{-1} yr^{-1})	Carbon foot print (kg CO_2eq kg^{-1} yr^{-1}
Wheat-maize	11800	0.85
Groundnut-wheat-maize	8532	0.76
Mustard-cotton-groundnut-wheat-maize	8324	0.68

Source: Yang *et al.* (2014)

In designing a diverse cropping system, there is need to examine the overall greenhouse gas emissions and the footprint of individual crop. Crops requiring low production inputs and those with a high yield of straw and roots for incorporation into the soil as carbon are keys to reducing the overall footprint of the system. Pathak *et al.* (2011) reported carbon sequestration potential of field crops under various cropping system (Table 5). The carbon build-up rate was maximum under jute-rice-wheat and maize-soybean-wheat cropping system.

Table 5: Potential of carbon sequestration of field crops under various crop rotation

Cropping system	C sequestration potential t C eq ha^{-1}	Reference
Jute-rice-wheat	1.45 – 3.33	Manna *et al.* (2005)
Maize-soybean-wheat	0.43 – 3.82	Hati *et al.* (2007)
Rice-wheat	0.41 – 1.87	Yadav *et al.* (2000)
Soybean-wheat	0.40 – 1.67	Behera *et al.* (2007)

Source: Pathak *et al.* (2011)

Intensifying crop rotations with less summer fallow frequencies

In arid and semiarid regions, the productivity of agroecosystems is often constrained by a low availability of water and nutrients (Rasouli *et al.*, 2014). One of the approaches employed to tackle these challenges is using summer fallow where the land is left unplanted for one growing season. During summer fallow, a proportion of the rainfall is conserved in the soil profile (Tanwar *et al.*, 2014), which is then available for crops grown the following year (Sun *et al.*, 2013). Additionally, summer fallowing encourages the release of N through the N mineralization of soil organic matter (Campbell *et al.*, 2008), thus increasing soil N availability and helping to reduce the amount of inorganic N fertilizer used in cropping (Koutika *et al.*, 2004). However, a number of studies have shown that the frequency of summer fallow in a crop rotation has a significant impact on the carbon footprint of the rotation (O'Dea *et al.*, 2013; Schillinger and Young, 2014). Crop intensification with reduced frequency of summer fallow in a rotation can increase crop production while reducing the carbon footprint. More intensified crop rotation systems with reduced frequency of summer fallow in the rotation can reduce the carbon footprint by as much as 250% (Liu *et al.*, 2016). Replacement of summer fallow with a forage or grain legume are more helpful in reducing the carbon footprint in addition to double farming profits compared with the system with a high frequency of summer fallow.

Integrating key cropping practices

Integrated cropping systems coupled with the adoption of best agronomic practices such as early sowing, optimum plant establishment methods (e.g. SRI in rice), use of recommended amount of fertilizer at right time using soil health card, and proper crop sequencing can increase crop productivity without increasing production inputs or GHG emissions. (Kirkegaard *et al.*, 2008; Miller *et al.*, 2003; Singh *et al.*, 2000). The carbon footprint of individual crop species is highly associated with crop biomass and the N concentration of plant parts like leaves, straw, stubbles, and roots. Integration of agronomical practices can significantly improve the net productivity of crops by improving the water and nitrogen use efficiencies. The increased net productivity in integrated cropping systems over monoculture systems is due to the improved diversity of the microbial populations and the function of microbial communities in the soil (Yang *et al.*, 2013; Cruz *et al.*, 2012). Leguminous-based cropping systems reduce the loss of soil organic carbon and nitrogen as compared to only cereal-based cropping systems (Gan *et al.*, 2014). A system that includes leguminous crop in the rotation has the lowest carbon footprint, regardless of the geographical locations (Fig. 3). Many studies across the world demonstrate that use of integrated agronomical practices can increase the system productivity by 15 to 50 %, reduce the carbon emissions associated with the crop inputs by 25 to 50 %, and lower the carbon footprint of cereal crops by 25 to 35 % (Berry *et al.*, 2008; Elsgaard *et al.*, 2013; Syp *et al.*, 2012; Brock *et al.*, 2012; Gan *et al.*, 2011).

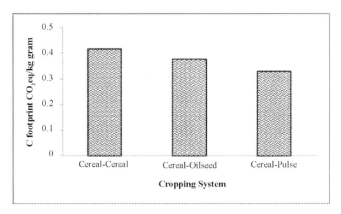

Fig. 3: Soil and crop management practices associated with carbon sequestration
Source: Gan *et al.* (2014)

Enhancing Soil Carbon Sequestration

The Intergovernmental Panel on Climate Change has identified one of the most important options for greenhouse gas mitigation is the sequestration of carbon into soils (IPCC, 2006). Several studies demonstrate that soil carbon sequestration plays a key role in reducing the carbon footprint of crop cultivation, because a per unit farmland GHG emission represents the balance between CO_2-eq emissions and carbon sequestration during the cultivation of crops per year. Modern soil and crop management practices have resulted in loss of SOC. Main agronomic practices for SOC loss are improper tillage operations, minimum crop rotations and residues management, imbalanced soil fertilization, and less use of organic compost or FYM. Hence, suitable crop and land management practices can be used to increase the amount of organic matter in the soil and thereby improve soil carbon sequestration, and mitigate GHG emissions to the atmosphere. About 1400 Pg of SOC is stored in surface soil (1 m depth), while 450 Pg SOC is stored above ground vegetation and dead organic matter. Increase in cropping intensity and reducing the frequency of fallow period in the crop rotation is an effective approach to improve biomass production and it also decreases organic matter decomposition rate and mineralization/oxidation of SOC (Dumanski *et al.*, 1998). GHG emissions associated with the crop production inputs can be offset by greater carbon conversion from atmospheric CO_2 into plant biomass and ultimately sequestered into the soil (Sainju *et al.*, 2010; Pinheiro *et al.*, 2015,

Hu *et al.*, 2015). Inclusion of intercropping in the crop rotation with reduced tillage coupled with stubble mulching can effectively lower carbon emissions and increase soil carbon sequestration within the small macro-aggregates and micro-aggregates of surface soil (Garcia-Franco *et al.*, 2015). Growing legumes as intercrop can substantially reduce the chemical N fertilizers application. Mixed cropping is also an effective approach in the intercropping system to balance the input and output of soil nutrients, suppress weeds and insects, control plant disease, and to increase the overall productivity with limited resources (Hirst 2009). Appropriate land/soil management programs are required to optimize carbon conversion from the atmosphere while minimizing carbon loss in the production of crops. Management of soil health friendly farming systems, minimization of soil and water loss by surface runoff and erosion, adoption of INM and organic amendment are found helpful in CO_2 sequestration (Singh and Lal, 2005). Reported data of C sequestration associated under soil and crop management practices are presented in Table 6.

Table 6: Soil and crop management practices associated with carbon sequestration

Management practice	Carbon sequestration ($*MMT C yr^{-1}$)
Crop rotation and cover crops	5.1 – 15.3
Crop residue management	11 – 67
Fertilizer management	6 – 18
Livestock manure (FYM)	3.6 – 9.0
Supplemental irrigation	1.0 – 3.2
Conservation tillage	17.8 – 35.7
Fallow reduction	1.4 – 2.7
MMTC: Million metric tons of carbon	

Source: Follett (2001)

Conclusion

Crop production practices which leads to less carbon emission are more desirable for sustainability and environmental safety from any production system. The transition to low-carbon agriculture requires identification of appropriate systems and management practices based on the farm resources available. Diversification of cropping systems, adoption of leguminous based crop rotation with reduced summer fallow, use of reduced tillage in combination with crop residue retention, improvement of nitrogen use efficiency, and enhancement of carbon sequestration into the soil together enhances agronomic productivity per unit consumption of C-based input. Adopting such holistic approach which decreases losses of crop, soil and water resources are important strategy to reduce the carbon footprint.

References

Banik A., Sen, M. and Sen, S.P. (1993). Methane emission from jute-retting tanks. *Ecological Engineering*, 2:73-79.

Behera, U., Sharma, A. and Pandey, H. (2007). Sustaining productivity of wheat–soybean cropping system through integrated nutrient management practices on the Vertisols of central India. *Plant and Soil*, 297:185–199.

Berry, P.M., Kindred, D.R., Paveley, N.D. (2008). Quantifying the effects of fungicides and disease resistance on greenhouse gas emissions associated with wheat production. *Plant Pathology*, 57:1000–1008.

Brock, P., Madden, P., Schwenke, G., Herridge, D. (2012). Greenhouse gas emissions profile for 1 tonne of wheat produced in Central Zone (East) New South Wales: a life cycle assessment approach. *Crop and Pasture Science*, 63:319–329.

Bumb, Balu L. and Baanante, Carlos A. (1996). *Policies to promote environmentally sustainable fertilizer use and supply to 2020.* 2020 Brief 40, International Food Policy Research Institute, Washington DC

Campbell, C.A., Lafond, G.P., Van den Bygaart A.J., Zentner, R.P., Lemke, R., May, W.E., Holzapfel, C.B. (2011). Effect of crop rotation, fertilizer and tillage management on spring wheat grain yield and N and P content in a thin Black Chernozem: a long-term study. *Canadian Journal of Plant Science*, 91:467–483.

Campbell, C.A., Zentner, R.P., Basnyat, P., DeJong, R., Lemke, R. and Desjardins, R., Reiter, M. (2008). Nitrogen mineralization under summer fallow and continuous wheat in the semi-arid Canadian prairie. *Canadian Journal of Soil Science*, 88:681–696.

Carbon Trust (2007). Carbon Footprint Measurement Methodology, Version 1.1. The Carbon Trust, London

Choudhary, M., Rana, K.S., Bana, R.S., Ghasal, P.C., Choudhary, G.L., Jakhar, P. and Verma, R.K. (2017). Energy budgeting and carbon footprint of pearl millet–mustard cropping system under conventional and conservation agriculture in rainfed semi-arid agro-ecosystem. *Energy*, 141:1052–1058.

Collins, N.E., Kimble, L.J. and Williams, T.H. (1976). Energy requirements for tillage on coastal plains soils. In: Lockeretz W, editor. Agriculture and energy. Academic Press, New York, p.233

Condron, L. M., Tiessen, H., Trasar-Cepada, C., Moir, J. O., and Stewart, J.W.B. (1993). Effects of liming on organic matter decomposition and phosphorous extractability in an acid humid Ranker soil from northwest Spain. *Biology and Fertility of Soils*, 15: 279–284.

Cruz, A.F., Hamel, C., Yang, C., Matsubara, T., Gan, Y., Singh, A.K., Kuwada, K. and Ishii, T. (2012). Phytochemicals to suppress Fusarium head blight in wheat-chickpea rotation. *Phytochemistry*, 78: 72–80.

Dhiman, S. D., Nandal, D. P., and Om, H. (2000). Productivity of rice (*Oryza sativa*)-wheat (*Triticum aestivum*) cropping system as affected by its residue management and fertility levels. *Indian Journal of Agronomy*, 45: 1–5.

Dumanski J., Desjardins, R.L., Tarnocai, C., Moreal, C., Gregorich, E.G., Kirkwood V. and Campbell C.A. (1998). Possibilities for future carbon sequestration in canadian agriculture in relation to land use changes. *Journal of Climate Research*, 40(1): 81-103.

Dwivedi, B.S. (2017). Revamping soil testing service: a pre-requisite for effective implementation of Soil Health Card Scheme. *Journal of Indian Society of Soil Science*, 65(S): 62-71.

Dyer, J.A. and Desjardins, R.L. (2003). The impact of farm machinery management on the greenhouse gas emissions from Canadian agriculture. *Journal of Sustainable Agriculture*, 22: 59-74.

Dyer, J.A. and Desjardins, R.L. (2006). Carbon dioxide emissions associated with the manufacturing of tractors and farm machinery in Canada. *Biosystems Engineering,* 93(1): 107-118

Elsgaard, L., Olesen, J.E., Hermansen, J.E., Kristensen, I.T., Børgesen, C.D. (2013). Regional greenhouse gas emissions from cultivation of winter wheat and winter rapeseed for biofuels in Denmark. *Acta Agriculturae Scandinavica, Section B,* 63:219–230.

FE (2014). Mineral fertiliser carbon footprint reference values. Fertilizers Europe, Brussels, Belgium

Follett, R.F. (2001). Soil management concepts and carbon sequestration in cropland soils. *Soil Tillage Research,* 61:77-92.

Forster, P., Ramaswamy, V., Artaxo, P., Bemsten, T., Betts, R., Fahey, D.W., Haywood, J., Lowe, D.C. and Myhre (2007). Changes in atmospheric constituents and in radiative forcing, In: Climate change 2007: The Physical Science Basis Contribution of Working Group I to the Fourth Assessment Report of the Intergovernmental Panel on Climate Change, pp 129-234.

Gan, Y., Liang, C., Wang, X. and McConkey, B. (2011). Lowering carbon footprint of durum wheat by diversifying cropping systems. *Field Crop Research,* 122:199–206.

Gan, Y., Liang, C., Chai, Q., Lemke, R.L., Campbell, C.A. and Zentner, R.P. (2014). Improving farming practices reduces the carbon footprint of spring wheat production. *Nature Communications,* 5: 5012–5012.

Gan, Y., Hamel, C., O'Donovan, J.T., Cutforth, H., Zentner, R.P., Campbell, C.A., Niu, Y., Poppy and L. (2015). Diversifying crop rotations with pulses enhances system productivity. *Nature Scientific Reports,* 5: 14625.

Garcia-Franco, N., Albaladejo, J., Almagro, M. and Martínez-Mena, M. (2015). Beneficial effects of reduced tillage and green manure on soil aggregation and stabilization of organic carbon in a Mediterranean agroecosystem. *Soil Tillage Research,* 153: 66–75.

Gathala, M.K., Kumar, V., Sharma, P.C., Saharawat, Y.S., Jat, H.S., Singh, M., Kumar, A., Jat, M.L., Humphreys, E., Sharma, D.K. and Sharma, S. (2013). Optimizing intensive cereal-based cropping systems addressing current and future drivers of agricultural change in the northwestern Indo-Gangetic Plains of India. *Agriculture Ecosystem and Environment,* 177: 85–97.

Grant Thornton (2016). Accelerating growth of Indian agriculture: Micro irrigation an efficient solution, Grant Thornton India LLP., New Delhi

Harker, K.N., O'Donovan, J.T., Irvine, R.B., Turkington, T.K., Clayton and G.W. (2009). Integrating cropping systems with cultural techniques augments wild oat (*Avena fatua*) management in barley. *Weed Science,* 57: 326–337.

Hati, K.K., Swarup, A., Dwivedi, A.K. (2007). Changes in soil physical properties and organic carbon status at the topsoil horizon of a Vertisol of central India after 28 years of continuous cropping, fertilization and manuring. *Agriculture, Ecosystem and Environment,* 119: 127–134.

Hirst, K.K. (2009) Mixed Cropping, Agricultural Technique Known as Mixed Cropping. http://archaeology.about.com/od/historyofagriculture/qt/mixed_cropping.html (access date: 3/28/2010)

Hu, F., Chai, Q., Yu, A., Yin, W., Cui, H., Gan, Y. (2015). Less carbon emissions of wheat–maize intercropping under reduced tillage in arid areas. *Agronomy for Sustainable Development,* 35: 701–711.

IPCC (2006). Intergovernmental panel on climate change. 2006 IPCC Guidelines for National Greenhouse Gas Inventories-Vol 4, Agriculture, Forestry and Other Land Use: Geneva 2, Switzerland

IRRI (1986). Annual Report. International Rice Research Institute, Los Banos, Philippines

Jat, H.S., Choudhary, K.M., Nandal, D.P., Yadav, A.K., Poonia, Tanuja, Singh, Yadvinder, Sharma, P.C. and Jat, M.L. (2020). Conservation Agriculture-based Sustainable Intensification of Cereal Systems Leads to Energy Conservation, Higher Productivity and Farm Profitability. *Environmental Management* https://doi.org/10.1007/s00267-020-01273-w

Jenny, H. and Raychaudhuri, S.P. (1960). Effect of climate and cultivation on nitrogen and organic matter resources in Indian soils. Indian Council of Agricultural Research, New Delhi

Kar, G., Singh, R., Kumar, A. and Sikka, A.K. (2014). Farm Level Water Footprints of Crop Production: Concept and Accounting. Bulletin No.-67. Directorate of Water Management, Indian Council of Agricultural Research, Chandrasekharpur, Bhubaneswar, India, p. 56

Karimi Poolad, Qureshi Asad Sarwar, Bahramloo Reza and Molden David (2012). Reducing carbon emissions through improved irrigation and groundwater management: A case study from Iran *Agricultural Water Management,* 108: 52-60.

Kirkegaard, J., Christen, O., Krupinsky, J., Layzell, D. (2008). Break crop benefits in temperate wheat production. *Field Crops Research,* 107: 185– 195.

Koutika, L.S., Ndango, R., Hauser, S. (2004). Nutrient concentrations and NH_4^+-N mineralization under different soil types and fallow forms in southern Cameroon. *Journal of Plant Nutrition and Soil Science,* 167: 591–595.

Kumar, V., Ghosh, B.C., and Bhat, K. (2000). Complementary effect of crop wastes and inorganic fertilizers on yield, nutrient uptake and residual fertility in mustarad (*Brassica juncea*) rice (*Oryza sativa*) cropping sequence. Indian *Journal of Agriculture Science,* 70: 69–72.

Kutcher, H.R., Brandt, S.A., Smith, E.G., Ulrich, D., Malhi, S.S. and Johnston, A.M. (2013). Blackleg disease of canola mitigated by resistant cultivars and four-year crop rotations in western Canada. *Canadian Journal of Plant Pathology,* 35: 209–221.

Ladd, J.N., van Gestel, M., Monrozier, L. J., and Amato, M. (1996). Distribution of organic ^{14}C and ^{15}N in particle-size fractions of soils incubated with ^{14}C, ^{15}N-labelled glucose, NH_4, and legume and wheat straw residues. *Soil Biology and Biochemistry,* 28: 893–905.

Lal, R. (2004). Soil carbon sequestration impacts on global climate change and food security. *Science,* 304: 1623–1627.

Lal R. (2004a). Carbon emission from farm operations. *Environment International,* 30: 981-990.

Liu, Chang, Cutforth, Herb, Chai, Qiang and Gan, Yantai (2016). Farming tactics to reduce the carbon footprint of crop cultivation in semiarid areas, A review. *Agronomy for Sustainability Development,* 36: 69.

Manna, M.C., Swarup, A. and Wajari, R.H. (2005). Long-term effect of fertilizer and manure application on soil organic carbon storage, soil quality and yield sustainability under sub-humid and semi-arid tropical India. *Field Crops Research,* 93: 264–280.

Mhango, W.G., Snapp, S.S. and Phiri, G.Y.K. (2013). Opportunities and constraints to legume diversification for sustainable maize production on smallholder farms in Malawi. *Renewable Agriculture and Food Systems,* 28: 234–244.

Miller, P.R., Gan, Y., McConkey, B.G., McDonald, C.L. (2003). Pulse crops for the northern Great Plains. *Agronomy Journal,* 95: 980–986.

Nelson, P. N., Ladd, J. N., and Oades, J. M. (1996). Decomposition of ^{14}C-labelled plant material in a salt-affected soil. *Soil Biology and Biochemistry,* 28: 433–441.

O'Dea, J.K., Miller, P.R., Jones, C.A. (2013). Greening summer fallow with legume green manures: on-farm assessment in North-Central Montana. *Journal of Soil Water Conservation,* 68: 270–282.

Pathak, H. and Wassmann, R. (2007). Introducing greenhouse gas mitigation as a development objective in rice-based agriculture: I. Generation of technical coefficients. *Agriculture System,* 94: 807-825.

Pathak, H., Byjesh, K., Chakrabarti, B. and Aggarwal, P.K. (2011). Potential and cost of carbon sequestration in Indian agriculture: Estimates from long-term field experiments. *Field Crops Research,* 120: 102–111.

Pinheiro, É.F.M., De Campos, D.V.B., De Carvalho, B.F., Dos Anjos, L.C. and Pereira, M.G. (2015). Tillage systems effects on soil carbon stock and physical fractions of soil organic matter. *Agricultural System,* 132: 35–39.

Rahman, M.A.T.M.T., Aktar, Z., Mondal, M.K. and Ahmed, T. (2012). Environmental friendly agricultural practice in the southwestern coastal zone of Bangladesh to adapt with climate change. *International Journal of Innovative Research Development,* 1(9): 33-44.

Rasouli, S, Whalen, J.K., Madramootoo, C.A. (2014). Review: reducing residual soil nitrogen losses from agroecosystems for surface water protection in Quebec and Ontario, Canada: best management practices, policies and perspectives. *Canadian Journal of Soil Science,* 94: 109–127.

Sainju, U.M., Jabro, J.D. and Caesar-TonThat, T. (2010). Tillage, cropping sequence, and nitrogen fertilization effects on dryland soil carbon dioxide emission and carbon content. *Journal of Environmental Quality,* 39: 935–945.

Schillinger, W.F., Young, D.L. (2014). Best management practices for summer fallow in the World's driest rainfed wheat region. *Soil Science Society of America Journal,* 78: 1707–1715.

Singh, A.K., Roy, M.L. and Ghorai, A.K. (2020). Optimize fertilizer use management through soil health assessment: Saves money and the environment. *International Journal of Current Microbiology and Allied Science,* 9(6): 322-330.

Singh A.K., Behera M.S., Mazumdar S.P. and Kundu D.K. (2019). Soil Carbon Sequestration in Long-Term Fertilization under Jute-Rice-Wheat Agro-Ecosystem. *Communications in Soil Science and Plant Analysis,* 50(6): 739-748.

Singh, A.K., Kumar, Mukesh and Mitra, S. (2018). Carbon footprint and energy use in jute and allied fibre production. *The Indian Journal of Agriculture Science,* 88(8): 1305-11.

Singh, Balram and Lal, Ratan (2005). The potential of soil carbon sequestration through improved management practices in Norway. *Environment, Development and Sustainability,* 7: 161-184.

Singh S., Singh S., Pannu C.J.S. and Singh J. (1999). Energy input and yield relations for wheat in different agro-climatic zones of the Punjab. *Applied Energy,* 63: 287–98.

Singh, Yadvinder, Singh, Bijay and Timsina, J. (2005). Crop Residue Management for Nutrient Cycling and Improving Soil Productivity in Rice-Based Cropping Systems in the Tropics. *Advances in Agronomy,* 85: 269–407.

Soltani, A., Rajabi, M.H., Zeinali, E. and Soltani, E. (2013). Energy inputs and greenhouse gases emissions in wheat production in Gorgan Iran. *Energy,* 50 (50): 54–61.

Sun, M., Gao, Z.Q., Zhao, W.F., Deng, L.F., Deng, Y., Zhao, H.M., Ren, A.X., Li, G., Yang, Z.P. (2013). Effect of Subsoiling in fallow period on soil water storage and grain protein accumulation of dryland wheat and its regulatory effect by nitrogen application. PLoS One 8

Syp, A., Jarosz, Z., Faber, A., Borzecka-Walker, M., Pudełko, R. (2012). Greenhouse gas emissions from winter wheat cultivation for bioethanol production in Poland. *Journal of Food, Agriculture and Environment,* 10: 1169–1172.

Tanwar, S.P.S., Rao, S.S., Regar, P.L., Datt, S., Praveen, K., Jodha, B.S., Santra, P., Kumar, R., Ram, R. (2014). Improving water and land use efficiency of fallow-wheat system in shallow lithic Calciorthid soils of arid region: introduction of bed planting and rainy season sorghum-legume intercropping. *Soil Tillage Research,* 138: 44–55.

Velayutham, M., Pal, D.K., and Bhattacharyya, T. (2000). Organic carbon stock in soils of India. In: Global Climatic Change and Tropical Ecosystems (eds. Lal R., Kimble J.M., and Stewart B.A.,). Advances in Soil Science, CRC Press, Boca Raton. pp. 71–95.

Verma, T. S., and Bhagat, R. M. (1992). Impact of rice straw management practices on yield, nitrogen uptake and soil properties in a wheat-rice rotation in northern India. *Fertilizer Research,* 33: 97–106.

Yadav, R.L., Dwivedi, B.S., Prasad, K., Tomar, O.K., Shurpali, N.J. and Pandey, P.S. (2000). Yield trends and changes in soil Organic-C and available NPK in a long-term rice–wheat system under integrated use of manures and fertilisers. *Field Crops Research,* 68: 219–246.

Yang, C., Hamel, C., Gan, Y., Vujanovic, V. (2013). Pyrosequencing reveals how pulses influence rhizobacterial communities with feedback on wheat growth in the semiarid prairie. *Plant Soil,* 367: 493–505.

Yang, X., Gao, W., Zhang, M., Chen, Y. and Sui, P. (2014). Reducing agricultural carbon footprint through diversified crop rotation systems in the North China Plain. *Journal of Clean Production,* 76:131–139.

29

Greenhouse Gas Estimation Techniques and Mitigation Technologies for Reducing Carbon Footprint in Agriculture

P. Bhattacharyya, S. R. Padhy and P. K. Dash

ICAR-National Rice Research Institute (NRRI), Cuttack, Odisha-753 006 India

Introduction

Precise and continuous measurement the greenhouse gases (GHGs) fluxes/ emissions in agricultural systems are required for accounting and budgeting of carbon. Further, the long-term measurements of GHGs fluxes are equally important as precise measurement. Therefore, high frequency and long-term measurement of GHGs across the cropping systems are essential for chalking out the GHGs emissions mitigation strategies. Various technologies with definite working principles are used for estimation of GHGs emissions in agriculture.

Greenhouse Gases Estimation Techniques

Two basic principles are followed to estimate greenhouse gases in agriculture. One is the collection of GHGs from the field and subsequent analysis of the concentration of gases through gas chromatography (GC). Another principle is based on sensors measurement with the help of infrared gas analyze (IRGA). The two techniques used for the measurement are (i) 'closed chamber method' (manual/automatic), and (ii) 'sensor-based method'. Generally, the GHGs fluxes in agricultural crop field are estimated by 'closed chamber method' for methane (CH_4) and nitrous oxide (N_2O). And for carbon dioxide (CO_2) measurement either soil respiratory chamber or canopy chamber is used that measures CO_2 with the help of infrared gas analyzer (IRGA).

Close Chambers Method

The gas samples are collected from the field with the help of chamber which closed in all side. So, the method is referred as "close chamber method'. The chamber is used only for collection of gases. The accessories required for gas-collections are base-plate, syringe, pump/fan, thermometer, septum or small pipes (Fig. 1). The GHGs samples collected by the chamber subsequently analysed in Gas chromatography (GC). In GC the concentration of gases are measured. The fluxes of the GHG are then estimated by using the concentration differences of particular gas per unit time per unit area considering the linear change in gaseous concentration during the study period (Bhattacharyya *et al.*, 2014, 2016, 2019).

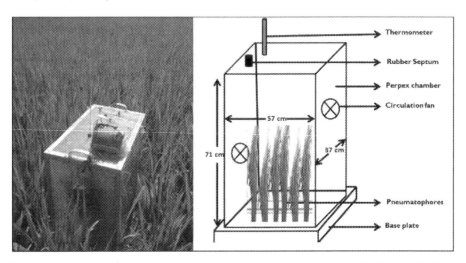

Fig. 1: Greenhouse gas collection by using 'close chamber' (with schematic diagram) in rice field

The base plates that supposed to be installed in the field previous to the sampling time should have sharp-edge base and smooth channel. It should be placed in the desired location in field prior to sowing or transplanting of crop and remains in the same place throughout the sampling period. It can be made up of good quality hard plastic pipes/ PVC pipes or aluminium frames. The base-plate and chamber base should be compatible. The bottom of the base plates are usually inserted at 6-12 cm into the soil. Importantly, the channels of base plate must be filled with desired quantity of water to make the chamber-base plate unit air-tight. There should be any leakage. Intrusion of outer air to the chamber and leakage of gas from the chamber to outside atmosphere should be strictly restricted during sampling time. In order to homogeneous mixing of gases inside the chamber during sampling, either fans or pump is necessary. Volume expansion/ contraction of gases due to temperature differences inside the chamber during sampling time must be corrected at the time of flux-estimation. So the temperature inside the chambers is monitored with the help of thermometer. The gas samples from the field or study area or different treatments must be collected with short interval (with high frequency) for getting precise emission. After collecting the GHGs samples from the field with the help of gas-chamber are analysed in GC. Gas samples should be kept in sealed (leak proof) syringe before analysis in GC. Minimum time lag (between gas sample collection and GC-analysis) and quick analysis in GC is recommended to check the loses of gas concentration.

Method of GHGs collection by chamber method

The step by step manual procedure of collection of gases by close chamber method is described below (Fig. 2)

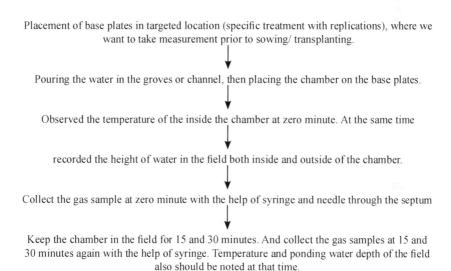

Fig. 2: Flowchart of the step by step procedures of gas collection from the field by the manual close chamber method.

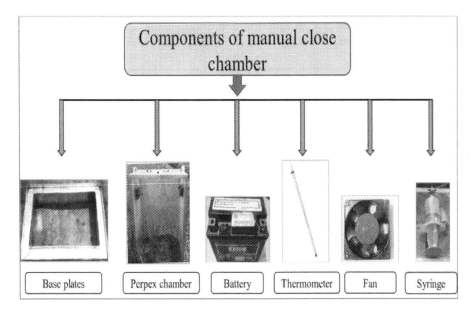

Fig. 3: Components of manual close chamber

The GHGs analysis through Gas Chromatography

The concentrations of GHGs, namely carbon dioxide, methane and nitrous oxide collected by chamber can be measured by gas chromatography. The detectors required for estimation of carbon dioxide, methane and nitrous oxide are flame ionization detector (FID) fitted with a methanizer, flame ionization detector, and electron capture detector (ECD), respectively (Fig. 4). The FID is generally detecting the hydrocarbons ions which generate when gas is heated with H_2-air flame. In GC, a Porapak-N / Porapak-Q column around 3-m-long nickel / stainless steel made (having ~3.18-3.20 mm outside diameter) is generally used. The common method which is generally used in GC for detection of CH_4 is,

(i) Column temperature: 70°C;

(ii) Helium, nitrogen / argon as carrier gas: Flow rate of around 20-30 cm^3 min^{-1};

(iii) Detector temperature: Around 250°C;

(iv) Hydrogen (H_2) gas flow rate: 30-40 ml min^{-1} for ignition.

Fig. 4: Gas Chromatography systems

However, different GC configuration's demands different methods that must be optimized according to its column material, column length, and the target gas to be analysed. The calibration is important aspect of GC analysis. It should be performed with the help of pure (99.999%) gas standards. For the measurement of N_2O concentration in GC the electron capture detector (ECD) is commonly used. In this detection, generally, a Porapak-Q column of 6 feet long with 1/8-inch outer diameter and SS (stainless steel) column (80/100 mesh) is used for separation. Only the electrophilic substances can be measured by ECD. The method generally used for detection of N_2O is,

(i) Column temperature: 60°C;

(ii) Nitrogen flow rate (as carrier gas): Around 15-18 ml min^{-1};

(iii) Injector temperature: Around 200°C;

(iv) Detector temperature: 340-350°C.

For the detection of nitrous oxide, the GC should be calibrated frequently by using ~ 100-110 ppb pure (99.999%) N_2O in N_2 as the primary standard. The concentration of secondary standards could be 300-320 and 390-400 ppb N_2O in N_2 (Bhattacharyya et al., 2016). The methanizer is used along with FID for estimation CO_2. In methanizer, basically, the CO and CO_2 in gases are converted to CH_4 and subsequently the CH_4 concentration is detected with the help of flame ionization detector in gas chromatography. The tubes of methanizer should be protected from the sulphur gas.

Calculation of GHGs flux

The amount of gas emitted per unit time per unit area is called flux. So, the flux is a vector and the SI unit of flux is g m^{-2} s^{-1}. The GHGs fluxes in agriculture

438   Conservation Agriculture and Climate Change

are either positive or negative. The positive flux refers to emission of GHGs from the agricultural field to atmosphere and opposite in case of negative flux.

Calculation of Methane (CH_4) Flux

$$CH_4 \ flux = \frac{(\Delta X \times EBV_{(STP)} \times 16 \times 10^3 \times 60)}{(10^6 \times 22400 \times T \times A)}$$

Where,

ΔX = Change in CH_4 concentration of zero and 30 minutes measured in GC;

$EBV_{(STP)}$ = Effective box volume at standard temperature and pressure;

T = Time of gas sample collection (in min, 15 / 30/ 60);

A = Area of the box (in m^2: length × breadth).

The $EBV_{(STP)}$ is calculated by the formula

$$\frac{P_1 V_1}{T_1} = \frac{P_2 V_2}{T_2}$$

Where,

P_1 = Barometric pressure at the time of sampling in mm Hg;

V_1 = EBV (Effective box volume);

T_1 = 273-K + temperature inside the box at the time of sampling in °C;

P_2 = Standard barometric pressure (760) in mm Hg;

V_2 = $EBV_{(STP)}$;

T_2 = 273°K;

The Effective box volume (EBV) is calculated by the formula given below:

EBV - Box [(H - h) × L × B] - V

Where,

H = Box height (cm); h = Height of the water in base plate (cm); L = Box length (cm); B = Box breadth (cm); V = Crop biomass volume (mL) inside the box

Calculation of N_2O flux

$$N_2O \ flux = \frac{(\Delta X \times EBV_{(STP)} \times 44 \times 10^3 \times 60)}{(10^6 \times 22400 \times T \times A)}$$

Where,

ΔX = Difference in concentrations of N_2O that is measured by GC; (30 and 0 minutes of gas sampling);

$EBV_{(STP)}$ = Effective box volume at standard temperature and pressure;

T = Flux time in min (15 or 30);

A = Crop area covered by the box in m^{-2};

The estimation of $EBV_{(STP)}$ and subsequent portion was same as mentioned above.

There are some limitations of close chamber method which is operated manually. The limitations include, (i) end point inhibition on fluxes due to accumulation of gases during sampling time; (ii) increase in temperature and pressure inside the chamber during sampling that change the gas volume/concentration; (iii) this method only give point data; (iii) getting diurnal and seasonal accounting of gases flux by this method requires high frequency sampling in field.

The gas sampling by the chamber method could be done through automatically instead of manually. The technique is also known as 'automatic chamber method'. The continuous gas collection and simultaneous analysis of GHG's concentration is the main difference from the manual method. The opening and closing of the lid of gas chamber is automatic and the chambers have to be kept in fixed position during the study period along with base plates. The gas sampling and subsequent analysis of gases in GC through purging are fully automatic and controlled by computer based software. So, it is not portable like 'manual chamber method'. Here, base plates and gas-chamber are attached together unlike the chamber used for manual gas collection. Additionally, in 'automatic chamber' has a suction pipe that pushed the collected gas to the analyser. The GHGs samples are collected at specific intervals with the help of suction pumps and then transferred it to the input port of GC for analysis (Fig 5) (Sarkar *et al.*, 2006, 2008; Bhattacharyya *et al.*, 2012, 2013, 2016, 2019). The step by step procedure of gas collection by this method and subsequent analysis of gases by analyser in presented in Fig. 6 for easy understanding.

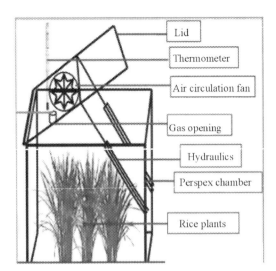

Fig. 5: Schematic diagram of automatic close chamber assembly for gas collection

Placement of 'base plate-automatic chamber' assembly in targeted location (specific treatment with replications), where we want to take measurement prior sowing transplanting.

↓

Checking the leakage of movable lid and joints of base plate-chamber assembly.

↓

Setting the method in computer software for gas sample collections cycle/intervals and put the automatic system on.

↓

The GC also must be on with control unit for real time analysis

↓

Recording the inside temperature of the chamber at each sampling cycle including at zero minute. At the same time recorded the depth of ponding in the field. That is the height of water in the field both inside and outside of the chamber.

↓

Ponding depth as well as the height of water in the field both inside and outside of the chamber should be recorded manually during sampling period.

↓

Gas sampling could be done throughout the day at specific intervals depending on the objective of the study and user requirement.

Fig. 6: The step by step procedure of estimation of GHGs fluxes by automatic chamber method.

The limitation of this method can be listed as, (i) the initial cost of the machine is relatively higher; (ii) it requires continuous power supply; (iii) it sometime underestimate or overestimate the fluxes due to temperature and moisture fluctuations inside the chamber.

Sensor Based Measure Eddy Covariance System for GHGs Estimation

The small three-dimensional pocket of wind is called eddy which that has magnitude as well as direction. The statistical covariance between vertical wind-component and concentration of scalar (desired GHG) gas at a specific time is the actual covariance-flux measured through eddy covariance system. This technique generally is used to estimate the turbulent GHGs fluxes, heat-fluxes and water vapour-fluxes in the atmospheric-boundary layer. This technique is widely known as 'Eddy Flux technique'. It provides continuous real time sensor-based measurement GHGs flux (Bhattacharyya *et al.*, 2016). The high frequency (10/ 20/ 40 Hz) time series data can be obtained by this technique. The positive 'eddy flux' means the emission of GHG from agricultural field to atmosphere while, the negative 'eddy flux' refers to the uptake or net downward movement of gases from atmosphere to crop field. It is highly precise system widely used globally for preparing real time flux database.

Basic components of eddy covariance system

The three basic unit of eddy covariance system comprises of three-dimensional sonic anemometer, carbon dioxide and H_2O gas analyser, and data logger. The other important accessories of the system are a power supply unit, flux tower or tripod and solar panel. (Fig. 7, 8). The data processing software is an important unit of the system as eddy covariance systems generate huge volume of real time 'time-series' that need to be analysed in user friendly manner. The software for processing the raw data should perform all necessary correction of the raw data before calculation of fluxes. It must have platform to import different time series data format, and capable to accommodate the data from auxiliary sensors.

Fig. 7: Open path Eddy Covariance System in rice field

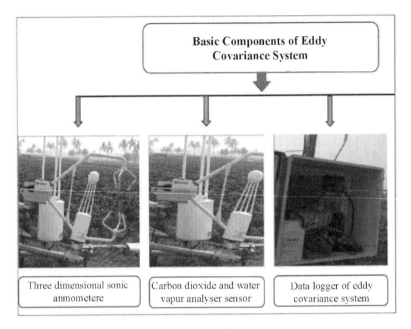

Fig. 8: Three basic components of eddy covariance system

Estimation of GHGs flux in Eddy Covariance system

The flux is estimated by calculating the covariance between vertical wind speed and GHGs concentration on time-series data. So, the net ecosystem exchange (CO_2 NEE) is estimated by adding the eddy CO_2 flux (F_C) and CO_2 storage (F_S). In majority of agricultural field, the FS is negligibly as canopy height are shorter. Therefore,

The mean vertical flux of CO_2 is,

$$F_C = \rho a * w' c'$$

Where, FC= mean CO_2 flux;

w= 30 minutes' covariance between vertical fluctuation;

c' is the CO_2 mixing ratio;

ρa is the air density ratio;

'Over bars' denotes 'time average' and primes refers to fluctuations from average value. Mostly, 30 minutes or 1 hr mean NEE is used for carbon foot print analysis.

This method is advanced and precise; however, there are also few limitations of this technique particularly in agricultural perspective. Those are, (i) it is costly instrument; (ii) it needs high frequency data measurement capability;

(iii) requires and un-interrupted power supply; (iv) expertise and skilled personnel are required for processing the huge real time data; (v) the system is not portable, so provides data is a single system at a particular time-frame; (vi) correction of raw data is pre-requisite; otherwise validity of data will significantly have hampered.

Mitigation Technologies of GHGs Emissions

The technologies and or management practices that reduce the greenhouse emissions from agriculture are referred as mitigation technologies. As a whole, agriculture acts as both sink and sources of GHGs. The net emissions of GHGs like CO_2, CH_4 and N_2O from agriculture are controlled by several factors viz. production potential, transport processes, fertilizer consumption/ application patterns, soil types, and intercultural operations performed. Lowland rice production systems, particularly plays important roles on greenhouse gas budget in agriculture. Lowland submerged rice paddies are major source of CH_4. However, under situations of intermittent flooding or alternate wetting and drying (AWD) condition favours N_2O emissions.

The mitigation options of GHGs emissions in agriculture can be classified into two different pathways like, supply-side and demand-side mitigation technologies. The supply-side mitigation technologies primarily focused on crop and livestock management, efficient land use changes (LUC) and carbon (C) sequestration. The manipulation of cropping sequence; manure-fertilizer management; soil amendment; water management etc., are few examples of crop management (Fig 9). Mitigation technologies in livestock sector includes, improved feeding and grazing for aquaculture and animal. Agroforestry is also an important climate change mitigation option. And in 'demand-side' mitigation options, we can have listed three main strategies, viz., reduction in food waste / loss, changing in dietary pattern and shortening the supply-chain (Smith *et al.*, 2014; IPCC, 2014, 2018)

In this chapter we will discuss only the supply-side mitigation options in crop management. So, the different mitigation and mitigation cum adaptation technologies for reduction of GHGs (CH_4, CO_2 and N_2O) emissions from agriculture are discussed in coming section.

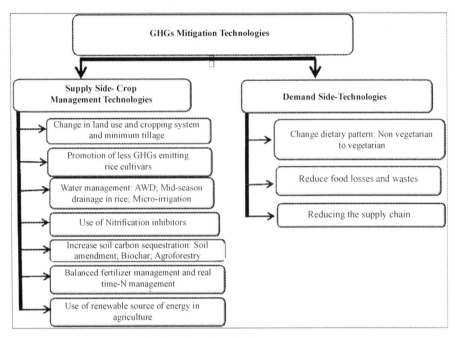

Fig. 9: The mitigation technologies of GHGs emissions.

Judicious water management

Continuous submergence in rice-paddies creates anaerobic (redox potential (Eh) below -200 mV) that is conducive of methanogenesis and methane production in soil. Therefore, alternate wetting and drying (AWD) practices in rice would increase the redox potential of soils and reduces methane production and subsequent emission to atmosphere. Alternative wetting and drying in the rice between maximum tillering (MT) to panicle initiation (PI) growth stages could mitigate CH_4 emissions from the rice field significantly. It has been found that only mid-season drainage can reduce the methane emissions from rice by 30-35%.

Similarly, micro-irrigation (drip or sprinkler) instead of flooded-irrigation in lowland agriculture curtail down the methane emissions. However, there exits trade-off between methane emission reduction and nitrous oxide flux-increment in agriculture related to water management.

Introduction of low methane emitting rice cultivars

The rice cultivars having less methane transmitting ability must be promoted. The less methane transmitting/ emission potential are manifested; either by producing less rhizodeposits / or root exudates; and or modified aerenchyma tissues that retards the transmission of gaseous CH_4 from rhizosphere to atmosphere; or by supplying higher oxygen to root zone through aerenchyma and facilitates oxidation at rhizosphere soil. Therefore, promotion of rice cultivars having low number of pore spaces in aerenchyma like, Kalinga 1, CR Dhan 201, 202 (aerobic rice cultivars) could reduce the methane emission by 8-10%.

Effective fertilizer management

The deep placement of urea super granules/ briquettes in lowland rice can reduce both CH_4 and N_2O emissions as compared to urea broadcasting. The balanced application of phosphorus and potassium fertilizers having the inherent sulphate (S) constituent can reduce CH_4 emissions. The sulphate has the inhibitory effect on methane production in soil. It competes with carbon as electron acceptor in reduced-soil condition. So, the application potassium as K_2SO_4 will reduce CH_4 emission. Apart from this, the proper potassium nutrition enhances the oxidizing power of rice root; inducing iron oxidation; and maintaining proper soil-pH in rhizosphere. So, the balanced application of potassium induces higher oxidizing conditions in soil and, thereby negatively affected the methanogenesis in rice-field and reducing CH_4 formation / emission.

Proper use of soil amendments

Soil amendments refers to organic and or inorganic substances that added to soil for improving its properties (physical, chemical and biological). The characteristics of soil amendments varied widely depending on their source of origin. The amendments generally used in agriculture are, phosphogypsum, basic-slag, flyash, (industrial origin), biochar, rice-straw (bio-origin), etc. As the phosphogypsum is rich in sulphate, so it helps to reduce the methane formation from soil (Ali *et al.*, 2012; Hussain *et al.*, 2015; Bhattacharyya *et al.*, 2016). On the other hand, the biochar could be used to sequester more carbon in recalcitrant soil C-pool that would subsequently reduce the carbon-footprint and mitigate climate change. The basic slag which is a by-product of steel industry contains significant amount of free iron and manganese oxides and silicate which compete with CO_2 and acetate as electron acceptors during methanogenesis and reduces methane emissions (Ali *et al.*, 2008, 2013). The combine application of urea with rice straw on a 1:1 nitrogen-basis can facilitate soil C-build-up, increase crop yield and lowers the N_2O emissions in lowland rice (Bhattacharyya *et al.*, 2012).

Real time nitrogen management

Crop demand based N management through leaf colour charts (LCC) or SPAD meter-reading can effectively reduce the nitrous oxide emissions by 5-15% and also increase the nitrogen use efficiency (NUE). Customized LCC like CLCC for rice, wheat and maize are available which very well judge the N-demand of crop on real time basis. Currently sensor based hyperspectral technology or remote-sensing-GIS platform also could be used very effectively for real time N-management in agriculture which has great potential to reduce GHGs emissions and carbon foot print in agriculture.

Nitrification inhibitor

The nitrification inhibitors of organic origin like Karanja oil and Nimin have found to reduce the N_2O emission from rice. The Dicynamide (DCD) is widely promoted nitrification inhibitor that can reduce the N_2O emission by 5-10%.

Modified cropping and tillage practices

Efficient cropping system can significantly reduce the GHGs emissions. For example, introduction of pulse in lowland rice-rice system could reduce the GHGs emission by 30-35% lowland (Neogi *et al.*, 2014). Zero and conservation tillage reduces the CO_2 and N_2O emissions and favours carbon sequestration. The zero/minimum tillage in DSR can mitigation CH_4 as well as N_2O emissions in rice-wheat cropping. (Pathak and Aggarwal, 2012; Bhattacharyya *et al.*, 2016). Further, the conservation agriculture where second-generation machinery like Zero-till seed drill, Happy seeder, are used along with *in-situ* residue management and crop diversification has the huge potential to reduce the GHGs emission and carbon footprint from agriculture.

References

Ali, M.A., Farouque, G., Haque, M., Kabir, A., (2012). Influence of soil amendments on mitigating methane emissions and sustaining rice productivity in paddy soil ecosystems of Bangladesh. *Journal of Environmental Science and Natural Resources,* 5: 179–185.

Ali, M.A., Hoque, M.A., Kim, P.J., (2013). Mitigating global warming potentials of methane and nitrous oxide gases from rice paddies under different irrigation regimes. *AMBIO,* 42: 357–358.

Ali, M.A., Lee, C.H., Kim, P.J., (2008). Effect of silicate fertilizer on reducing methane emission during rice cultivation. *Biology and Fertility of Soils,* 44: 597–604.

Bhattacharyya, P, Neogi, S., Roy, K. S., Dash, P. K., Tripathi, R. and Rao, K. S. (2013). Net ecosystem CO_2 exchange and carbon cycling in tropical lowland flooded rice ecosystem. *Nutrient Cycling in Agroecosystems,* 95: 133-144.

Bhattacharyya, P., Dash, P.K., Swain, C.K., Padhy, S.R., Roy, K.S., Neogi, S., Berliner, J., Adak, T., Pokhare, S.S., Baig, M.J. and Mohapatra, T., (2019). Mechanism of plant mediated methane emission in tropical lowland rice. *Science of The Total Environment,* 651: 84-92.

Bhattacharyya, P., Munda, S., Dash, P.K. (2019) Climate change and greenhouse gas emission. New India Publishing Agency, New Delhi, 110088, India. ISBN: 978-93-895-7175-2; p. 195

Bhattacharyya, P., Neogi, S., Roy, K.S., Dash, P.K., Nayak, A.K. and Mohapatra, T., (2014). Tropical low land rice ecosystem is a net carbon sink. *Agriculture, Ecosystems & Environment*, 189: 127-135.

Bhattacharyya, P., Roy, K.S., Nayak, A.K., (2016). Greenhouse Gas Emmission from Agriculture: Monitiring, Quantification & Mitigation. Narendra Publishing House. New Delhi. ISBN 13-9789384337964.

Bhattacharyya, P., Roy, K.S., Neogi, S., Adhya, T.K., Rao, K.S. and Manna, M.C., (2012). Effects of rice straw and nitrogen fertilization on greenhouse gas emissions and carbon storage in tropical flooded soil planted with rice. *Soil and Tillage Research*, 124: 119-130.

Hussain, S., Peng, S., Fahad, S., Khaliq, A., Hunag, J., Ciu, K., Nie, L. (2015). Rice management interventions to mitigate greenhouse gas emission: a review. *Environmental Science and Pollution Research*, 22: 3342–3360. http://dx.doi.org/10.1007/s11356-014-3760-4.

Intergovernmental Panel on Climate Change. (2014). Mitigation of climate change. *Contribution of working group III to the fifth assessment report of the intergovernmental panel on climate change*. IPCC, 2014

Intergovernmental Panel on Climate Change. (2018). *Global Warming of 1.5° C: An IPCC Special Report on the Impacts of Global Warming of 1.5° C Above Pre-Industrial Levels and Related Global Greenhouse Gas Emission Pathways, in the Context of Strengthening the Global Response to the Threat of Climate Change, Sustainable Development, and Efforts to Eradicate Poverty*. Intergovernmental Panel on Climate Change.

IPCC 2014. Summary for policymakers. Climate Change (2014). Impacts, Adaptation, andVulnerability. Part A: Global and Sectoral Aspects. Contribution of Working Group IIto the Fifth Assessment Report of the Intergovernmental Panel on Climate Change.Cambridge University Press, Cambridge, United Kingdom and New York, NY, USA, pp. 1–32.

Neogi, S., Bhattacharyya, P., Roy, K.S., Panda, B.B., Nayak, A.K., Rao, K.S. and Manna, M.C. (2014). Soil respiration, labile carbon pools, and enzyme activities as affected by tillage practices in a tropical rice–maize–cowpea cropping system. *Environmental Monitoring and Assessment*, 186(7): 4223-4236.

Pathak, H. and Aggarwal, P.K. (2012). Low carbon Technologies for Agriculture: A study on Rice and wheat Systems in the Indo-Gangetic Plains. *Indian Agricultural Research Institute, p. xvii, 78*.

Sarkar, R. K., Panda, D., Reddy, J. N., Patnaik, S. S. C., Mackill, D. J., & Ismail, A. M. (2009). Performance of submergence tolerant rice (Oryza sativa) genotypes carrying the Sub1 quantitative trait locus under stressed and non-stressed natural field conditions. *Indian Journal of Agricultural Sciences*, 79(11): 876-883.

Sarkar, R. K., Reddy, J. N., Sharma, S. G., & Ismail, A. M. (2006). Physiological basis of submergence tolerance in rice and implications for crop improvement. *Current Science*, 899-906.

Smith, P., and Coauthors (2014). Agriculture, Forestry and Other Land Use (AFOLU). Climate Change 2014: Mitigation of Climate Change. Contribution of Working Group III to the Fifth Assessment Report of the Intergovernmental Panel on Climate Change, O. Edenhofer *et al.*, Eds., Cambridge University Press, Cambridge, United Kingdom and New York, NY, USA.http://www.ipccnggip.iges.or.jp/public/2006gl/pdf/4_ Volume4/V4_04_Ch4_ Forest_Land.pdf.

30

Prospects and Constraints in Adoption of Conservation Agricultural Practices

Shamna. A, S.K. Jha, S. Kumar and M.L. Roy

ICAR-Central Research Institute for Jute and Allied Fibres Barrackpore West Bengal – 700 121, India

Introduction

Persistent income from agriculture is most important for sustaining the farming sector. Conservation agriculture (CA) is a resource-saving agricultural production system that aims to achieve production intensification and high yields while enhancing the natural resource base through compliance with three interrelated principles, along with other good production practices of plant nutrition and pest management (Abrol and Sangar, 2006). Conservation agriculture is largely promoted as one of the few win–win technologies affordable to farmers, in the sense that potentially it improves farmers' yields (in the long term) at the same time conserving the environment. CA is described by FAO (http://www.fao.org.ag/ca) as a concept for resource saving agricultural crop production which is based on enhancing the natural and biological processes above and below the ground. As per FAO definition CA is to a.) Achieve acceptable profits, b.) High and sustained production levels, and c.) Conserve the environment. It aims at reversing the process of degradation inherent to the conventional agricultural practices like intensive agriculture, burning/removal of crop residues. Shifting from tillage-based agriculture to no-tillage CA systems removes unsustainable elements in the current tillage based systems and replaces them with CA elements that make the production systems profitable and ecologically sustainable. Conservation agriculture offers an opportunity for arresting and reversing the downward spiral of resource degradation, decreasing cultivation costs and making agriculture more resource – use-efficient, competitive and sustainable.

Adoption of RCTs/CA

To bring a visible change both socially and economically, in the system through introduction of resource conservation techniques require a wider adoption of the same. Adoption of any technology depend on attributes of technology like relative advantage, observability, visibility, profitability and trialability. Adoption also depend on an array of personal, socio economic and cultural factors of the respondents of a particular social system. Innovations are more likely to be adopted when they have a high 'relative advantage' (perceived superiority to the idea or practice that it supersedes), and when they are readily trialable (easy to test and learn about before adoption). Non-adoption or low adoption of a number of conservation practices is readily explicable in terms of their failure to provide a relative advantage (particularly in economic terms) or a range of difficulties that landholders may have in trialing them. Full adoption of resource conservation techniques is very rare. Farmers try different technologies in combination in a bid to draw maximum profitability with minimum possible time. Farmers involvement in participatory research and technology development and other demonstration trials can increase the adoption rate of CA. Adoption of RCT depends on personal, social, environmental, institutional and political factors inaddition to the technology characters. Pascal and Josef, (2007) have suggested hypothetical pathways towards adopting conservation agriculture. (Fig. 1)

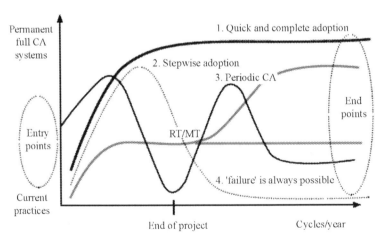

Fig. 1: Hypothetical pathway towards adopting conservation agriculture.

There are following possible four pathways towards adopting conservation agriculture:

1. Quick and complete adoption of conservation agriculture in its fullest form

2. Stepwise adoption of conservation agriculture practices, which may or may not lead to complete adoption over time (RT = reduced tillage, MT = minimum tillage)

3. Conservation agriculture practised during some cycles but not others

4. Use of conservation agriculture practices stops soon after the end of the project, perhaps because incentives are no longer available.

CA is both management and knowledge intensive and complex to practice, requiring more planning than tillage-based systems. It cannot be reduced to a technology package, adoption requiring both change and adaptation based on experiential learning (Derpsch, 2008; Friedrich & Kassam (2009).

Kassam *et al.* (2014) have elaborated the following necessary conditions for the introduction of CA and transformation of tillage-based systems.

a. Reliable local individual and institutional champions

b. Dynamic institutional capacity to support CA

c. Engaging with farmers

d. The Importance of farmers' organizations

e. Providing knowledge, education and learning services

f. The need for scientists and extension agents to recognise and characterise the problems related to CA adoption and facilitate problem solving

g. The need to build up a nucleus of knowledge and learning system for CA in the farming, extension and scientist community

h. Mobilizing input supply and output marketing sectors for CA

i. Accessibility and affordability of required inputs and equipment

The support to foster these necessary conditions must be mobilised at the individual, group, institutional and policy levels within the private, public and civil sectors so that the behaviour patterns of all stakeholders involved in the CA innovation system are mutually reinforcing to induce the development of the sufficient conditions, or the enabling environment, for adoption and spread. In cases where the learning process is missing or the benefits to the farmer are not obvious, then non-adoption or disadoption can occur.

Prospects of Conservation Agriculture

The promotion of CA under Indian/Asian context has the following prospects (Bhan and Behera, 2014)

- *Reduction in cost of production* – This is a key factor contributing to rapid adoption of zero-till technology. Most studies showed that the cost of wheat production is reduced by Rs. 2,000 to 3,000 ($ 33 to 50) per hectare (Malik *et al.*, 2005; RWC-CIMMYT, 2005). Cost reduction is attributed to savings on account of diesel, labour and input costs, particularly herbicides.

- *Reduced incidence of weeds* – Most studies tend to indicate reduced incidence of Phalaris minor, a major weed in wheat, when zero-tillage is adopted resulting in reduced in use of herbicides.

- *Saving in water and nutrients* – Limited experimental results and farmers experience indicate that considerable saving in water (up to 20% – 30%) and nutrients are achieved with zero-till planting and particularly in laser levelled and bed planted crops. De Vita *et al.* (2007) stated that higher soil water content under no-till than under conventional tillage indicated the reduced water evaporation during the preceding period.

- *Increased yields* – In properly managed zero-till planted wheat, yields were invariably higher compared to traditionally prepared fields for comparable planting dates. CA has been reported to enhance the yield level of crops due to associated effects like prevention of soil degradation, improved soil fertility, improved soil moisture regime (due to increased rain water infiltration, water holding capacity and reduced evaporation loss) and crop rotational benefits.

- *Environmental benefits* – Conservation agriculture involving zero-till and surface managed crop residue systems are an excellent opportunity to eliminate burning of crop residue which contribute to large amounts of greenhouse gases like CO_2, CH_4 and N_2O.

- *Crop diversification opportunities* – Adopting Conservation Agriculture systems offers opportunities for crop diversification. Cropping sequences/rotations and agroforestry systems when adopted in appropriate spatial and temporal patterns can further enhance natural ecological processes

- *Resource improvement* – No tillage when combined with surface management of crop residues begins the processes whereby slow decomposition of residues results in soil structural improvement and increased recycling and availability of plant nutrients. Surface residues acting as mulch, moderate soil temperatures, reduce evaporation, and improve biological activity.

Singh *et al.* (2011) evaluated superiority of RCTs over the conventional practices in terms of cost saving and efficient inputs-use. The zero tillage provided maximum saving in input cost on human and machine labour in Uttar Pradesh; it was Rs. 862/ha, i.e. around 22 per cent over conventional tillage. In Haryana and Punjab, this saving was around 21 per cent each, followed by Bihar (~20%). Besides labour saving, the adoption of RCTs also helped in saving to the extent of expenditures on seed (2-3%), chemical plant nutrients (7-10%) and irrigation (7-15%), thereby resulting in a total input cost saving of around 8-9 per cent over the non-adopters. The variation in cost of saving across states could be attributed to the methods of their application. These advantages are quite attractive and can serve as incentives to a farmer to switch over to RCTs. The comparative study on adopters and nonadopters of 'Resource Conservation Technologies' in the rice-wheat cropping system has clearly indicated the superiority of RCT over conventional practices in terms of cost saving and more efficient use of inputs. In order to enhance the productivity, profitability and sustainability of rice-wheat cropping system, the tillage technologies developed at the research farms need to be transferred and fine-tuned at farmer's field through on-farm participatory research.

Jat *et al.* (2014) reported that during the initial 2–3 years, the benefits of CA based rice production system were not visible but subsequently it was more productive and profitable than CT based system. In case of wheat, yield and economical benefit of CA based production systems were apparent right from beginning. Moreover, the wheat yields were constrained by conventional tillage based management in preceding rice crop. The results showed that CA based systems became economically profitable than CT based RW production systems.

Promoting Conservation Agriculture among Farmers

The followings are effective sequence of actions that could constitute initial interventions for promoting the transformation towards CA systems (Kassam *et al.*, 2014)

1. Identify what are the limiting factors to farmers making improvements to their livelihoods to catch their attention.

2. Identification of factors limiting crop yields and what could be done to alleviate these.

3. Identify one or more farmers already undertaking CA and demonstrating its agronomic, financial and/or livelihood benefits, and set up study visits.

4. Set up demonstration for researchers and advisory staff and farmers' groups leaders, to catch their interest.

5. Initiate 'learning by doing' e.g., through participatory forms of investigation and learning. Gain insight into what farmers know already and how they would tackle the apparent problems in the light of new knowledge introduced.

6. Determine what are optimum means of achieving CA's benefits for different situations of farm size, resource-endowments, through on-station and on farm research and benchmark demonstration, observation, FFS etc. and Field Days on farms already attempting CA. Record-keeping, analysis and feedback loops, Operational Research, are all important

7. Importing suitable samples of equipment (e.g., direct seeders for animal or tractor power, knife rollers, walking tractors with no-till seeded attachments, etc.) to be able to demonstrate their use at the beginning.

8. Interact with any already-established farmers' groups, e.g., co-operatives, to gain interest and support.

Socio Economic Aspects in Adoption of CA/RCTs

When the technology reaches the farmers field and wider diffusion of the technology occur, the socio economic impact can be studied precisely. The socio economic aspects should have special focus on the small and marginal farmers and agriculture labourers. Socio-economic and Process aspects of adoption of RCTs ponder in to the following factors (Kumar, 2014)

- Has the adoption of conservation agriculture practices changed the workload and division or sharing for labour between men and women (also seasonality)? How?

- How do those who adopted conservation agriculture practices (men and women) benefit economically in the process?

- Have the adopted conservation agriculture practices helped families to cope better with adverse situations?

- How do the landless daily-wage labourers and other marginalized groups benefit from conservation agriculture practices?

- How did the access to distribution and control of key resources change for smallholder and marginal farmers (men and women) as a result of conservation agriculture adoption?

- Have land tenure systems played a role in adopting conservation agriculture?
- Have certain government policies like subsidy or promotion influenced conservation agriculture?
- How do those who adopted conservation agriculture practices (men and women) benefit economically in the process?

Constraints for Adoption of Conservation Agriculture

CA has the potential to improve the socio economic status of the farmers but it has to be customised under the existing farming situation. The following are a few important constraints which impede broad scale adoption of CA.

- Knowledge gaps among the stakeholders
- Lack of mechanisation
- Lack of training on RCT to small and medium scale farmers
- Social and cultural barriers
- Limited or no access to inputs and output markets
- Use of crop residues for livestock feed and fuel
- Burning of crop residues
- Lack of knowledge about the potential of CA to agriculture leaders, extension agents and farmers
- Lack of skilled and scientific manpower

Conclusion

For wider adoption of Resource conservation technologies, there is a need to educate the farmers and develop more appropriate and labour saving implements. Participatory technology development can increase the rate of adoption of RCT. The impact of the Technology will be realised when its adoption occurs in a large scale. The role of scientists and extension agents are immense for increasing the agriculture output sustainably by providing training and support to farmers in Conservation Agriculture leading to the adoption of RCTs.

References

Abrol, I. P. and Sangar, S. (2006). Sustaining Indian agriculture-conservation agriculture the way forward. *Current Science*, 91(8): 1020-2015.

Derpsch, R. (2008). Critical Steps in No-till Adoption. In No-Till Farming Systems, eds. T. Goddard, M.A. Zoebisch, Y.T. Gan, W. Ellis, A. Watson and S. Sombatpanit, 479-495. Special Publication No. 3. Bangkok: World Association of Soil and Water Conservation (WASWC).

Friedrich, T. and Kassam, A. H. (2009). Adoption of Conservation Agriculture Technologies: Constraints and Opportunities. Invited paper at the IV World Congress on Conservation Agriculture. 4-7 February 2009, New Delhi, India.

Pascal Kaumbutho, Kienzle Josef, eds. (2007). Conservation agriculture as practised in Kenya: two case studies. Nairobi. African Conservation Tillage Network, Centre de Coopération Internationale de Recherche Agronomique pour le Dévelopment, Food and Agriculture Organization of the United Nations.

Malik, R. K., Gupta, R. K., Singh, C. M., Yadav, A., Brar, S. S., Thakur, T. C., Singh, S. S., Singh, A. K., Singh, R., & Sinha, R. K. (2005). Accelerating the Adoption of Resource Conservation Technologies in Rice Wheat System of the Indo-Gangetic Plains. Proceedings of Project Workshop, Directorate of Extension Education, Chaudhary Charan Singh Haryana Agricultural University (CCSHAU), June 1-2, 2005. Hisar, India: CCSHAU.

RWC-CIMMYT. (2005). Agenda Notes. 13th Regional Technical Coordination Committee Meeting. RWC-CIMMYT, Dhaka, Bangladesh.

Bhan, S. and Behera, U.K (2014) .Conservation agriculture in India – Problems, prospects and policy issues. *International Soil and Water Conservation Research*, 2(4): 1-12.

Jat, R.K, Sapkota, T.B, Singh, R.G, Jat, M.L., Kumar, M and Gupta, R.K (2014). Seven years of conservation agriculture in a rice–wheat rotation of Eastern Gangetic Plains of South Asia: Yield trends and economic profitability. *Field Crops Research*, 164: 199–210.

Kumar, R. (2014). Socio economic assessment of conservation agriculture in India. In book: Conservation Agriculture for Carbon Sequestration and Sustaining Soil Health, Chapter: Socio-economic assessment of conservation agriculture in India, Publisher: New India Publishing Agency, New Delhi, Editors: J. Somasundaram, R.S. Chaudhary, A. Subba Rao, K.M. Hati, N.K. Sinha, M. Vassanda Coumar, p. 517.

Singh, N.P., Singh, R.P. Ranjit Kumar, Vashist, A.K., Farida Khan and Nisha Varghese., (2011). Adoption of Resource Conservation Technologies in Indo-Gangetic Plains of India: Scouting for Profitability and Efficiency. *Agricultural Economics Research Review*, 24(1): 15-24.

Kassam,A., Friedrich,T., Shaxson, Bartz,H., Mello, I., Kienzle,J and Pretty,J (2014). The spread of Conservation Agriculture: policy and institutional support for adoption and uptake, Field Actions Science Reports [Online], Vol. 7 | 2014, URL:http://journals.openedition.org/factsreports/3720.

31

Retrospect and Prospects of Resource Conservation Technologies in Indo-Gangetic Plains

S.K. Jha, S. Kumar, Shamna A., M.L. Roy and T. Samajdar

ICAR-Central Research Institute for Jute and Allied Fibres
Barrackpore, West Bengal – 700 121, India

Introduction

The resource conservation technologies (RCTs) is primarily focussed on resource savings through minimal tillage, ensuring soil nutrients and moisture conservation through crop residues and growth of cover crops, and adoption of spatial and temporal crop sequencing. These practices have long been practised by the farmers in the Indo Gangetic Plains (IGP) but it got eroded in recent times due to various geo-socio economic- political reasons. This region (10.5 million ha) comprises of Punjab, Haryana, Uttar Pradesh, Himachal Pradesh, Bihar and West Bengal and have major share in production of rice and wheat due to adoption of green revolution technologies followed by area expansion. It has been reported that rice-wheat system has strained the natural resources of this region (Swarup and Singh, 1989; Kumar and Yadav, 1993; Lal *et al.*, 2004). In present scenario there is little scope to meet future target by earlier approaches like area expansion and intensive use of natural resources. Therefore, RCTs can be regarded as energy, water and labour efficient system to sustain soil health and production environment and produce more at less cost (Jat *et al.*, 2012; Gathala *et al.*, 2011b).

Major emphasis is given upon the RCTs components like zero/reduced tillage, direct seeded rice, crop residue management and crop diversification. However, other RCTs like site specific nutrient management, leaf colour chart, laser land leveller, nutrient use efficient genotypes, integrated crop and pest management provide options to the farmers as per their resource endowment.

Impact of Various CA/ RCTs

Zero tillage

The resource conserving technology that has been most successful in the western IGP is zero where it helps in sowing of wheat after paddy (tractor-drawn ZT seed drill). This specialized seeding implement allows wheat seed to be planted directly into unploughed fields with a single pass of the tractor, often with simultaneous basal fertilizer application. In contrast, conventional tillage practices for wheat in these systems involve multiple passes of the tractor to accomplish ploughing, harrowing, planking and seeding operations. Experimental evidence of Uttar Pradesh has shown that ZT reduced irrigation requirements in wheat compared to conventional (Bhushan *et al.*, 2007).

A study in Tarai-Teesta flood plain of eastern IGP of West Bengal showed (Fig. 1) that there was 11.54% increase in productivity in ZT wheat over conventional tilled (CT) wheat, where maximum saving took place in land preparation.

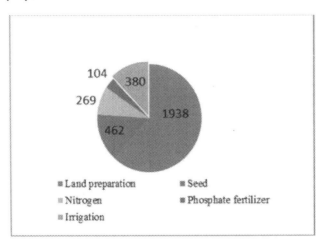

Fig. 1: Savings (Rs. /ha) in wheat through ZT

Besides, availability of zero-till seeder and its servicing, skilled operator and sometimes reluctance of local tiller operator for apprehension of lower earnings from single tillage pass are also revealed as the factors of non-adoption (Biswas, 2016). In similar kind of study, among the components of cost effects (pooled over the locations), saving on land preparation was the highest (86%) followed by seed (20%) and irrigation (17%). The combination of a significant "yield effect" and "cost-saving effect" makes ZT adoption for wheat worthwhile. These factors were more crucial for adopters in Punjab (Farooq *et al.*, 2007) and Bihar (Keil *et al.*, 2015).

An adoption pattern study of resource conservation technologies in wheat (Kumar *et al.*, 2017) of Haryana showed that majority of the farmers used laser land leveller (88.33 %) followed zero tillage (84.17 %) and rotary tillage (38.33 %). In case of application of these technologies in combination; zero tillage + laser land leveller was used by 72.5 % respondents followed by rotary tillage + laser land leveller (34.17 %) and zero tillage + rotary tillage (26.67 %). One more thing has been observed that 22.50 % of the respondents used more than two technologies (zero tillage + rotary tillage + laser land leveller). Impact of the rotary tillage on wheat showed (n=46) that (Fig. 2 and 3) it saved time (40-90 minutes/acre) and cost of cultivation. At the same time there was increase in yield, organic carbon content of soil, soil moisture and fertility of the soil.

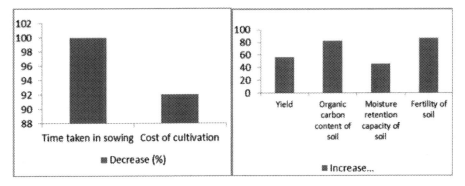

Fig. 2: Decrease due to rotary tillage   **Fig. 3:** Increase due to rotary tillage

Like, rotary tillage sowing of wheat, majority of farmers reported saving of time under zero tillage. They observed reduction in overall weed population, cost saving, harvesting similar yield, increased soil fertility, organic carbon in the soil and moisture retention capacity of soil. Besides, they reported less lodging and avoiding of terminal heat in wheat.

In West Bengal (Mandal *et al.*, 2014) found that ZT was useful in almost all types of soil except very heavy clay soils. It had the potential to save the cost of wheat cultivation by at least Rs. 7,500/ha and around 17% more yield as compared to the traditional one. Advantage of ZT technology (Nora *et al.*, 2005) in arsenic contaminated soil reduced risk of arsenicosis to a large extent by 20-40% lesser ground water utilization. ZT wheat field needed irrigation for only 1.5-2.0 hrs. (one acre of land) against 4.0-5.0 hrs for conventional tillage.

Crop residues

Crop residues are the parts or portion of a plant or crop left in the field after harvest or that part of the crop which is not used domestically or sold commercially or discarded during processing. More than 340 Mt of crop residues from various crops are produced annually of which major quantity is contributed by rice and wheat (nearly 240 Mt). There is vast potential to efficiently recycle crop residues, especially in rice-wheat belt of Punjab, Haryana and western Uttar Pradesh, where it is burnt *in situ*. It has been proved that crop residue recycling in rice-wheat is able to increase rice and wheat yield by 13 and 8%, decrease cost effectiveness by 5 and 3% and energy efficiency by 13 and 6% (PDFSR, 2011). From farmer's point of view, recycling of rice residue is problematic to succeeding wheat crop because of slow rate of decomposition of rice straw (due to high silica) coupled with shorter period available for decomposition (long duration maturing variety like Pusa 44 give a very small gap of 5-10 days for harvesting of paddy and sowing of wheat/other *Rabi* crops).

The Government of India in April, 2018 launched a scheme on "Promotion of Agricultural Mechanization for in-situ management of crop residue in the States of Punjab, Haryana, Uttar Pradesh and National Capital Territory of Delhi". It provided subsidy to farmers for purchasing a bouquet of machineries (ex- Super Straw Management System, Happy Seeder, straw chopper, shrub master, zero till drill, rotavator etc.) for in-situ crop residue management. In all the State Governments i.e. Punjab, Haryana and Uttar Pradesh have distributed 32570 different machines to the farmers on individual ownership basis and established 7960 Custom Hiring Centres.

Overall, there was about 15% and 41% reduction in number of burning events in current year (2018) as compared to that in 2017 and 2016, respectively. Till date more than 21,000 farmers have been trained and more than 6,400 demonstrations (ha) has been conducted. In Punjab, KVKs have converted 25 villages (2017) as Zero Stubble Burning villages whereas in 2018 it rose to 76.

Crop diversification

In Upper Gangetic Plains of Punjab and Haryana crop diversification have not been very successful because diversified agriculture was incompatible with commercial farming, stability of paddy and wheat yields coupled with strong domestic demand and good export prospect and so on and undermine the merits of diversified crops. However, diversification is much needed to minimize yield/market risk, improve biodiversity and diversify income source and enhance resource sustainability. It will be a key strategy for future gains in crop production.

In lower Gangetic plain of West Bengal and Bihar, sub-merged rice field offer an opportunity to diversify farm income of small marginal farmers. Crop diversification in rice field using woven and non-woven jute fabrics/gunny bags makes effective utilization of resources applied in rice field, reduces the irrigation requirement of long duration vegetables due to its long association (2-3 months) with dwarf and semi dwarf rice field (Annual Report DARE, 2012-13). This intervention by ICAR-CRIJAF, Barrackpore (CRIJAF, 2014-15) at Makaltala and Farmania villages of North 24 Parganas ensured the harvest of one ton vegetables/ha in addition to 5.4 q rice substantiated the income of poor farmers.

Direct seeded rice (DSR)

Direct seeded rice (DSR) provides a viable option for saving of labour as well as irrigation requirement, according to Pandey and Velasco (2005), low wages and adequate availability of water favour transplanting, whereas high wages and low water availability favour direct seeded rice in IGPs. A study with regard to utilization of DSR method of sowing in rice-wheat systems in Uttarakhand, Uttar Pradesh and Bihar (Malabayabas *et al.*, 2011) showed that the average yield of DSR accross sample farms in all the three states was 5 percent less than that of transplanted rice. Farmers following DSR got wheat yield increased by 9 percent. The corresponding net present values (m USD in Fig. 4) and benefit-cost ratios in the states were respectively 1.46, 1.36, and 1.50. Hence, the greater proportion of benefits from DSR adoption was derived from the change in rice production. Further it was found that in labour shortage areas DSR is advantageous because it reduced labor inputs for crop establishment, land preparation, and irrigation. It can be maximized in areas where irrigation water is scarce.

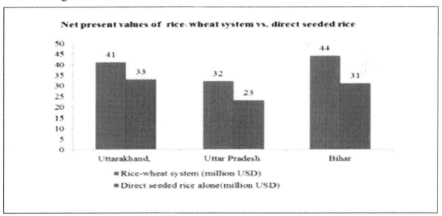

Fig. 4: Net present value (m USD) of rice-wheat vs. direct seeded rice

In spite of potential benefits, DSR suffers from inherent constraints like potential threat from high weed infestation and weed species. The DSR, therefore, requires the firsthand knowledge about weeds for wiser and safer use of herbicide as per the weed tag/population. Sowing of DSR crop in early June due to heavy and frequent rains in some parts of the IGP, small and fragmented holdings, appears to be the most important constraint in the adoption of tractor drawn zero till sowing of DSR.

Site specific nutrient management (SSNM)

Site specific nutrient management (SSNM) provide an approach to "feeding crops" with nutrients as and when they are needed and hence making synergy for nutrient demand and supply under a certain production system. This makes the efficient utilization of nutrients by crop plants and avoids the wastages of fertilizers. A field evaluation of Nutrient Expert for wheat at 173 locations across four districts of Bihar has shown increase in crop yield by over 13.9% (Shahi *et al.*, 2018). For efficient and effective SSNM, the use of soil and plant nutrient status sensing devices, remote sensing, GIS, decision support systems, simulation models and machines for variable application of nutrients play an important role.

Leaf colour chart

The concept for the use of leaf colour chart, an indicator to apply N in rice was crystallized during 1990s. The International Rice Research Institute and the Philippine Rice Research Institute developed a leaf colour chart (LCC) that helps in guiding farmers for real-time nitrogen management in rice farming. The technology is inexpensive, and easily affordable by most resource poor rice farmers. An adoption and impact study in Nadia district of West Bengal (Islam *et al.*, 2007) showed LCC was adopted by 57–63% (on an average) farmers. This technique, saved N by 25 kg/ha as well as insecticide application by 50%.

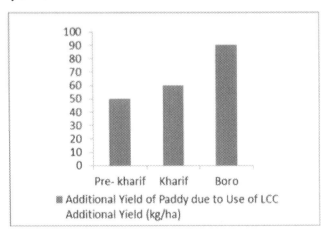

Fig 5. Additional paddy yield/ha due to use of LCC

In the *boro* season the highest saving of N was 31.4 kg/ ha (Fig. 5) giving additional yield of 90 kg/ha. The additional yield of paddy in pre Kharif and Kharif season was 50 and 60 kg/ha. In sampled area, farmer's age (negative influence), education (years of schooling) and farming experience were the major determinants of LCC adoption. In Punjab, LCC was beneficial for higher fertilizer N-use efficiency in wet direct seeded rice (Singh *et al.*, 2006).

Laser land leveller

Given the existing consumption pattern, India will need to produce at least 37% more rice and wheat by 2025 as compared to the year 2000, with nearly 10% less water available for irrigation (Jat *et al.*, 2006). In rice-wheat system of IGPs about 10-25% irrigation water is lost due to poor water management and uneven field (Kahlown *et al.*, 2000). Farmers' participatory trials of laser land levelling (LLL) saved 20-25% of irrigation water (Ravi Gopal *et al.*, 2010). Therefore, applying LLL can reduce the use of irrigation water and save energy through reduced duration of irrigation (Jat *et al.*, 2009, Jat *et al.*, 2006, Jat *et al.*, 2011). Almost all the farmers (Fig. 6) reported saving in water (30-40%) and time (one third) required for irrigation in laser levelled wheat fields (average time for levelling 1.86 hours/acre) of Haryana (Kumar *et al.*, 2017).

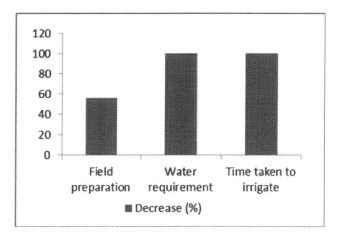

Fig. 6: Effect of laser land leveler

A study in Punjab and Haryana (during 2011) confirmed that the adoption of laser land levelling has direct implications for food security through increased local food production, water resource conservation through reduced duration of irrigation, and also reduced energy use due to fewer hours of irrigation required in rice wheat systems. The majority of farmers hired laser leveling services from service providers (rental charges USD 10/h (one USD= Rs. 50 during the time of survey) and total cost of leveling of one hectare of land was

USD 50. It included the entire laser leveling package comprising the laser leveler, driver and fuel. There was reduction in irrigation time in rice field by 47-69 hour/ha/season and improved yield by 7% (approx.) compared with traditionally levelled field; whereas, in wheat irrigation time was reduced by 10-12 hour/ha/season and yield increased by 7-9% in laser levelled field. The farmers benefitted by an additional USD 143.5 /ha/year through increase yield and reduced electricity consumption for irrigation (Aryal *et al.*, 2015).

Nutrient use efficient genotypes

During the last three decades, much research has been conducted to identify and/or breed nutrient efficient plant species or genotypes/cultivars within species and to further understand the mechanisms of nutrient efficiency in crop plants. However, success in releasing nutrient efficient cultivars has been limited. Some of the rice varieties like Rasi, Vikas, RP 5929, some of rice hybrids, etc are reported to be efficient utilizers of nutrients (Srinivasarao *et al.*, 2017).

Integrated crop and pest management (ICPM)

Plant health is governed by several biotic and abiotic factors like diseases, insects and weeds. There is considerable damage to potential agricultural production due to insect pests, diseases and weed interactions. Evidences indicate that pests cause 25 percent loss in rice, 5-10 percent in wheat, 30 percent in pulses, 35 percent in oilseeds, 20 percent in sugarcane and 50 percent in cotton (Dhaliwal and Arora, 1996). These losses though cannot be eliminated altogether; these can be reduced.

Standardized crop production and protection technologies using flora and fauna for commercial application to provide better pest control and crop economics than the conventional chemical control with other pest control measures is often referred as 'Integrated Crop and Pest Management'. India has successfully reduced pesticide consumption (404 g/ha in 1990-91 and 265 g/ha in 1998-99) without adversely affecting the agricultural productivity (Birthal, 2003). This was facilitated by appropriate policies that discouraged pesticide use and favoured IPM application. Despite this, adoption of IPM is low owing to a number of socio-economic, institutional and policy constraints (such as lack of understanding of farmers regarding biological processes of pests/ and their predators, methods of application of new technology, poor resources of farmers, community vs. individual approach)

However, empirical evidences of Integrated Pest Management (IPM) programme in Punjab, has demonstrated acceptance by the farmers. Only few components of the IPM technology resulted in decline of 10 to 15 percent pesticides application by the respondents. It is assumed, once the farmers start using full package of practices of IPM technology, the pesticide consumption may go down still further. The results, however, did not provide conclusive evidence of the financial superiority of IPM technology over the traditional methods of pest control. The farmers also, perceived an improvement in quality of crop, soil and human health as a result of use of IPM technology (Malik, 2004).

Similar kind of findings from Haryana showed that IPM was effective against major insect pests of sugarcane and paddy crops. In cotton crop need-based pesticide applications along with other alternatives such as mechanical and bio-agents has been found economical. With IPM, sugarcane yield was higher by 16 percent over the traditional method, and without any additional cost. Higher benefit cost ratio was observed under IPM practice in both cotton and paddy. The awareness among farmers regarding ill effects of pesticides on human and animal health was also high. However, they were not much aware of their effects on natural resources like soil and water (Dixit and Rai,2004).

Conclusion

The adoption of RCTs is expected to yield benefits to the farmers in terms of reduced losses due to soil erosion, saving of energy and irrigation costs, savings on labour input, increased productivity and water-use efficiency, reduced pumping of groundwater, increased nutrient-use efficiency and adoption of new crop rotations. Although considerable debate is currently going on these issues, but based on the above mentioned research evidences, the planners and policymakers can make suitable policies for popularization/increased adoption of RCTs to substantiate the income of poor farmers.

References

Annual Report (2012-13) DARE. Crop Management: Crop diversification in waterlogged rice field. p. 44.

Aryal, J.P., Mehrotra, B.M., Jat, M.L. and Sidhu, H.S. (2015). Impacts of laser land levelling in rice-wheat systems of the north-western indo-gangetic plains of India. *Food Security,* 7(3). DOI: 10.1007/s12571-015-0460-y·

Bhushan, L., Ladha, J.K., Gupta, R.K., Singh, S, Tirol-Padre, A., Saharawat, Y.S., Gathala, M. and Pathak, H. (2007). Saving of water and labour in a Rice-Wheat System with No-Tillage and Direct Seeding Technologies. *Agronomy Journal,* 99: 1288-1296.

Birthal, P.S. (2003). Economic potential of biological substitutes for agrochemicals. NCAP Policy Paper 18. National Centre for Agricultural Economics and Policy Research, New Delhi.

Biswas, B. (2016). On-farm impact analysis of resource conservation technology on wheat at Tarai-Teesta Flood plain of Eastern Indo-Gangetic Plain (IGP). *Journal of Applied and Natural Science,* 8(2): 833-839.

CRIJAF (2014). Central Research Institute for Jute and Allied Fibres. *Annual Report,* 2014-15.

466 Conservation Agriculture and Climate Change

Dixit, A.K. and Rai, K.N. (2004). Adoption and Impact of Integrated Pest Management in Important Crops in Haryana, NCAP Proceedings 11, In: Birthal P. S. and Sharma O.P. (Eds.) Integrated Pest Management in Indian Agriculture, National Centre for Agricultural Economics and Policy Research, New Delhi, India pp.161-174.

Farooq, U., Sharif, M. and Erenstein, O. (2007). Adoption and impacts of zero tillage in the rice–wheat zone of irrigated Punjab, Pakistan. Research Report. CIMMYT India &RWC, New Delhi, India.

Gathala, M., Ladha, J.K., Balyan, V., Saharawat, Y.S., Kumar, V. and Sharma, P.K. (2011b). Effect of tillage and crop establishment methods on physical properties of a medium-textured soil under 7-year rice–wheat rotation. *Soil Science Society of America Journal*, 75: 1-12.

Islam Z., Bagchi, B. and Hossain M. (2007). Adoption of leaf colour chart for nitrogen use efficiency in rice: Impact assessment of a farmer-participatory experiment in West Bengal, India. *Field Crop Research*, 103(1): 70-75.

Jat, M.L., Chandna, P., Gupta, R., Sharma, S.K. and Gill, M.A. (2006). Laser land leveling: A precursor technology for resource conservation. New Delhi: Rice-Wheat Consortium for the Indo-Gangetic Plains. Rice-Wheat Consortium Technical Bulletin Series 7.

Jat, M.L. (2012). Laser Land Leveling in India: A Success. In Presentation given at a conference on BLessons Learned from Postharvest and Mechanization Projects, and Ways Forward. Asian Development Bank's Postharvest Projects' Post-Production Workgroup of the Irrigated Rice Research Consortium (IRRC), heldat the International Rice Research Institute, Los Banos, Manila, Philippines, May 22-24.

Jat, M.L., Gupta, R., Saharawat, Y. and Khosla, R. (2011). Layering precision land leveling and furrow irrigated raised bed planting: productivity and input use efficiency of irrigated bread wheat in Indo-Gangetic plains. *American Journal of Plant Sciences*, 2(4): 578–588.

Kahlown, M.A., Raoof, A. and Hanif, M. (2000). Rice yield as affected by plant densities. Mona Reclamation Experimental Project, WAPDA, Bhalwal, Report No. 238.

Jat, M.L., Gathala, M.K., Ladha, J.K., Saharawat, Y.S., Jat, A.S., Kumar, V. and Gupta, R. (2009). Evaluation of precision land leveling and double zero-till systems in the rice–wheat rotation: Wateruse, productivity, profitability and soil physical properties. *Soil and Tillage Research*, 105(1):112–121.

Jat, M.L. Malik, R.K., Saharawat, Y.S., Gupta, R., Bhag, M. and Paroda, R. (2012). Proceedings of regional dialogue on conservation agricultural in South Asia, New Delhi, India, APAARI, CIMMYT, ICAR, p. 32.

Keil, A.K., D'Souza, A. and Mc Donald, A. (2015). Zero -tillage as a pathway for sustainable wheat intensification in the Eastern Indo-Gangetic Plains: does it work in farmers' fields? *Food Security*, 7(5): 983-1001.

Kumar, A., Singh, R., Singh, S., Sendhil, R., Chand, R. and Pandey, J.K. (2017). Impact of Resource Conservation Technologies in Haryana. *Journal of Community Mobilization and Sustainable Development*, 12(2): 257-264.

Kumar, A. and Yadav, D.S. (1993). Effect of long term fertilization on soil fertility and yield on rice –wheat cropping system. *Journal of Indian Society of Soil Science*, 41: 178-180

Lal R. Hansen, Hobbes, D.O., P.R. and Uphoff N. (2004). Sustainable Agriculture and the International Rice-Wheat System (mimeo), pp. 495-512.

Malabayabas, A.J., Julia, A., Templeton, D. and Singh, P. (2012). "Ex-ante Impact of Direct Seeding of Rice as an Alternative to Transplanting Rice in the Indo-Gangetic Plain," *Asian Journal of Agriculture and Development*, Southeast Asian Regional Centre for Graduate Study and Research in Agriculture (SEARCA), 9(2): 1-17.

Malik, R.P.S. (2011). Economics of Integrated Pest Management in Rice and Cotton in Punjab. NCAP Proceedings 11 In: Birthal P.S. and Sharma O.P. (Eds.) Integrated Pest Management in Indian Agriculture, National Centre for Agricultural Economics and Policy Research (NCAP) New Delhi, India pp .145-159.

Mandal B., Ray D.P. and Hansda A.K. (2014). Application of Resource Conserving Technology (RCT) in Wheat: A Promising Better Option Particularly During Winter Season in West Bengal. *International Journal of Bioresource Science,* 1(1): 7-17.

PDFSR (2011). Project Directorate for Farming System Research, *Annual Report,* 2011-12.

Ravi, G., Jat, R.K., Malik, R.K., Kumar, V, Alam, M.M., Jat, M.L., Mazid, M.A., Sahrawat, Y.S., Mc Donald, A. and Gupta, R. (2010). Direct dry seeded rice production technology and weed management in rice based systems. Technical bulletin. International Maize and Wheat Improvement Centre, New Delhi India. p. 28..

Shahi, V. B., Dutta, S. K and Majumdar, K. (2018) Nutrient management in high yield wheat system in Bihar using nutrient expert tool. *Journal of Pharmacognosy and Phytochemistry,* 7(6): 459-463.

Singh, B., Gupta, R.K., Singh, Y., Gupta, S.K., Singh, J., Bains, J.S. and Vashishta, M. (2006). Need-Based Nitrogen Management Using Leaf Color Chart in wet direct- seeded in Northwestern India. *Journal of New Seeds,* 8(1): 35-47.

Srinivasarao, Ch., Srinivas, K., Sharma, K.L. and Kundu, S. (2017). Soil Health Improving Strategies for Resilient Rice Based Cropping Systems of India. *Journal of Rice Research,* 10 (1): 54-63.

Swarup, A. and Singh, K.N. (1989). Effect of 12 years rice-wheat cropping and fertilizer use on soil properties and crop yields in a sodic soil. *Field Crop Research,* 21: 277-287.

Printed in the United States
by Baker & Taylor Publisher Services